KLINCK MEMORIAL LIBRARY
Concordia University
River Forest, IL 60305-1499

ANNUAL REVIEW OF EARTH AND PLANETARY SCIENCES

KLINCK MEMORIAL LIBRARY
Concordia University
River Forest, IL 60305-1499

EDITORIAL COMMITTEE (1990)

ARDEN L. ALBEE
ROBIN BRETT
KEVIN C. BURKE
RAYMOND JEANLOZ
GERARD V. MIDDLETON
FRANCIS G. STEHLI
M. NAFI TOKSÖZ
GEORGE W. WETHERILL

Responsible for the organization of Volume 18
(Editorial Committee, 1988)

ARDEN L. ALBEE
ROBIN BRETT
WILLIAM W. HAY
RAYMOND JEANLOZ
GERARD V. MIDDLETON
FRANCIS G. STEHLI
M. NAFI TOKSÖZ
GEORGE W. WETHERILL
GEORGE A. THOMPSON (Guest)
TJEERD H. VAN ANDEL (Guest)

Production Editor KEITH DODSON
Subject Indexer CHERI D. WALSH

ANNUAL REVIEW OF EARTH AND PLANETARY SCIENCES

VOLUME 18, 1990

GEORGE W. WETHERILL, *Editor*
Carnegie Institution of Washington

ARDEN L. ALBEE, *Associate Editor*
California Institute of Technology

FRANCIS G. STEHLI, *Associate Editor*
DOSECC Science Advisory Committee

KLINCK MEMORIAL LIBRARY
Concordia University
River Forest, IL 60305-1499

ANNUAL REVIEWS INC 4139 EL CAMINO WAY P.O. BOX 10139 PALO ALTO, CALIFORNIA 94303-0897

ANNUAL REVIEWS INC.
Palo Alto, California, USA

COPYRIGHT © 1990 BY ANNUAL REVIEWS INC., PALO ALTO, CALIFORNIA, USA. ALL RIGHTS RESERVED. The appearance of the code at the bottom of the first page of an article in this serial indicates the copyright owner's consent that copies of the article may be made for personal or internal use, or for the personal or internal use of specific clients. This consent is given on the condition, however, that the copier pay the stated per-copy fee of $2.00 per article through the Copyright Clearance Center, Inc. (21 Congress Street, Salem, MA 01970) for copying beyond that permitted by Sections 107 or 108 of the US Copyright Law. The per-copy fee of $2.00 per article also applies to the copying, under the stated conditions, of articles published in any *Annual Review* serial before January 1, 1978. Individual readers, and nonprofit libraries acting for them, are permitted to make a single copy of an article without charge for use in research or teaching. The consent does not extend to other kinds of copying, such as copying for general distribution, for advertising or promotional purposes, for creating new collective works, or for resale. For such uses, written permission is required. Write to Permissions Dept., Annual Reviews Inc., 4139 El Camino Way, P.O. Box 10139, Palo Alto, CA 94303-0897 USA.

International Standard Serial Number: 0084-6597
International Standard Book Number: 0-8243-2018-2
Library of Congress Catalog Card Number: 72-82137

Annual Review and publication titles are registered trademarks of Annual Reviews Inc.

∞ The paper used in this publication meets the minimum requirements of American National Standard for Information Sciences—Permanence of Paper for Printed Library Materials, ANSI Z39.48-1984.

Annual Reviews Inc. and the Editors of its publications assume no responsibility for the statements expressed by the contributors to this *Review*.

TYPESET BY AUP TYPESETTERS (GLASGOW) LTD., SCOTLAND
PRINTED AND BOUND IN THE UNITED STATES OF AMERICA

Annual Review of Earth and Planetary Sciences
Volume 18, 1990

CONTENTS

SOME RELATED ARTICLES IN OTHER *ANNUAL REVIEWS*

From the *Annual Review of Astronomy and Astrophysics*, Volume 27 (1989):

The Orion Molecular Cloud and Star-Forming Region, Reinhard Genzel and Jürgen Stutzki

A New Component of the Interstellar Matter: Small Grains and Large Aromatic Molecules, J. L. Puget and A. Léger

Interaction Between the Solar Wind and the Interstellar Medium, Thomas E. Holzer

T Tauri Stars: Wild as Dust, Claude Bertout

Classification of Solar Flares, T. Bai and P. A. Sturrock

From the *Annual Review of Ecology and Systematics*, Volume 20 (1989):

Environmental Stresses and Conservation of Natural Populations, P. A. Parsons

The Biological Homology Concept, G. P. Wagner

Nutrient Dynamics and Food-Web Stability, D. L. De Angelis, P. J. Mulholland, A. V. Palumbo, A. D. Steinman, M. A. Huston, and J. W. Elwood

Analyzing Body Size as a Factor in Ecology and Evolution, Michael LaBarbera

Behavioral Environments and Evolutionary Change, William T. Wcislo

Landscape Ecology: The Effect of Pattern on Process, Monica Goigel Turner

Phenotypic Plasticity and the Origins of Diversity, Mary Jane West-Eberhard

The Role of Optimizing Selection in Natural Populations, Joseph Travis

The Importance of Fossils in Phylogeny Reconstruction, Michael J. Donoghue, James A. Doyle, Jacques Gauthier, Arnold G. Kluge, and Timothy Rowe

From the *Annual Review of Fluid Mechanics*, Volume 22 (1990):

Wave Loads on Offshore Structures, O. M. Faltinsen

Boundary Layers in the General Ocean Circulation, Glenn R. Ierley

Parametrically Forced Surface Waves, John Miles and Diane Henderson

Wave–Mean Flow Interactions in the Equatorial Ocean, M. J. McPhaden and P. Ripa

From the *Annual Review of Materials Science*, Volume 19 (1989):

Ceramic Materials Science in Society, W. David Kingery

Dynamical Diffraction Imaging (Topography) With X-Ray Synchrotron Radiation, M. Kuriyama, B. W. Steiner, and R. C. Dobbyn

From the *Annual Review of Nuclear and Particle Science*, Volume 39 (1989):

Highest Energy Cosmic Rays, J. Wdowczyk and A. W. Wolfendale

Solar Neutrinos, Raymond Davis, Jr., Alfred K. Mann, and Lincoln Wolfenstein

From the *Annual Review of Physical Chemistry*, Volume 40 (1989):

Physical Chemistry at Ultrahigh Pressures and Temperatures, Raymond Jeanloz

ANNUAL REVIEWS INC. is a nonprofit scientific publisher established to promote the advancement of the sciences. Beginning in 1932 with the *Annual Review of Biochemistry*, the Company has pursued as its principal function the publication of high quality, reasonably priced *Annual Review* volumes. The volumes are organized by Editors and Editorial Committees who invite qualified authors to contribute critical articles reviewing significant developments within each major discipline. The Editor-in-Chief invites those interested in serving as future Editorial Committee members to communicate directly with him. Annual Reviews Inc. is administered by a Board of Directors, whose members serve without compensation.

1990 Board of Directors, Annual Reviews Inc.

J. Murray Luck, Founder and Director Emeritus of Annual Reviews Inc.
Professor Emeritus of Chemistry, Stanford University
Joshua Lederberg, President of Annual Reviews Inc.
President, The Rockefeller University
James E. Howell, Vice President of Annual Reviews Inc.
Professor of Economics, Stanford University
Winslow R. Briggs, *Director, Carnegie Institution of Washington, Stanford*
Sidney D. Drell, *Deputy Director, Stanford Linear Accelerator Center*
Sandra M. Faber, *Professor of Astronomy, University of California, Santa Cruz*
Eugene Garfield, *President, Institute for Scientific Information*
William Kaufmann, *President, William Kaufmann, Inc.*
D. E. Koshland, Jr., *Professor of Biochemistry, University of California, Berkeley*
Donald A. B. Lindberg, *Director, National Library of Medicine*
Gardner Lindzey, *Director, Center for Advanced Study in the Behavioral Sciences, Stanford*
William F. Miller, *President, SRI International*
Charles Yanofsky, *Professor of Biological Sciences, Stanford University*
Richard N. Zare, *Professor of Physical Chemistry, Stanford University*
Harriet A. Zuckerman, *Professor of Sociology, Columbia University*

Management of Annual Reviews Inc.

John S. McNeil, Publisher and Secretary-Treasurer
William Kaufmann, Editor-in-Chief
Mickey G. Hamilton, Promotion Manager
Donald S. Svedeman, Business Manager
Ann B. McGuire, Production Manager

ANNUAL REVIEWS OF
Anthropology
Astronomy and Astrophysics
Biochemistry
Biophysics and Biophysical Chemistry
Cell Biology
Computer Science
Earth and Planetary Sciences
Ecology and Systematics
Energy
Entomology
Fluid Mechanics
Genetics
Immunology
Materials Science
Medicine
Microbiology
Neuroscience
Nuclear and Particle Science
Nutrition
Pharmacology and Toxicology
Physical Chemistry
Physiology
Phytopathology
Plant Physiology and Plant Molecular Biology
Psychology
Public Health
Sociology

SPECIAL PUBLICATIONS

Excitement and Fascination of Science, Vols 1, 2 and 3

Intelligence and Affectivity, by Jean Piaget

A detachable order form/envelope is bound into the back of this volume.

James A. Van Allen

Annu. Rev. Earth Planet. Sci. 1990. 18:1–26
Copyright © 1990 by Annual Reviews Inc. All rights reserved

WHAT IS A SPACE SCIENTIST? AN AUTOBIOGRAPHICAL EXAMPLE

James A. Van Allen

Department of Physics and Astronomy, University of Iowa, Iowa City, Iowa 52242

INTRODUCTION

Space science is not a professional discipline in the usual sense of that term as exemplified by the traditional terms astronomy, geology, physics, chemistry, and biology. Rather, it is a loosely defined mixture of all of these fields plus an exotic and expensive operational style. The distinctive features of space science are the use of rocket vehicles for propelling scientific equipment through and beyond the appreciable atmosphere of the Earth; the rigorous mechanical, electrical, and thermal requirements of such equipment; and (usually) the remote control of the equipment and the radio transmission of data from distant points in space to an investigator at a ground laboratory. Space science is primarily observational and interpretative; it is directed toward the investigation of natural conditions and natural phenomena. But it can be and sometimes is experimental in the sense that artificial conditions are created and the consequences observed. Most space science has been and will continue to be conducted by unmanned, automated, commandable spacecraft. But some is conducted by human flight crews performing direct hands-on manipulation of equipment. The latter mode of operation is of dubious efficacy and, in any case, will probably be the technique of choice only in specialized subfields involving preliminary laboratory-type experiments under free-fall or low-*g* conditions.

The personal and professional backgrounds of space scientists are diverse, as is commonly the case in new and interdisciplinary fields. In accepting the invitation of Editor Wetherill of the *Annual Review of Earth*

0084–6597/90/0515–0001$02.00

and Planetary Sciences to write an autobiographical account of my career as a space scientist, I did so with a full realization of the diversity and individualism of those who belong to the fraternal order of space scientists. My account is a personal one and does not include references to primary sources, as would a proper scholarly paper. Some of this account is abridged from my monograph *Origins of Magnetospheric Physics* (Smithsonian Institution Press, 1983), but most of it is not.

PARENTAGE, BOYHOOD, AND EARLY EDUCATION

I was born to Alfred Morris and Alma Olney Van Allen on the 7th of September 1914, the second of their four sons, in Mount Pleasant (population then about 3000), Iowa, the county seat of Henry County. My mother grew up near Eddyville, Iowa, on a small farm that her father had inherited from his father, who had moved from Ohio to Iowa in the mid-1840s. My paternal grandfather, George Clinton Van Allen, was one of 11 children of Cornelius and Lory Ann Van Allen, the former a shipbuilder in Pillar Point, New York, at the eastern end of Lake Ontario. He attended Wesleyan University in Connecticut for two years and later studied law and became proficient in land titles and surveying. He passed through Mount Pleasant in 1862 as a member of the survey party that was laying out the route of the Burlington and Missouri River Railroad, which later became part of the Chicago, Burlington and Quincy (now the Burlington Northern) Railroad. (Several of my prized possessions are the magnetic compass and the drafting instruments that he used.) In late 1862, with his wife of five years, Jennie, he settled in Mount Pleasant, built a small house, and established a law office. My father, an only child, was born in Mount Pleasant in 1869. He attended the local public schools and Iowa Wesleyan College, and then studied law at the University of Iowa in Iowa City, receiving an LLB degree in 1892. He joined his father as a practicing lawyer and continued to practice law for the remainder of his life.

My boyhood activities were centered within our closely knit family, which had a strong resemblance to earlier pioneer families.

The virtues of frugality, hard work, and devotion to education were enforced rigorously and on a daily basis, especially by my father. My mother exemplified the pioneer qualities of affection and nurture for her husband and their children and of comprehensive self-reliance: cooking all meals from scratch, baking delicious bread twice a week, washing clothes with a washboard and tub, maintaining a meticulous standard of household cleanliness, canning large quantities of fruits and vegetables, and, most important of all, ministering to her children through health and frequent sickness during the many epidemics of those days. Before her marriage, she had taught in one-room country schools near Eddyville and had attended the Iowa Wesleyan Academy for two years.

My first clear recollection is waving a tremulous farewell to her as I set off on foot to kindergarten a few days after my fourth birthday. Two months later, my older brother and I went with my father to the public square in Mount Pleasant to witness the celebration of the armistice of World War I by a horde of raucous and exuberant people of all ages. The culmination of this celebration was the burning of a huge straw-filled effigy of Kaiser Wilhelm.

I enjoyed school work greatly under the guidance of devoted teachers, most of whom were unmarried woman who had gone into teaching as a durable profession. Our father read to my brothers and me for about an hour after supper nearly every evening—from the *Book of Knowledge*, *An Illustrated History of the Civil War*, the *National Geographic* magazine, and, occasionally, from the *Atlantic Monthly*. Then he shooed us off to our respective corners to do our homework for two or three hours. Our chores varied with the seasons. We raised a large flock of chickens year-round. In the summer we planted and cultivated a one-acre vegetable garden and a large apple orchard, and in the winter we split wood for the cook stove, shoveled snow, ran errands, fired the furnace, and tried to keep warm. We had a car but seldom used it, even during the summer. During the winter, the car was set up on wooden blocks in the barn to "save the tires." For the most part, we walked everywhere.

I was intensely interested in mechanical and electrical devices. *Popular Mechanics* and *Popular Science* were my favorite magazines. I built elementary electrical motors, primitive (crystal) radios, and other devices described therein. Two highlights were the construction of a Tesla coil that produced, to my mother's horror, foot-long electrical discharges and caused my hair to stand on end; and the complete disassembly and reassembly of those mysterious "black boxes"—the engine and planetary transmission of an ancient Model T Ford that my older brother and I had bought for $25 (later recovered on resale).

In high school my favorite subjects were mathematics (including solid geometry), Latin, grammar, and manual training (woodworking). As a senior in 1930–31, I had my first course in physics, with many opportunities for laboratory work, a memorable experience. During the same year I edited the senior annual *The Target*. I graduated from Mount Pleasant High School in June 1931 as class valedictorian. My valedictory oration was entitled "Pax Romana—Pax Americana," based on my study of Roman history in school and on my father's tutelage. The thesis of this oration was that America, by virtue of its economic, cultural, and military stength, would dominate world affairs and enforce world peace for a limited period of history but would then lose its influence because of its preoccupation with "bread and the circus games."

COLLEGE AND GRADUATE WORK

Throughout my boyhood, there was never any doubt that my three brothers and I would go to college and have an opportunity to "amount to something." The matter was not subject to discussion. In the autumn of 1931, following in the footsteps of my father, mother, and older brother George, I entered Iowa Wesleyan College in Mount Pleasant. The tuition was $45 per semester, and I lived at home. The academic work was demanding, and I took all the courses offered there in physics, chemistry, and mathematics (four years of each), a summer field course in geology, and the one available course in astronomy (using Moulton's 1933 *Astronomy*, which I still have), the only formal course in astronomy that I ever took. Professor Thomas Poulter in physics and Professor Delbert Wobbe in chemistry were my principal inspirations. Each was the one-man faculty of his respective department. I wavered between choosing physics or chemistry as my major but decided on physics after Poulter offered me a part-time student assistantship. I worked in his high-pressure research laboratory and learned to blow glass, to run a metal-turning lathe and a milling machine, and to braze, silver-solder, and weld. More importantly, I came to have an almost worshipful regard for his mechanical ingenuity, his intuitive use of physics and chemistry as a way of life, and his devotion to experimental research. Poulter was in the process of preparing for his role as chief scientist of the Second Byrd Antarctic Expedition, a part of the Second International Polar Year. Following my freshman year I became a part of those preparations. I helped build a simple seismograph and was entrusted with checking out a field magnetometer on loan from the Department of Terrestrial Magnetism of the Carnegie Institution of Washington (DTM/CIW), one of the most beautiful instruments that I have ever seen. In the autumn of 1932 I used this instrument to make precision measurements of the geomagnetic field at three ad hoc locations in Henry County. The measurements involved also the determination of latitude and longitude by observation of the Sun with the theodolite on the magnetometer. All of this was done by carefully following the third edition of Daniel L. Hazard's *Directions for Magnetic Measurements* (US Department of Commerce, Serial Number 166, 1930). I copied my field notes onto clean forms and mailed them proudly to John A. Fleming, then director of DTM/CIW, as a modest contribution to the world survey that was underway. I received a prompt acknowledgment from him that concluded by making it clear that only raw field notes could be accepted as valid. I then sent him those, thereby learning a durable lesson about the sanctity of raw data.

My other introduction to geophysical research was serving as an observer of meteor trails during the Perseid shower of August 1932. Arrangements for the observations were worked out between Poulter and astronomy professor C. C. Wylie of the University of Iowa; sky "reticles" devised and built by Poulter from welding rods were used to carry out the observations. These 6-ft-long conical devices with an eye-ring at the vertex and a coordinate system of radial and circular rods at the other end were mounted on fixed stands. One was located in my backyard in Mount Pleasant and the other in Iowa City, 50 miles to the north. The conical fields of view were positioned so that they included a common volume of the atmosphere spanning the estimated altitude range of meteoric luminosity. During the early morning hours of 22 August, Raymond Crilley manned the Iowa City reticle, and I the Mount Pleasant one, using accurate watches for coordination. Each of us observed about 20 bright meteor trails. Of these, Wylie identified seven as identical cases. He later published the calculated altitudes of the beginning and end points of each of these trails. At the time, I had the impression that this was the first successful attempt to make such measurements, and the impression provided part of the thrill of making them. Later, I learned that my impression was not true. During the ensuing Antarctic expedition Poulter used this system to obtain one of the world's most comprehensive sets of observations of meteor trails. Also, he made extensive use of the DTM/CIW magnetometer and the seismograph that I had helped construct. The 1935 graduation ceremony at Iowa Wesleyan College included a public parade honoring Poulter and Admiral Richard E. Byrd. The latter gave the commencement address. I graduated summa cum laude and was the first student to walk across the platform. Poulter moved forward to congratulate me, but I was so flustered that I scurried past him, clutching my diploma.

During the summer of 1934, I went by automobile to California with my mother, father, and two of my brothers to visit prospective graduate schools in the west. Two of my most pleasurable recollections were visits to the laboratories of Jesse Du Mond at Caltech and Paul Kirkpatrick at Stanford. My eyes popped at the elegance and scope of their laboratories, and I was deeply grateful for the careful explanations of their research that they gave me, a young kid who had dropped in uninvited. But in the end I followed my family's tradition of attending the University of Iowa. In 1935, the faculty of its Department of Physics numbered five: George W. Stewart (head of the department since 1909), John A. Eldridge, Edward

P. T. Tyndall, Claude J. Lapp, and Alexander Ellett. The latest addition occurred in 1928 with Ellett's arrival. My assigned advisor was Tyndall, a warm-hearted and spirited individual with a PhD from Cornell University. My central preoccupation was with introductory graduate-level courses based on Slater and Frank's *Introduction to Theoretical Physics*, Abraham and Becker's *Classical Electricity and Magnetism*, and Pauling and Wilson's *Introduction to Quantum Mechanics*; on instructors' original lectures on classical mechanics, statistics, and partial differential equations; and on lectures and laboratory studies in atomic physics. I found the work to be rigorous and demanding.

I was eager to start research, and soon after my arrival Tyndall introduced me to the art of growing large single crystals of spectroscopically pure zinc and of measuring their physical properties. I completed an MS degree in June 1936 with an original experimental thesis, "A Sensitive Apparatus for Determining Young's Modulus at Small Tensional Strains." By that time Ellett, who formerly worked with atomic beams, was actively converting his research interests to the new field of experimental nuclear physics. I decided to join in this work. Together with Robert Huntoon, a more senior graduate student, and others, I helped build a copy of the famous Cockroft-Walton high-voltage power supply and accelerator. Our capacitors were made of plates of window glass on which we glued aluminum foil; the rectifiers and the accelerator tube used glass cylinders from a local company that supplied them to service stations for the then prevalent model of gasoline pumps. Everything was improvisation. Central elements of the measuring equipment were an ionization chamber and a Dunning-type pulse amplifier with a voltage gain of about one million, built with vacuum tubes of course and a nightmare to shield adequately against pickup of AC ripple and coronal discharges, of which we had a plethora. Because of the absence of air conditioning or any effective humidity control, operation during the summer was impossible. But on a good day in the autumn of 1938, we finally got an ion beam of a few microamperes with an accelerating potential of 400 kV. My objective was to measure the absolute cross section of the reaction

$$H^2+H^2 \rightarrow H^1+H^3$$

over as great a range of bombarding energy as possible. The novel feature of my experiment was the use of a gaseous (i.e. infinitesimally thin) target, which involved the controlled flow of deuterium gas through the custom-built reaction chamber. After several months of fixing leaks in the vacuum system, replacing burnt-out filaments in the rectifiers, repairing damage from high-voltage spark-overs, etc, etc, I finally got everything to work at the same time. With the help of a fellow graduate student, I then made a continuous run of 40 hr, being unwilling to turn off anything because of the well-founded expectation that many weeks might be required to restore full operation. However, with good luck, I was able to make a confirmatory run two weeks later. These two runs provided the basis of my PhD dissertation, which, with Ellett's approval, I then wrote up under the title "Absolute Cross-Section for the Nuclear Disintegration $H^2 + H^2 \rightarrow H^1 + H^3$ and its Dependence on Bombarding Energy" (50–380 keV). I defended my work successfully before the examining committee and received the degree in June 1939.

Following an oral paper that I gave at the spring 1939 American Physical Society meeting, Hans Bethe expressed a keen interest in the results but found that the trend of my curve of cross section vs. bombarding energy was impossible to believe at the lower energies because of basic quantum-mechanical theory. This criticism was unsettling to put it mildly. Ellett and I went over the entire matter critically and eventually realized that my method of measuring the beam current through the reaction chamber was faulty. I had collected the ion beam in a Faraday cup *after* it had passed through the chamber and had measured the charge collected per unit time there. I failed to take account of the partial neutralization of the beam by charge exchange in the target gas, an effect of increasing importance at the lower energies. As a result, the measured current was too low and the calculated cross section was correspondingly too large. A follow-on experiment by Stanley Atkinson, using the same apparatus, established the magnitude of this effect and corrected my results.

Many years later, the cross section of the deuteron-deuteron reaction at much lower energies became a matter of importance in the development of equipment for the current major effort to achieve controlled fusion in the laboratory.

DEPARTMENT OF TERRESTRIAL MAGNETISM OF THE CARNEGIE INSTITUTION OF WASHINGTON

Concurrently with the early nuclear physics work at Iowa, Merle Tuve, Lawrence Hafstad, and Odd Dahl had built a Van de Graaff (electrostatic) power supply and an ion accelerator tube at DTM and had succeeded in getting a stable beam at bombarding energies up to 1 MeV. The principal emphasis of their early work, under the urging of theoretician Gregory Breit, was the careful measurement of the proton-proton scattering cross section, then regarded as one of the most fundamental problems in nuclear physics. Norman Heydenberg, also one of Ellett's former students at Iowa, was one of Tuve's principal collaborators. In the spring of

1939 Ellett recommended me to Tuve, and I received a Carnegie Research Fellowship to work at DTM.

Earlier, in late 1938, Otto Hahn and Fritz Strassman in Germany had discovered nuclear fission. The DTM laboratory was converted to confirmatory experiments, which were successful. More importantly, Richard Roberts discovered the delayed emission of neutrons from fission products. This discovery provided the basis for the control of nuclear fission in all subsequently developed nuclear power plants.

My own work at DTM during 1939–40 was the measurement of the absolute cross section for photodisintegration of the deuteron by 6.2-MeV gamma rays from protons on fluorine. This was done in collaboration with Nicholas Smith, another Carnegie fellow, formerly at the University of Chicago. In addition, Norman Ramsey, yet another Carnegie fellow, and I measured neutron-proton cross sections using a small proportional counter that I had devised for observing the recoil protons.

Of much greater importance to my future career was my crossing of the culture gap at DTM from nuclear physics to the department's traditional research in geomagnetism, cosmic rays, auroral physics, and ionospheric physics. I was impressed especially by the work of Scott Forbush and Harry Vestine. Also, there were occasional visits by Sydney Chapman and Julius Bartels, who were then completing their great two-volume treatise *Geomagnetism*. As a result, my interest in low-energy nuclear physics dwindled, and I resolved to make geomagnetism, cosmic rays, and solar-terrestrial physics my fields of research—at some unidentified future date.

PROXIMITY FUZES

By late 1939 the war in Europe was already several months old, and Tuve foresaw the inevitable involvement of the United States. He abandoned experimental work and turned his remarkable talents to the problem of what scientists in the United States should be doing to help remedy the desperately inadequate quality of our military establishment. He made intensive inquiries, especially among high-ranking naval officers, and returned to DTM with a vivid impression of the ineffectiveness of anti-aircraft guns and with full knowledge of the embryonic British work on proximity fuzes for eliminating the range error of time-fuzed projectiles. He seized on this as *the* matter to which he would devote his own staff and, by recruitment, other physicists and engineers of kindred inclination—including Ellett from Iowa and Charles Lauritsen, his son Thomas, and William Fowler from Caltech. As a Carnegie fellow, I was apart from these early efforts, but by the summer of 1940 I asked to become a part of this enterprise and was appointed to a staff position in Section T (for Tuve) of the National Defense Research Council (NDRC) of the newly created Office of Scientific Research and Development, headed by Vannevar Bush.

I worked first on a photoelectric proximity fuze and succeeded in solving the basic problem of making a circuit such that the fuze would have equal sensitivity over a large range of ambient light levels. My circuit gave an output approximately proportional to the logarithm of the current from a photoelectric cell by using a fundamental characteristic of a vacuum-tube diode. My demonstration of a breadboard of this circuit to Charlie Lauritsen and Willy Fowler showed that I got the same size pulse by waving my hand in front of a photocell when illuminated by full sunlight as I got in a darkened room. Their exuberant response not only made my day, but it also propelled the photoelectric fuze into the realm of serious consideration.

But soon thereafter, I was transferred to work on the radio proximity fuze. Dick Roberts had built a simple self-excited rf oscillator operating at about 70 MHz after the fashion of the one that the British called an autodyne circuit. In brief, the plate current of the one-tube oscillator with a short antenna was affected by the reflected signal from a nearby conductor. The basic scheme was that the transient pulse as a fuze passed an aircraft could be amplified so as to trigger a gaseous tube (thyratron) to fire the detonator of the projectile. This device became the focus of a truly huge development.

For the first time in my life I worked under conditions in which urgency was the motto, multiple approaches to a problem were fostered, money was no object, and the first approximation to a solution was the prime objective. As Tuve put it, " I don't want you to waste your time saving money."

THE APPLIED PHYSICS LABORATORY OF JOHNS HOPKINS UNIVERSITY

The radio proximity fuze group soon outgrew the capacity of DTM, and thus Tuve negotiated an arrangement with Johns Hopkins University (JHU) such that JHU would assume contractual oversight of the project. In early 1942 JHU rented a large Chevrolet garage in Silver Spring, Maryland, and established the Applied Physics Laboratory (APL). Along with other members of the group, I was transferred to APL/JHU in April 1942, thereby qualifying as a plank-owner, as that term is used in the navy for a member of the crew who places a new ship on commission.

My own work was principally on developing what was termed a rugged vacuum tube, i.e. one that would survive acceleration of some 20,000 g as it was propelled through the barrel of a 5″/38 navy gun. The starting point was the miniature vacuum tubes that had been developed for use in electronic hearing aids by the Raytheon and Sylvania companies. I worked principally with tube engineer Ross Wood of Raytheon in the trial-and-error process of remedying the numerous shortcomings of the early tubes. I conducted field tests of each batch of tubes by putting them in a small cylinder that was mounted in a projectile. These projectiles were then fired vertically by a converted 10-pounder gun at a test site in southern Maryland along the Potomac River. We recovered the projectiles with a posthole digger and returned the tubes to the laboratory for detailed scrutiny. (In July 1942, I was commissioned a deputy sheriff of Montgomery County in order to legally carry a loaded revolver for coping with hypothetical hijackers on our daily expeditions to and from the test site.) I would then report the results to Ross by phone or, if we had important conclusions, by personal visit by train to Newton, Massachusetts, where he operated a pilot line. On most of these trips I would return to Silver Spring with a batch of improved tubes. One of the most nagging problems was the breakage of the fine filaments. I reasoned that distortion of the structure that supported the filaments was the cause of the failure. In a moment of inspiration I sketched out a scheme for a minute coil spring (wrapped around a mandrel), to the free end of which one end of the filament would be welded. My hope was that the spring would maintain nearly constant tension on the filament during acceleration in the barrel of the gun, and also that the tension would be such as to tune microphonics outside of the frequency pass-band of the amplifier. Wood executed this idea using the skills of the women who built these tubes with the aid of microscopes. The scheme worked and became an essential feature of the millions of tubes that were manufactured during the three subsequent years of World War II.

By late autumn 1942, the first of the Section T radio proximity fuzes were coming off the production line. Realistic and extensive testing at the Dahlgren Proving Ground over the Potomac River ("airbursts" as the projectile approached the water) and past an aircraft suspended between two towers at Jack Workman's test facility near Socorro, New Mexico, had been conducted. Despite numerous duds and premature bursts, it was estimated that the effectiveness of naval antiaircraft fire would be increased by a factor of order five if the proximity fuzes were substituted for the time fuzes then in use throughout the fleet.

In early November 1942, the Naval Bureau of Ordnance (Bu Ord) determined that the fuzes were ready for issue to the Pacific Fleet. Neil Dilley, Robert Peterson, and I were given spot commissions as United States Naval Reserve line officers with the rank of lieutenant junior grade.

Our job was to assist Commander William S. ("Deke") Parsons, United States Navy (USN), principal liason officer from Bu Ord during the development work, in introducing this new fuze to gunnery officers of combatant ships in the South Pacific. Parsons (later the weaponier on the *Enola Gay*, which dropped the first atomic bomb at Hiroshima) flew ahead to an unrevealed location in the Pacific theater. Dilley, Peterson, and I oversaw the loading of the first secret issue of some 5000 carefully counted, proximity-fuzed (also called VT fuzes to disguise their nature) 5″/38 projectiles into the hold of a troop ship at Mare Island near San Francisco. Within a week of receiving our commissions, signed personally by Frank Knox, Secretary of the Navy, we were at sea en route to a secret destination. The ship traveled without escort. I was able to keep track of our progress in latitude by elementary celestial observations and in longitude by the progressive change in mean time between sunrise and sunset and the occasional one-hour changes in ship's time.

About two weeks later we arrived in Nouméa, New Caledonia, headquarters of the Commander of the South Pacific Fleet (COMSOPAC). Parsons had already laid the groundwork and assigned us to various segments of the fleet. I was assigned as assistant gunnery officer on the staff of Rear Admiral Willis A. Lee, a task group commander of Task Force 38 (commanded by Admiral William F. Halsey) and Task Force 58 (commanded by Admiral Raymond A. Spruance). Admiral Lee was also type commander of battleships in the Pacific Fleet (COMBATPAC) with headquarters on the USS *Washington*. I arrived on the *Washington* only about two weeks after her celebrated role in the major engagement with a Japanese task force in the strait between Tulagi and Guadalcanal, thereafter called iron-bottom bay. Lee was the informal president of the Navy "gun club" and was acknowledged to be one of the leading gunnery officers of the US Navy. He was thoroughly familiar, both theoretically and practically, with the fundamental ineffectiveness of antiaircraft weapons and of the often fatal fallacy of supposing that an attacking aircraft could be stopped by "filling the air with shrapnel." He was deeply impressed by my briefings on the VT fuzes and immediately recognized their potential in quantitative terms. I gave him a clear statement on the necessity of a clear field of fire (not over our own ships), of the expectation of at least 15% duds and premature bursts (which posed no hazard to the firing ship), and of the airbursts that occurred as projectiles approached the sea at the end of flight. Also, I informed him of the then prevailing doctrine that despite the potential effectiveness of proximity-fuzed projectiles for shore bombardment, such usage was forbidden on the security ground that duds might be recovered by the enemy and either duplicated by them or used as a basis for countermeasures, i.e. "jamming" by radio transmitters so as to cause premature bursts. He endorsed my written description of the properties of the new ammunition and immediately ordered a pro rata distribution of the available supply to all combatant ships of his task group. My job was to effect this distribution and to brief gunnery officers and commanding officers on their proper use. I encountered a wide range of understanding and lack of understanding of the range-error problem and varying degrees of acceptance. The toughest operational problem was the restriction on firing over other ships of the task group under the complex conditions of actual air attack.

After eight months of sea duty on the *Washington* and other ships I was ordered back to Bu Ord to serve as liasion officer with APL/JHU and to read and summarize combat reports from ships using the VT fuze against attacking aircraft. Finding such desk work onerous, I requested transfer back to the Pacific Fleet to help remedy the grave shortcomings of the fuzes—most notably, the large percentage of duds that were occurring as the useful shelf life of their batteries expired during their long period of transport, usually at elevated temperature conditions, by cargo ships from the states to combatant ships. I then made contact again with Admiral Lee on the *Washington* and with Commander Lloyd Muston, COMSOPAC staff gunnery officer, in Nouméa and engaged in setting up rebattering stations at ammunition depots at Nouméa, Espiritu Santo, Tulagi, Guadalcanal, and Manus Island, and on ammunition barges at Eniwetok Atoll, Kwajalein, and Ulithi. I also had temporary duty on a succession of destroyers to instruct gunnery officers and conduct tests of the fuzes. And I made frequent reports to Bu Ord on the status of the work and (usually urgent) requests for fresh batteries, tools, and equipment—by air transport, if possible, to try to maintain the fleet's supply of workable fuzes. During this period I was on the *Washington* as assistant staff gunnery officer during the Battle of the Philippines Sea, in which the ship successfully defended herself against kamikaze attack. In March 1945 I returned to duty at Bu Ord and as liaison officer at APL/JHU until my transfer to the inactive reserve as a lieutenant commander in March 1946, after the end of World War II hostilities.

The period 1940–45 was a part of my life totally foreign to my previous aspiration to become an academic physicist. But I lost no energy grieving over the turn of events. On the contrary, I plunged into "the war effort" with the patriotic fervor of those days and with the exhilaration of applying my knowledge of physics and mathematics and my laboratory skills to solving difficult problems of practical importance and national urgency. My service as a naval officer was, far and away, the most broadening experience of my lifetime. I had considerable responsibility in the real world of life-or-death, and for the first time, I dealt with a vertical cross section of the human race on a one-to-one basis, from apprentice seamen to admirals. I was deeply impressed by every such relationship, by the code of honor of the navy, and by the validity of military protocol. I gained a profound respect for the raw power and grandeur of the sea and a corresponding respect for seamen. Much of my boyhood reading was in that vein. As a high-school senior, I had hoped for an appointment to the US Naval Academy, and our US congressman, a close friend and former classmate of my father's in college and law school, nominated me subject to passing the academic and physical examinations. But I failed the latter. Eleven years later I received a spot commission as a lieutenant junior grade in the Naval Reserve under the relaxed wartime standards.

Among other things that I learned in the navy by close observation of my peers and superiors was how to make a sound decision when the basis for a decision was diffuse, inadequate, and bewildering. This lesson has served me well. Another strong and durable impression was the great gap between the life of a bureaucrat in Washington and the real situation on a combatant ship.

HIGH-ALTITUDE RESEARCH

While still on terminal leave from the navy, I was rehired as a physicist at APL and encouraged to organize a research group to engage in high-altitude research based on the prospectively available opportunity to conduct experiments with captured and refurbished German V-2 rockets. I had earned my spurs by my wartime work, and Tuve gave me a free hand and ample financing to develop this field as I saw fit. My interest in geophysics stemmed from Tom Poulter's work and from my association with the " old-line" geophysicists at DTM. This line of scientific interest and my laboratory experience in nuclear physics and with rugged electronic devices and high-performance ordnance combined to lay the groundwork for my future research career. I was eager to attack problems of the primary cosmic radiation, the ionosphere, and geomagnetism by rocket techniques, which promised direct observation of many phenomena that had been previously a matter of conjecture, albeit sophisticated conjecture.

I gathered together a spirited group of like-minded individuals—Robert Peterson, Lorence Fraser, Howard Tatel, Clyde Holliday, John Hopfield, and several others—and got to work. Parallel efforts were underway at several other military or quasi-military laboratories. Of these, the group at the Naval Research Laboratory (NRL), inspired by Ed Hulburt, their long-time leader in atmospheric and ionospheric physics, was the most noteworthy. We adopted NRL as our principal competitor and sometime collaborator.

The opportunity to use V-2's for scientific work was provided by the Army Ordnance Department by virtue of the foresight and broad vision of Colonel Holger N. Toftoy.

Under the leadership of Ernst Krause of NRL, a small and highly informal group of prospective participants in this effort was assembled to maximize the scientific work and to allocate flight opportunities in an equitable manner. I was a member of this group, which called itself the V-2 Rocket Panel. We had no formal organization, no official authority, and no budget. Nonetheless, we oversaw, in effect, the entire national effort in this field for over a decade. Krause was the original chairman, but he left the NRL in 1946 and I was chosen to succeed him, continuing thereafter as chairman until the effective termination of our functions in 1958.

In 1946, with the support of Merle Tuve, I initiated and supervised the development of a high-performance American sounding rocket, the Aerobee, to be used exclusively for scientific purposes. This rocket soon joined the V-2 as a basic vehicle for high-altitude research. During the period 1946–51, payloads of scientific instruments were carried by 48 V-2's and 30 Aerobees.

The emphasis of our APL work was in the fields of cosmic rays, solar ultraviolet, high-altitude photography, atmospheric ozone, and ionospheric current systems. The site for most of the launchings was the White Sands Proving Ground (later renamed White Sands Missile Range) near Las Cruces, New Mexico. But in 1949 and 1950, I organized successful Aerobee-firing expeditions on the USS *Norton Sound* to the equatorial Pacific and the Gulf of Alaska, respectively. (As of 17 January 1985, a total of 1037 Aerobees had been fired for a wide variety of investigations in atmospheric physics, cosmic rays, geomagnetism, astronomy, and other fields.)

The national effort in high-altitude research during those early free-wheeling and spirited days was characterized by many failures and many noteworthy successes. Substantial advances in knowledge were achieved in atmospheric structure, ionospheric physics, cosmic rays, high-altitude photography of large areas of the cloud cover and surface of the Earth, geomagnetism, and the ultraviolet and X-ray spectra of the Sun.

The V-2 Rocket Panel [later renamed the Upper Atmosphere Rocket Research Panel (UARRP) and, still later, the Rocket and Satellite Research Panel] presided over the entire effort. Beginning in the mid-1950s, the panel spawned one of the important components of our national participation in the 1957–58 International Geophysical Year (IGY). Its members became influential in the planning of the IGY, actively promoted the adoption of scientific satellites of the Earth as an element of the IGY program, and laid the foundations for the scientific program of the

National Aeronautics and Space Administration, a major agency of the federal government created in 1958.

RETURN TO THE UNIVERSITY OF IOWA

In 1950, and despite the flourishing of our high-altitude work, the new director of APL, R. C. Gibson, split my assignment so as to include supervision of the residual proximity fuze group. I was competent to provide such supervision but had no interest in pursuing further developmental work on fuzes. I did the job but interpreted the split assignment as foreshadowing the termination of academic-style research in geophysics at APL. A few months later I received a telephone call from Professor Tyndall, my former research mentor at the Universtity of Iowa. He informed me that Louis A. Turner had resigned as head of the Department of Physics after four years and that he (Tyndall) had suggested me as a possible successor. I was thrilled by this prospect and soon thereafter made a short visit to Iowa City for interviews and a departmental colloquium. Several weeks dragged on after I returned to Silver Spring, with no news. I finally received a letter from Tyndall advising me that they had offered the position to the individual who was their first choice and were awaiting his response. Another few weeks of suspense came to an end when Tyndall called to offer me the position, which would also carry the rank of full professor. At that time Abigail, my wife of five years, had been west of the Mississippi only once and considered Iowa to be terra incognito from the cultural point of view. Nonetheless, she agreed to support my decision whatever it might be. I then accepted the offer but told Tyndall that I would need six months to wind up my obligations at APL. This was agreed.

On a very cold first of January 1951, my wife and I with our two young daughters arrived in Iowa City in our old station wagon, pulling an even older trailer containing most of our earthly possessions. We plowed through the snow to move into a "barracks apartment," one of a cluster of small metal-sheathed buildings that had been erected during the war as temporary quarters for naval cadets and other personnel associated with the university. The sole source of heat was a cast-iron stove that was fed fuel oil from an external 55-gal drum by gravity flow through a small copper tube. The small living room could be made comfortably warm, but the remainder of the apartment presented a challenging problem in heat transfer. However, the monthly rent was only $35.

I entered my new duties with enthusiasm and dedication. I had no research budget, but the department had an excellent machine shop and two skilled instrument makers, as well as a large stock of more-or-less obsolete but still usable electrical instruments.

With the help of George W. Stewart I got a small but very important grant from the private Research Corporation as seed money and started research on cosmic rays using balloon-borne equipment; I also recruited several able graduate students as collaborators. Soon thereafter, I wrote a proposal to the US Office of Naval Research (ONR) for measuring the primary cosmic-ray intensity at high latitudes above the appreciable atmosphere, using small military-surplus rockets carried to an altitude of about 50,000 ft by a balloon and launched from that starting point to reach a summit altitude of some 250,000 ft. By this inexpensive technique, I hoped to resume high-altitude research on a low budget. The proposal was accepted. Support by the ONR has continued without a break for the subsequent 38 years and has provided the base for all of my research during this period.

In the summer of 1952, two of my students, Leslie Meredith and Gary Strein, our lab technician Lee Blodgett, and I made our first rockoon (rocket-balloon combination) expedition to measure the cosmic-ray intensity above the atmosphere in the Arctic. We traveled on the Coast Guard icebreaker USCGC *Eastwind*, whose primary mission was the resupply of the weather station at Alert at the shore of the Arctic Ocean on northeastern Ellesmere Island. We released balloon-borne Deacon rockets from the heliocopter deck whenever we could persuade the captain to steam downwind for an hour while we inflated and released the balloon under zero relative wind conditions. After several failures we diagnosed and cured the problem and got a succession of successful flights to altitudes of about 200,000 ft at locations off the coast of Greenland—the first research rocket flights ever made at such high geomagnetic latitudes. All of the instrumentation, including the telemetry transmitters and nose cones, were built in our own shop; a single Geiger-Mueller (G-M) tube was the radiation sensor.

We reported our results with pride to the UARRP and set to work to refine the instrumentation using ionization chambers and scintillation counters as well as G-M tubes.

During succeeding summers, Arctic rockoon expeditions on various ships were the heart of our work. These were led by Melvin Gottlieb and Frank McDonald. The 1953 expedition yielded a remarkable new finding—namely, the first direct detection of the electrons that, we surmised, were the primaries for producing auroral luminosity.

For a 15-month period in 1953–54, my family and I took a leave from the University of Iowa to join Lyman Spitzer at Princeton University in an experimental program to investigate the confinement of hot plasma by a magnetic field in a twisted figure-eight-shaped tube that he called a stellerator—so named with the hopeful prospect of providing a demonstration of controlled thermonuclear fusion in deuterium and eventually in a mixture of deuterium and tritium. All of this recalled my PhD thesis. I built and operated a crude model of such a machine, called the Model B-1 stellerator. (Model A was a previously built smaller device of table-top size.) I demonstrated the validity of Spitzer's rotational transform of magnetic fields in the twisted toroid with a miniature electron gun and fluorescent screen and got plasma confinement times of a few milliseconds in a hydrogen plasma. The difficulty of making a much larger machine of this nature and, as it appeared to me, the remote prospect for achieving self-sustained fusion on a reasonable time scale convinced me to return to Iowa and resume my high-altitude research, which was already yielding significant original results. This I did in August 1954.

Plans for the International Geophysical Year were by then being formulated, and my colleagues and I on the UARRP were eager to add investigations with rocket-borne equipment to these plans. I proposed to continue the rockoon program with further auroral, cosmic-ray, and magnetic-field measurements in the Arctic, equatorial latitudes, and the Antarctic. This program was funded by the National Science Foundation within its special IGY program. Our field work culminated in 1957 with two shipboard expeditions that I led. The first was aboard the USS *Plymouth Rock* from Norfolk to northwestern Greenland; the second, aboard the large icebreaker USS *Glacier* from Boston via the Panama Canal to the Central Pacific and thence to Antarctica. Of the 30 rockoon flights that Laurence Cahill and I attempted during these two expeditions, which ranged from 79°N to 75°S latitude within a four-month period, 20 were successful in yielding high-altitude cosmic-ray, auroral particle, and magnetic-field data.

EARLY SATELLITE WORK

In 1956 I made a formal proposal to the IGY directorate for a simple but globally comprehensive cosmic-ray investigation using one of the early US Earth satellites as a powerful follow-on to my previous and planned work with rockets. In addition, I proposed study of the auroral primary radiation, also on a global basis, if and when orbits of sufficiently high inclination were available.

The cosmic-ray proposal was given a favorable rating, placed in the pool of experiments to be conducted by one of the early IGY satellites, and funded by the National Science Foundation. The development of the instrument was in the capable hands of George H. Ludwig, a former Air Force pilot and a graduate student at Iowa. He introduced many novel features, including the use of then new transistors throughout the electronics and the design and construction of a miniature magnetic tape recorder.

Following the Soviet's successful flights of the first Earth satellite *Sputnik I* and then *Sputnik II*, the Army's rocket vehicle Jupiter C was adopted as a US alternative to the planned but faltering Vanguard vehicle for placing an early US payload into Earth orbit. By virtue of preparedness and good fortune, the Iowa cosmic-ray instrument was selected as the principal element of the payload of the first flight of a four-stage Jupiter C, launched on 31 January 1958 (1 February GMT).

Both the vehicle and our instrument worked. The data from the single Geiger-Mueller tube on *Explorer I* (as the payload was called) yielded the discovery of the radiation belt of the Earth—a huge region of space populated by energetic charged particles (principally electrons and protons), trapped within the external geomagnetic field. The attempted launch of *Explorer II* was a vehicular failure, but the launch of *Explorer III* in 26 March 1958, with an augmented version of the Iowa instrument, was successful. The *Explorer III* data provided massive confirmation of our earlier discovery and clarified many features of the earlier body of data.

Soon thereafter we were invited to provide radiation-detecting instruments for two satellites that were to observe the effects of several nuclear bombs to be detonated after delivery to high altitudes by rockets. On a time scale of less than three months, Carl McIlwain, George Ludwig, and I designed and built the radiation packages for these satellites—using much smaller and more discriminating detectors, chosen for the first time with knowledge of the existence of the natural radiation belts and the enormous intensity of charged particles therein.

Explorer IV was launched successfully on 26 July 1958. Our apparatus operated as planned and provided the principal body of observations of the artificial radiation belts that were produced by the three high-altitude nuclear bursts—called Argus I, II, and III. The back-up launch of our apparatus on *Explorer V* was a vehicular failure. Analysis of our *Explorer IV* data on the natural radiation belt as well as on the artificial radiation belts from the Argus bursts propelled the entire subject to a new level of understanding and broad scientific interest.

The first Soviet confirmation of the existence of natural radiation belts came from *Sputnik III*, launched in May 1958.

Late in 1958, the Iowa group supplied radiation detectors on two missions, *Pioneer I* and *Pioneer III*, that were intended to impact the Moon. The lunar objective was not achieved, but our data established the large-scale structure and radial dimensions of the region containing geomagnetically trapped radiation. Another lunar flight, *Pioneer IV* (also unsuccessful in reaching the Moon), carried our apparatus in early 1959 and provided a valuable body of confirmatory data.

SPACE RESEARCH UNDER THE AUSPICES OF THE NATIONAL AERONAUTICS AND SPACE ADMINISTRATION AND THE OFFICE OF NAVAL RESEARCH

Our early satellite work was done under the auspices of the IGY/National Science Foundation program, the Office of Naval Research, the Army Ballistic Missile Agency, the Jet Propulsion Laboratory, and the Air Force.

The creation of the National Aeronautics and Space Administration (NASA) as a new, civilian agency of the federal government was formalized by President Eisenhower's signature on the enabling legislation on 10 October 1958. As NASA got itself organized, it moved toward becoming the central national agency for the planning and support of space science and applications. Nonethless, a substantial effort in these areas continued under the auspices of the military departments of the Department of Defense, our principal relationship therein being with the Office of Naval Research.

The creation of NASA led to a dramatic change in space research in the United States. Whereas previously it had been performed by only a small cadre of individuals who might well be described as members of the UARRP and their immediate associates, principally in military and quasi-military laboratories, it then assumed national scope and became "civilianized." In anticipation of the NASA legislation, the National Academy of Sciences established the Space Science Board (SSB) in the early summer of 1958 to advise the federal government on the conduct of scientific research in space. This board was chaired during its important and most influential period by Lloyd Berkner. I was an original member of the SSB and served from 1958 to 1970 and again from 1980 to 1983. A major planning study was conducted under my chairmanship during a two-month period in the summer of 1962 at the University of Iowa. This study yielded a classical document and became the prototype for subsequent summer studies by the SSB. Space science in the United States benefited greatly from the close relationship and mutual respect between Berkner and James Webb, the second administrator of NASA and an especially effective one during the period of great growth of the agency. Indeed, members of the SSB had the heady, but only partially true, perception that they were writing the national scientific program in space in the form of well-considered advice from the interested segment of the scientific community.

Space exploration was transformed from being an arcane field with only a handful of participants to an activity of high national visibility. Members of Congress vied for membership on freshly created space committees.

Senator Lyndon B. Johnson and Congressman John W. McCormack led the legislative drive to put the United States in space on a scale adequate to restore national pride and international prestige following what was perceived as national humiliation by the early successes of the Soviet Union, previously thought by most Americans to be a technologically backward nation. The political scene culminated in the hesitant but eventually dramatic decision by President John F. Kennedy to undertake the landing of a man on the Moon and his safe return to the Earth. His formal public announcement of the Apollo project was made on 25 May 1961. The politics of this decision has been discussed voluminously by many, many others, and I have nothing to add.

In parallel with the Apollo project, programs of space science and the numerous practical applications of space technology were also flourishing. These had much less public visibility but in the long run have proven to be of far greater importance and durability.

Our research at Iowa centered on expanding our knowledge of the energetic particle population of the Earths's external magnetic field and the multifold physics therof. In 1959 Thomas Gold suggested the term " magnetosphere" for the region around the Earth in which the geomagnetic field has a controlling influence on the motion of charged particles, and the term "magnetospheric physics" was widely adopted. Magnetosphere joined the already established list of "spheres"—atmosphere, ionosphere, mesosphere, thermosphere, etc—as a geophysical term.

Much earlier, the presence of thermal and quasi-thermal plasma (ionized gas) in the Earth's magnetic field had been established by Owen Storey and others in the interpretation of "whistlers," a low-frequency electromagnetic phenomenon resulting from lightning strokes.

In situ measurements of this plasma became a central objective of magnetospheric physics using Earth satellites. Another central objective was the investigation of the physics of the aurorae, geomagnetic storms, and the ring current of the Earth. Also, space technology opened up new fields of investigation of cosmic rays, energetic particles from the Sun, solar X-ray flares, and the detailed nature of the interplanetary medium.

The Iowa group played an imprtant role in these developments. Louis Frank made a marked advance in our understanding of the plasma physics of the magnetosphere by developing and flying, first on the *NASA/Orbiting Geophysical Observatory II* in 1965, a low-energy proton electron differential energy analyzer (LEPEDEA) that was sensitive to particles having energies as low as 1 keV.

With the support of the Office of Naval Research and later of NASA we developed and built complete satellites with full complements of scientific instrumentation, thus becoming the first university to succeed in this comprehensive undertaking. Indeed, at one point in time, we had built and flown more satellites than the combined number built by all foreign nations, excepting the Soviet Union. The Injun series of Iowa satellites comprised *Injun I* (launched 29 June 1961), *Injun II* (not placed in orbit because of a launch vehicle failure on 24 January 1962), *Injun III* (launched 12 December 1962), *Injun IV* (launched 21 November 1964), and *Injun V* (launched 8 August 1968). These were placed in low-altitude, high-inclination orbits and had investigation of the aurorae as one of their primary objectives. A notable advance was made by Donald Gurnett in devising and successfully flying a VLF (very low frequency) radio receiver on *Injun III*. A large variety of plasma/wave phenomena were observed, and this new field of investigation began to assume an important role in magnetospheric physics.

We also provided, under increasingly competitive circumstances, instruments as part of the scientific payload of NASA spacecraft: *OGO I, II, III*, and *IV*; *Explorers VII, XII*, and *XIV*; and *IMP*'s (Interplanetary Monitoring Platforms) *-D, -E*, and *-F* (*Explorers 33, 35*, and *34*). *IMP-D* was placed into a very eccentric orbit of Earth with apogee beyond the Moon's orbit, and *IMP-E* was injected into a durable orbit around the Moon. Both of these spacecraft were exceedingly fruitful in studying solar X rays and solar energetic particles and in exploring the outer fringes of the magnetosphere—especially the magnetotail, which had been discovered by other groups using *Explorer VI* and studied further by us and others with *Explorer XII* and *Explorer XIV* in very eccentric orbits.

The latest of the Universtiy of Iowa's small satellites was *Hawkeye I*, which was placed in an eccentric, 90° inclination orbit in 1974 and continued to operate properly until its reentry into the atmosphere nearly four years later. It yielded important results on the configuration of the bow shock and magnetopause and on the topology of the geomagnetic field at large radial distances over the northern polar cap and in the vicinity of the polar cusp—a special feature of central importance in the entry of solar plasma into the magnetosphere.

Other Earth satellite missions in which the University of Iowa has had an important role have included *IMP-G, IMP-I, IMP-H, IMP-J, UK-4* (*United Kingdom-4*), S^3 (*Small Standard Satellite*), *DE-1* (*Dynamics Explorer I*), and *ISEE-1, -2*, and *-3* (*International Sun Earth Explorers 1, 2*, and *3*). Frank's imaging camera on *DE-1* has provided a large and classical album of global images of the aurorae and other faint light features of the Earth under lighting conditions previously thought to make such images impossible. One of my great regrets has been that Sydney Chapman did not live to see these, inasmuch as he often expressed such an aspiration when he lectured on auroral physics at the University of Iowa in 1963. Meanwhile, the investigations of Gurnett and his colleagues with their VLF receivers on *ISEE-3* have been exceedingly fruitful, as have the plasma measurements on *ISEE-1* and *-2* by Frank et al. The imaginative scheme of Robert Farquhar at the Goddard Space Flight Center for diverting *ISEE-3* from its original station at the Ll Lagrangian libration point of the Earth-Sun system to an orbit enabling it to fly by Comet Giacobini-Zinner, the spacecraft being then renamed *ICE* (*International Comet Explorer*), resulted in the first in situ observation of dust and plasma physical phenomena associated with a comet (September 1985).

PLANETARY EXPLORATION

As early as 1960 and in parallel with the activities just sketched, one of my driving aspirations was to push on with magnetospheric studies of the other planets.

The emphasis at NASA in the early 1960s was on manned lunar flights. But the Jet Propulsion Laboratory (JPL) and other groups had already made extensive general studies of the ballistics of flight to other planets—especially Venus and Mars. The interest in Mars was driven by the desire for geological studies of its surface and, perhaps more importantly, by the desire to search for any form of biological activity there. Also, Mars and Venus were ballistically much easier to reach than was Mercury or the outer planets. The first planetary target to be adopted by JPL/NASA was Venus. I proposed a simple radiation detector for the first mission with the purpose of searching for the existence of a Venusian radiation belt and the consequent inferences on the magnetization of this planet, then completely unknown. My instrument was selected and was incorporated into the payload of *Mariner I*, an early in-flight failure, and of *Mariner II*, launched successfully on 27 August 1962. The cruise phase was quite successful, yielding, most importantly, the first continuous measurements of the solar wind by Conway Snyder and Marcia Neugebauer and of the interplanetary magnetic field by Paul Coleman et al, as well as the detection of numerous solar energetic particle events by my apparatus and a companion instrument of Hugh Anderson and Victor Neher. *Mariner II* passed by Venus on 14 December 1962 at a radial miss distance of 41,000 km. In our measurements there was not the slightest indication of the presence of the planet, thereby implying an upper limit on its magnetic moment as 0.18 that of the Earth, its "sister" planet. A casual, and perhaps even correct, interpretation of this result is that Venus is simply rotating too slowly (period 243 days) to drive an internal self-excited dynamo

I had an improved model of the radiation instrument on *Mariner III* (launch failure) and *Mariner IV* (launched successfully of 28 November 1964), which made the first-ever encounter with Mars on 15 July 1965. The interplanetary data yielded a nearly continuous record of solar X-ray flares and of the presence of energetic solar particles, including the discovery of energetic solar electrons and, in stereoscopic combination with data from *Explorer 35* in lunar orbit, the first determination of the altitude in the Sun's atmosphere at which 2–10 Å X rays are emitted. At Mars, as at Venus, we got a null result and inferred an upper limit on Mars' magnetic moment as 0.001 that of the Earth.

Meanwhile, as a member of the Space Science Board and, from 1966–70, of the Lunar and Planetary Missions Board, I had adopted the role of being a special and unremitting advocate of missions to the outer planets—especially Jupiter. The first fruits of these efforts were the adoption by NASA of two missions to Jupiter—later called *Pioneer 10* and *Pioneer 11*—with emphasis on energetic particle and magnetic-field measurements. A special motivation for this emphasis was the radio-astronomical evidence that Jupiter has a huge radiation belt whose population of relativistic electrons emits the observed synchrotron radiation in the decimetric wavelength range. The *Pioneer 10/11* project was managed by the Ames Research Center of NASA with a keen concern for optimizing the scientific yield of the mission. The spacecraft were built by the TRW Company. My proposal for an energetic particle instrument was accepted, after some reduction in its scope, during the vigorous national competition for payload space. *Pioneer 10* was launched successfully on 3 March 1972 and *Pioneer 11* on 6 April 1973. For the subsequent 17 years, the inflight data from these two spacecraft have been a central part of my research life and that of several of my students and associates. *Pioneer 10* made the first-ever encounter (December 1973) with Jupiter and yielded a large body of new knowledge, most especially on its magnetosphere. *Pioneer 11* encountered Jupiter a year later along a different trajectory and confirmed and substantially expanded the earlier findings. *Pioneer 10* has continued on an escape trajectory out of the solar system and, at the date of writing, is about 47 AU from the Sun (nearly 7 billion km)—the most remote man-made object in the Universe. It is still working well and providing daily data on cosmic rays and the interplanetary medium in the outer heliosphere. After its close encounter with Jupiter, *Pioneer 11* made the first-ever encounter with Saturn in September 1979, discovering its magnetosphere and yielding a rich body of new information on the planet itself and its system of rings and satellites. This spacecraft is now also on a solar-system-escape trajectory at a current distance of over 28 AU, and it too is transmitting data on a daily basis.

In the late 1960s and early 1970s a Grand Tour of the Outer Planets was being advocated by the Jet Propulsion Laboratory, in particular, and by other planetary enthusiasts who were advising NASA on new programs. JPL had shown that the forthcoming configuration of the outer planets Jupiter, Saturn, Uranus, and Neptune (a once-in-179-year phenomenon) would make it ballistically feasible to have a single spacecraft fly by all four of these remote planets. The Grand Tour, as such, was a budgetary casualty of late 1970. Soon, thereafter, I was asked by JPL to chair a Science Working Group to develop a more modest-sounding mission, tentatively called MJS (Mariner/Jupiter Saturn). The two-spacecraft mission that we developed was eventually approved and came to life in 1974. It was later renamed Voyager. Although the term Grand Tour was now eschewed in polite conversation, it did not escape our attention that the configuration of the outer planets was independent of budgetary-political considerations in the White House and the Congress.

The successes of the two Pioneer missions produced a greatly enhanced interest in the Voyager missions, as well as in ground-based study of the outer planets. Competition for payload space brought forth a wealth of proposals of new and sophisticated instruments and eventual selection of an excellent complement.

Both Voyagers were launched successfully in the late summer of 1977. Each flew by Jupiter and Saturn and provided great advances, most notably in high-resolution imaging of the atmospheres of the planets, of their satellites, and of their rings and in our understanding of the plasma physics of their magnetospheres. Since its Saturn encounter, *Voyager 1* is on a solar-system-escape trajectory, but *Voyager 2* made the first-ever encounter with Uranus in early 1986 and is now approaching Neptune for a 25 August 1989 encounter—thus prospectively achieving the objectives of the Grand Tour as visualized at the outset of this program.

I have had no part in the execution of the Voyager program but have been a guilty bystander, so to speak, and one of its enthusiastic fans.

In 1976–77, I chaired still another JPL/Ames Research Center science working group, called JOP/SWG [Jupiter Orbiter with (Atmospheric) Probe/Science Working Group]. Our purpose was to develop a follow-on Jupiter mission of more advanced capability than the Pioneers and the Voyagers. This program would have a deep atmospheric entry probe and an orbiter having a useful lifetime of at least two years, in contrast to the limited period (days to weeks) of nearby observation available on a fly-by.

The mission, renamed *Galileo*, has suffered a plethora of delays—financial, political, and technical—principally as the result of the inadequacies and defaults of the shuttle launching system, which had been adopted by NASA in the late 1970s and 1980s. The launch of *Galileo* is now scheduled for October 1989, but many uncertainties remain. Also, because of the less than originally planned capability of the shuttle, it has been necessary to adopt an ingenious but very long flight path to Jupiter, requiring over 6 years vs. the 20-month flights of *Pioneers 10/11*. I recognize that the probability of my own survival to 1995 is substantially less than unity.

Nonetheless, I still hope to function in my interpretative role as an Interdisciplinary Scientist beginning after *Galileo*'s scheduled entry into orbit around Jupiter in that year.

CONCLUDING COMMENTS

In the period 1951–85, I served as head of the Department of Physics (which became the Department of Physics and Astronomy in 1969) of the University of Iowa. My formal teaching involved a full gamut of courses: General Physics, General Astronomy, Electricity and Magnetism, Introduction to Modern Physics, Radio Astronomy, Intermediate Mechanics, and a specialized course in Solar-Terrestrial Physics. Perhaps my favorite was General Astronomy, an introductory but rigorous course on the solar system, with laboratory, which I taught for 17 years.

My closest working relationships with students involved ones at the graduate level. The following are those who finished advanced degrees under my guidance. The first date in the parentheses after each name is the date of an MS degree, the second of a PhD degree. Some students who did their MS work with me later earned a PhD elsewhere or under another advisor at Iowa. Others terminated their graduate work at the MS level. Every one was a collaborator in the fullest sense of the word, a fact that is amply represented in authorship or coauthorship of published work.

JOHARI BIN ADNAN	(1983, —)
SIXTEN INGVAR ÅKERSTEN	(1969, —)
HUGH RIDDELL ANDERSON	(1958, —)
THOMAS PEYTON ARMSTRONG	(1964, —)
DANIEL N. BAKER	(1973, 1974)
KENNETH E. BUTTREY	(1955, —)
LAURENCE JAMES CAHILL, JR.	(1956, 1959)
CHARLES P. CATALANO	(— , 1971)
JAMES R. CESSNA	(1965, —)
PHILLIP CHANG	(1962, —)
TSAN-FU CHEN	(1973, 1978)
JOHN D. CRAVEN	(1964, —)
JOSÉ M. da COSTA	(1971, —)
JERRY F. DRAKE	(1967, 1970)
ROBERT A. ELLIS, JR.	(— , 1954)
R. WALKER FILLIUS	(1963, 1965)
HERBERT R. FLINDT	(1968, —)
LOUIS A. FRANK	(1961, 1964)
JOHN W. FREEMAN	(1961, 1963)
THEODORE A. FRITZ	(1964, 1967)
SISTER JEAN GIBSON, O.S.B.	(— , 1969)

CYNTHIA LEE GROSSKREUTZ	(1982, —)
DONALD A. GURNETT	(— , 1965)
ROLLIN CHARLES HARDING	(1966, —)
H. KENT HILLS	(1964, —)
WILLIAM G. INNANEN	(— , 1972)
ROBERT CHANDLER JOHNSON	(1957, —)
JOSEPH E. KASPER	(1955, —)
STAMATIOS MIKE KRIMIGIS	(1963, 1965)
CURTIS D. LAUGHLIN	(1960, —)
WEI CHING LIN	(1961, 1963)
THOMAS A. LOFTUS	(1969, —)
GEORGE HARRY LUDWIG	(1959, 1960)
CARL E. McILWAIN, JR.	(1956, 1960)
LESLIE H. MEREDITH	(1952, 1954)
RAYMOND F. MISSERT	(1955, 1957)
STEVEN R. MOSIER	(1967, —)
MICHAEL O'CONNOR	(1968, —)
MELVIN N. OLIVEN	(1966, 1970)
MARK E. PESSES	(1976, 1979)
GUIDO PIZZELLA	(— , 1962)
ROBERT C. PLACIOUS	(1953, —)
RICHARD LOUIS RAIRDEN	(1981, —)
BRUCE A. RANDALL	(1969, 1972)
JOANNA M. RANKIN	(— , 1970)
ERNEST C. RAY	(1953, 1956)
NICOLAOS A. SAFLEKOS	(— , 1975)
EMMANUEL T. SARRIS	(— , 1973)
MELVIN SCHWARTZ	(— , 1958)
DAVIS D. SENTMAN	(— , 1976)
STANLEY D. SHAWHAN	(1965, —)
HAROLD E. TAYLOR	(— , 1966)
JAMES DENNIS THISSEL	(1963, —)
MICHELLE F. THOMSEN	(1974, 1977)
WILLIAM R. WEBBER	(1955, 1957)
JOEL M. WEISBERG	(1975, 1978)
CHARLES D. WENDE	(1966, 1968)
MICHAEL JAMES WIEMER	(1964, —)
SAIYED MASOOD ZAKI	(1964, —)

In addition, I have benefited from the highly competent efforts of an uncounted number of members of our technical staff, of whom the following are representative: William A. Whelpley, Roger F. Randall, Robert B. Brechwald, Evelyn D. Robison, John E. Rogers, Joseph G. Sentinella, Donald C. Enemark, W. Lee Shope, Edmund Freund, and Robert Markee.

Finally, and most importantly, I am indebted to my wife of 44 years, Abigail, and our five children, who have provided the circumstances under which sustained and intensive professional work has seemed worthwhile.

Annu. Rev. Earth Planet. Sci. 1990. 18:27–53

THE VALLES/TOLEDO CALDERA COMPLEX, JEMEZ VOLCANIC FIELD, NEW MEXICO[1]

Grant Heiken, Fraser Goff, Jamie N. Gardner, and W. S. Baldridge

Earth and Space Sciences Division, Los Alamos National Laboratory, Los Alamos, New Mexico 87545

J. B. Hulen and Dennis L. Nielson

University of Utah Research Institute, 391 Chipeta Way, Suite C, Salt Lake City, Utah 84108

David Vaniman

Earth and Space Sciences Division, Los Alamos National Laboratory, Los Alamos, New Mexico 87545

INTRODUCTION

Valles caldera is famous as an example of a resurgent caldera and for its high-temperature geothermal system of volcanic origin (Smith & Bailey 1968, Dondanville 1978). Although the caldera and its hydrothermal system are the products of events that have occurred during the last 1.12 Myr, the Jemez volcanic field has experienced continuous volcanism for the last 13 Myr and is known to have had at least three major periods of hydrothermal activity (Gardner et al 1986, WoldeGabriel & Goff 1989). This paper does not review all aspects of research concerning the Jemez

[1] The US Government has the right to retain a nonexclusive royalty-free license in and to any copyright covering this paper.

volcanic field, only the significant discoveries that have been made since 1980.

The Jemez volcanic field (Figures 1, 2) consists of a diverse suite of basaltic through rhyolitic rocks that were erupted from >13 to 0.13 Ma, although the field is best known for the Bandelier Tuff and the Toledo and Valles calderas (Smith & Bailey 1966, 1968, Gardner et al 1986, Self et al 1986). The volcanic field overlies the western edge of the Rio Grande rift at its intersection with the Jemez volcanic lineament. The lineament is a chain of Miocene to Quaternary volcanic centers stretching from southeastern Colorado across New Mexico to central Arizona (Mayo 1958, Laughlin et al 1976, Aldrich 1986). The largest volume and diversity of volcanic rocks erupted along the Jemez volcanic lineament are in the Jemez

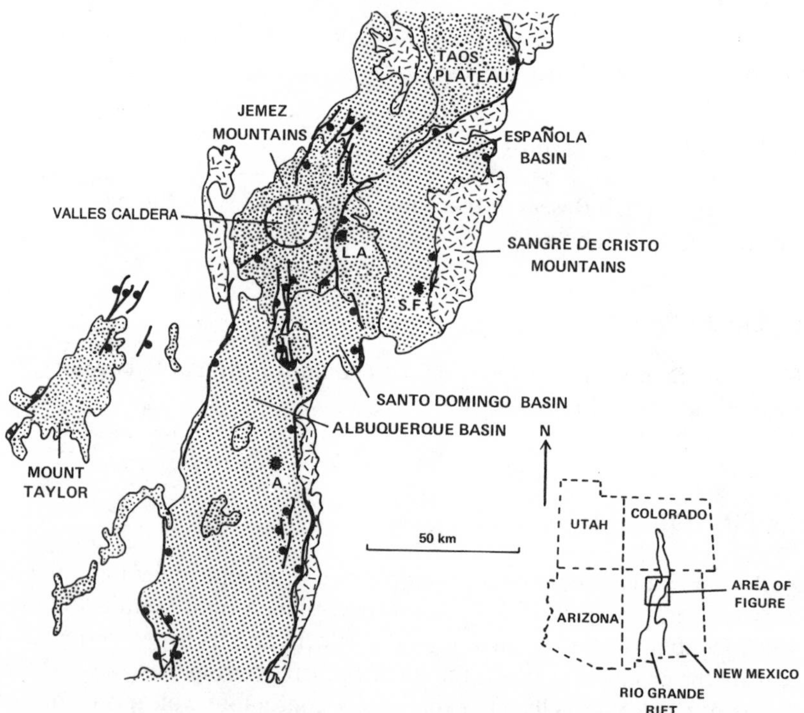

Figure 1 Regional map showing generalized relations of the Jemez volcanic field (Jemez Mountains) to basins of the north-central Rio Grande rift and to the Jemez volcanic lineament; major fault zones of the region are also shown. LA, SF, and A are the cities of Los Alamos, Santa Fe, and Albuquerque, respectively (from Self et al 1986). Dashes = Precambrian metamorphic/plutonic basement; regular stipple = sedimentary fill of the Rio Grande rift; irregular stipple = volcanic rocks.

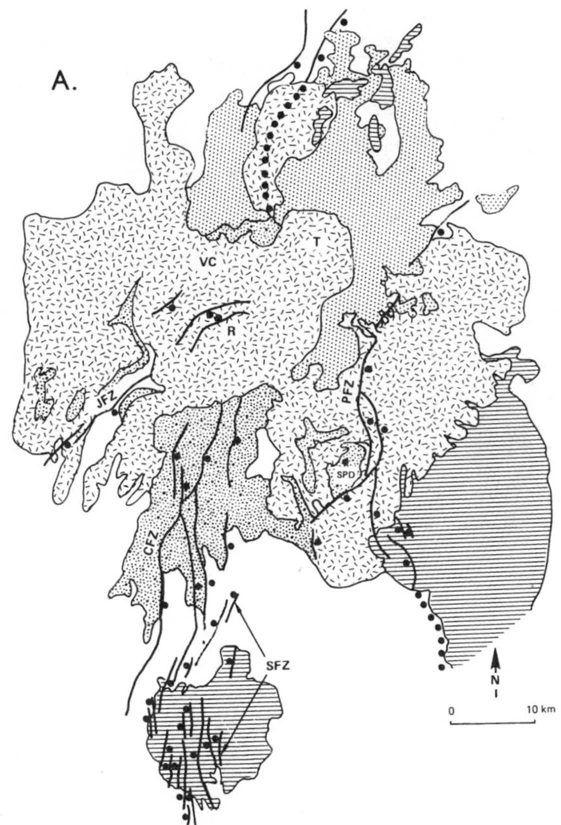

Figure 2 (*A*) Generalized geologic map of the Jemez volcanic field, showing distribution of major stratigraphic groups. Irregular stipple = Keres Group formations; coarse, regular stipple = Polvadera Group formations; random dash = Tewa Group formations; horizontally ruled pattern = young basalt fields. Major fault zones (dotted where inferred) are the Jemez fault zone (JFZ), Santa Ana Mesa fault zone (SFZ), Cañada de Cochiti fault zone (CFZ), and Pajarito fault zone (PFZ). Other abbreviations: VC = Valles caldera, R = resurgent dome of caldera, T = Toledo embayment, SPD = St. Peter's Dome (from Gardner & Goff 1984). (*B*) Basic stratigraphy of the Jemez volcanic field (Bailey et al 1969) compared with revised stratigraphy of Gardner et al (1986); patterns are the same as those in (*A*). Dashed lines indicate uncertainty, and more revisions may be forthcoming. The revised stratigraphy diagram is also a schematic south-to-north (bottom-to-top) section through the volcanic field.

[*continued overleaf*]

KLINCK MEMORIAL LIBRARY
Concordia University
River Forest, IL 60305-1499

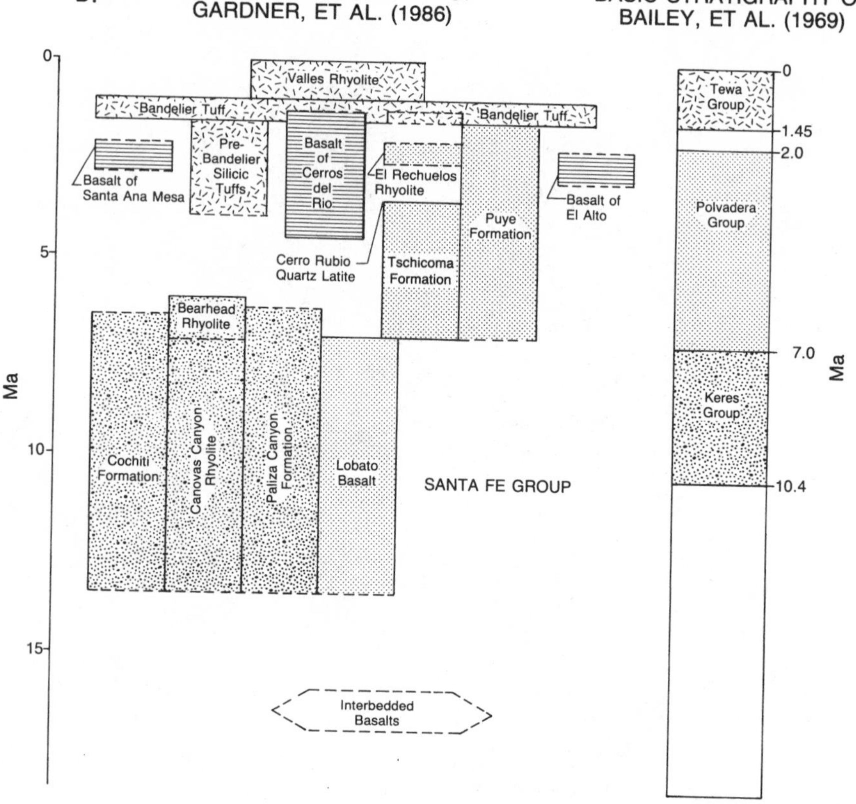

Figure 2 (continued)

Mountains. Tertiary basin-fill rocks of the Española and Santo Domingo Basins of the Rio Grande rift underlie volcanic rocks on the eastern side of the volcanic field, whereas Paleozoic to Mesozoic sedimentary rocks and Precambrian crystalline rocks of the Colorado Plateau underlie volcanic rocks on the western flank.

Early reconnaissance studies of the geology and petrology of the Jemez Mountains were conducted by J. W. Powell and J. P. Iddings (Iddings 1890). Comprehensive investigations of the geology, volcanism, and hydrology of the Jemez Mountains began in the 1920s and intensified during the 1940s by members of the US Geological Survey. These studies continued through the 1960s and, in limited fashion, to the present day (Ross & Smith 1961, Smith et al 1961, 1970, Griggs 1964, Doell et al 1968,

Bailey et al 1969, Smith 1979). Doell & Dalrymple (1966) identified the Jaramillo normal polarity magnetic event at Jaramillo Creek, in the Valles caldera.

SIGNIFICANT GEOLOGIC RESEARCH IN THE JEMEZ VOLCANIC FIELD SINCE 1980

Starting in 1980, diverse research in the Valles caldera region began to appear in the geological literature. In addition to graduate thesis projects (e.g. Stein 1983, Gardner 1985), research was stimulated by geologic investigations of the US Department of Energy (USDOE) Hot Dry Rock geothermal program (Kolstad & McGetchin 1978, Laughlin et al 1983, Heiken & Goff 1983) and by data and samples released from the "Baca" project of Union Oil Company of California (UNOCAL) (Faust et al 1984, Nielson & Hulen 1984). The largest single source of funding for new Valles research has come through the Continental Scientific Drilling Program (CSDP) of the USDOE. The Valles caldera has been discussed for nearly 20 years as a high-priority site for the investigation of fundamental processes in magmatism, hydrothermal systems, and ore deposition. The USDOE Office of Basic Energy Sciences has sponsored scientific investigations, task groups, and workshops to identify critical data gaps and to identify drilling objectives and sites (e.g. Luth & Hardee 1980, Taschek 1981, Heiken 1985). The support of the CSDP has resulted in continuous coring of three scientific core holes in the Valles and Toledo calderas (VC-1, VC-2A, and VC-2B; Goff et al 1986, 1987, Goff & Nielson 1986).

The extensive subsurface data base at Valles caldera (nearly 40 deep wells) makes it the best explored Quaternary caldera complex in the USA and allows for precise planning and realistic expectations in meeting scientific goals. For example, the known composite stratigraphic section in the Valles region is nearly 7 km thick. Collectively, wells within and adjacent to the caldera complex have penetrated 1950 m of Quaternary caldera fill, over 300 m of Tertiary precaldera volcanic and sedimentary rocks, over 800 m of Paleozoic sedimentary rocks, and 3700 m of Precambrian basement (Laughlin et al 1983, Nielson & Hulen 1984).

Tectonic History

Rifting in the vicinity of the Jemez Mountains began no later than 16.5 Ma. The tectonic history of this region has been analyzed by Gardner & Goff (1984) and Self et al (1986). From 13 to 10 Ma, tectonic activity was accompanied by eruptions of basalt and rhyolite along the Cañada de Cochiti fault zone (Figure 2*A*); these volcanic rocks are interbedded with immature, basin-fill conglomerates and laharic breccias of the Cochiti Formation. During the period 10 to 7 Ma, basalt, rhyolite, and basaltic differentiates were erupted along the Cañada de Cochiti fault zone and now form half of the volume of the volcanic field. From 7 to 6 Ma, eruptions of basaltic differentiates and rhyolite ceased, and new eruptions of dominantly mixed magma began, but with a sharp reduction in volcanic activity. From 7 to 4 Ma, a lull in basaltic volcanism occurred, and the Cañada de Cochiti fault zone became inactive.

The volcanic events from 7 to 4 Ma may have been a response to decreasing tectonic activity in the vicinity of the Jemez Mountains. Instead of magma being vented along active faults, as was the case earlier in the history of the volcanic field, pockets of these mantle- and lower-crust-derived magmas coalesced to form the hybrid dacites of the Tschicoma Formation (Polvadera Group). The Puye Formation, a broad volcaniclastic alluvial fan shed off the Polvadera Group's volcanic highland, was deposited between 7 and 4 Ma.

Major tectonic activity shifted from the Cañada de Cochiti to the Pajarito fault zone between 5 and 4 Ma. This shift was accompanied by renewed basaltic activity at 4 Ma, peripheral to and east of the volcanic field (in the rift). Tectonic activity along the Pajarito and Jemez fault zones continued through the Quaternary (Gardner & House 1987). Vertical displacement of the Tshirege Member of the Bandelier Tuff, dated at 1.12 Ma, is 100 to 200 m along the Pajarito fault zone and is 50 m along the Jemez fault zone. Numerous localities show faulting of Quaternary units younger than the Bandelier Tuff along both of these faults (Gardner & House 1987, Goff & Shevenell 1987).

Stratigraphy

The formalized stratigraphy for volcanic and volcaniclastic rocks in the Jemez volcanic field was developed by Bailey et al (1969) and Smith et al (1970). These authors divided the volcanic field into the Keres, Polvadera, and Tewa Groups (oldest to youngest). The Keres Group consists of mostly two-pyroxene andesites and interbedded basalts, the Polvadera Group is mainly dacite, and the Tewa Group is almost all high-silica rhyolites. Bailey, Smith, and coworkers proposed that each group represents a discrete time interval and petrologic entity. After further geologic mapping, radiometric dating, and petrologic analysis, it is now evident that temporal and petrologic overlap occurs among all groups and their many formations (Gardner et al 1986, Loeffler et al 1988).

The following is a summary of major periods of intrusive, volcanic, and hydrothermal activity:

1. Volcanism began at about 16.5 Ma (alkalic basalts in the Santa Fe

Group), but the oldest rocks included as formal units of the volcanic field were erupted between 14 and 13 Ma.

2. Hydrothermally altered volcanic and hypabyssal rocks of the Cochiti mining district (the Bland Group of Stein 1983) are not Eocene or Oligocene as previously thought. They are the cores of Keres Group andesitic volcanoes that were intruded by the Bearhead Rhyolite and later by epithermal quartz veins. Ages of all units within this altered sequence range from 11.2 to 5.6 Ma (Stein 1983, Wronkiewicz et al 1984, Gardner et al 1986, WoldeGabriel & Goff 1989).
3. The Basalt of Chamisa Mesa (10.4 Ma) is chemically, temporally, and petrographically indistinguishable from basalts of the Paliza Canyon Formation.
4. The Cochiti Formation is included within the Keres Group because of intimate spatial, temporal, and tectonic relations to formations in the group.
5. The Lobato Basalt occupies a much wider time range (14–7 Ma) than was previously thought ($\sim$8–7 Ma). Lobato basalts are petrographically and chemically similar to basalts of the Paliza Canyon Formation, and the age range of the two units is the same, indicating that the only distinction between them is geographical.
6. Dacitic rocks of the Tschicoma Formation (7–3 Ma) are chemically and petrographically similar to dacites in the Paliza Canyon Formation. On the other hand, the Tschicoma Formation, originally defined by Griggs (1964) to consist of dacite only, contains large unmapped areas of two-pyroxene andesite grossly similar to the andesites in the Paliza Canyon Formation. Overlap in chemistry, petrography, and age of these formations is significant, and the age break between them acknowledges the geographical distinction built into the formal stratigraphy of Bailey et al (1969).
7. The El Rechuelos Rhyolite, on the northern side of the volcanic field, is restricted to two domes and a smaller vent, all dated at 2 Ma. Other rhyolites in the northern Jemez Mountains are older (7.5–5.8 Ma) (Loeffler et al 1988). The older rhyolites resemble the Canovas Canyon and Bearhead Rhyolites of the southern Jemez volcanic field.
8. The Tewa Group contains recently characterized ignimbrites that predate the Bandelier Tuff and temporally overlap the waning stages of Polvadera Group volcanic activity (Self et al 1986).
9. The Cerro Rubio Quartz Latite, located in the Toledo embayment, chemically and temporally resembles Tschicoma Formation dacites and is now included in the Tschicoma Formation (Heiken et al 1986, Gardner et al 1986).
10. Post-Toledo caldera moat volcanism is more complicated than pre-

viously thought (e.g. Smith 1979). Several rhyolites formerly thought to be post-Valles are now realized to be post-Toledo in age (Heiken et al 1986). Chemical and petrologic variations within the Cerro Toledo Rhyolite define subgroups of rhyolite types and show the restoration of compositional zonation of the Bandelier magma chamber between two caldera-forming eruptions (Stix et al 1988).

11. Post-Valles caldera moat volcanoes were formed by eruption of several distinct magma batches and are grouped as the Valles Rhyolite (Gardner et al 1986, Spell 1987, Self et al 1988).

Volcanic Petrology

With few exceptions, the rocks of the volcanic field are subalkaline and are of the calc-alkaline series. In fact, most andesites of the Jemez Mountains satisfy the criteria suggested by Gill (1981) for high-potassium orogenic andesites. Additionally, many units that have been described previously as dacite, latite, quartz latite, and rhyodacite are all dacites.

Eruptions of mantle-derived basalt span the entire history of the volcanic field, except during a tectonic lull from 7 to 4 Ma. Magmas derived by differentiation of mantle-derived material were erupted early in the history of the field, from 10 to 7 Ma. These magmas, represented primarily by two-pyroxene andesites and by some dacites of the Paliza Canyon Formation, make up over half of the original volume of the volcanic field ($\sim$2000 km^3).

High-silica rhyolitic magmas, derived from partial melts of lower crust, were erupted from 13 to 6 Ma, with pronounced activity between 7 and 6 Ma (Gardner et al 1986). These magma types are represented by the Canovas Canyon and Bearhead Rhyolites of the Keres Group. On the other hand, the "early rhyolite" of Loeffler et al (1988) (fomerly called El Rechuelos) was derived primarily through fractional crystallization of basalt with addition of about 5% of lower crustal rock.

Dacites that were formed by mixing of basalt or basaltic andesite and high-silica rhyolite began erupting at about 7 Ma. Eruptions of these mixed magmas began during the waning phases of the basaltic and high-silica rhyolitic activity discussed above. Most of the mixed magmas were erupted as dacites of the Tschicoma Formation, but some occur in the Paliza Canyon Formation.

Based upon the chemical and petrologic variations within the outflow sheets of the Bandelier ignimbrites (mostly the Tshirege Member), Smith & Bailey (1966) proposed that the Bandelier magma chamber was compositionally zoned. Smith (1979) suggested that the Bandelier pluton evolved from a (Tschicoma) dacitic parent magma. This concept was also proposed by Gardner (1985), using trace-element and rare-earth-element data for all Jemez volcanic rocks. Stix et al (1988) found that the post-Toledo caldera (Cerro Toledo) rhyolites were generated by fractional crystallization of Bandelier magma and not by diffusive processes or by wall-rock assimilation.

The first post-Valles caldera moat rhyolites also appear to have been derived from the Bandelier magma chamber (Spell 1987). The latest eruptions of moat rhyolites are more mafic in composition, have $^{87}Sr/^{86}Sr$ ratios of about 0.705 (versus about 0.710 for the Bandelier Tuff), and could have been derived from a mixture of high-silica rhyolitic (Bandelier) and dacitic magmas (Gardner et al 1986, Self et al 1988).

The volatile content of silicate melt inclusions in quartz phenocrysts of the Tshirege Member of the Bandelier Tuff indicates that the depth to the top of the magma chamber is about 5 km ($\sim$1.5 kbar; Sommer 1977). Warshaw & Smith (1988) investigated the pyroxene, fayalite, and iron-titanium oxide compositions in both members of the Bandelier Tuff. They report that the basal ash fall of the Tshirege (Tsankawi Pumice Bed) was erupted from a magma chamber with a temperature of $\sim$700°C, while the uppermost Tshirege ignimbrite was erupted at a temperature of $\sim$850°C.

Volcanic Activity of the Toledo/Valles Caldera Complex

Ignimbrite plateaus surrounding the Jemez volcanic field are mostly the products of two caldera-forming eruptions that occurred during the last 1.45 Myr. With an estimated minimum cumulative volume of 600 km^3, the two members of the Bandelier Tuff are the most evident products of silicic volcanism of the central Jemez Mountains volcanic field. Concurrent with the eruption of these two batches of rhyolitic tephra was the formation, by collapse, of the Toledo and Valles calderas. Many modern concepts of pyroclastic flow deposits and facies, caldera formation, caldera resurgence, and zoned magma chambers were developed here (Ross & Smith 1961, Smith & Bailey 1966, 1968, Smith 1979, Wright et al 1981).

SAN DIEGO CANYON IGNIMBRITES Underlying the Bandelier Tuff and extending southwest of the present-day rim of the Valles caldera are erosional remnants of two ignimbrites; these ignimbrites were previously designated "pre-Bandelier ignimbrite A" and "pre-Bandelier ignimbrite B" (Self et al 1986) and are now called the San Diego Canyon ignimbrites (Turbeville & Self 1988). The full extent of these deposits is unknown, but they are estimated to have a volume of >5 km^3 (Figure 3*A*; Turbeville & Self 1988). The San Diego Canyon ignimbrites have K/Ar (sanidine) ages of 2.84 ± 0.07 Ma ("B") and 3.64 ± 1.64 Ma ("A"). Chemically, they are similar to, but slightly more mafic than, the two members of the Bandelier Tuff (Self et al 1986, Turbeville & Self 1988).

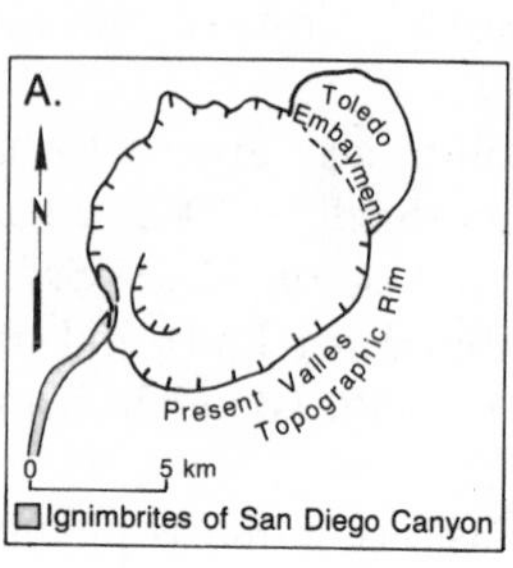

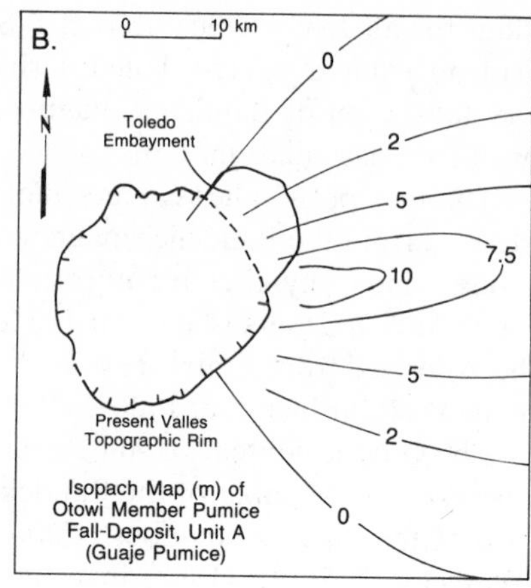

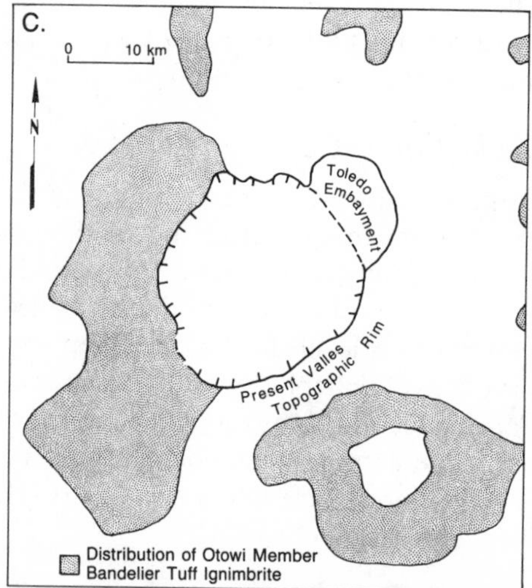

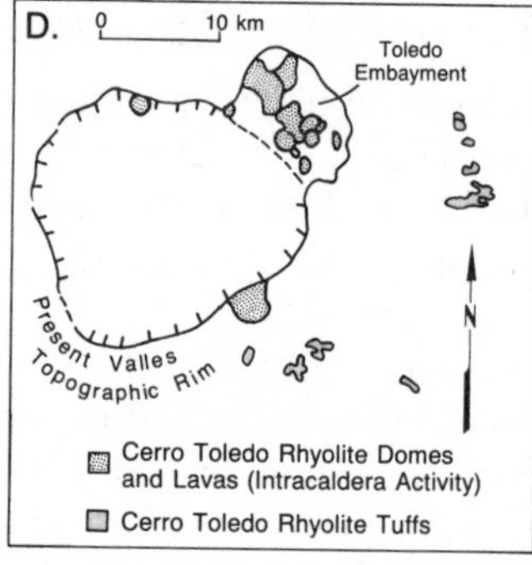

Figure 3 Simplified maps showing distribution of major silicic tuffs and lava domes erupted in the central Jemez Mountains during the last 3 Myr. For reference, the present topographic rim of the Valles caldera and the Toledo embayment is shown on each map. (*A*) Distribution of San Diego Canyon ignimbrites (from Turbeville & Self 1988). (*B*) Isopach map, in meters, of unit A of the Otowi Member, Bandelier Tuff pumice-fall deposit (Guaje Pumice; from Self et al 1986). (*C*) Distribution of the Otowi Member, Bandelier Tuff ignimbrite (from Smith et al 1970). (*D*) Distribution of Cerro Toledo Rhyolite domes, lava flows, and tuffs (from Heiken et al 1986). (*E*) Isopach map of fall unit A, Tshirege Member, Bandelier Tuff pumice-fall deposit (Tsankawi Pumice; from Self et al 1986). (*F*) Distribution of the Tshirege Member, Bandelier Tuff ignimbrite (from Smith et al 1970). (*G*) Distribution of the silicic domes, lava flows, and pyroclastic rocks of the Valles Rhyolite (from Smith et al., 1970) and the El Cajete pumice-fall deposit (from Self et al 1988).

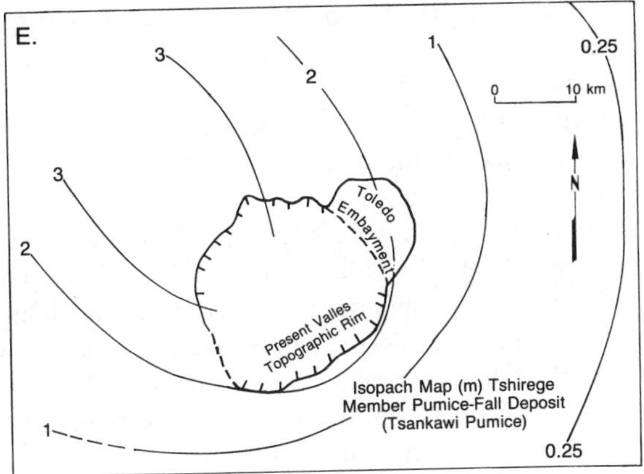

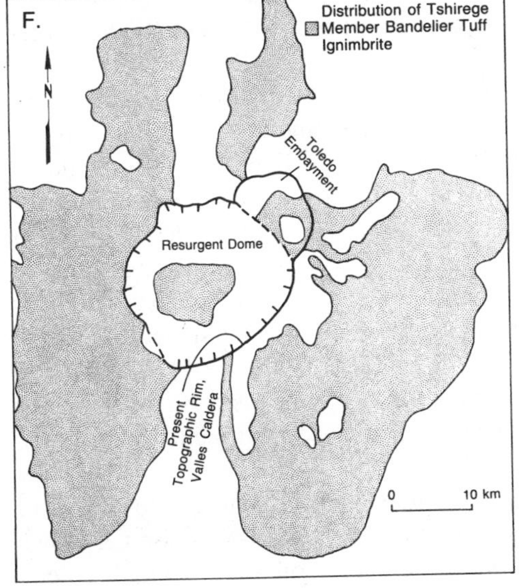

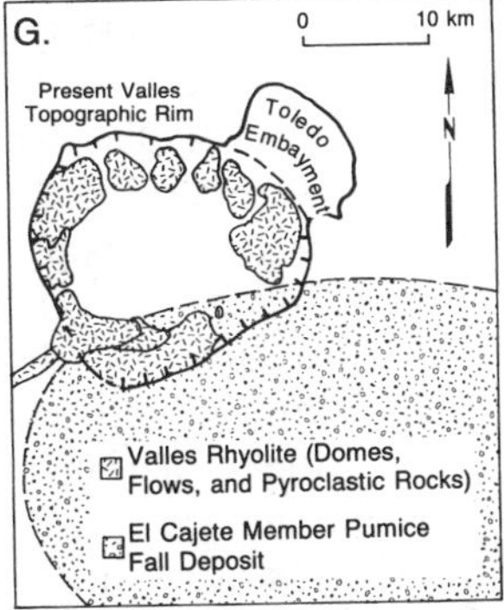

Figure 3 (continued)

Field evidence indicates that these ignimbrites were once extensive, were erupted from what is now the western margin of the Valles caldera, and may have formed a small caldera, which is now hidden by the present-day caldera complex.

OTOWI MEMBER OF THE BANDELIER TUFF AND THE TOLEDO CALDERA The Otowi Member of the Bandelier Tuff consists of a Plinian pumice-fall deposit (Guaje Pumice Bed), which is overlain by thin surge beds and massive pyroclastic flow units. The age of this eruption, based upon several K/Ar dates on sanidine, is 1.45 Ma (Doell et al 1968). The Guaje pumice-fall deposits are distributed to the east of the caldera complex (Figure 3*B*). The five fall units defined cumulatively have a maximum thickness of over 8 m (Self et al 1986).

Surge beds and massive ignimbrites of the Otowi Member are distributed symmetrically about the Valles caldera (Figure 3*C*). Multiple flow units have cumulative thicknesses outside the caldera of as much as 180 m and thicknesses within the caldera of <60 m to 800 m (possibly twice that in the eastern, undrilled half of the caldera complex). The orangish-tan, nonwelded to densely welded ignimbrite contains abundant lithic clasts and pumice clasts in a vitric-crystal or crystal-vitric ash matrix (30–35% phenocrysts of mostly sanidine and quartz). Lithic clasts, which make up from a trace to 30% of the tuffs, are mostly derived from older volcanic rocks but include rare fragments of Precambrian basement (Eichelberger & Koch 1979).

The distribution and thickness of the Otowi Member ignimbrite, the distribution of lag breccias, the flow direction indicators such as aligned pumices (Potter & Oberthal 1983), and the K-Ar dates on an intracaldera arc of four domes (Goff et al 1984) all support the interpretation that the Otowi Member was erupted from ring fractures more or less coincident with those of the younger Valles caldera. These features do not support the interpretation that the Otowi Member of the Bandelier Tuff was erupted from the Toledo embayment, a 9-km-diameter structure located on the northeast margin of the Valles caldera (Figure 2*A*). The Toledo embayment may be of tectonic origin, or it may be a small crater formed at a slightly earlier time; in either case, it is now mostly filled with Cerro Toledo Rhyolite domes.

Within the caldera, the Otowi Member has a maximum thickness of 833 m in well Baca-12 and thins to 177 m in well Baca-4 (Nielson & Hulen 1984). Thus, the unit thins toward the center of the proposed Toledo caldera. This may be an indicator that the caldera did not subside as a massive piston, but rather as a mosaic of blocks, which create a hummocky caldera floor. However, the data suggest that maximum subsidence was centered over the eastern ring-fracture system.

INTRACALDERA VOLCANIC ACTIVITY, TOLEDO CALDERA The Cerro Toledo Rhyolite is a post-Toledo-caldera-collapse sequence of rhyolitic domes, flows, and associated pyroclastic deposits (Izett et al 1981, Heiken et al 1986). Exposed Cerro Toledo Rhyolite domes are clustered mostly within the Toledo embayment and along the north rim and just beyond the southeast rim of the Valles caldera (Figure 3*D*). K-Ar ages of these domes range from 1.5 Ma to 1.2 Ma, placing them in time between eruptions of the two members of the Bandelier Tuff. The domes consist primarily of gray, flow-banded to massive rhyolite, which contains small amounts of sanidine and rare quartz phenocrysts. Contemporaneous with the Cerro Toledo domes are tuffs and epiclastic rocks; most of these consist of very fine-grained phreatomagmatic ash, overlain by interbedded pumice-fall beds and fine-grained ash-fall beds. The tuffs are nearly aphyric, with only traces of K-feldspar and plagioclase.

Using cuttings from deep wells, Nielson & Hulen (1984) have identified a clastic deposit within the caldera fill, the S_3 Sandstone, as being correlative with the Cerro Toledo tuffs. The S_3 Sandstone reaches a maximum thickness of 70 m and separates the Otowi and Tshirege Members of the Bandelier Tuff. This unit appears to be composed of interbedded tuffs and epiclastic sandstones.

TSHIREGE MEMBER OF THE BANDELIER TUFF AND THE VALLES CALDERA The Valles caldera was formed during eruption of the Tshirege Member of the Bandelier Tuff at 1.12 Ma (Doell et al 1968). The eruption sequence consists of a basal pumice-fall deposit overlain by thin surge beds and pyroclastic flow units that make up the ignimbrite (Figure 3*E*) (Self et al 1986).

The basal Plinian pumice-fall deposit, the Tsankawi Pumice Bed, consists of four pumice-fall units and two finer grained ash-fall units, with composite maximum thickness of 3.5 m. Each fall unit has a different dispersal axis, ranging to the south, west, and north (Figure 3*E*).

The Tshirege Member ignimbrite is exposed on nearly all sides of the volcanic field and forms plateaus flanking the central Jemez Mountains (Figure 3*F*). It also forms thick sections of caldera fill, which have been sampled during geothermal drilling. The orangish-tan tuffs of the ignimbrite are nonwelded to densely welded, crystal-vitric to vitric-crystal tuff ($\sim$32% phenocrysts of mostly sanidine and quartz, with traces of hornblende and magnetite). Thicknesses outside the caldera range from 15 m to over 270 m, whereas within the caldera, densely welded tuffs range from 400 m to over 1100 m thick.

THE VALLES RHYOLITE: INTRACALDERA DOMES, LAVA FLOWS, PYROCLASTIC DEPOSITS, AND SEDIMENTARY ROCKS After formation of the Valles caldera, less explosive activity continued within the caldera until about 130 ka. Rhyolitic domes and lava flows were erupted along the caldera ring-fracture system, from the massive Cerro del Medio in the eastern moat to the very small Cerro la Jara dome on the southern margin (Figure 3*G*; Spell 1987). Some domes are up to 700 m thick. Several extensive pumice-fall deposits and small ignimbrites are associated with dome growth. Coarse breccias, gravels, and lacustrine mudstones are interbedded with tuffs and rhyolite lavas.

The youngest (130 ka) intracaldera eruption is that of the Banco Bonito obsidian flow (Marvin & Dobson 1979), which erupted along the southwest caldera ring fracture. The Battleship Rock ignimbrite (well exposed in San Diego Canyon) and the widespread El Cajete pumice-fall bed may have been explosive precursors to eruption of the 30- to 70-m-thick Banco Bonito lava flow (Self et al 1988).

Subsurface Structure and Resurgence of the Valles/Toledo Calderas

The structural framework of the Valles caldera is dominated by high-angle normal faulting (including ring faulting) and resurgent domal uplift of the subsided cauldron block. The Jemez volcanic field, including the calderas, is located at the intersection of northeast-trending faults of the Jemez volcanic lineament with north-trending faults at the western edge of the Rio Grande rift (Aldrich & Laughlin 1984, Gardner & Goff 1984). Reactivation of these earlier structures during caldera formation and subsequent resurgent doming is responsible for much of the complex fault pattern presently observed in caldera-fill rocks. For example, the northeast-trending apical graben of the caldera's resurgent dome is on strike with the Jemez fault zone to the southwest (Figure 2*A*). Ring-fracture and apical graben, high-angle normal faults are believed to be the principal permeable conduits for circulation of the high-temperature Valles hydrothermal system.

The structure of Valles caldera has been discussed by Goff et al (1985), Nielson & Hulen (1984), Heiken et al (1986), and Wilt & Vonder Haar (1986). These interpretations rely on the gravity data acquired by UNOCAL during geothermal exploration and on the drilling results of several organizations. There is a pronounced circular, negative Bouguer gravity anomaly of as much as -35 mgal that coincides with the location of the caldera depression (Segar 1974). The gravity low, however, is asymmetric and is greatest in the eastern caldera, below the Valle Grande.

A model of structural resurgence is constrained by drilling results in the southwestern quadrant of the caldera. Depth to basement increases eastward across faults; it is 1555 m (5100 ft) beneath Sulphur Springs in core hole VC-2B (J. N. Gardner and J. B. Hulen, unpublished data, 1988) and 3110 m (10,200 ft) in well Baca-12 in the Redondo Creek graben (Nielson & Hulen 1984). According to the gravity model, depth to basement is approximately 4570 m (15,000 ft), or about 1525 m (5000 ft) below sea level, in the vicinity of Valle Grande. Such gross asymmetry prompted

Nielson & Hulen (1984) and Heiken et al (1986) to propose that the Valles caldera is a "trap door" caldera (Figure 4).

The intracaldera Tshirege Member of the Bandelier Tuff, 1156 m thick in well Baca-5 (Nielson & Hulen 1984), was structurally uplifted by more than 1000 m during the 100,000-yr interval following caldera collapse. During uplift, extension of the upper resurgent dome resulted in normal faulting and an apical graben.

Magma chamber depth is estimated at between 5 and 7 km (1.5 to 2.0 kbar). Nielson & Hulen (1984) used a structural model to estimate the depth to the top of the (crystallized) Bandelier pluton to be about 4.7 km. Because this estimate accounts for posteruption resurgence, the depth to the magma chamber before eruption of the Bandelier tuff may have been between 5 and 6 km.

Nielson & Hulen (1984) have suggested a domed plate as a model for deformation of the Redondo dome. The model has two important implications: (*a*) On the basis of the amplitude and diameter of the dome, the top of the presumed causative magma body is at a depth of about 5 km; and (*b*) the widespread normal faulting presently observed at the crest of the dome is predicted by the model to have formed during extension above a neutral plane situated about halfway between the dome crest and the top of the magma body (compression prevails beneath this neutral plane). This relation implies that fracture-controlled hydrothermal circulation is confined principally to the upper half of the domed sequence, a conclusion supported by drilling (Hulen & Nielson 1986a).

THERMAL HISTORY Much attention has been focused on the Quaternary thermal regime beneath Fenton Hill, on the southwestern caldera margin (Heiken & Goff 1983). Kolstad & McGetchin (1978) used heat-flow data and dates of major events in the caldera history to derive a model of conductive thermal cooling of the Bandelier pluton after formation of Valles caldera. Heat flow beneath Fenton Hill averages about 4 HFU, but the thermal gradient increases from 60°C km^{-1} to 90°C km^{-1} with depth in the Precambrian section. Although most estimates of the age of the most recent thermal heating at the Fenton Hill site vary from 4 to 1 Ma, Harrison et al (1986) used transient analyses on the thermal gradients to suggest that the heating event may be as young as 10 ka. Sasada (1988) also recognizes a young (postcaldera) heating event from detailed examination of fluid inclusions in Fenton Hill core samples. The source of this heating is not entirely clear (the youngest volcanism in the Valles region is 130 ka), but it must be related to magmatic/hydrothermal activity associated with the southwestern caldera (see also Sass & Morgan 1988).

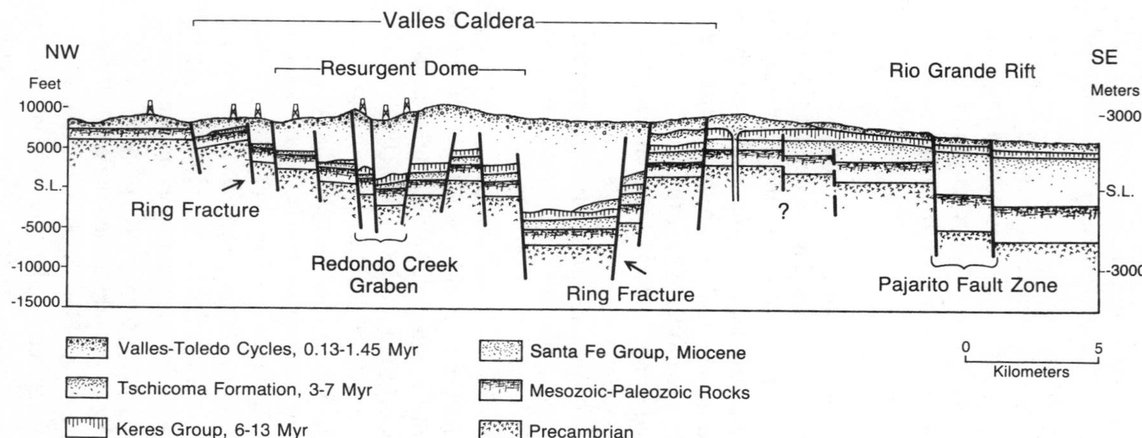

Figure 4 Schematic northwest-southeast cross section of the Valles caldera, based on the stratigraphy of many geothermal wells projected into the plane of the section and on the gravity interpretation of Segar (1974); the detailed stratigraphy of intracaldera volcanic rocks are omitted for clarity.

There is a broad area in the southwestern caldera where shallow gradients exceed 450°C km^{-1}, including the Redondo Creek and the Sulphur Springs geothermal reservoirs (Goff & Gardner 1980). These areas are known upflow zones for thermal fluids and hot-water production. Shallow heat flow in the caldera locally exceeds 10 HFU, caused by convective processes (Reiter et al 1975, Sass & Morgan 1988), whereas heat flow along the western Rio Grande rift averages about 2.8 HFU (Reiter et al 1976). Deep magnetotelluric investigations of the Valles caldera and adjacent Rio Grande rift (Hermance & Pedersen 1980) recognized a widespread electrical conductor at 15-km depth but were not able to discriminate unique (magmatic?) sources beneath the caldera or the rift. Shallower electrical and telluric techniques were integrated by Wilt & Vonder Harr (1986) to define a broad conductive anomaly beneath the southwestern caldera at 1- to 3-km depth. Parts of this anomaly correlate with known fluid production in the Redondo Creek and Sulphur Springs areas. Depending on interpretation, the areal extent of the geothermal reservoir(s) in the caldera varies from 10 to 30 km^2.

Geophysical Evidence for a Magma Body

Attenuation of seismic waves beneath the caldera complex was first recognized by Suhr (1981). Olsen et al (1986) studied the complex, shallow crustal structure beneath the Jemez volcanic field using a time-term technique to process first-arrived time data from a network of temporary seismic stations called the Caldera and Rift Deep Seismic Experiment (CARDEX). Station-time terms are correlated to gravity anomalies and generally support models for which crystalline basement drops eastward to depths of 600 m below sea level beneath the axial basins of the Rio Grande rift. Although Olsen et al (1986) imply that the large gravity low of the caldera and associated time-term delays are caused by residual melt in the silicic Bandelier pluton, other workers have suggested that the gravity low can be modeled solely as a deep asymmetric hole filled with low-density caldera fill (Segar 1974, Nielson & Hulen 1984, Wilt & Vonder Haar 1986).

Ankeny et al (1986) applied a simultaneous inversion of earthquake and travel-time data to derive a velocity model of the upper 10 km of crust beneath the Jemez volcanic field. The model displays a prominent, low-velocity, cylindrically shaped body beneath the caldera that is 15 km in diameter and 12–15 km high. This low-velocity body, centered beneath the southwestern sector of the caldera complex, is interpreted as resulting from the combined effects of a silicic magma body and the high temperatures of surrounding rock.

Configuration of the Hydrothermal System

The Valles/Toledo caldera complex possesses a diverse suite of thermal waters that are typical of those existing at many geothermal areas around the world (Goff & Grigsby 1982, Trainer 1984). Acid-sulfate waters with associated mud pots and fumaroles discharge in the central and western resurgent dome areas, particularly at Sulphur Springs (Goff et al 1985). These acidic springs result from condensation of steam and oxidation of H_2S to form natural sulfuric acid that mixes with near-surface groundwaters. Thermal meteoric waters occur at isolated spots throughout the western ring-fracture zone of the caldera. They appear to be mostly dilute groundwaters heated by the relatively high subsurface temperatures present at shallow depths. Deep reservoir waters are encountered by wells in the Redondo Creek and Sulphur Springs areas, beneath the acid-sulfate/vapor zone that caps the liquid-dominated part of the hydrothermal system. They are neutral-chloride in character, with anomalously high concentrations of As, B, Br, Cs, Li, Rb, and other trace elements. Hot springs derived, in part, from deep-reservoir water are encountered outside the southwest caldera margin in San Diego Canyon (along the Jemez fault zone) and in wells drilled in the flanking plateaus. They appear to be mixtures of reservoir water and groundwater (Goff et al 1981, 1988).

Stable isotope and tritium analyses of over 100 thermal and nonthermal waters in the Valles/Toledo caldera complex are the basis for the interpretation that recharge to the geothermal reservoir comes from meteoric precipitation and slow infiltration of cool groundwater to depth, particularly from the basins of the northern and eastern caldera moat (Vuataz & Goff 1986).

Meteoric precipitation recharges the hydrothermal system, which equilibrates at depths of 2–3 km and at temperatures of about 300°C, in caldera-fill tuffs and precaldera volcanic rocks. Thermal waters rise convectively to depths of roughly 500 to 600 m before flowing laterally to the southwest toward the caldera wall. A vapor zone that contains steam, CO_2, H_2S, and other volatile components has formed above a boiling interface with liquids at about 200°C. This interface is the upper surface of a convecting liquid-dominated system. The lateral flow system crosses the southwestern caldera wall above Precambrian basement and then travels along the Jemez fault zone and through semipermeable Paleozoic strata. Reservoir water and groundwater mix along the lateral flow path and form derivative waters that issue as hot springs or that flow in subsurface aquifers southwest of the caldera. An extremely good correlation is found between the $^{87}Sr/^{86}Sr$ ratio of hydrothermal fluids and that of coexisting rocks in the geothermal reservoir and its lateral flow system (Vuataz et al 1988).

Chronology of the Hydrothermal System(s)

Hydrothermally altered rocks are present in precaldera volcanic rocks throughout the central and southeastern Jemez Mountains. Wronkiewicz et al (1984) and Gardner et al (1986) concluded that the gold-bearing quartz veins and hydrothermally altered rocks of the Cochiti mining district were formed about 6 Ma. Recent K-Ar dates by WoldeGabriel & Goff (1989) on hydrothermal illites identify this event at 6.5–5.6 Ma. Wronkiewicz et al (1984) have demonstrated from fluid-inclusion studies that the Cochiti veins were formed at temperatures of about 195–375°C and that the fluids had salinities of about 0–4.9 wt% equivalent NaCl.

Hydrothermally altered precaldera volcanic rocks along the northern and western Valles caldera wall (Gardner et al 1986) and K-Ar dates on hydrothermal illites from altered Paleozoic rocks in CSDP corehole VC-1 (Ghazi & Wampler 1987) suggest that other hydrothermal events may have occurred between 6 and 1.12 Ma. These events were associated with volcanism of the Polvadera Group or with formation of the Toledo caldera.

Much more is known about the evolution of the hydrothermal system associated with the Valles caldera (Table 1). Goff & Shevenell (1987) applied U-Th disequilibrium and U-U dating techniques to the present and ancient travertine deposits at the Soda Dam hot springs to show that the hydrothermal system was initiated at about 1 Ma. Ghazi & Wampler (1987) obtained a K-Ar date of 1.0 ± 0.1 Ma on hydrothermal illite from Paleozoic limestone at about 500-m depth in corehole VC-1. Geissman (1988) obtained a reverse magnetic polarity of unique magnetic character on hydrothermally altered Paleozoic rocks from core hole VC-1 and concluded that high-temperature (300°C) fluids permeated the rocks at ≥ 0.98 Ma. WoldeGabriel & Goff (1989) dated hydrothermal illite throughout the VC-2A core by K-Ar techniques and obtained ages of 0.83–0.66 Ma. Because temperatures in VC-2A are still 200°C at only 400-m depth, these dates may have been slightly reset and could reflect the minimum age of initial hydrothermal activity.

The age of the vapor cap above the liquid-dominated reservoir is less constrained. Goff & Shevenell (1987) suggested that the vapor cap began to form at about 0.5 Ma. This age corresponds to cessation of travertine deposition of the oldest large travertine deposit at Soda Dam and coincides with the age of breaching of the southwest caldera wall and draining of intracaldera lakes (Doell et al 1968, Nielson & Hulen 1984). It is postulated that the draining of the lakes and the resulting loss of hydraulic head on the hydrothermal system caused the maximum elevation of the hydrothermal fluids in the geothermal reservoir to drop (Trainer 1984).

Table 1 Geochronology of volcanic, hydrothermal, and geomorphic events associated with Valles caldera, New Mexico (Goff et al 1989)

Event	Age	Method	Reference
1. Eruption of Tshirege Member of the Bandelier Tuff; formation of Valles caldera	1.12 Ma	K-Ar	Doell et al 1968, Izett et al 1981
2. Uplift of resurgent dome; eruption of early rhyolites	~1.0 Ma	Inference	Smith & Bailey 1968
3. Eruption of northern arc of postcaldera moat rhyolites	1.04–0.45 Ma	K-Ar	Doell et al 1968
4. Initial formation of Valles hydrothermal system and voluminous travertine deposit at Soda Dam	~1.0 Ma	U-U	Goff & Shevenell 1987
	1.0 Ma	K-Ar	Ghazi & Wampler 1987
	>0.98 Ma	Paleomagnetism	Geissman 1988
5. Formation of Sulphur Springs subsystem of Valles hydrothermal system	≥0.83 Ma	K-Ar	WoldeGabriel & Goff 1989
6. Formation of Sulphur Springs molybdenite deposit	≥0.66 Ma	K-Ar	WoldeGabriel & Goff 1989
7. Breaching of southwest caldera wall; deep erosion of southwest caldera moat zone	~0.5 Ma	Inference	Doell et al 1968, Nielson & Hulen 1984
	~0.5 Ma	K-Ar	Hulen & Nielson 1988a
8. Cessation of voluminous travertine deposition at Soda Dam	0.48 Ma	U-U	Goff & Shevenell 1987
9. Initial formation of vapor zone above liquid-dominated hydrothermal system	≤0.5 Ma	Inference	Goff & Shevenell 1987
10. Eruption of southern cluster of postcaldera moat rhyolites	0.49–0.13 Ma	K-Ar, fission-track, U-Th	Doell et al 1968, Marvin & Dobson 1979, Gardner et al 1986, Self et al 1988
11. Partial filling of southwest caldera breach	≤0.65 Ma	Paleomagnetism	Geissman 1988
	<0.5 Ma	K-Ar	Hulen & Nielson 1988a
12. Formation of hydrothermal calcite veins along Jemez fault zone beneath southwest caldera moat	>400–95 ka	U-Th	Sturchio & Binz 1988
13. Late hydrothermal fluorite vug in Sulphur Springs subsystem	~150 ka?	U-Th	N. Sturchio, unpublished data
14. Second period of travertine deposition at Soda Dam	110–60 ka	U-Th	Goff & Shevenell 1987
15. Last pulse of thermal activity at Fenton Hill	40–10 ka	Transient analysis	Harrison et al 1986
16. Final period of travertine deposition at Soda Dam	5 ka to present	U-Th	Goff & Shevenell 1987

Intracaldera Hydrothermal Alteration

Well-zoned hydrothermal alteration and metallic mineral assemblages recognized in deep geothermal wells have proved useful not only for locating permeable fluid channels in the Valles hydrothermal system but also for understanding how that system has evolved with time. For example, we know that phyllic (quartz-sericite-pyrite) alteration in the Redondo Creek area is commonly associated with widely spaced but active thermal fluid entries, and that similar but impermeable phyllic zones represent analogous channels now hydrothermally sealed (Hulen & Nielson 1986a,b). The hydrothermal alteration in the Redondo Creek sector of the active Valles system (the "Baca" geothermal system) is similar in many respects to that in fossil systems that have formed certain epithermal precious metal ores.

Hydrothermal alteration of the rocks penetrated by CSDP core hole VC-2A is the most intense and pervasive yet encountered in any Valles borehole. The alteration is separable into an intense, upper phyllic zone (0–163 m depth) and a subjacent, mostly moderate-intensity chlorite-sericite zone (163–528 m; Hulen et al 1987). The high-level phyllic zone is notable in containing a unique occurrence of poorly crystalline molybdenite, locally reaching concentrations of 0.56 wt% in scattered breccia zones. The molybdenite occurs in veinlets and cemented breccia intergrown with quartz, illite or sericite, fluorite, and pyrite, and locally with traces of rhodochrosite, sphalerite, and chalcopyrite. Fluid-inclusion homogenization temperatures obtained for quartz and fluorite intergrown with the molybdenite suggest that the molybdenite was deposited from a dilute liquid at temperatures of ~200°C (Hulen et al 1987, Sasada 1987, Gonzalez & McKibben 1987).

Textural evidence from the VC-2A core suggests that the high-temperature alteration postdated or accompanied the waning stages of resurgent doming. Well Baca-8, located north of VC-2A, has also shown cooling and a drop in the water table (Hulen & Nielson 1986a), as have wells in the Redondo Creek area, where cooling of 50–100°C relative to past activity has been observed; this temperature decline has been attributed to draining of an intracaldera lake (Hulen & Nielson 1988b).

CONCLUSIONS

1. The Jemez volcanic field has been continuously active over the last 13 Myr, and volcanic rocks are predominantly calc-alkaline in composition. Although basaltic to rhyolitic rocks have been erupted, most of the activity produced two-pyroxene andesites (erupted from ~10 to

~7 Ma), dacites (~7 to 3 Ma), and high-silica rhyolite (~1.45 to ~0.30 Ma). The older volcanic rocks were derived mostly by fractional crystallization, although magma mixing and melting of lower crust may have been responsible for some of the more silicic volcanic rocks. Volcanism is related to tectonic activity along the Rio Grande rift and the Jemez volcanic lineament.

2. High-silica rhyolite ignimbrites from the evolving Bandelier magma chamber were erupted as early as 3.64 Ma. It is now known that the Valles caldera complex contains at least two small and two large calderas, but the earlier collapse features were probably obliterated by collapse of the Toledo (1.45 Ma) and Valles (1.12 Ma) calderas. The Bandelier magma chamber may have been compositionally zoned, but processes within the chamber (estimated volume is >3000 km^3) are not resolved. The roof of the Bandelier magma chamber was at a depth of 5–6 km when the Tshirege Member was vented, and the eruption temperature varied from 700°C to 850°C.
3. Structure of the caldera complex is controlled by northeast-trending structures associated with the earlier Rio Grande rift and the Jemez volcanic lineament. Drill holes throughout the southwestern sector of the caldera and gravity models of the entire caldera complex indicate that the caldera-fill sequence is very thick (perhaps over 3 km) and that depths to the caldera floor are asymmetric (relatively shallow in the western half and deep in the eastern half). The resurgent dome structure is very complex. Drilling results have not verified that resurgence occurred after formation of the Toledo caldera. Structural resurgence was caused by buoyant rise of the Bandelier magma chamber after formation of the Valles caldera. Caldera-fill deposits are now exposed at the summit of the Redondo resurgent dome, 1 km above the caldera moat.
4. The highest convective heat flow and subsurface seismic attenuation in the caldera complex occur in the southwestern sector and are coincident with the main zones of hydrothermal upflow and the youngest moat rhyolite eruptions. Two different investigations of the Precambrian rocks in the western caldera margin indicate that the subsurface temperatures beneath the caldera may be increasing. If so, the "reheating" must be caused by evolution of the new batch of rhyolitic magma below the southwestern caldera moat.
5. The Valles hydrothermal system consists of local meteoric recharge, equilibration at temperatures approaching 300°C, convective upflow along faults and fractures, and lateral flow along structures that cut the southwestern caldera wall (Jemez fault zone). A vapor zone and acid-sulfate condensation zone originate from subsurface boiling (200°C)

above the liquid-dominated reservoir in the caldera. A lateral flow system produces a subsurface tongue of mixed reservoir water and cooler groundwaters in Paleozoic rocks of the southwestern caldera margin. The areal extent of the geothermal reservoir inside the caldera is 10–30 km^2, but permeable zones are limited.

6. Episodic hydrothermal events have occurred in the Jemez volcanic field during the last 8 Myr (e.g. the Cochiti mining district, with an age range from 6.5 to 5.6 Ma). The Valles hydrothermal system has been continuously active for the last 1 Myr, but the vapor zone first formed about 0.5 Ma. Creation of the vapor zone is apparently linked to breaching of the southwestern caldera wall by the ancestral Jemez River, draining of intracaldera lakes, and resulting loss of hydraulic head on the liquid-dominated hydrothermal reservoir.
7. Hydrothermal alteration and ore mineralization in Valles caldera resemble those found at many fossil calderas hosting economic ore deposits. Alteration style ranges from advanced argillic to anhydrous calc-silicate, depending on depth and location. Secondary mineral assemblages and fluid-inclusion studies indicate temperatures of formation and salinities similar to those presently occurring in the hydrothermal system, although a complex evolutionary history can be unraveled. Sub-ore-grade molybdenite (up to 0.56 wt% MoS_2) is the principal ore mineral found so far, but Cu, Pb, Zn, Mn, Bi, Te, and Ag minerals have also been recently identified.

Acknowledgments

This review is condensed from a field guide prepared by the authors for the 1989 General Assembly of the International Association of Volcanology and Chemistry of the Earth's Interior (Goff et al 1989). Many thanks to all the explorers, developers, researchers, and graduate students, past and present, who have contributed to the general knowledge of the Jemez Mountains region. Much of the information incorporated into this paper was obtained from the geothermal efforts of UNOCAL (Santa Rosa, California), GEO Operator Corp. (San Mateo, California), and SUNEDCO (Dallas, Texas), and from the research and development efforts of the Hot Dry Rock project funded by the US Department of Energy. The cooperation of the US Forest Service and of the Baca Land and Cattle Company and other landowners in the Jemez Mountains is appreciated. This review was funded by the US Department of Energy, Office of Basic Energy Sciences, through the helpful efforts of George Kolstad and by an Institutional Supporting Research grant from Los Alamos National Laboratory.

Literature Cited

Aldrich, M. J. 1986. Tectonics of the Jemez lineament in the Jemez Mountains and the Rio Grande rift. *J. Geophys. Res.* 91: 1753–62

Aldrich, M. J., Laughlin, A. W. 1984. A model for the tectonic development of the southeastern Colorado Plateau boundary. *J. Geophys. Res.* 89: 10,207–18

Ankeny, L. A., Braile, L. W., Olsen, K. H. 1986. Upper crustal structure beneath the Jemez Mountains volcanic field, New Mexico, determined by three-dimensional simultaneous inversion of seismic refraction and earthquake data. *J. Geophys. Res.* 91: 6188–98

Bailey, R. A., Smith, R. L., Ross, C. S. 1969. Stratigraphic nomenclature of volcanic rocks in the Jemez Mountains, New Mexico. *US Geol. Surv. Bull. 1274-P.* 19 pp.

Doell, R. R., Dalrymple, G. B. 1966. Geomagnetic polarity epochs: A new polarity event and the age of the Brunhes-Matuyama boundary. *Science* 152: 1060–72

Doell, R. R., Dalrymple, G. B., Smith, R. L., Bailey, R. A. 1968. Paleomagnetism, potassium-argon ages, and geology of rhyolites and associated rocks of the Valles caldera, New Mexico. *Geol. Soc. Am. Mem.* 116: 211–48

Dondanville, R. F. 1978. Geologic characteristics of the Valles caldera geothermal system. *Geotherm. Resour. Counc. Trans.* 2: 157–60

Eichelberger, J. C., Koch, F. G. 1979. Lithic fragments in the Bandelier Tuff, Jemez Mountains, New Mexico. *J. Volcanol. Geotherm. Res.* 5: 115–34

Faust, C. R., Mercer, J. W., Thomas, S. D. 1984. Quantitative analysis of existing conditions and production strategies for the Baca geothermal system, New Mexico. *Water Resour. Res.* 20: 601–18

Gardner, J. N. 1985. *Tectonic and petrologic evolution of the Keres Group: implications for the development of the Jemez volcanic field, New Mexico.* PhD thesis. Univ. Calif., Davis. 295 pp.

Gardner, J. N., Goff, F. 1984. Potassium-argon dates from the Jemez volcanic field: implications for tectonic activity in the north-central Rio Grande rift. *Guideb. N. M. Geol. Soc. Field Conf., 35th*, pp. 75–81

Gardner, J. N., House, L. 1987. Seismic hazards investigations at Los Alamos National Laboratory, 1984 to 1985. *Rep. LA-11072-MS*, Los Alamos Natl. Lab., Los Alamos, N. Mex. 76 pp.

Gardner, J. N., Goff, F., Garcia, S., Hagan, R. C. 1986. Stratigraphic relations and lithologic variations in the Jemez volcanic field, New Mexico. *J. Geophys. Res.* 91: 1763–78

Geissman, J. W. 1988. Paleomagnetism and rock magnetism of Quaternary volcanic rocks and late Paleozoic strata, VC-1 core hole, Valles caldera, New Mexico, with emphasis on remagnetization of late Paleozoic strata. *J. Geophys. Res.* 93(B6): 6001–26

Ghazi, A., Wampler, J. 1987. Potassium-argon dates of clays from brecciated and hydrothermally-altered rocks from VC-1, Valles caldera, New Mexico. *Eos, Trans. Am. Geophys. Union* 68: 1515 (Abstr.)

Gill, J. B. 1981. *Orogenic Andesites and Plate Tectonics.* New York: Springer-Verlag. 390 pp.

Goff, F., Gardner, J. N. 1980. Geologic map of the Sulphur Springs area, Valles caldera geothermal system, New Mexico. *Los Alamos Sci. Lab. Map LA-8634-MAP*, scale 1 : 5000. 2 sheets

Goff, F., Grigsby, C. 1982. Valles caldera geothermal systems, New Mexico, USA. *J. Hydrol.* 56: 119–36

Goff, F., Nielson, D. L., eds. 1986. *Caldera Processes and Magma-Hydrothermal Systems, Continental Scientific Drilling Program–Thermal Regimes, Valles Caldera Research, Scientific and Management Plan. Rep. LA-10737-OBES*, Los Alamos Natl. Lab., Los Alamos, N. Mex. 163 pp.

Goff, F., Shevenell, L. 1987. Travertine deposits of Soda Dam, New Mexico and their implications for the age and evolution of the Valles caldera hydrothermal system. *Geol. Soc. Am. Bull.* 99: 292–302

Goff, F., Grigsby, C., Trujillo, P., Counce, D., Kron, A. 1981. Geology, water geochemistry, and geothermal potential of the Jemez Springs area, Cañon de San Diego, New Mexico. *J. Volcanol. Geotherm. Res.* 10: 227–44

Goff, F., Heiken, G., Tamanyu, S., Gardner, J., Self, S., et al. 1984. Location of Toledo caldera and formation of the Toledo embayment, Jemez Mountains, New Mexico. *Eos, Trans. Am. Geophys. Union* 65: 1145 (Abstr.)

Goff, F., Gardner, J. N., Vidale, R., Charles, R. 1985. Geochemistry and isotopes of fluids from Sulphur Springs, Valles caldera, New Mexico. *J. Volcanol. Geotherm. Res.* 23: 2734–97

Goff, F., Rowley, J., Gardner, J., Hawkins, W., Goff, S., et al. 1986. Initial results from VC-1: first Continental Scientific Drilling Program core hole in the Valles caldera, New Mexico. *J. Geophys. Res.* 91: 1742–52

Goff, F., Nielson, D., Gardner, J., Hulen, J.,

Lysne, P., et al. 1987. Scientific drilling at Sulphur Springs, Valles caldera, New Mexico: corehole VC-2A. *Eos, Trans. Am. Geophys. Union* 68: 649, 661–62 (Abstr.)

Goff, F., Shevenell, L., Gardner, J. N., Vuataz, F.-D., Grigsby, C. O. 1988. The hydrothermal outflow plume of Valles caldera, New Mexico, and a comparison with other outflow plumes. *J. Geophys. Res.* 93(B6): 6041–58

Goff, F., Gardner, J. N., Baldridge, W. S., Hulen, J. B., Nielson, D. L., et al. 1989. Volcanic and hydrothermal evolution of the Valles caldera and Jemez volcanic field. *N. Mex. Bur. Mines Miner. Resour. Mem.* 46: 381–434

Gonzalez, C. M., McKibben, M. A. 1987. Thermal history of vein mineralization in CSDP corehole VC-2A, Valles caldera, New Mexico. *Geol. Soc. Am. Abstr. With Programs* 19: 679 (Abstr.)

Griggs, R. L. 1964. Geology and groundwater resources of the Los Alamos area, New Mexico. *US Geol. Surv. Water-Supply Pap. 1753*. 107 pp.

Harrison, T. M., Morgan, P. M., Blackwell, D. D. 1986. Constraints on the age of heating at the Fenton Hill site, Valles caldera, New Mexico. *J. Geophys. Res.* 91: 1899–1908

Heiken, G., ed. 1985. *Workshop on Recent Research in the Valles Caldera. Rep. LA-10339-C.* Los Alamos Natl. Lab., Los Alamos, N. Mex. 68 pp.

Heiken, G., Goff, F. 1983. Hot dry rock geothermal energy in the Jemez volcanic field, New Mexico. *J. Volcanol. Geotherm. Res.* 15: 223–46

Heiken, G., Goff, F., Stix, J., Tamanyu, S., Shafiqullah, M., et al. 1986. Intracaldera volcanic activity, Toledo caldera and embayment, Jemez Mountains, New Mexico. *J. Geophys. Res.* 91: 1799–1815

Hermance, J. F., Pedersen, J. 1980. Deep structure of the Rio Grande rift: a magnetotelluric interpretation. *J. Geophys. Res.* 85: 3899–3912

Hulen, J. B., Neilson, D. L. 1986a. Hydrothermal alteration in the Baca geothermal system, Redondo dome, Valles caldera, New Mexico. *J. Geophys. Res.* 91: 1867–86

Hulen, J. B., Nielson, D. L. 1986b. Stratigraphy and hydrothermal alteration in borehole Baca-8, Sulphur Springs area, Valles caldera, New Mexico. *Geotherm. Resour. Counc. Trans.* 10: 187–92

Hulen, J. B., Nielson, D. L. 1988a. Hydrothermal brecciation in the Jemez fault zone, Valles caldera, New Mexico: results from Continental Scientific Drilling Program core hole VC-1. *J. Geophys. Res.* 93: 6077–90

Hulen, J. B., Nielson, D. L. 1988b. Clay mineralogy and zoning in Continental Scientific Drilling Program corehole VC-2A, Valles caldera, New Mexico. *Geotherm. Resour. Counc. Trans.* 12: 291–98

Hulen, J., Nielson, D. L., Goff, F., Gardner, J. N., Charles, R. 1987. Molybdenum mineralization in an active geothermal system, Valles caldera, New Mexico. *Geology* 15: 748–52

Iddings, J. P. 1890. On a group of rocks from the Tewan Mountains, New Mexico, and on the occurrence of primary quartz in certain basalts. *US Geol. Surv. Bull. 66.* 34 pp.

Izett, G. A., Obradovich, C. W., Naeser, C. W., Cebula, C. T. 1981. Potassium-argon and fission-track zircon ages of Cerro Toledo rhyolite tephra in the Jemez Mountains, New Mexico. *US Geol. Surv. Prof. Pap. 1199-D*, pp. 37–43

Kolstad, C. D., McGetchin, T. R. 1978. Thermal evolution models for the Valles caldera with reference to a hot-dry-rock geothermal experiment. *J. Volcanol. Geotherm. Res.* 3: 197–218

Laughlin, A. W., Brookins, D. G., Damon, P. E. 1976. Late-Cenozoic basaltic volcanism along the Jemez zone of New Mexico and Arizona. *Geol. Soc. Am. Abstr. With Programs* 8: 598 (Abstr.)

Laughlin, A. W., Eddy, A. C., Laney, R., Aldrich, M. J. 1983. Geology of the Fenton Hill, New Mexico, hot dry rock site. *J. Volcanol. Geotherm. Res.* 15: 21–41

Loeffler, B. M., Vaniman, D. T., Baldridge, W. S., Shafiqullah, M. 1988. Neogene rhyolites of the northern Jemez volcanic field, New Mexico. *J. Geophys. Res.* 93(B6): 6157–68

Luth, W., Hardee, R. 1980. Comparative assessment of five potential sites for hydrothermal-magma system: a summary. *US Dep. Energy Rep. DOE/TIC-11303.* 51 pp.

Marvin, R. F., Dobson, S. W. 1979. Radiometric ages: compilation B. *US Geol. Surv. Isochron/West* 26: 3–32

Mayo, E. G. 1958. Lineament tectonics and some ore districts of the southwest. *Min. Eng. (N.Y.)* 10: 1169–75

Nielson, D. L., Hulen, J. B. 1984. Internal geology and evolution of the Redondo dome, Valles caldera, New Mexico. *J. Geophys. Res.* 89: 8695–8711

Olsen, K. H., Braile, L. W., Stewart, J. N., Daudt, C. R., Keller, G. R., et al. 1986. Jemez Mountains volcanic field, New Mexico: time term interpretation of the CARDEX seismic experiment and comparison with Bouguer gravity. *J Geophys. Res.* 91: 6175–87

Potter, D. B., Oberthal, C. M. 1983. Vent

sites and flow directions of the Otowi ash flows (lower Bandelier Tuff), New Mexico. *Geol. Soc. Am. Bull.* 98: 66–76

Reiter, M., Edwards, C. L., Hartman, H., Weidman, C. 1975. Terrestrial heat flow along the Rio Grande rift, New Mexico and southern Colorado. *Geol. Soc. Am. Bull.* 86: 811–18

Reiter, M., Weidman, C., Edwards, C. L., Hartman, H. 1976. Sub-surface temperature data in Jemez Mountains, New Mexico. *N. Mex. Bur. Mines Miner. Resour. Circ. 151.* 16 pp.

Ross, C. S., Smith, R. L. 1961. Ash-flow tuffs: their origin, geologic relations and identification. *US Geol. Surv. Prof. Pap. 366.* 81 pp.

Sasada, M. 1987. Fluid inclusions from VC-2A corehole in Valles caldera, New Mexico—evidence for transition from hot-water-dominated to vapor-dominated system. *Geotherm. Resour. Counc. Jpn., Annu. Meet., Abstr. With Program* (Abstr.)

Sasada, M. 1988. Fluid inclusion evidence for recent temperature rising at Fenton Hill hot dry rock test site west of the Valles caldera, New Mexico, USA. In *Exploration and Development of Geothermal Resources. Int. Symp. Geotherm. Energy, Proc. Geotherm. Res. Soc. Jpn.*, pp. 526–29

Sass, J. H., Morgan, P. 1988. Conductive heat flux in VC-1 and the thermal regime of Valles caldera, Jemez Mountains, New Mexico. *J. Geophys. Res.* 93(B6): 6027–39

Segar, R. L. 1974. Quantitative gravity interpretation, Valles caldera area, Sandoval and Rio Arriba counties, New Mexico. *Earth Sci. Lab. Open File Rep. NM/BACA-27,* Univ. Utah Res. Inst., Salt Lake City. 12 pp.

Self, S., Goff, F., Gardner, J. N., Wright, J. V., Kite, W. M. 1986. Explosive rhyolite volcanism in the Jemez Mountains: vent locations, caldera development and relation to regional structure. *J. Geophys. Res.* 91: 1779–98

Self, S., Kircher, D. E., Wolff, J. A. 1988. The El Cajete Series, Valles caldera, New Mexico. *J. Geophys. Res.* 93(B6): 6113–28

Smith, R. L. 1979. Ash-flow magmatism. *Geol. Soc. Am. Spec. Pap. 180*, pp. 5–28

Smith, R. L., Bailey, R. A. 1966. The Bandelier Tuff—a study of ash-flow eruption cycles from zoned magma chambers. *Bull. Volcanol.* 29: 83–104

Smith, R. L., Bailey, R. A. 1968. Resurgent cauldrons. *Geol. Soc. Am. Mem.* 116: 613–62

Smith, R. L., Bailey, R. A., Ross, C. S. 1961. Structural evolution of the Valles caldera, New Mexico and its bearing on the emplacement of ring dikes. *US Geol. Surv. Prof. Pap. 424-D*, pp. D145–49

Smith, R. L., Bailey, R. A., Ross, C. S. 1970. Geologic map of the Jemez Mountains, New Mexico. *US Geol. Surv. Misc. Geol. Invest. Map I-571*, scale 1 : 125,000, 1 sheet

Sommer, M. A. 1977. Volatiles H_2O, CO_2, and CO in silicate melt inclusions in quartz phenocrysts from the rhyolitic Bandelier air-fall and ash-flow tuff, New Mexico. *J. Geol.* 85: 423–32

Spell, T. L. 1987. *Geochemistry of Valles Grande Member ring fracture rhyolites, Valles caldera, New Mexico.* MS thesis, N. Mex. Inst. Min. Technol., Socorro. 212 pp.

Stein, H. L. 1983. *Geology of the Cochiti mining district, Jemez Mountains, New Mexico.* MS thesis. Univ. N. Mex., Albuquerque. 122 pp.

Stix, J., Goff, F., Gorton, M. P., Heiken, G., Garcia, S. R. 1988. Restoration of compositional zonation in the Bandelier silicic magma chamber between two caldera-forming eruptions: geochemistry and origin of the Cerro Toledo Rhyolite, Jemez Mountains, New Mexico. *J. Geophys. Res.* 93(B6): 6129–47

Sturchio, N. C., Binz, C. M. 1988. Uranium-series age determination of calcite veins, VC-1 drill core, Valles caldera, New Mexico. *J. Geophys. Res.* 93(B6): 6097–6102

Suhr, G. 1981. Seismic crust anomaly under the Valles caldera in New Mexico, USA. Unpubl. Consult. Rep. PRAKLA-SEISMOS GMBH, Hannover, Fed. Rep. Germ. 56 pp.

Taschek, R. F. 1981. Scientific and operational plan for drilling in the Valles caldera, New Mexico. Doc. 0491A prepared for Cont. Sci. Drill. Comm., Natl. Sci. Found. 53 pp.

Trainer, F. W. 1984. Thermal springs in Cañon de San Diego as a window into Valles caldera, New Mexico. *Guideb. N. Mex. Geol. Soc. Field Conf., 35th*, pp. 249–55

Turbeville, B. N., Self, S. 1988. San Diego Canyon ignimbrites: pre-Bandelier Tuff explosive rhyolite volcanism in the Jemez Mountains, New Mexico. *J. Geophys. Res.* 93(B6): 6148–56

Vuataz, F.-D., Goff, F. 1986. Isotope geochemistry of thermal and nonthermal waters in the Valles caldera, Jemez Mountains, northern New Mexico. *J. Geophys. Res.* 91: 1835–53

Vuataz, F.-D., Goff, F., Fouillac, C., Calvez, J. Y. 1988. A strontium isotope study of the VC-1 core hole and associated hydrothermal fluids and rocks from Valles caldera, Jemez Mountains, New Mexico. *J. Geophys. Res.* 93(B6): 6059–67

Warshaw, C. M., Smith, R. L. 1988. Pyroxenes and fayalites in the Bandelier Tuff, New Mexico: temperatures and comparison with other rhyolites. *Am. Mineral.* 73: 1025–37

Wilt, M., Vonder Haar, S. 1986. A geologic and geophysical appraisal of the Baca geothermal field, Valles caldera, New Mexico. *J Volcanol. Geotherm. Res.* 27: 349–70

WoldeGabriel, G., Goff, F. 1989. Temporal relationships of volcanism and hydrothermal systems in two areas of the Jemez volcanic field, New Mexico. *Geology*. In press

Wright, J. V., Smith, A. L., Self, S. 1981. A terminology for pyroclastic deposits. In *Tephra Studies*, ed. S. Self, R. S. J. Sparks, pp. 457–63. Boston: Reidel

Wronkiewicz, D., Norman, D., Parkinson, G., Emanuel, K. 1984. Geology of the Cochiti mining district, Sandoval County, New Mexico. *Guideb. N. Mex. Geol. Soc. Field Conf., 35th*, pp. 219–22

Annu. Rev. Earth Planet. Sci. 1990. 18:55–99
Copyright © 1990 by Annual Reviews Inc. All rights reserved

CRITICAL TAPER MODEL OF FOLD-AND-THRUST BELTS AND ACCRETIONARY WEDGES

F. A. Dahlen

Department of Geological and Geophysical Sciences,
Princeton University, Princeton, New Jersey 08544

INTRODUCTION

The fold-and-thrust belts and submarine accretionary wedges that lie along compressive plate boundaries are one of the best understood deformational features of the Earth's upper crust. Although there is considerable natural variation among the many fold-and-thrust belts and accretionary wedges that have been recognized and explored, several features appear to be universal. In cross section, fold-and-thrust belts and accretionary wedges occupy a wedge-shaped deformed region overlying a basal detachment or décollement fault; the rocks or sediments beneath this fault show very little deformation. The décollement fault characteristically dips toward the interior of the mountain belt or, in the case of a submarine wedge, toward the island arc; the topography, in contrast, slopes toward the toe or deformation front of the wedge. Deformation within the wedge is generally dominated by imbricate thrust faults verging toward the toe and related fault-bend folding.

Two North American fold-and-thrust belts that exhibit these features are shown in Figure 1. Neither of these two examples is tectonically active today; the southern Canadian fold-and-thrust belt was active during the late Jurassic and Cretaceous (150–100 Ma), whereas the southern Appalachians were deformed during the late Carboniferous to Permian Alleghenian orogeny (300–250 Ma). Figure 2 shows two examples that are currently active: the Taiwan fold-and-thrust belt, produced by the subduction of the Eurasian plate beneath the Philippine Sea plate (Suppe 1981, 1987); and the Barbados accretionary wedge, produced by the sub-

0084–6597/90/0515–0055$02.00

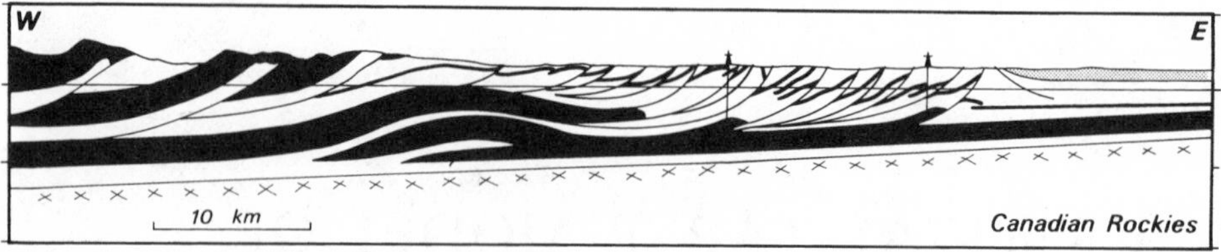

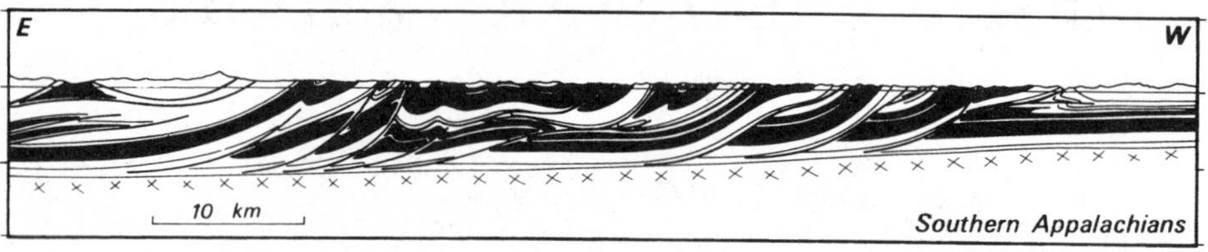

Figure 1 Cross sections of two foreland fold-and-thrust belts. (*Top*) Canadian Rockies (Bally et al 1966). (*Bottom*) Southern Appalachians (Roeder et al 1978). No vertical exaggeration.

duction of the North American plate beneath the Caribbean plate (Westbrook 1975, 1982). These are in a sense two end members of a spectrum, since Taiwan is moderately tapered and rapidly accreting and eroding, whereas Barbados is narrowly tapered, slowly accreting, and noneroding. Evidence that thin-skinned folding and thrusting was a common phenomenon much farther back in the Earth's history is shown in Figure 3. This cross section of the 1900-Ma Asiak fold-and-thrust belt in the northwest Canadian shield was constructed without benefit of seismic or drilling data, by downplunge projection of geological maps (Hoffman et al 1988). Every structural detail has been trod upon by a field geologist's boot.

Mechanically, a fold-and-thrust belt or accretionary wedge is analogous to a wedge of sand in front of a moving bulldozer. The sand, rock, or sediment deforms until it develops a constant critical taper; if no fresh material is encountered at the toe, the wedge then slides stably without further deformation as it is pushed. The magnitude of the critical taper is governed by the relative magnitudes of the frictional resistance along the base and the compressive strength of the wedge material. An increase in the sliding resistance increases the critical taper, since it is the drag on the base that is fundamentally responsible for the deformation. An increase in the wedge strength, on the other hand, decreases the critical taper, since a stronger wedge can be thinner and still slide stably over a rough base

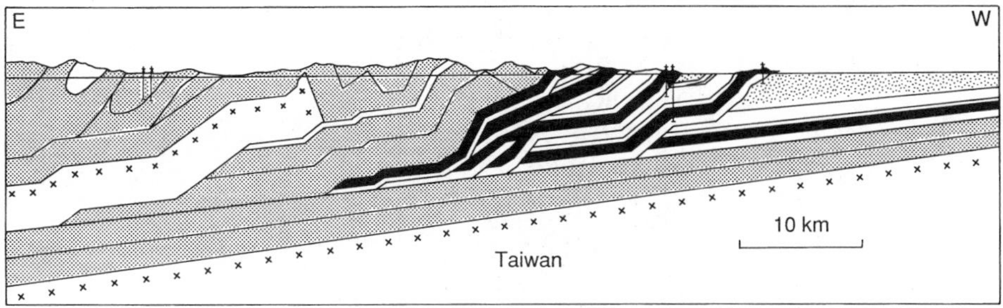

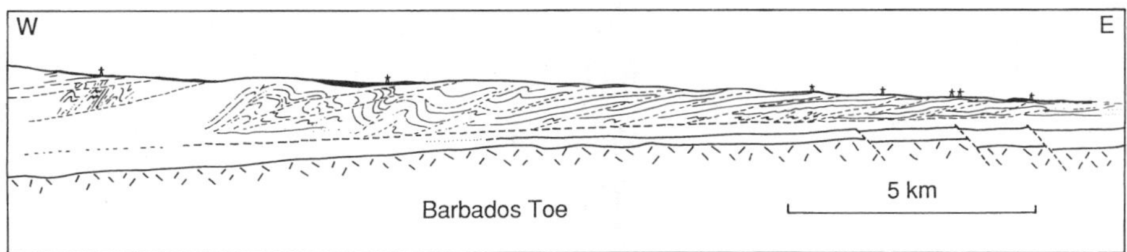

Figure 2 (*Top*) Cross section of the active fold-and-thrust belt in northern Taiwan (Suppe 1980). (*Bottom*) Cross section of the frontal region of the Barbados accretionary wedge near 15°30′N latitude (Behrmann et al 1988). Locations of Deep Sea Drilling Project and Ocean Drilling Program drill sites are indicated. No vertical exaggeration.

without deforming. The state of stress within a critically tapered wedge in the upper crust is everywhere on the verge of Coulomb failure, since the taper is attained by a process of continued brittle frictional deformation.

This paper describes an idealized mechanical model of a fold-and-thrust belt or accretionary wedge, based on this bulldozer analogy. The first analyses in this spirit were developed by Elliott (1976) and Chapple (1978); their ideas were later refined and extended to incorporate a brittle frictional rheology by Davis et al (1983). For the most part, we consider only the simplest possible version of the model, which ignores cohesion and assumes that the material properties within the wedge and on the basal décollement fault are spatially uniform (Dahlen 1984). A more general approximate analysis is, however, also discussed. Special attention is paid to the effects of pore fluids, since elevated pore-fluid pressures play such an important role in the mechanics of overthrust faulting (Hubbert & Rubey 1959). Pore-fluid pressure effects were accounted for by Davis et al (1983) and Dahlen (1984); however, a significant point was not spelled out clearly, and that is rectified here.

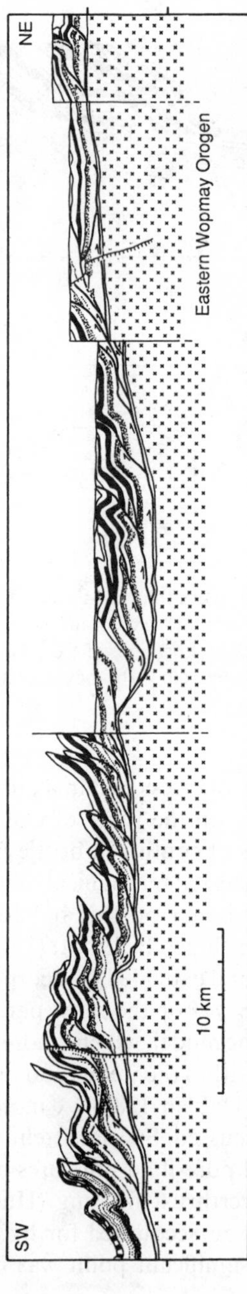

Figure 3 Cross section of the Asiak fold-and-thrust belt, eastern Wopmay orogen, in the northwest Canadian shield (Hoffman et al 1988). No vertical exaggeration.

MECHANICS OF A BULLDOZER WEDGE

Kinematics

We begin with a simplified discussion of the mechanics of a bulldozer wedge. Suppose a rigid hillside of slope β is covered with a uniform layer of dry sand of thickness h (Figure 4). If at time $t = 0$ a bulldozer begins moving uphill at a uniform velocity V, scraping up sand, a critically tapered wedge of deformed sand will form in front of the moving bulldozer. Let α denote the surface slope of this deformed wedge; the critical taper is the angle at the toe, $\alpha+\beta$. The mass flux per unit length along strike into the toe of the wedge is ρhV, where ρ is the sand density. We ignore compaction and assume that ρ is a constant. The growth of the wedge with time is described by the mass conservation law

$$\frac{\mathrm{d}}{\mathrm{d}t}\left[\frac{1}{2}\rho W^2 \tan(\alpha+\beta)\right] = \rho hV, \tag{1}$$

where W is the wedge width. Since $\alpha+\beta$ does not change with time, Equation (1) reduces to

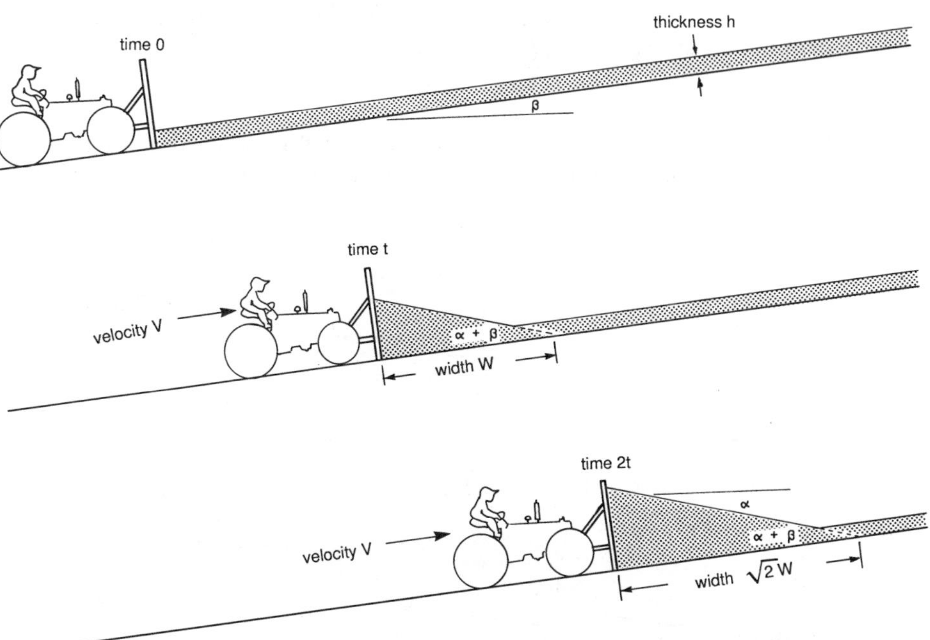

Figure 4 Cartoon depicting the self-similar growth of a bulldozer wedge.

$$W\frac{\mathrm{d}W}{\mathrm{d}t}=\frac{hV}{\tan(\alpha+\beta)}. \tag{2}$$

This has the solution

$$W=\left[\frac{2hVt}{\tan(\alpha+\beta)}\right]^{1/2}\approx\left[\frac{2hVt}{\alpha+\beta}\right]^{1/2}. \tag{3}$$

The final approximation is valid for a wedge of narrow taper, $\alpha+\beta \ll 1$, where α and β are measured in radians. Because the critical taper is governed only by the unvarying strength of the sand and the basal friction, both the width and the height of a bulldozer wedge grow like $t^{1/2}$. The growth is self-similar in the sense that the wedge at time $2t$ is indistinguishable from the wedge at time t, magnified $2^{1/2}$ times.

An eroding wedge will attain a dynamic steady state when the accretionary influx rate of fresh material into the toe is balanced by the erosive efflux (Figure 5). The steady-state width of a uniformly eroding wedge is given by the flux balance condition

$$\dot{e}W\sec(\alpha+\beta)\approx\dot{e}W=hV, \tag{4}$$

where $\dot{e}$ is the rate of erosion. A steady-state wedge must continually deform both to accommodate the influx of fresh material into its toe and to maintain its critical taper against erosion.

Critical Taper

Let (x, z) be a system of Cartesian coordinates with x aligned along the top of the wedge and z pointing down (Figure 6). To determine the critical

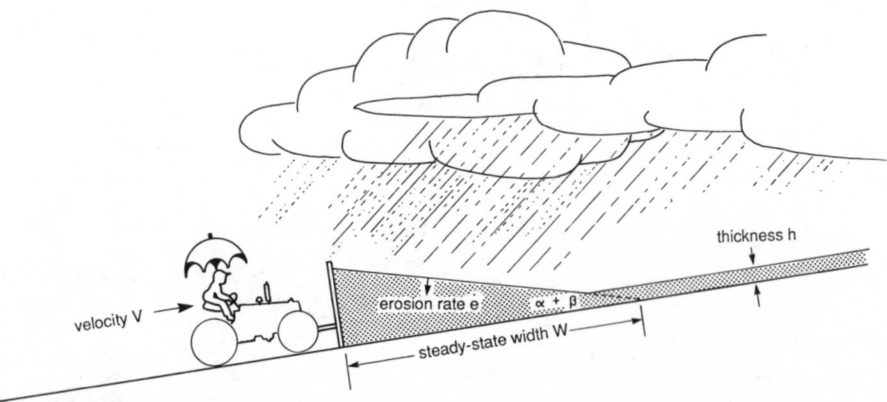

Figure 5 An eroding wedge attains a dynamic steady-state width given by $\dot{e}W = hV$.

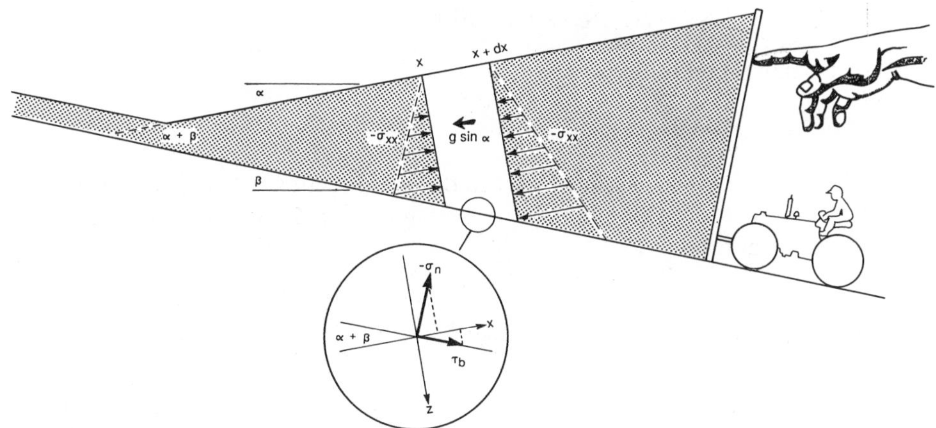

Figure 6 Schematic diagram illustrating the horizontal balance of forces on an element of a bulldozer wedge.

taper, we consider the balance of forces on an infinitesimal segment of the wedge lying between x and $x+\mathrm{d}x$; it suffices to consider the forces acting in the $\pm x$ direction. First, there is a gravitational body force whose x component, per unit length along strike, is

$$F_{\mathrm{g}} = -\rho g H \sin\alpha\,\mathrm{d}x, \tag{5}$$

where g is the acceleration of gravity, and H is the local wedge thickness. Second, there is the net force exerted by the compressive tractions σ_{xx} acting on the sidewalls at x and $x+\mathrm{d}x$; if we adopt the convention that a compressive stress is negative, this force is given by

$$F_{\mathrm{s}} = \int_0^H [\sigma_{xx}(x+\mathrm{d}x, z) - \sigma_{xx}(x, z)]\,\mathrm{d}z. \tag{6}$$

Third, and finally, there is the surface force exerted on the base; this is given in terms of the local shear and normal tractions τ_{b} and σ_{n} by

$$F_{\mathrm{b}} = [\tau_{\mathrm{b}}\cos(\alpha+\beta) - \sigma_{\mathrm{n}}\sin(\alpha+\beta)]\,\mathrm{d}x. \tag{7}$$

We assume that the base is governed by a frictional sliding condition

$$\tau_{\mathrm{b}} = -\mu_{\mathrm{b}}\sigma_{\mathrm{n}}, \tag{8}$$

where μ_{b} is the coefficient of basal friction; the basal traction then reduces to

$$F_b = -\sigma_n[\mu_b \cos(\alpha+\beta) + \sin(\alpha+\beta)]\,dx. \tag{9}$$

The balance condition is

$$F_g + F_s + F_b = 0. \tag{10}$$

The first two forces, F_g and F_s, act in the $-x$ direction, whereas F_b acts in the $+x$ direction. Upon taking the limit as $dx \to 0$, Equation (10) reduces to the exact result

$$-\rho g H \sin\alpha - \sigma_n[\mu_b \cos(\alpha+\beta) + \sin(\alpha+\beta)] + \frac{d}{dx}\int_0^H \sigma_{xx}\,dz = 0. \tag{11}$$

For $\alpha \ll 1$ and $\beta \ll 1$ we employ the approximations $\sin\alpha \approx \alpha$, $\sin(\alpha+\beta) \approx \alpha+\beta$, $\cos(\alpha+\beta) \approx 1$, and $\sigma_n \approx -\rho g H$. This reduces Equation (11) to

$$\rho g H(\beta+\mu_b) + \frac{d}{dx}\int_0^H \sigma_{xx}\,dz \approx 0. \tag{12}$$

The failure criterion for noncohesive dry sand can be written in the form

$$\frac{\sigma_1}{\sigma_3} = \frac{1+\sin\phi}{1-\sin\phi}. \tag{13}$$

Here σ_1 and σ_3 are the greatest and least principal compressive stresses, respectively, and ϕ is the angle of internal friction (Jaeger & Cook 1969, pp. 87–91). In a narrow taper ($\alpha \ll 1$ and $\beta \ll 1$), the principal stresses are approximately horizontal and vertical, that is

$$\sigma_{zz} \approx \sigma_3 \approx -\rho g z, \tag{14a}$$

$$\sigma_{xx} \approx \sigma_1 \approx -\left(\frac{1+\sin\phi}{1-\sin\phi}\right)\rho g z. \tag{14b}$$

The sidewall traction term in Equation (12) reduces in this approximation to

$$\frac{d}{dx}\int_0^H \sigma_{xx}\,dz \approx -\left(\frac{1+\sin\phi}{1-\sin\phi}\right)\rho g H(\alpha+\beta), \tag{15}$$

where we have used the relation $dH/dx \approx \alpha+\beta$. Upon inserting Equation (15) into (12), we obtain the approximate critical taper equation for a dry sand wedge in front of a bulldozer:

$$\alpha+\beta \approx \left(\frac{1-\sin\phi}{1+\sin\phi}\right)(\beta+\mu_b). \tag{16}$$

Discussion

Equation (16) shows that the critical taper $\alpha+\beta$ is increased by an increase in the coefficient of basal friction μ_b, whereas it is decreased by an increase in the internal friction angle ϕ. For $\phi = 30°$, a typical value for dry sand, the critical surface slope is given by $\alpha \approx \frac{1}{3}(\mu_b - 2\beta)$. Idle conjecture might have led to the conclusion that the surface slope α of a bulldozer wedge is at the angle of repose ($\phi = 30°$), but in fact the state of stress and therefore the slope are completely different.

Equation (12) describes the quasi-static balance of forces in any thin-skinned wedge being pushed up a frictional incline. To make use of this result, it is necessary to relate the horizontal compressive stress σ_{xx} to the stress due to the lithostatic overburden, $\sigma_{zz} \approx -\rho g z$. In a critically tapered wedge, σ_{xx} is related to σ_{zz} by the Coulomb failure law:

$$\sigma_{xx}^{\text{failure}} \approx -\left(\frac{1+\sin\phi}{1-\sin\phi}\right)\rho g z. \tag{17}$$

A thinner (subcritical) wedge being pushed up the same incline has σ_{xx} greater than $\sigma_{xx}^{\text{failure}}$; such a wedge fails and increases its taper until it becomes critical. A thicker (supercritical) wedge has σ_{xx} less than $\sigma_{xx}^{\text{failure}}$, so it can be pushed up the incline without deforming if no fresh material is encountered at the toe. In determining the critical taper, we have solved a stability problem, since a subcritical wedge is unstable, and a supercritical wedge stable, when pushed up the same incline. Any wedge that is formed by offscraping and the progressive failure of the material within it should have $\sigma_{xx} \approx \sigma_{xx}^{\text{failure}}$—this is the essential premise of the critical taper model.

BALANCE OF FORCES IN A POROUS MEDIUM

The brittle frictional strength of rocks in the upper crust is significantly affected by the presence of water and other interstitial pore fluids. The important role played by pore-fluid pressure in overthrust faulting was first pointed out in the classic and influential paper of Hubbert & Rubey (1959). Their discussion is extremely lucid and well worth reading over 30 years later. One aspect that led to some controversy following the original publication is their calculation of the force exerted by a pore fluid on a porous solid (Laubscher 1960, Moore 1961, Hubbert & Rubey 1960, 1961). This is a subtle issue that has been overlooked in previous critical taper analyses, and we address it in some detail here.

Microscopic Equations

We model a porous medium as a solid skeleton or matrix whose pore spaces are completely filled by a homogeneous incompressible fluid of constant density ρ_f and constant viscosity ν. The density $\rho_s(\mathbf{x})$ of the solid is regarded as a function of position $\mathbf{x}$ to allow for density variations from grain to grain. We denote the stress tensors within the fluid and solid, respectively, by $\boldsymbol{\sigma}_f(\mathbf{x})$ and $\boldsymbol{\sigma}_s(\mathbf{x})$. The fluid stress is related to the fluid velocity $\mathbf{u}(\mathbf{x})$ within the pore spaces by the Newtonian constitutive equation

$$\boldsymbol{\sigma}_f = -p_f\mathbf{I} + \nu[\nabla_x\mathbf{u} + (\nabla_x\mathbf{u})^T], \tag{18}$$

where $\mathbf{I}$ is the identity tensor, and T denotes the transpose. We affix a subscript $\mathbf{x}$ to the gradient operator ∇ to emphasize that it describes the change in a quantity due to a change in the microscopic position variable $\mathbf{x}$. The quantity $p_f(\mathbf{x})$ is the microscopic pore-fluid pressure. The pointwise momentum balance equations within the solid and fluid are

$$\nabla_x \cdot \boldsymbol{\sigma}_s + \rho_s\mathbf{g} = \mathbf{0}, \tag{19a}$$

$$\nabla_x \cdot \boldsymbol{\sigma}_f + \rho_f\mathbf{g} = \mathbf{0}, \tag{19b}$$

where $\mathbf{g}$ is the acceleration of gravity. Inertial forces $\rho_f(\partial_t\mathbf{u} + \mathbf{u}\cdot\nabla_x\mathbf{u})$ have been ignored in writing Equation (19b), since the flow is assumed to be in the creeping (low-Reynolds-number) regime. Equations (18) and (19b) together can be written in the form

$$-\nabla_x p_f + \nu\nabla_x^2\mathbf{u} + \rho_f\mathbf{g} = \mathbf{0}. \tag{20}$$

This is the well-known Navier-Stokes equation, with inertial forces ignored (Batchelor 1967, pp. 216–17).

Volume Averaging

Equations (19) and (20) are far too complicated to use directly because of the rapid variation from fluid to solid on the microscopic scale. We seek instead a system of averaged equations that are valid on the macroscopic scale. The procedure of averaging a system of microscopic equations to obtain a simpler system of macroscopic equations is a common one; in electromagnetism it is the basis for extending Maxwell's equations to dielectric and magnetic materials (Jackson 1962, pp. 103–8, 150–54). Averaging has been used to obtain the macroscopic equations governing a porous medium by several authors, including Whitaker (1969), Saffman (1971), Slattery (1972, pp. 191–215), Gray & O'Neill (1976), and Lehner (1979).

Consider an averaging volume V centered on an arbitrary point $\bar{\mathbf{x}}$, as shown in Figure 7 (*left*). Let V_s and V_f be the volumes within V occupied by the solid and fluid, respectively. The porosity $\eta(\bar{\mathbf{x}})$ is defined by

$$\eta = V_f/V. \tag{21}$$

In order for the averaging to be meaningful, the size of the averaging volume must be much smaller than a typical macroscopic scale length but large enough to average over many solid grains and many pore spaces. In this limit, it is immaterial whether we regard $\bar{\mathbf{x}}$ as the centroid of the whole volume V, or as the centroid of the solid matter contained within V_s, or as the centroid of the pore space V_f. We regard the porosity η and all other macroscopic variables as continuous functions of the macroscopic position variable $\bar{\mathbf{x}}$.

Upon averaging Equation (19a) over V_s and Equation (19b) over V_f, we obtain

$$\frac{1}{V_s}\int_{V_s} \nabla_{\mathbf{x}}\cdot\boldsymbol{\sigma}_s\,\mathrm{d}V + \bar{\rho}_s\mathbf{g} = \mathbf{0}, \tag{22a}$$

$$\frac{1}{V_f}\int_{V_f} \nabla_{\mathbf{x}}\cdot\boldsymbol{\sigma}_f\,\mathrm{d}V + \rho_f\mathbf{g} = \mathbf{0}. \tag{22b}$$

The quantity $\bar{\rho}_s(\bar{\mathbf{x}})$ is the macroscopic solid density, given by

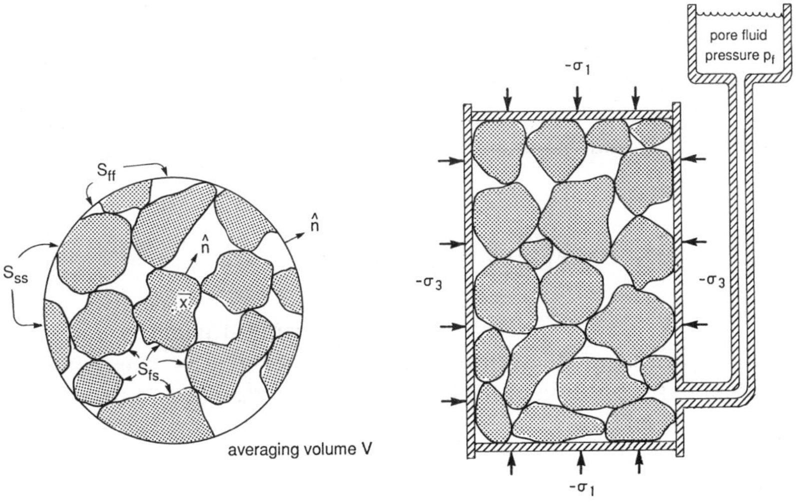

Figure 7 (*Left*) Schematic diagram of a portion of a porous medium within a representative averaging volume V. The interface between the solid grains and the pore spaces is denoted by S_{fs}. (*Right*) Schematic diagram of a typical rock-mechanics laboratory friction or fracture experiment.

$$\bar{\rho}_s = \frac{1}{V_s}\int_{V_s} \rho_s \, \mathrm{d}V. \tag{23}$$

Equations (22) can alternatively be written in the form

$$\frac{1}{V}\int_{V_s} \nabla_{\mathbf{x}} \cdot \boldsymbol{\sigma}_s \, \mathrm{d}V + (1-\eta)\bar{\rho}_s \mathbf{g} = \mathbf{0}, \tag{24a}$$

$$\frac{1}{V}\int_{V_f} \nabla_{\mathbf{x}} \cdot \boldsymbol{\sigma}_f \, \mathrm{d}V + \eta \rho_f \mathbf{g} = \mathbf{0}. \tag{24b}$$

We define the macroscopic solid and fluid stresses $\bar{\boldsymbol{\sigma}}_s(\bar{\mathbf{x}})$ and $\bar{\boldsymbol{\sigma}}_f(\bar{\mathbf{x}}) = -\bar{p}_f(\bar{\mathbf{x}})\mathbf{I}$ by

$$\bar{\boldsymbol{\sigma}}_s = \frac{1}{V_s}\int_{V_s} \boldsymbol{\sigma}_s \, \mathrm{d}V, \tag{25a}$$

$$\bar{\boldsymbol{\sigma}}_f = -\bar{p}_f \mathbf{I} = \frac{1}{V_f}\int_{V_f} \boldsymbol{\sigma}_f \, \mathrm{d}V. \tag{25b}$$

Note that the macroscopic fluid stress is considered to be isotropic, even though viscous shear stresses may be comparable to the dynamical pressure fluctuations on the microscopic scale.

The boundary of the averaging volume V has portions S_{ss} in the solid grains and portions S_{ff} in the fluid-filled pore spaces, as shown in Figure 7 (*left*). We denote the fluid-solid interface situated within the volume V by S_{fs} and use $\hat{\mathbf{n}}$ to denote the unit normal that points out of the averaging volume on S_{ss} and S_{ff} and out of the solid grains on S_{fs}. Consider the quantity

$$\nabla_{\bar{\mathbf{x}}} \cdot [(1-\eta)\bar{\boldsymbol{\sigma}}_s] = \frac{1}{V}\nabla_{\bar{\mathbf{x}}} \cdot \int_{V_s} \boldsymbol{\sigma}_s \, \mathrm{d}V. \tag{26}$$

Physically, $\nabla_{\bar{\mathbf{x}}}$ describes the change in an averaged quantity due to an infinitesimal shift in the centroid $\bar{\mathbf{x}}$ of the averaging volume. Only the variation in the position of the boundary of V contributes to this change, hence

$$\nabla_{\bar{\mathbf{x}}} \cdot \int_{V_s} \boldsymbol{\sigma}_s \, \mathrm{d}V = \int_{S_{ss}} \hat{\mathbf{n}} \cdot \boldsymbol{\sigma}_s \, \mathrm{d}A. \tag{27}$$

By Gauss' theorem, the right side of Equation (27) can be written in the form

$$\int_{S_{ss}} \hat{\mathbf{n}} \cdot \boldsymbol{\sigma}_s \, dA = \int_{V_s} \nabla_{\mathbf{x}} \cdot \boldsymbol{\sigma}_s \, dV - \int_{S_{fs}} \hat{\mathbf{n}} \cdot \boldsymbol{\sigma}_s \, dA. \tag{28}$$

Thus, it follows that

$$\frac{1}{V} \int_{V_s} \nabla_{\mathbf{x}} \cdot \boldsymbol{\sigma}_s \, dV = \nabla_{\bar{\mathbf{x}}} \cdot [(1-\eta)\bar{\boldsymbol{\sigma}}_s] + \frac{1}{V} \int_{S_{fs}} \hat{\mathbf{n}} \cdot \boldsymbol{\sigma}_s \, dA. \tag{29}$$

Equation (29), which relates the average of a divergence to the divergence of an average, is known as the Slattery-Whitaker averaging theorem. A similar result applies to the fluid stress, namely

$$\frac{1}{V} \int_{V_f} \nabla_{\mathbf{x}} \cdot \boldsymbol{\sigma}_f \, dA = -\nabla_{\bar{\mathbf{x}}}(\eta \bar{p}_f) - \frac{1}{V} \int_{S_{fs}} \hat{\mathbf{n}} \cdot \boldsymbol{\sigma}_f \, dA. \tag{30}$$

The sign in front of the surface integral over S_{fs} differs from that in Equation (29) because of the convention that $\hat{\mathbf{n}}$ points out of the solid grains into the pore spaces.

The final macroscopic solid and fluid equations, obtained by inserting Equations (29) and (30) into (24), are

$$\nabla_{\bar{\mathbf{x}}} \cdot [(1-\eta)\bar{\boldsymbol{\sigma}}_s] + (1-\eta)\bar{\rho}_s \mathbf{g} + \frac{1}{V} \int_{S_{fs}} \hat{\mathbf{n}} \cdot \boldsymbol{\sigma} \, dA = \mathbf{0}, \tag{31a}$$

$$-\nabla_{\bar{\mathbf{x}}}(\eta \bar{p}_f) + \eta \rho_f \mathbf{g} - \frac{1}{V} \int_{S_{fs}} \hat{\mathbf{n}} \cdot \boldsymbol{\sigma} \, dA = \mathbf{0}. \tag{31b}$$

On the fluid-solid interface S_{fs} there is continuity of traction ($\hat{\mathbf{n}} \cdot \boldsymbol{\sigma}_s = \hat{\mathbf{n}} \cdot \boldsymbol{\sigma}_f$), so we have just written $\hat{\mathbf{n}} \cdot \boldsymbol{\sigma}$ in the surface integrals. By adding Equations (31) we obtain the simple result

$$\nabla_{\bar{\mathbf{x}}} \cdot \bar{\boldsymbol{\sigma}} + \bar{\rho} \mathbf{g} = \mathbf{0}. \tag{32}$$

The quantities

$$\bar{\boldsymbol{\sigma}} = (1-\eta)\boldsymbol{\sigma}_s - \eta \bar{p}_f \mathbf{I}, \tag{33a}$$

$$\bar{\rho} = (1-\eta)\bar{\rho}_s + \eta \rho_f, \tag{33b}$$

are the aggregate stress and density, respectively, of the fluid-filled porous medium; these macroscopic aggregate variables satisfy the same static equilibrium equation as the microscopic solid and fluid variables.

Force Exerted by the Fluid on the Solid

The quantity $\mathbf{F}(\bar{\mathbf{x}})$, defined by

$$\mathbf{F} = \frac{1}{V}\int_{S_{\mathrm{fs}}} \hat{\mathbf{n}}\cdot\boldsymbol{\sigma}\,\mathrm{d}A, \tag{34}$$

appears explicitly as an additional apparent body force in the macroscopic solid balance equation (31a). Physically, $\mathbf{F}$ is the macroscopic force per unit volume exerted by the pore fluid on the solid matrix; the solid exerts an equal and opposite force on the fluid, and this appears in Equation (31b). It is straightforward to evaluate $\mathbf{F}$ if the fluid is in a hydrostatic rest state:

$$\mathbf{u} = \mathbf{0}, \tag{35a}$$

$$\boldsymbol{\sigma}_{\mathrm{f}} = -(p_0 + \rho_{\mathrm{f}}\mathbf{g}\cdot\mathbf{x})\mathbf{I}, \tag{35b}$$

where p_0 is a constant reference pressure. By Gauss' theorem, we have

$$\begin{aligned}\mathbf{F}_{\mathrm{hydro}} &= -\frac{1}{V}\int_{V_{\mathrm{f}}} \nabla_{\bar{\mathbf{x}}}\cdot\boldsymbol{\sigma}_{\mathrm{f}}\,\mathrm{d}V + \frac{1}{V}\int_{S_{\mathrm{ff}}} \hat{\mathbf{n}}\cdot\boldsymbol{\sigma}_{\mathrm{f}}\,\mathrm{d}A \\ &= \frac{1}{V}\int_{V_{\mathrm{f}}} \rho_{\mathrm{f}}\mathbf{g}\,\mathrm{d}V + \frac{1}{V}\int_{S_{\mathrm{ff}}} \hat{\mathbf{n}}\cdot\boldsymbol{\sigma}_{\mathrm{f}}\,\mathrm{d}A \\ &= \eta\rho_{\mathrm{f}}\mathbf{g} - \frac{1}{V}\int_{S_{\mathrm{ff}}} \hat{\mathbf{n}}p_0\,\mathrm{d}A - \frac{1}{V}\int_{S_{\mathrm{ff}}} \hat{\mathbf{n}}(\rho_{\mathrm{f}}\mathbf{g}\cdot\mathbf{x})\,\mathrm{d}A.\end{aligned} \tag{36}$$

Upon evaluating the first surface integral in Equation (36) using the Slattery-Whitaker averaging theorem, we obtain

$$\frac{1}{V}\int_{S_{\mathrm{ff}}} \hat{\mathbf{n}}p_0\,\mathrm{d}A = p_0\nabla_{\bar{\mathbf{x}}}\left[\frac{1}{V}\int_{V_{\mathrm{f}}} \mathrm{d}V\right] = p_0\nabla_{\bar{\mathbf{x}}}\eta. \tag{37}$$

A similar manipulation of the second surface integral gives

$$\begin{aligned}\frac{1}{V}\int_{S_{\mathrm{ff}}} \hat{\mathbf{n}}(\rho_{\mathrm{f}}\mathbf{g}\cdot\mathbf{x})\,\mathrm{d}A &= \nabla_{\bar{\mathbf{x}}}\left[\frac{1}{V}\int_{V_{\mathrm{f}}} \mathbf{x}\,\mathrm{d}V\right]\cdot\rho_{\mathrm{f}}\mathbf{g} = \nabla_{\bar{\mathbf{x}}}(\eta\bar{\mathbf{x}})\cdot\rho_{\mathrm{f}}\mathbf{g} \\ &= \eta(\rho_{\mathrm{f}}\mathbf{g}) + (\rho_{\mathrm{f}}\mathbf{g}\cdot\bar{\mathbf{x}})\nabla_{\bar{\mathbf{x}}}\eta,\end{aligned} \tag{38}$$

by definition of the centroid $\bar{\mathbf{x}}$. Combining Equations (36)–(38), we find that

$$\mathbf{F}_{\mathrm{hydro}} = -(p_0 + \rho_{\mathrm{f}}\mathbf{g}\cdot\bar{\mathbf{x}})\nabla_{\bar{\mathbf{x}}}\eta = -\bar{p}_{\mathrm{f}}\nabla_{\bar{\mathbf{x}}}\eta. \tag{39}$$

A hydrostatic pore fluid thus exerts no net force on a constant-porosity solid; more generally, $\mathbf{F}_{\text{hydro}}$ is in the direction of decreasing porosity. The above analysis shows that there is no Archimedean buoyancy force $-(1-\eta)\rho_f\mathbf{g}$ on a constant-porosity solid; this point is obscured in the discussion of Hubbert & Rubey (1959) because they do not calculate the physically relevant quantity $\mathbf{F}_{\text{hydro}}$.

The macroscopic moment balance equation (31b) in the fluid can be rewritten in the form

$$\mathbf{F}-\mathbf{F}_{\text{hydro}} = -\eta(\nabla_{\bar{x}}\bar{p}_f-\rho_f\mathbf{g}). \tag{40}$$

The quantity $-\eta(\nabla_{\bar{x}}\bar{p}_f-\rho_f\mathbf{g})$ is thus the additional force per unit volume on the solid due to the motion of the fluid. It is customary to write this so-called seepage force in the form (Bear 1972, pp. 184–89)

$$\mathbf{F}-\mathbf{F}_{\text{hydro}} = \eta\nu\mathbf{K}^{-1}\cdot\bar{\mathbf{u}}, \tag{41}$$

where $\mathbf{K}(\bar{\mathbf{x}})$ is the permeability tensor, and $\mathbf{K}\cdot\mathbf{K}^{-1}=\mathbf{K}^{-1}\cdot\mathbf{K}=\mathbf{I}$. The quantity $\bar{\mathbf{u}}(\bar{\mathbf{x}})$ is the macroscopic fluid velocity or averaged fluid flux per unit area (note the division by V instead of V_f):

$$\bar{\mathbf{u}} = \frac{1}{V}\int_{V_f}\mathbf{u}\,dV. \tag{42}$$

Equation (41) is a constitutive relation governing the macroscopic flow; the linear relation between $\mathbf{F}-\mathbf{F}_{\text{hydro}}$ and $\bar{\mathbf{u}}$ is a consequence of the linearity of the Navier-Stokes equation (20), which governs the flow on the microscopic scale (Neumann 1977). Inserting Equation (41) into Equation (40) reduces the macroscopic fluid equation to

$$\bar{\mathbf{u}} = -\nu^{-1}\mathbf{K}\cdot(\nabla_{\bar{x}}\bar{p}_f-\rho_f\mathbf{g}). \tag{43}$$

This is the usual form of Darcy's law (Batchelor 1967, pp. 223–24; Bear 1972, pp. 119–25).

Simplified Notation

Once the macroscopic equations have been derived, it is convenient to simplify the notation by dispensing with the subscripts on ∇ and the overbars used to denote averaged quantities. Accordingly, we rewrite Equation (43) in the form

$$\mathbf{u} = -\nu^{-1}\mathbf{K}\cdot(\nabla p_f-\rho_f\mathbf{g}). \tag{44}$$

Fluid flow within a noncompacting porous medium is determined by solving Equation (44) together with the macroscopic incompressibility condition $\nabla\cdot\mathbf{u}=0$. In ground-water hydrology, it is common to rewrite Equation (44) in the form $\mathbf{u} = -\nu^{-1}\rho_f g \mathbf{K}\cdot\nabla\Phi$, where $\Phi = (\rho_f g)^{-1}(p_f - \rho_f \mathbf{g}\cdot\mathbf{x})$ is the piezometric head (Bear 1972, pp. 122–23). The stress $\boldsymbol{\sigma}_s$ in the solid is related to that in the fluid by

$$\nabla\cdot[(1-\eta)\boldsymbol{\sigma}_s]+(1-\eta)\rho_s\mathbf{g} = \nabla(\eta p_f)-\eta\rho_f\mathbf{g}, \tag{45}$$

whereas the aggregate or porous medium stress $\boldsymbol{\sigma}$ satisfies

$$\nabla\cdot\boldsymbol{\sigma}+\rho\mathbf{g} = \mathbf{0}. \tag{46}$$

Equations (44)–(46) are valid at every point $\mathbf{x}$ in the macroscopic medium.

Effective Stress

The Coulomb failure criterion, which we consider next, depends on the effective stress in the porous medium, defined by

$$\boldsymbol{\sigma}^* = \boldsymbol{\sigma}+p_f\mathbf{I} = (1-\eta)(\boldsymbol{\sigma}_s+p_f\mathbf{I}). \tag{47}$$

Equations (44)–(46) are readily combined to yield

$$\nabla\cdot\boldsymbol{\sigma}^*+(1-\eta)(\rho_s-\rho_f)\mathbf{g}+\nu\mathbf{K}^{-1}\cdot\mathbf{u} = \mathbf{0}, \tag{48}$$

or, alternatively,

$$\nabla\cdot\boldsymbol{\sigma}^*+(1-\eta)\rho_s\mathbf{g}-(1-\eta)\nabla p_f+\eta\nu\mathbf{K}^{-1}\cdot\mathbf{u} = \mathbf{0}. \tag{49}$$

These equations involving the effective stress $\boldsymbol{\sigma}^*$ are commonly employed in soil mechanics and slope stability problems (e.g. Iverson & Major 1986). The second term $[(1-\eta)\rho_s\mathbf{g}]$ in Equation (49) is the gravitational attraction on the solid matrix, and the final term $[\eta\nu\mathbf{K}^{-1}\cdot\mathbf{u}]$ is the seepage force due to the motion of the fluid through the porous medium. The third term $[-(1-\eta)\nabla p_f]$ is frequently interpreted as an Archimedean buoyancy force acting on the solid matrix (Bear 1972, pp. 184–89), since that is what it reduces to if the fluid is in a hydrostatic rest state.

Discussion

In enumerating the body forces acting on a porous medium, it is necessary to distinguish which of the three stresses $\boldsymbol{\sigma}$, $\boldsymbol{\sigma}_s$, or $\boldsymbol{\sigma}^*$ is being considered—failure to do this was the cause of much of the controversy initiated by Hubbert & Rubey (1959). The aggregate stress $\boldsymbol{\sigma}$ satisfies Equation (46); the only body force in this case is the direct attraction of gravity on the porous medium, $\rho\mathbf{g}$. Equation (46) is valid even if the porosity is spatially variable and if there are nonhydrostatic pressure gradients causing fluid to percolate through the medium. If, instead, we wish to solve directly for the solid stress $\boldsymbol{\sigma}_s$, we must employ Equation (45). If the posority is uniform, (45) reduces to

$$\nabla \cdot \boldsymbol{\sigma}_s + \rho_s \mathbf{g} - \eta(1-\eta)^{-1}(\nabla p_f - \rho_f \mathbf{g}) = \mathbf{0}. \tag{50}$$

This has the same form as (46) if the pore-fluid pressure is hydrostatic; more generally, however, we must add $(1-\eta)^{-1}$ times the seepage force $\eta v \mathbf{K}^{-1} \cdot \mathbf{u}$. Finally, if we wish to solve directly for the effective stress $\boldsymbol{\sigma}^*$, we must employ Equation (49); in this case it is necessary to account for the Archimedean term $-(1-\eta)\nabla p_f$ as well as for the seepage force. Once any of the three stresses $\boldsymbol{\sigma}$, $\boldsymbol{\sigma}_s$, or $\boldsymbol{\sigma}^*$ has been determined, the others may be found subsequently; we are free to adopt the most convenient strategy.

COULOMB FAILURE CRITERION

There are a number of equivalent prescriptions of the Coulomb criterion; in reviewing these here, we account for cohesion as well as pore-fluid pressure. From a fundamental point of view, the strength of brittle materials is not well understood; in particular, there is no satisfactory explanation for the dependence on effective stress $\boldsymbol{\sigma}^*$. The Coulomb law is regarded here as a strictly empirical relation; laboratory data supporting its validity for rocks are reviewed by Jaeger & Cook (1969, pp. 136–82, 210–12) and Paterson (1978, pp. 16–50, 71–87). The effective stress principle was first stated for soils by Terzaghi (1923); a historical review is given by Skempton (1960).

Formulation in Terms of Principal Stresses

The right side of Figure 7 shows a sketch of an idealized laboratory rock mechanics fracture experiment. Three variables can be controlled independently: the applied axial stress σ_1, the confining stress σ_3, and the pore-fluid pressure p_f. The observed relationship between these three quantities at failure is

$$\sigma_1 + p_f = B(\sigma_3 + p_f) - C. \tag{51}$$

The constant B is related to the internal frictional angle ϕ by

$$B = \frac{1+\sin\phi}{1-\sin\phi}. \tag{52}$$

The constant C is called the uniaxial compressive strength.

The tractions exerted by the pistons and confining walls are transmitted to both the solid grains and the pore spaces, as shown in Figure 7 (*right*). The extent to which the fluid shares the load depends on the ratio A_f/A, where A is the total area of the boundary and A_f is the fraction of that area lying within the pore spaces. This so-called areal porosity is equal to the volumetric porosity η if the sample is homogeneous and isotropic.

With that proviso, the experimentally applied tractions may be interpreted as the macroscopic principal stresses in the porous aggregate:

$$\sigma_1 = (1-\eta)\sigma_{1s} - \eta p_f, \tag{53a}$$

$$\sigma_3 = (1-\eta)\sigma_{3s} - \eta p_f. \tag{53b}$$

The Coulomb law (51) can be rewritten in the form

$$\sigma_1^* = \left(\frac{1+\sin\phi}{1-\sin\phi}\right)\sigma_3^* - C, \tag{54}$$

where the quantities

$$\sigma_1^* = \sigma_1 + p_f = (1-\eta)(\sigma_{1s} + p_f), \tag{55a}$$

$$\sigma_3^* = \sigma_3 + p_f = (1-\eta)(\sigma_{3s} + p_f) \tag{55b}$$

are the effective principal stresses in the porous aggregate. With the sign convention adopted here, both σ_1^* and σ_3^* are negative in a typical experiment or in an active fold-and-thrust belt or accretionary wedge; failure occurs in the shaded region shown on the left in Figure 8.

Alternative Formulations

Let ψ be the counterclockwise angle from the x-axis to the local axis of greatest principal stress in a material, as shown in the center of Figure 8. Any two-dimensional state of stress can be written in the form

$$\sigma_{xx} = -p - R\cos 2\psi, \tag{56a}$$

$$\sigma_{zz} = -p + R\cos 2\psi, \tag{56b}$$

$$\sigma_{xz} = R\sin 2\psi, \tag{56c}$$

where

$$R = [\tfrac{1}{4}(\sigma_{zz} - \sigma_{xx})^2 + \sigma_{xz}^2]^{1/2}, \tag{57a}$$

$$p = -\tfrac{1}{2}(\sigma_{xx} + \sigma_{zz}). \tag{57b}$$

The quantity R is the radius of the Mohr circle, and p is called the mean aggregate stress. The aggregate principal stresses σ_1 and σ_3 are given in terms of p and R by

$$\sigma_1 = -p - R, \tag{58a}$$

$$\sigma_3 = -p + R. \tag{58b}$$

Alternatively, they may be written in terms of σ_{zz}, σ_{xx}, and ψ in the form

$$\sigma_1 = \sigma_{zz} - \tfrac{1}{2}(\sigma_{zz} - \sigma_{xx})(1 + \sec 2\psi), \tag{59a}$$

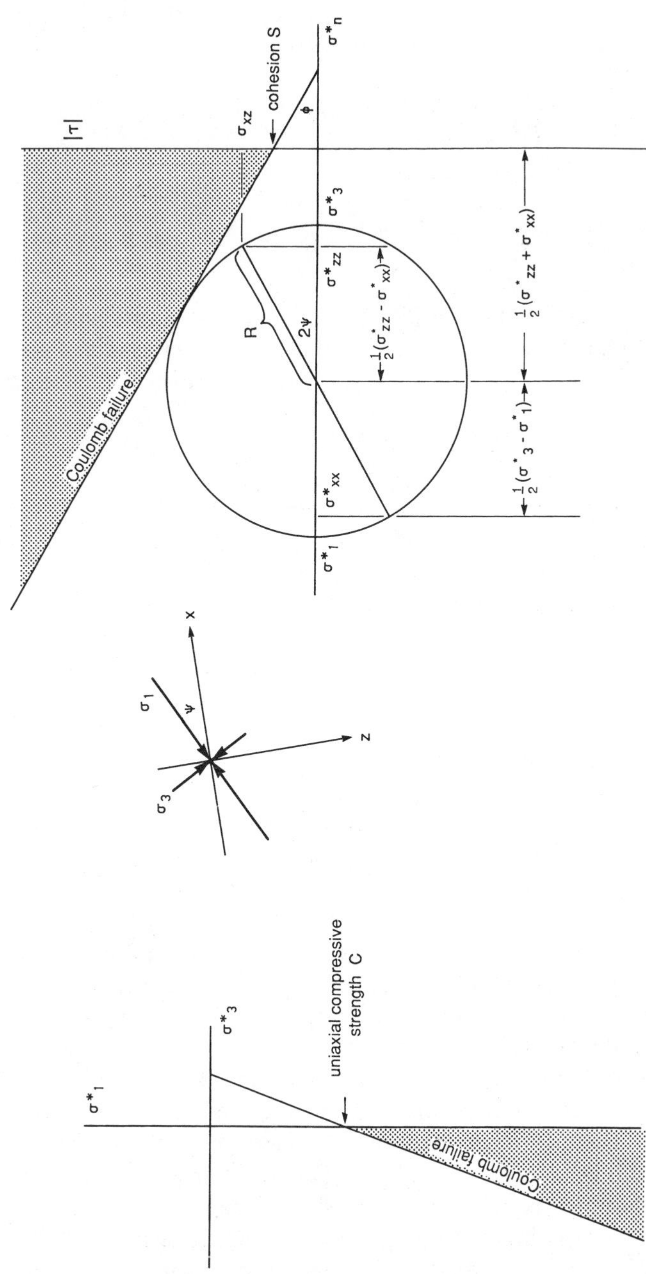

Figure 8 (*Left*) Relation between the effective principal stresses σ_1^* and σ_3^* in a noncohesive Coulomb material. (*Middle*) Orientation of the principal stresses σ_1 and σ_3 with respect to the coordinate axes x, z. (*Right*) Mohr diagram depicting the state of effective stress within a noncohesive Coulomb wedge.

$$\sigma_3 = \sigma_{zz} - \tfrac{1}{2}(\sigma_{zz} - \sigma_{xx})(1 - \sec 2\psi). \tag{59b}$$

The Coulomb criterion (54) can be written in terms of R, p, and p_f in the form

$$R = S\cos\phi + (p - p_f)\sin\phi = S\cos\phi + p^*\sin\phi. \tag{60}$$

The quantity S is the cohesion, given by

$$S = \tfrac{1}{2}C\left(\frac{1-\sin\phi}{\cos\phi}\right) = \tfrac{1}{2}C\left(\frac{1-\sin\phi}{1+\sin\phi}\right)^{1/2}. \tag{61}$$

Shear failure within an idealized Coulomb material occurs on conjugate surfaces oriented at angles $\pm\tfrac{1}{2}(90° - \phi)$ with respect to the axis of greatest principal stress σ_1. The shear traction $|\tau|$ on these planes is related to the effective normal traction $\sigma_n^* = \sigma_n + p_f = (1-\eta)(\sigma_{ns} + p_f)$ by

$$|\tau| = S - \mu\sigma_n^*, \tag{62}$$

where $\mu = \tan\phi$ is the coefficient of internal friction. The right side of Figure 8 shows a Mohr circle representation of the state of stress in a porous medium at failure. The quantities $\tfrac{1}{2}(\sigma_{zz} - \sigma_{xx}) = \tfrac{1}{2}(\sigma_{zz}^* - \sigma_{xx}^*)$ and $\sigma_{xz} = \sigma_{xz}^*$ can be written in terms of $\sigma_{zz}^* = \sigma_{zz} + p_f$ and ψ in the form

$$\tfrac{1}{2}(\sigma_{zz}^* - \sigma_{xx}^*) = \frac{S\cot\phi - \sigma_{zz}^*}{\csc\phi\sec 2\psi - 1}, \tag{63a}$$

$$\sigma_{xz}^* = \frac{\tan 2\psi\,[S\cot\phi - \sigma_{zz}^*]}{\csc\phi\sec 2\psi - 1}. \tag{63b}$$

Equations (63) are the most convenient form of the Coulomb failure criterion to use in the critical taper analysis that follows.

Noncohesive Approximation

Coefficients of internal friction measured in laboratory fracture experiments are in the range $\mu = 0.6$–1.0 for virtually all rocks; the corresponding internal friction angles are in the range $\phi = 30$–$45°$. Cohesion S varies much more widely, from nearly zero up to 150 MPa, with a strong dependence on porosity, cementation, mineralogy, and other factors. The shale and sandstone sedimentary rocks that are the predominant constituents of fold-and-thrust belts and accretionary wedges generally have $S = 5$–10 MPa (Hoshino et al 1972). Equations (62) and (63) show that such low values of cohesion are only important at shallow depths or where the pore-fluid pressure is very high; this suggests that a noncohesive critical taper model should be a reasonably good approximation for many geological applications. Equations (63) reduce, in the absence of cohesion, to

$$\tfrac{1}{2}(\sigma_{zz}^* - \sigma_{xx}^*) = \frac{-\sigma_{zz}^*}{\csc\phi \sec 2\psi - 1}, \tag{64a}$$

$$\sigma_{xz}^* = \frac{-\tan 2\psi\, \sigma_{zz}^*}{\csc\phi \sec 2\psi - 1}. \tag{64b}$$

It is noteworthy that these equations are satisfied by $\boldsymbol{\sigma}_{\mathrm{s}} + p_{\mathrm{f}}\mathbf{I}$ as well as by $\boldsymbol{\sigma}^*$, since a multiplicative factor of $1-\eta$ can be canceled on both sides.

Friction

Cohesion is also negligible in the case of frictional sliding on preexisting faults. Byerlee (1978) has shown that the laboratory coefficient of friction is remarkably uniform for a wide variety of rock types: $|\tau| = -0.85\sigma_{\mathrm{n}}^*$ for $|\sigma_{\mathrm{n}}^*| < 200$ MPa. Clay-rich fault gouges are characterized by lower coefficients, in the range 0.3–0.5 (Morrow et al 1981, Logan & Rauenzahn 1987).

NONCOHESIVE COULOMB WEDGE

Theory

Consider a submarine wedge with a planar upper surface, as shown in Figure 9; the results for a subaerial wedge can be recovered by setting the fluid density ρ_{f} equal to zero wherever it appears in a numbered equation below. We adopt a system of Cartesian coordinates, with x lying along the top of the wedge and z pointing obliquely down. Equation (46), the static equilibrium equation in terms of the aggregate stress $\boldsymbol{\sigma}$, becomes

$$\frac{\partial \sigma_{xx}}{\partial x} + \frac{\partial \sigma_{xz}}{\partial z} - \rho g z \sin\alpha = 0, \tag{65a}$$

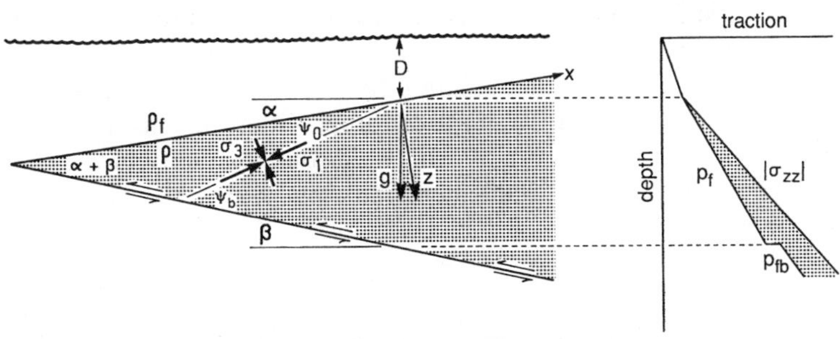

Figure 9 Idealized cross section of a submarine noncohesive critical wedge, showing the coordinate axes x, z and the angles α, β, ψ_0, and ψ_b. Strength in the wedge is proportional to the effective stress $\sigma_{zz}^* = \sigma_{zz} + p_{\mathrm{f}}$, shown schematically by the shaded area on the right.

$$\frac{\partial \sigma_{xz}}{\partial x}+\frac{\partial \sigma_{zz}}{\partial z}+\rho g z \cos \alpha=0. \tag{65b}$$

The boundary conditions on the upper surface of the wedge, $z=0$, are

$$\sigma_{xz}=0, \qquad \sigma_{zz}=-\rho_{\mathrm{f}} g D, \tag{66}$$

where D is the water depth. It is convenient to define the generalized Hubbert & Rubey (1959) pore-fluid to lithostatic pressure ratio by

$$\lambda=\frac{p_{\mathrm{f}}-\rho_{\mathrm{f}} g D}{-\sigma_{zz}-\rho_{\mathrm{f}} g D}. \tag{67}$$

We assume that λ, ρ, and the coefficient of internal friction μ are constant. Recall that ρ is the macroscopic density of the porous aggregate, given by $\rho=(1-\eta)\rho_{\mathrm{s}}+\eta\rho_{\mathrm{f}}$; if the wedge material is homogeneous, so that ρ_{s} is constant, then constancy of ρ implies constancy of the porosity η. Let ψ_0 be the angle between the x-axis and the axis of greatest principal stress σ_1, as shown in Figure 9. Equations (64)–(66) are then satisfied by

$$\sigma_{xz}=(\rho-\rho_{\mathrm{f}}) g z \sin \alpha, \tag{68a}$$

$$\sigma_{zz}=-\rho_{\mathrm{f}} g D-\rho g z \cos \alpha, \tag{68b}$$

$$\sigma_{xx}=-\rho_{\mathrm{f}} g D-\rho g z \cos \alpha\left[\frac{\csc \phi \sec 2\psi_0-2\lambda+1}{\csc \phi \sec 2\psi_0-1}\right], \tag{68c}$$

provided that

$$\frac{\tan 2\psi_0}{\csc \phi \sec 2\psi_0-1}=\left(\frac{1-\rho_{\mathrm{f}}/\rho}{1-\lambda}\right)\tan \alpha. \tag{69}$$

Equation (69) relates the stress orientation angle ψ_0 to the surface slope α; we have assumed that ψ_0 is constant and have made use of the relation $\mathrm{d}D/\mathrm{d}x=-\sin \alpha$.

Equations (68) are an exact solution for the state of stress in a sloping half-space on the verge of Coulomb failure. All that remains is to satisfy the basal boundary condition. We allow for the possibility of a different pore-fluid regime on the décollement fault by writing the basal sliding condition in the form

$$\tau_{\mathrm{b}}=-\mu_{\mathrm{b}}(\sigma_{\mathrm{n}}+p_{\mathrm{fb}}). \tag{70}$$

The quantity p_{fb} is the pore-fluid pressure on the base, and μ_{b} is the basal coefficient of friction. Both μ_{b} and the basal pore-fluid to lithostatic pressure ratio defined by

$$\lambda_b = \frac{p_{fb} - \rho_f g D}{-\sigma_{zz} - \rho_f g D} \tag{71}$$

are assumed to be constant. In reality, the pore-fluid pressure cannot exhibit a jump discontinuity such as that shown in Figure 9, but the introduction of two constant Hubbert-Rubey ratios λ and λ_b provides a simple means of allowing for elevated pore-fluid pressures in the décollement zone. In order for a critical wedge to exist, its base must be a zone of weakness, i.e.

$$0 \leq \mu_b(1-\lambda_b) \leq \mu(1-\lambda). \tag{72}$$

The shear stress and normal stress on a surface whose dip is β are given as usual by (Malvern 1969, pp. 102–11)

$$\tau_b = \tfrac{1}{2}(\sigma_{zz} - \sigma_{xx}) \sin 2(\alpha+\beta) + \sigma_{xz} \cos 2(\alpha+\beta), \tag{73a}$$

$$\sigma_n = \sigma_{zz} - \sigma_{xz} \sin 2(\alpha+\beta) - \tfrac{1}{2}(\sigma_{zz} - \sigma_{xz})\,[1 - \cos 2(\alpha+\beta)]. \tag{73b}$$

Equations (68) and (73) are used to determine the dip of the surface on which the frictional sliding condition (70) is satisfied. We find, after some algebra, that β is given by

$$\alpha + \beta = \psi_b - \psi_0, \tag{74}$$

where

$$\frac{\tan 2\psi_b}{\csc\phi \sec 2\psi_b - 1} = \mu_b\left(\frac{1-\lambda_b}{1-\lambda}\right). \tag{75}$$

Equation (74) is the exact critical taper equation for a homogeneous noncohesive Coulomb wedge. It can be regarded as an equation of the form $\alpha+\beta = \psi_b - \psi_0(\alpha)$, which implicitly gives the surface slope α in terms of the basal dip β, the density ratio ρ_f/ρ, and the strength parameters μ, λ, μ_b, and λ_b. The quantity ψ_b is the angle between the axis of greatest principal stress σ_1 and the base of the wedge; thus, Equation (74) expresses an elementary relation between two internal angles and the opposite external angle of a triangle, as shown in Figure 9. Since the orientation of the principal stresses is everywhere the same, a noncohesive critical wedge is self-similar in the sense that a magnified version of any portion of it near the toe is indistinguishable from the wedge as a whole; this is a consequence of the absence of an inherent length scale in the equations of equilibrium and in the boundary and failure conditions. The exact critical taper equation involves only the angles α and β and the dimensionless parameters ρ_f/ρ, μ, λ, μ_b, and λ_b.

Equations (69) and (75) may be rewritten in the explicit form

$$\psi_0 = \tfrac{1}{2}\arcsin\left(\frac{\sin\alpha'}{\sin\phi}\right) - \tfrac{1}{2}\alpha', \tag{76a}$$

$$\psi_b = \tfrac{1}{2}\arcsin\left(\frac{\sin\phi'_b}{\sin\phi}\right) - \tfrac{1}{2}\phi'_b. \tag{76b}$$

The quantity α' is a modified surface slope angle defined by

$$\alpha' = \arctan\left[\left(\frac{1-\rho_f/\rho}{1-\lambda}\right)\tan\alpha\right], \tag{77}$$

and ϕ'_b is an effective basal friction angle defined by

$$\phi'_b = \arctan\left[\mu_b\left(\frac{1-\lambda_b}{1-\lambda}\right)\right]. \tag{78}$$

The multivalued nature of the arcsin functions in Equations (76) gives rise to a multiplicity of solutions with both compressional and extensional states of stress within the wedge and both possible orientations of the shear stress τ_b on the basal décollement fault (Dahlen 1984). The solution applicable to an active fold-and-thrust belt or accretionary wedge is obtained by choosing both ψ_0 and ψ_b to be positive acute angles, as shown in Figure 9. A different but equivalent form of the general solution (for the special case of a noncohesive dry subaerial wedge) dates back to Coulomb's (1773) analysis of the load exerted on a rough retaining wall. Dahlen (1984) was unaware of this venerable result and rediscovered it in the present context. A more systematic derivation, which exploits the observed scale invariance, is given by Barcilon (1987). Lehner (1986) has shown how the various multiple solutions can be obtained by a graphical construction method on the Mohr diagram.

Step-Up of Thrusts From the Basal Décollement Fault

Because of the self-similarity of a homogeneous noncohesive wedge, the failure surfaces oriented at $\pm\tfrac{1}{2}(90^\circ-\phi)$ with respect to the axis of greatest compressive stress σ_1 have the same dip everywhere. Forward-verging thrusts step up from the basal décollement fault at an angle

$$\delta_b = \tfrac{1}{2}(90^\circ-\phi)-\psi_b, \tag{79}$$

whereas the conjugate back-thrusts step up at a steeper angle

$$\delta'_b = \tfrac{1}{2}(90^\circ-\phi)+\psi_b. \tag{80}$$

The idealized geometry of these thrust faults and a Mohr diagram illustrating the basal state of effective stress in a noncohesive Coulomb wedge are shown in Figure 10. The failure stress $|\tau|$ on both the forward and backward thrusts at the point where they step up into the wedge is given by

$$|\tau| = (\rho - \rho_f) gz \sin\alpha \left(\frac{\cos\phi}{\sin 2\psi_0} \right). \tag{81}$$

The frictional resistance on the basal décollement fault at the same point is

$$\tau_b = (\rho - \rho_f) gz \sin\alpha \left(\frac{\sin 2\psi_b}{\sin 2\psi_0} \right). \tag{82}$$

The quantity $\tau_b/|\tau| = \sin 2\psi_b/\cos\phi$ is a depth-independent measure of the ratio of décollement fault strength to wedge strength that must be in the range $0 \leq \tau_b/|\tau| \leq 1$ for every active fold-and-thrust belt or accretionary wedge. The quantity $\rho gz \sin\alpha$ has been employed in glaciology for at least 40 years to estimate the traction acting on the bed of a glacier (Orowan 1949). It has also been used by Elliott (1976) and others to estimate the traction at the base of a subaerial thrust sheet. For a noncohesive wedge on the verge of Coulomb failure everywhere, this glacial rule-of-thumb is biased low by a factor $\sin 2\psi_b/\sin 2\psi_0$, as shown by Equation (82).

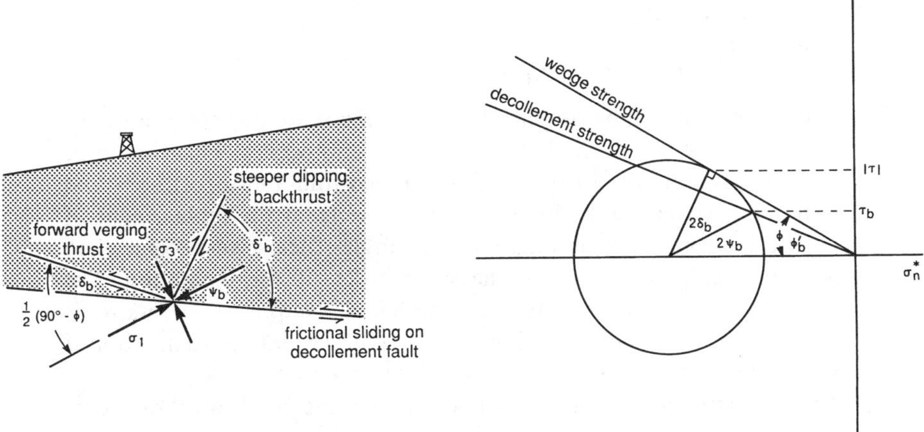

Figure 10 (*Left*) Geometry of self-similar thrust fault orientation within a critical noncohesive Coulomb wedge. Forward-verging thrusts exhibit a shallower dip due to the inclination between σ_1 and the base. (*Right*) Mohr diagram illustrating the basal state of effective stress.

Fluid Pressure Distribution and Fluid Flow

The pore-fluid pressure distribution within a noncohesive critical wedge is given by

$$p_f = \lambda \rho g z \cos\alpha. \tag{83}$$

In general, active fold-and-thrust belts and accretionary wedges are observed to be overpressured, i.e. the fluid-pressure ratio λ exceeds the hydrostatic value $\lambda_{\mathrm{hydro}} = \rho_f/\rho$. Such elevated pore-fluid pressures will give rise to fluid flow within the wedge, and these percolating fluids will exert a seepage force on the solid wedge material, as discussed above. If the permeability of the wedge material is isotropic ($\mathbf{K} = k\mathbf{I}$), then the flow in a submarine wedge will be everywhere upward and normal to the upper surfaces:

$$\mathbf{u} = -v^{-1}k\rho g \cos\alpha(\lambda - \rho_f/\rho)\hat{\mathbf{z}} = -v^{-1}k\rho g \cos\alpha(\lambda - \lambda_{\mathrm{hydro}})\hat{\mathbf{z}}. \tag{84}$$

A subaerial fold-and-thrust belt has an additional small downslope component driven by the topography:

$$\mathbf{u} = -v^{-1}k\rho_f g \sin\alpha\hat{\mathbf{x}} - v^{-1}k\rho g \cos\alpha(\lambda - \lambda_{\mathrm{hydro}})\hat{\mathbf{z}}. \tag{85}$$

In a steady-state wedge with a constant permeability k, the fluid flux $\mathbf{u}$ is pervasive and uniform. The source of the upward-flowing fluids is the dewatering of the sediments subducted beneath the wedge (Westbrook et al 1982, Moore 1989).

Small-Angle Approximation

The above results can be simplified by specializing to the case of a wedge having a narrow taper: $\alpha \ll 1$, $\beta \ll 1$, $\psi_0 \ll 1$, and $\psi_b \ll 1$. This should be a useful approximation for many thin-skinned fold-and-thrust belts and accretionary wedges. The simplifications arise from the replacement of sines and tangents of small angles by the angles themselves; cosines and secants are replaced by one to the same order of approximation. The specification $\psi_b \ll 1$ implies that the principal compressive stress σ_1 is quasi-horizontal; strictly speaking, this is only a valid approximation if the basal décollement fault is very weak $[\mu_b(1-\lambda_b) \ll \mu(1-\lambda)]$.

The approximate critical taper equation takes the purely algebraic form

$$\alpha + \beta \approx \frac{(1-\rho_f/\rho)\beta + \mu_b(1-\lambda_b)}{(1-\rho_f/\rho) + 2(1-\lambda)\left(\dfrac{\sin\phi}{1-\sin\phi}\right)}. \tag{86}$$

This is the generalization of Equation (16) for a submarine wedge with pore-fluid pressure effects taken into account.

Laboratory Sandbox Models

Davis et al (1983) tested the predictions of the critical taper theory using an extremely simple laboratory model. Their apparatus consisted of a bottomless box containing well-sorted sand resting on a sheet of Mylar; the Mylar was supported by a flat rigid base whose dip was adjustable (Figure 11). The process of plate subduction was mimicked by slowly pulling the Mylar sheet beneath the sand; the frictional drag on the base induced deformation within the sand. Frictional drag on the transparent sidewalls was minimized by coating them with graphite before the sand was emplaced. The sand was stratified with passive black marker beds to allow the deformation within the wedge to be observed during an experimental run. Typically, the deformation was dominated by motion along a few discrete forward- and backward-verging thrust faults, as shown in Figure 12. The first faults formed near the rigid buttress at the back of the initially untapered wedge, and the locus of active faulting then moved toward the toe; deformation ceased once the critical taper was attained.

The loosely packed dry sand employed by Davis et al (1983) had a coefficient of internal friction $\mu \approx 0.6$ ($\phi \approx 30^\circ$). The measured coefficient of friction between sand and Mylar was $\mu_b \approx 0.3$. The left side of Figure 13 compares the theoretical and observed critical tapers for four values of the basal dip β. Both the exact relationship between α and β and the small-angle approximation (16) are shown. Clearly, in this case, the small-angle approximation is valid; more importantly, the agreement of both the exact

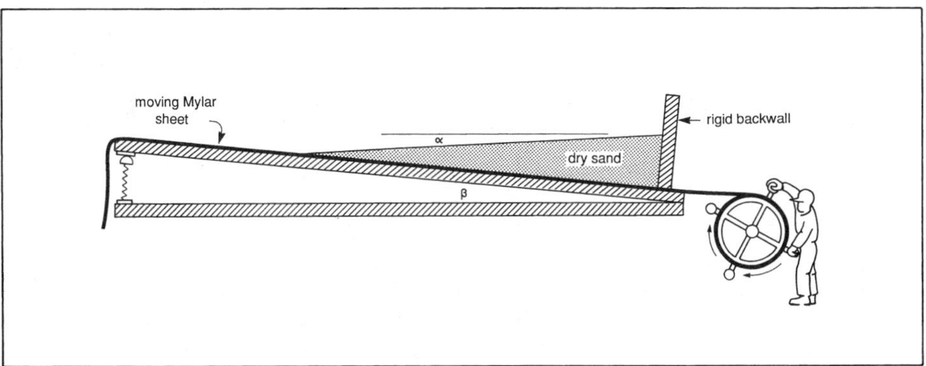

Figure 11 Schematic diagram of laboratory sandbox model. Actual length of rigid base is approximately 1 m. Figure in cap (Dan Davis) not shown to scale.

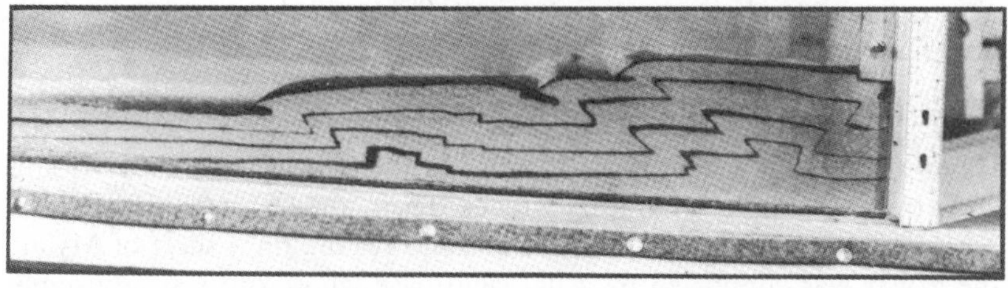

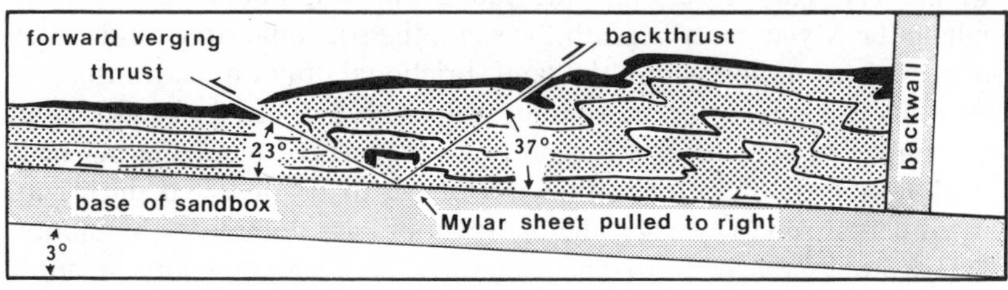

Figure 12 Photographic side view of a deforming sand wedge, showing discrete forward- and backward-verging thrust faults.

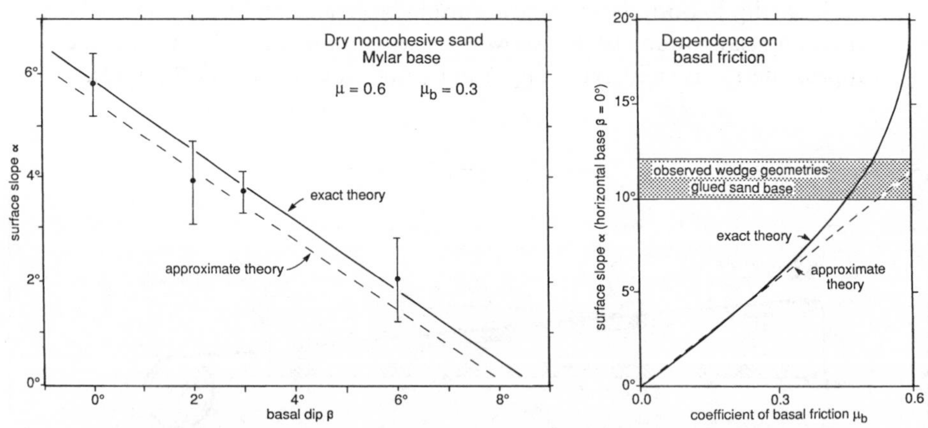

Figure 13 (*Left*) Comparison of theoretical critical surface slope with measured slopes in sandbox experiments (Davis et al 1983, Dahlen et al 1984). Dashed line is the small-angle approximation $\alpha = 5.7° - \frac{2}{3}\beta$. Dots represent the average of 8 experimental runs at $\beta = 0°$, 2 at $\beta = 2°$, 14 at $\beta = 3°$, and 9 at $\beta = 6°$. Bars denote the standard deviation. (*Right*) Theoretical dependence of α (for $\beta = 0°$) on the coefficient of basal friction μ_b. Dashed line is the small-angle approximation $\alpha = \frac{1}{3}\mu_b$.

and approximate results with the data is well within the experimental uncertainty. The observed basal step-up angles of the forward and backward thrust faults were compared with Equations (82) and (83) by Dahlen et al (1984). The angles were measured off photographs such as that in Figure 12; only freshly formed faults undistorted by subsequent faulting or rotation were included in the observations. The predicted step-up angles for $\mu = 0.6$, $\mu_b = 0.3$, and $\lambda = \lambda_b = 0$ are $\delta_b = 21^\circ$ and $\delta_b' = 38^\circ$. These agree very well with the observed values $\delta_b = 22 \pm 2^\circ$ and $\delta_b' = 38 \pm 4^\circ$.

To test the predicted dependence on the coefficient of basal friction μ_b, Goldberg (1982) conducted experimental runs using Mylar coated with sand as a base; the sand was glued to the Mylar base. The resulting surface slopes were significantly steeper than those produced on uncoated Mylar, in the range $\alpha = 10$–12° compared with $\alpha = 5$–7° for $\beta = 0^\circ$. The right side of Figure 13 shows the predicted dependence on μ_b for $\mu = 0.6$ and $\lambda = \lambda_b = 0$. The increasing discrepancy between the exact results and the small-angle approximation as μ_b approaches μ is evident; the approximation is accurate to within 10% provided the basal friction is 10–15% less than the internal friction. Goldburg (1982) did not attempt to measure μ_b for a glued sand base directly; the observed wedge geometries suggest, however, that $\mu_b \approx 0.5$.

Mulugeta (1988) has recently performed more sophisticated modeling experiments using a motorized Plexiglass squeeze-box in a centrifuge. Experiments using sand with rounded grains gave results in good agreement with critical taper theory, but those using more angular sand did not. This was attributed to the greater degree of compaction of the angular sand; an alternative explanation might, however, be the extreme sensitivity of the taper to μ_b when $\mu_b \approx \mu$ (Figure 13).

GEOLOGICAL APPLICATIONS

Both the success of the critical taper model on the laboratory scale and the scale invariance of the noncohesive theory encourage applications to more complicated and less well-constrained geological situations. We apply the model here to the two end-member wedges illustrated in Figure 2.

Taiwan

The island of Taiwan is the site of an ongoing collision between the Luzon island arc situated on the Philippine Sea plate and the stable continental margin of China situated on the Eurasian plate. The divergence between the strike of the arc and the margin results in a southward-propagating collision that began about 4 Ma in the north and is occurring now at the southern tip of the island (Suppe 1981). Farther to the south, the oceanic crust of the South China Sea is subducting beneath the Luzon arc along the Manila Trench, forming a submarine accretionary wedge on the east side of the trench. The subaerial fold-and-thrust belt that comprises more than half of the island of Taiwan forms by an expansion of this accretionary wedge as the arc encounters the thick sedimentary deposits on the Chinese continental slope and shelf. At the southernmost tip of the island, the mountains have just risen above sea level; to the north, the wedge grows in both height and width to become the Western Foothills, Hsuehshan Range, and Central Mountains of Taiwan.

This growth cannot continue unobstructedly because of the rapid tropical erosion. Between 23°N and 25°N latitude, the mountains exhibit a remarkably uniform surface slope and width $W = 90$ km. This morphology suggests strongly that the wedge has attained a dynamic steady state, with the accretionary influx balanced by the erosive efflux. The local rate of convergence between the Eurasian and Philippine Sea plates is $V = 70$ km Myr^{-1} (Seno 1977, Ranken et al 1984), and the thickness of the incoming sediments at the deformation front is estimated by drilling and seismic data to be $h = 7$ km (Suppe 1981). The average erosion rate, determined from both hydrological and geomorphological studies (Li 1976, Peng et al 1977), is $\dot{e} = 5.5$ km Myr^{-1}. The observed balance $hV = \dot{e}W$ confirms the steady-state nature of the deformation. North of 25°N, the mountains are being rifted on their eastern flank by the back-arc spreading in the Okinawa Trough; this major change in tectonic conditions is caused by the southward propagation of a reversal in the polarity of subduction as the Philippine Sea plate subducts beneath the Eurasian plate to the northeast of Taiwan (Suppe 1984).

Taiwan is an ideal natural laboratory for testing the validity of the critical taper model because of the abundant geophysical data acquired during petroleum exploration. The steady-state region is characterized by a regional surface slope $\alpha = 3°$ and a regional basal dip $\beta = 6°$ (Davis et al 1983). The basal dip is best determined in the front 30 km of the wedge by seismic reflection profiling, deep drilling, and the construction of retrodeformable cross sections (Suppe 1980). Pore-fluid pressures are also known in the front 30 km of the wedge from formation pressure tests and geophysical logging of numerous deep wells. The measured pore-fluid to lithostatic pressure ratio is $\lambda = \lambda_b = 0.67$ (Suppe & Wittke 1977, Davis et al 1983).

The critical taper model can be used to determine the range of basal and internal friction values consistent with the observed wedge geometry and pore-fluid pressures. The left side of Figure 14 shows the theoretical relation between α and β for various values of μ_b and μ. The geometry of the Taiwan wedge is consistent with the model, but the parameters cannot

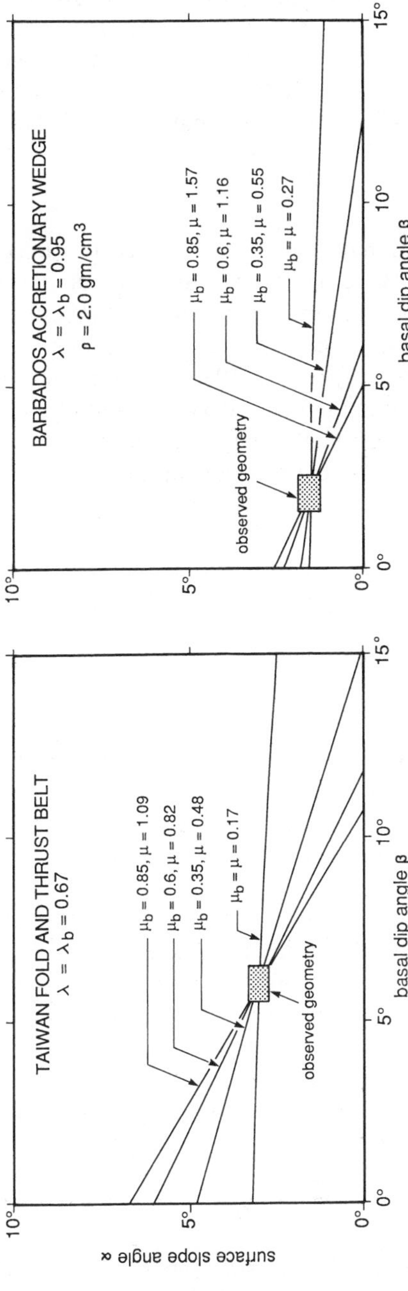

Figure 14 Theoretical surface slope α versus basal dip β for the Taiwan fold-and-thrust belt (*left*) and the Barbados accretionary wedge (*right*). Many possible combinations of μ_b and μ are consistent with the observed geometries and pore-fluid pressures.

be well constrained; the coefficient of friction μ_b on the décollement fault can be as low as 0.17 or in excess of 1. If μ_b is in the range of typical laboratory measurements for rocks and clay-rich fault gouges (0.3–0.85), then μ must exceed μ_b by 20–30%. This is an indication that the Taiwan wedge is not so pervasively fractured that frictional sliding is possible on surfaces of optimum orientation everywhere.

Barbados

The Barbados accretionary complex is located to the east of the Lesser Antilles island arc, where 65 Ma oceanic crust of the North American plate is subducting beneath the Caribbean plate (Westbrook 1975, 1982). The local rate of convergence between the two plates is $V = 2$ km Myr^{-1} (Minster & Jordan 1978, Sykes et al 1982). The thickness h of the incoming sediments varies by more than an order of magnitude, ranging from 7 km of mostly terrigenous sediments near the delta of the Orinoco River in the south to less than 1 km of hemipelagic muds in the north. The continuous accretion of these sediments since the Eocene has produced an accretionary prism of unusually great width ($\approx$300 km).

The geometry and internal structure of the Barbados wedge are particularly well known in the vicinity of 15°30′N latitude, where there has been a concentrated program of seismic reflection profiling and drilling (Westbrook et al 1982, 1989, Moore et al 1982, 1988). Only the uppermost two thirds of the thin incoming sediments in this region are being accreted at the deformation front; the underlying sediments are being underthrust beneath the toe, essentially undeformed (see Figure 2). The dip of the basal décollement fault is extremely shallow ($\beta = 2°$). The upper surface exhibits a slightly convex shape with a mean slope $\alpha = 1.5°$.

The entering sediments have initial porosities of 50–60%, but the porosity decreases both with depth and with distance to the deformation front as a result of compaction and dewatering (Bray & Karig 1985). If the average porosity for the wedge as a whole is $\eta = 30\%$, then the average value of the aggregate density of the porous medium is $\rho = 2000$ kg m^{-3}. Pore-fluid pressures are not nearly as well known in Barbados or any other accretionary wedge as they are in Taiwan. A single estimate obtained inadvertently at Deep Sea Drilling Project Site 542 during an attempt to free stuck drill string suggests that the pore-fluid to lithostatic pressure ratio near the deformation front is very high ($\lambda \approx \lambda_b \approx 0.9$–1). The ubiquitous presence of mud volcanic activity (Westbrook & Smith 1983) as well as the narrowness of the taper provide additional evidence for nearly lithostatic fluid pressures.

The right side of Figure 14 shows the theoretical relation between α and β for submarine wedges having $\rho = 2000$ kg m^{-3} and $\lambda = \lambda_b = 0.95$. As in the case of Taiwan, the Barbados observations are consistent with the critical taper model, but they place very little constraint on the friction parameters μ_b and μ. The constraint that μ_b must be greater than 0.27 is not very robust because the pore-fluid pressure data are so circumstantial.

Heat Flow Constraint on Friction in Taiwan

An increase in μ_b increases the critical taper, whereas an increase in μ decreases it; this is the reason that the observed or estimated geometries and pore-fluid pressures in Taiwan and Barbados cannot constrain the level of friction. Twenty years ago, Brune et al (1969) pointed out that the level of frictional stress on the San Andreas fault could be determined by measuring the heat flow anomaly produced by frictional heating. Since that time, an extensive program of heat flow measurements has failed to reveal the expected narrow anomaly around the fault (Henyey & Wasserburg 1971, Lachenbruch & Sass 1973, 1980, 1988). In Taiwan, there is a substantial heat flow anomaly that can be attributed to brittle frictional heating; Barr & Dahlen (1989a,b) have used this anomaly to infer the effective coefficient of basal friction. A steady-state thermal model of the Taiwan fold-and-thrust belt was developed for this purpose; both shear heating on the décollement fault and internal strain heating within the deforming brittle wedge were incorporated in a mechanically consistent manner.

The left side of Figure 15 shows a contour map of the observed surface heat flow in Taiwan; the data were collected by Lee & Cheng (1986) from more than 100 oil wells, geothermal wells, and shallow boreholes. Prior to contouring, a robust smoother was applied to suppress spurious observations and enhance the regional pattern. In the steady-state region between 23°N and 25°N the smoothed contours are roughly parallel to the strike of the fold-and-thrust belt; the heat flow increases from its tectonically undisturbed value of 100 mW m^{-2} at the deformation front to more than 240 mW m^{-2} at the rear. The smoothed data between 23°N and 25°N are shown projected along strike on the right side of Figure 15. Theoretical heat flow curves for various values of the quantity $\mu_b(1-\lambda_b)$ are shown superimposed for comparison; in each case, the corresponding values of the wedge strength parameters μ and λ are chosen to be consistent with the observed Taiwan geometry. The heat flow data are best fit by $\mu_b(1-\lambda_b) = 0.16$. The effective coefficient of basal friction $\mu_b(1-\lambda_b)$ has been chosen as the fitting parameter rather than μ_b, since the basal pore-fluid ratio λ_b is only well determined in the vicinity of the deformation front. If the value measured there prevails along the entire décollement fault, the inferred coefficient of sliding friction is $\mu_b = 0.5$. This is lower than Byerlee's universal value of 0.85 for most rocks; it is, however, within the range of measured friction values for clay-rich fault gouges.

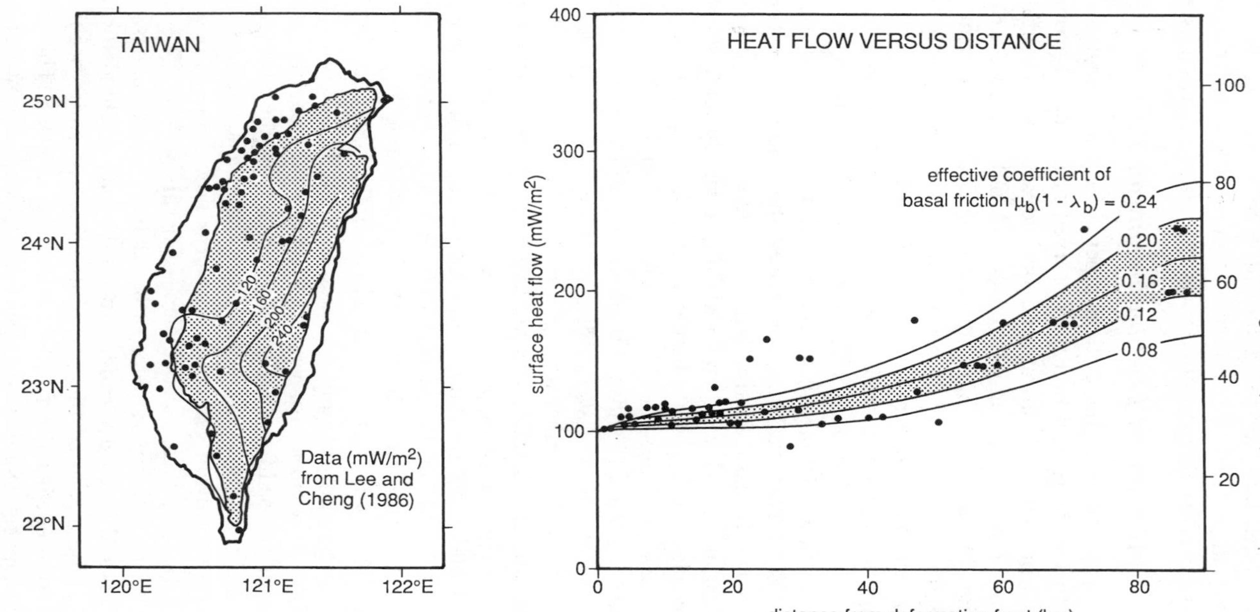

Figure 15 (*Left*) Smoothed surface heat flow on the island of Taiwan. (*Right*) Theoretical heat flow versus distance from the deformation front for various values of the effective coefficient of basal friction. Values in the shaded range $[\mu_b(1-\lambda_b) = 0.12\text{–}0.20]$ provide acceptable fits to the smoothed data (shown as dots).

APPROXIMATE GENERAL THEORY FOR A THIN-SKINNED WEDGE

The exact critical taper model described above is based on a number of extreme simplifying assumptions; in addition to ignoring cohesion, we have assumed that all the parameters ρ, λ, μ, λ_b, and μ_b are constant. Clearly, such simplifications are not realistic; on the other hand, there has been little motivation to develop a more general model in view of the almost complete lack of deep in situ measurements of pore-fluid pressure and other variables in active wedges. This situation is likely to change during the next decade as the geophysical exploration and characterization of active fold-and-thrust belts and accretionary wedges continue. Seismic data have already been used to infer the porosity distribution of two active wedges, Barbados (Westbrook et al 1989) and the Nankai Trough off southwestern Japan (Karig 1986). In this section, we extend the critical taper model to take into account spatial variations in porosity, pore-fluid pressure, and other parameters. For simplicity, we assume the principal stresses σ_1 and σ_3 are nearly horizontal and vertical, respectively, and consistently invoke small-angle approximations; the resulting theory for a thin-skinned wedge is quasi-analytical.

Theory

We consider a geometrically irregular submarine wedge, as shown in Figure 16; the case of a subaerial wedge can be recovered by setting the fluid density ρ_f equal to zero, as before. To begin with, we employ x- and z-axes that are locally aligned with the top of the wedge, as shown in the top sketch in Figure 16. The small-angle equations of static equilibrium within the wedge are then

$$\frac{\partial \sigma_{xx}}{\partial x} + \frac{\partial \sigma_{xz}}{\partial z} - \rho g \alpha \approx 0, \tag{87a}$$

$$\frac{\partial \sigma_{xz}}{\partial x} + \frac{\partial \sigma_{zz}}{\partial z} + \rho g \approx 0. \tag{87b}$$

The quantity $\partial \sigma_{xz}/\partial x$ can be ignored to first order in Equation (87b); the vertical (principal) stress due to the overlying water and porous wedge material is then found by integration to be

$$\sigma_{zz} \approx \sigma_3 \approx -\rho_f g D - g \int_0^z \rho \, dz. \tag{88}$$

The pore-fluid pressure p_f within the wedge is written in the form

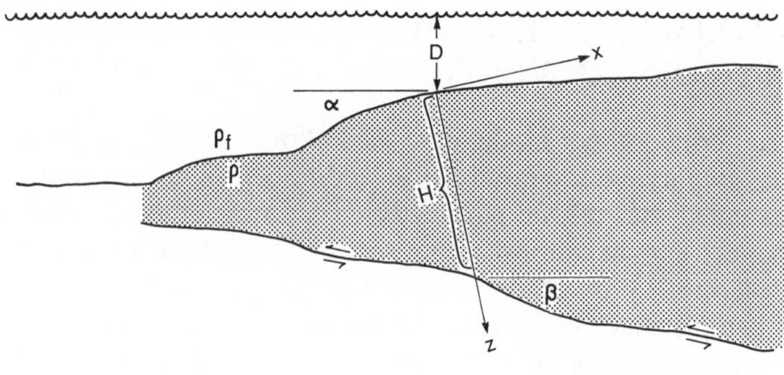

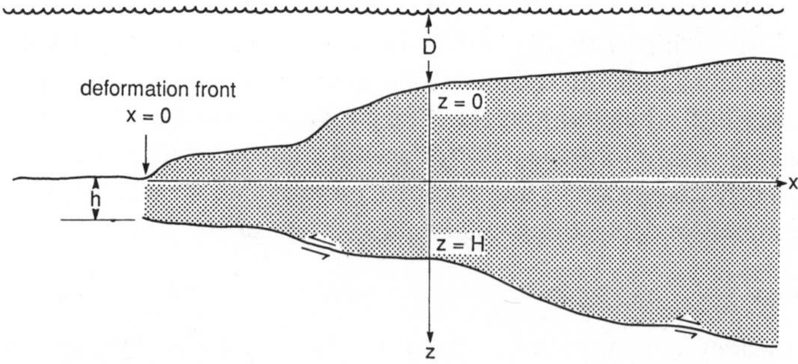

Figure 16 (*Top*) Schematic cross section of a geometrically irregular thin-skinned wedge with spatially variable mechanical properties. (*Bottom*) In applying the approximate critical taper equation (98), it is permissible to employ horizontal and vertical x- and z-axes.

$$p_{\mathrm{f}} = \rho_{\mathrm{f}} g D + \lambda g \int_0^z \rho \, \mathrm{d}z. \tag{89}$$

Equation (89) simply serves to define the quantity λ; both λ and the aggregate density $\rho = (1-\eta)\rho_{\mathrm{s}} + \eta\rho_{\mathrm{f}}$ are regarded as functions of x and z within the wedge. In the small-angle approximation, it is also permissible to ignore the quantity σ_{xz}^2 in Equation (57a); the cohesive Coulomb failure criterion (60) thus takes the form

$$\tfrac{1}{2}(\sigma_{zz} - \sigma_{xx}) \approx S \cos\phi - \tfrac{1}{2}(\sigma_{xx} + \sigma_{zz} + 2p_{\mathrm{f}}) \sin\phi. \tag{90}$$

Equation (90) can be solved for the horizontal (principal) stress:

$$\sigma_{xx} \approx \sigma_1 \approx -\rho_{\mathrm{f}} g D - C - g\Lambda \int_0^z \rho \, \mathrm{d}z, \tag{91}$$

where

$$\Lambda = 1+2(1-\lambda)\left(\frac{\sin\phi}{1-\sin\phi}\right). \tag{92}$$

The uniaxial compressive strength C and the angle of internal friction ϕ are also allowed to vary arbitrarily with position (x, z) inside the wedge.

The shear traction on the basal décollement fault, in the small-angle approximation, is

$$\tau_b \approx (\sigma_3-\sigma_1)(\alpha+\beta)+\sigma_{xz}. \tag{93}$$

To find σ_{xz} on $z = H$, we integrate Equation (87a); this gives

$$\sigma_{xz} \approx g\alpha\int_0^H (\rho-\rho_f)\,dz+\int_0^H (\partial C/\partial x)\,dz+g\int_0^H \frac{\partial}{\partial x}\left(\Lambda\int_0^z \rho\,dz\right)dz. \tag{94}$$

In writing Equation (94), we have used the fact that $dD/dx \approx -\alpha$. The final result for τ_b is most succinctly stated in terms of the depth-averaged quantities

$$\bar{\rho} = \frac{1}{H}\int_0^H \rho\,dz, \tag{95a}$$

$$\bar{C} = \frac{1}{H}\int_0^H C\,dz, \tag{95b}$$

$$\bar{\Gamma} = 2H^{-2}\int_0^H \Lambda\int_0^z \rho\,dz\,dz = 2H^{-2}\int_0^H \rho\int_0^z \Lambda\,dz\,dz. \tag{95c}$$

We find, after some reduction,

$$\tau_b \approx (\bar{\rho}-\rho_f)gH\alpha+\bar{C}(\alpha+\beta)+(\bar{\Gamma}-\bar{\rho})gH(\alpha+\beta) + H(d\bar{C}/dx)+\tfrac{1}{2}gH^2(d\bar{\Gamma}/dx). \tag{96}$$

The basal boundry condition requires that τ_b be equal to the coefficient of basal friction times the effective normal traction; we allow also for the possibility of a basal cohesion or plasticity S_b by writing

$$\tau_b \approx S_b+\mu_b(1-\lambda_b)\bar{\rho}gH. \tag{97}$$

Upon equating Equations (96) and (97), we obtain the approximate critical taper equation:

$$\alpha+\beta \approx \frac{[(1-\rho_f/\bar{\rho})\beta+\mu_b(1-\lambda_b)]\bar{\rho}gH+S_b-[d\bar{C}/dx+\frac{1}{2}gH(d\bar{\Gamma}/dx)]H}{(\bar{\Gamma}-\rho_f)gH+\bar{C}}. \tag{98}$$

Once this result has been obtained, it is permissible to ignore the slight variable tilt of the x- and z-axes at different points in the wedge, as shown in Figure 16 (*bottom*). At every horizontal location x, we use Equations (95) to calculate the vertical averages $\bar{\rho}$, $\bar{C}$, and $\bar{\Gamma}$. The critical taper equation (98) then relates the local taper $\alpha+\beta$ to the local properties $\bar{\rho}$, $\bar{C}$, $\bar{\Gamma}$, S_b, $\mu_b(1-\lambda_b)$, and the local thickness H. The approximations that have been employed are valid as long as the base is relatively weak and the horizontal gradients in the properties are small. Since $\alpha+\beta \approx dH/dx$, we can regard Equation (98) as a first-order ordinary differential equation of the form $dH/dx \approx F(x, H)$. This equation can be integrated to find the wedge thickness $H(x)$, given the thickness $H(0) = h$ of the entering sediments at the deformation front $x = 0$.

Special Cases

If all the properties are constant within the wedge, Equation (98) reduces to

$$\alpha+\beta \approx \frac{(1-\rho_f/\rho)\beta+\mu_b(1-\lambda_b)+S_b/\rho gH}{(1-\rho_f/\rho)+2(1-\lambda)\left(\dfrac{\sin\phi}{1-\sin\phi}\right)+C/\rho gH}. \tag{99}$$

This is the generalization of Equation (86) to a wedge with cohesion and basal plasticity. In the lithostatic limit $\lambda = \lambda_b = 1$, Equation (99) becomes

$$\alpha+\beta \approx \frac{(1-\rho_f/\rho)\beta+S_b/\rho gH}{(1-\rho_f/\rho)+C/\rho gH}. \tag{100}$$

This is the small-angle solution for a perfectly plastic wedge obtained by Chapple (1978).

It is very common for pore-fluid pressures to be hydrostatic at shallow depths and overpressured at deeper depths. If cohesion and basal plasticity are ignored and both ρ and ϕ are assumed to be constant, Equation (98) reduces to

$$\alpha+\beta \approx \frac{(1-\rho_f/\rho)\beta+\mu_b(1-\lambda_b)+H\left(\dfrac{\sin\phi}{1-\sin\phi}\right)\left(\dfrac{d\bar{\lambda}}{dx}\right)}{(1-\rho_f/\rho)+2(1-\bar{\lambda})\left(\dfrac{\sin\phi}{1-\sin\phi}\right)}, \tag{101}$$

where

$$\bar{\lambda} = 2H^{-2} \int_0^H z\lambda \, dz. \tag{102}$$

Equation (102) shows how vertical variations in the pore-fluid pressure ratio are averaged in such a way that the bottom of the column is more strongly weighted than the top; intuitively, it makes sense that the deeper portions of a wedge should be more important in determining the overall wedge strength.

To model the observed convexity of the Barbados wedge, Zhao et al (1986) considered a linear increase in cohesion with depth, $S = Kz$. Equation (98) reduces in this case to

$$\alpha+\beta \approx \frac{(1-\rho_f/\rho)+\mu_b(1-\lambda_b)-\left(\dfrac{H}{\rho g}\dfrac{dK}{dx}\right)\left(\dfrac{\cos\phi}{1-\sin\phi}\right)}{(1-\rho_f/\rho)+2(1-\lambda)\left(\dfrac{\sin\phi}{1-\sin\phi}\right)+\left(\dfrac{K}{\rho g}\right)\left(\dfrac{\cos\phi}{1-\sin\phi}\right)}. \tag{103}$$

Equation (103) assumes that there is no basal plasticity and that all the wedge properties except the cohesion are constant. Zhao et al (1986) estimated the horizontal variation of K by a piecewise application of an exact critical taper model having $S = Kz$; such a procedure ignores the term involving the derivative dK/dx in Equation (103). Fletcher (1989) derived Equation (103) and showed that the neglect of dK/dx could lead to nonnegligible errors. Another defect of the Zhao et al (1986) analysis is their neglect of density variations; they treated ρ as constant in spite of the pronounced porosity variations.

Décollement Fault in Salt

Stresses in geological materials are limited by brittle fracture and frictional sliding at low temperatures, but by thermally activated processes (especially dislocation climb) at high temperatures (Brace & Kohlstedt 1980). Figure 17 illustrates the idealized dependence of typical rock strengths on depth; brittle frictional behavior is seen to prevail in the upper 10–15 km of the crust except in the case of evaporites, which flow plastically at substantially shallower depths. This extreme weakness of salt at upper crustal pressures and temperatures has an important effect on the mechanics of fold-and-thrust belts above salt, as noted by Davis & Engelder (1985).

The approximate critical taper of a fold-and-thrust belt whose décollement fault is in a salt horizon is given by Equation (99) with $\mu_b(1-\lambda_b) = 0$. The depth-independent plastic strength of salt is $S_b \approx 1$ MPa; fold-and-thrust belts overlying salt have extremely narrow tapers ($\alpha+\beta \approx 1^\circ$), since

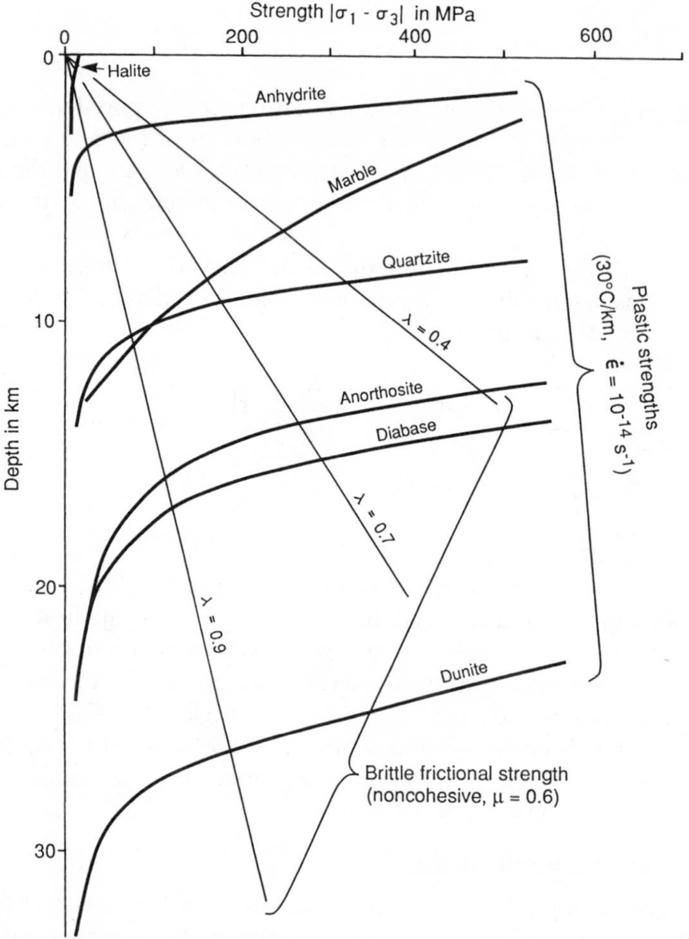

Figure 17 Idealized strength of rocks as a function of depth. Brittle frictional strength is largely independent of rock type, whereas plastic strengths vary considerably. This leads to a corresponding variation in depth to the brittle-plastic transition: 0–2 km for evaporites, 10–15 km for quartzofeldspathic rocks, and 25–30 km for olivine-rich rocks. Plastic strengths are extrapolated from laboratory data by assuming a typical geologic strain rate and geothermal gradient.

$S_b/\rho gH \ll 1$ if H is greater than a few kilometers. A cross section of the Zagros fold-and-thrust belt overlying the Hormuz salt in western Iran is shown in Figure 18; the extremely subdued topography is shown to true scale at the top.

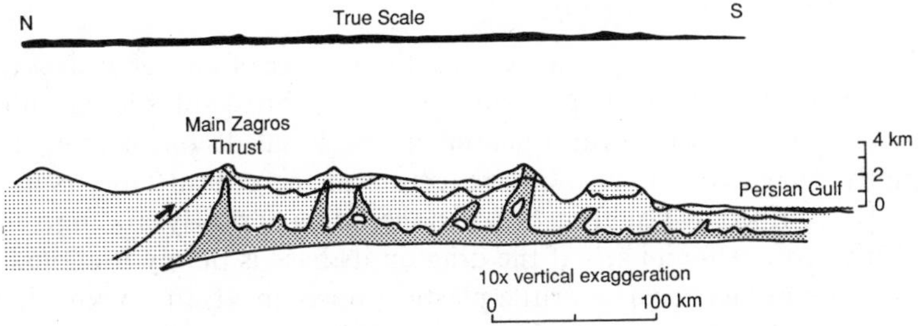

Figure 18 Cross section of the Zagros fold-and-thrust belt (Farhoudi 1978). Broad symmetrical anticlines are cored by diapiric salt, denoted by the dark stipple.

Forward-verging thrust faults generally predominate in fold-and-thrust belts not underlain by salt, since their shallower dip allows them to accommodate a greater amount of horizontal shortening than the steeper dipping back-thrusts for a given increase in gravitational potential energy. Salt-based fold-and-thrust belts, in contrast, generally lack a dominant sense of vergence; they are commonly characterized by broad, relatively symmetric folds (Davis & Engelder 1985). The theoretical step-up angles of forward- and backward-verging thrust faults are very nearly equal ($\delta_b' \approx \delta_b$) because the principal compressive stress σ_1 is almost parallel to the base (Figure 19). Simply stated, a fold-and-thrust belt riding on salt cannot tell from which direction it is being pushed.

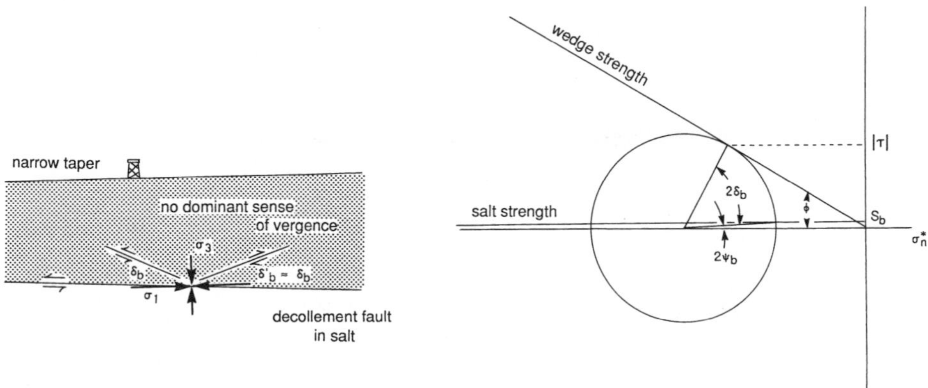

Figure 19 (*Left*) Schematic cross section illustrating the narrow taper and nearly symmetrical forward- and backward-verging thrust faults that characterize a fold-and-thrust belt overlying salt. (*Right*) Mohr diagram illustrating the basal state of effective stress. Compare with Figure 10.

Brittle-Plastic Transition

Basal plasticity can be important even in the absence of salt for wedges that have grown large enough to protrude beneath the brittle-plastic transition. Once this occurs, the shear traction on the basal décollement fault is abruptly decreased; the resulting reduction in the critical taper is again roughly described by Equation (99). A noneroding or slowly eroding wedge can only grow self-similarly if the drag on its base is purely frictional. It is thus the existence of a brittle-plastic transition at 10–15 km depth that ultimately limits the height of very wide fold-and-thrust belts and accretionary wedges. This provides a natural explanation for the break in topographic slope of the Higher Himalaya at the edge of the Tibetan Plateau (Figure 20).

CONCLUSION

The critical taper model is a useful paradigm for understanding the large-scale mechanics of fold-and-thrust belts and accretionary wedges. The theoretical formulation for a homogeneous noncohesive wedge is exact and entirely analytical. If one eschews elegance, the model can be generalized to take into account the spatial variation of mechanical properties and irregular geometries. The approximate general theory discussed here is limited to narrow tapers; exact, purely numerical models can, however, be developed using plastic slip-line theory (Hill 1950, pp. 128–60, 294–300; Stockmal 1983) or the finite element method. Hydrological models of pore-fluid pressures and flow rates in accretionary wedges have recently been developed (Shi & Wang 1988, Screaton et al 1989), and a logical next step would be to combine these with the critical taper model. The principal

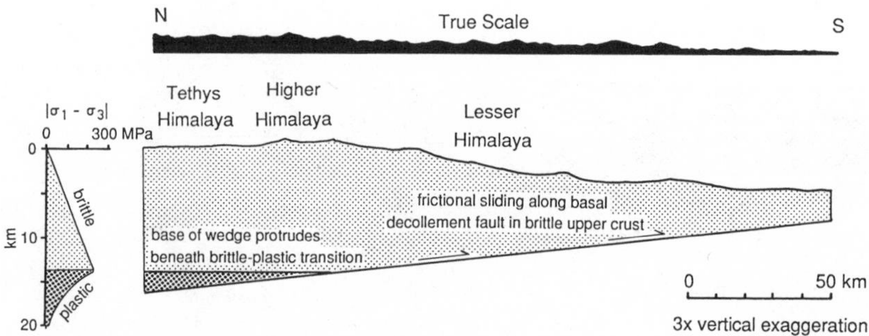

Figure 20 Schematic cross section of the Himalayan fold-and-thrust belt, showing the break in topographic slope in the Higher Himalaya at the edge of the Tibetan Plateau. A typical north-south topographic profile (along 82°30′E) is shown to true scale at the top.

obstacle at the present time is the meagerness of available data; the development of more detailed models must be preceded by the acquisition of structural, pore-fluid pressure, and heat flow data from active wedges.

ACKNOWLEDGMENTS

The critical Coulomb taper model has been developed in collaboration with Terence Barr, Dan Davis, John Suppe, and Wu-Ling Zhao. I thank Nancy Breen and Dan Orange for calling my attention to the seepage force and prompting me to think more carefully about the effect of pore fluids. Financial support has been provided by the National Science Foundation under grant EAR-8804098.

Literature Cited

Bally, A. W., Gordy, P. L., Stewart, G. A. 1966. Structure, seismic data and orogenic evolution of southern Canadian Rocky Mountains. *Bull. Can. Pet. Geol.* 14: 337–81

Barcilon, V. 1987. A note on "Noncohesive critical Coulomb wedges: An exact solution" by F. A. Dahlen. *J. Geophys. Res.* 92: 3681–82

Barr, T. D., Dahlen, F. A. 1989a. Brittle frictional mountain building. 2. Thermal structure and heat budget. *J. Geophys. Res.* 94: 3923–47

Barr, T. D., Dahlen, F. A. 1989b. Constraints on friction and stress in the Taiwan fold-and-thrust belt from heat flow and geochronology. *Geology*. In press

Batchelor, G. K. 1967. *An Introduction to Fluid Dynamics*. Cambridge: Univ. Press

Bear, J. 1972. *Dynamics of Fluids in Porous Media*. New York: Elsevier

Behrmann, J. H., Brown, K., Moore, J. C., Mascle, A., Taylor, E., et al. 1988. Evolution of structures and fabrics in the Barbados accretionary prism: insights from Leg 110 of the Ocean Drilling Program. *J. Struct. Geol.* 10: 577–91

Brace, W. F., Kohlstedt, D. L. 1980. Limits on lithospheric stress imposed by laboratory experiments. *J. Geophys. Res.* 85: 6248–52

Bray, C. J., Karig, D. E. 1985. Porosity of sediments in accretionary prisms and some implications for dewatering processes. *J. Geophys. Res.* 90: 768–78

Brune, J. N., Henyey, T. L., Roy, R. F. 1969. Heat flow, stress, and rate of slip along the San Andreas fault, California. *J. Geophys. Res.* 74: 3821–27

Byerlee, J. D. 1978. Friction of rocks. *Pure Appl. Geophys.* 92: 3681–82

Chapple, W. M. 1978. Mechanics of thin-skinned fold-and-thrust belts. *Geol. Soc. Am. Bull.* 89: 1189–98

Coulomb, C. A. 1773. Sur une application des règles de maximis et minimis à quelques problèmes de statique relatifs à l'architecture. *Acad. R. Sci. Mém. Math. Phys.* 7: 343–82

Dahlen, F. A. 1984. Noncohesive critical Coulomb wedges: an exact solution. *J. Geophys. Res.* 89: 10,125–33

Dahlen, F. A., Suppe, J., Davis, D. M. 1984. Mechanics of fold-and-thrust belts and accretionary wedges: cohesive Coulomb theory. *J. Geophys. Res.* 89: 10,087–10,101

Davis, D. M., Engelder, T. 1985. The role of salt in fold-and-thrust belts. *Tectonophysics* 119: 67–88

Davis, D., Suppe, J., Dahlen, F. A. 1983. Mechanics of fold-and-thrust belts and accretionary wedges. *J. Geophys. Res.* 88: 1153–72

Elliott, D. 1976. The motion of thrust sheets. *J. Geophys. Res.* 81: 949–63

Farhoudi, G. 1978. A comparison of Zagros geology to island arcs. *J. Geol.* 86: 323–34

Fletcher, R. C. 1989. Approximate analytical solutions for a cohesive fold-and-thrust wedge: some results for lateral variation in wedge properties and for finite wedge angle. *J. Geophys. Res.* 94: 10,347–54

Goldburg, B. 1982. *Formation of critical taper wedges by compression in a sandbox model*. BS thesis. Princeton Univ., Princeton, N.J. 70 pp.

Gray, W. G., O'Neill, K. 1976. On the general equations for flow in porous media and their reduction to Darcy's law. *Water Resour. Res.* 12: 148–54

Henyey, T. L., Wasserburg, G. J. 1971. Heat

flow near major strike-slip faults in California. *J. Geophys. Res.* 76: 7924–46

Hill, R. 1950. *The Mathematical Theory of Plasticity*. Oxford: Clarendon

Hoffman, P. F., Tirrul, R., King, J. E., St.-Onge, M. R., Lucas, S. B. 1988. Axial projections and modes of crustal thickening, eastern Wopmay orogen, northwest Canadian shield. In *Processes in Continental Lithospheric Deformation. Geol. Soc. Am. Spec. Pap.*, ed. S. P. Clark, B. C. Burchfiel, J. Suppe, 218: 1–29

Hoshino, K., Koide, H., Inami, K., Iwamura, S., Mitsui, S. 1972. Mechanical properties of Japanese Tertiary sedimentary rocks. *Rep. 244*, Geol. Surv. Jpn., Kawasaki. 200 pp.

Hubbert, M. K., Rubey, W. W. 1959. Role of fluid pressure in mechanics of overthrust faulting. 1. Mechanics of fluid-filled porous solids and its application to overthrust faulting. *Geol. Soc. Am. Bull.* 70: 115–66

Hubbert, M. K., Rubey, W. W. 1960. Role of fluid pressure in mechanics of overthrust faulting. Reply to discussion by H. P. Laubscher. *Geol. Soc. Am. Bull.* 71: 617–28

Hubbert, M. K., Rubey, W. W. 1961. Role of fluid pressure in mechanics of overthrust faulting. Reply to discussion by W. L. Moore. *Geol. Soc. Am. Bull.* 72: 1587–94

Iverson, R. M., Major, J. J. 1986. Groundwater seepage vectors and the potential for hillslope failure and debris flow mobilization. *Water Resour. Res.* 22: 1543–48

Jackson, J. D. 1962. *Classical Electrodynamics*. New York: Wiley

Jaeger, J. C., Cook, N. G. W. 1969. *Fundamentals of Rock Mechanics*. London: Methuen

Karig, D. E. 1986. Physical properties and mechanical state of accreted sediments in the Nankai Trough, S.W. Japan. In *Structural Fabrics in Deep Sea Drilling Project Cores From Forearcs. Geol. Soc. Am. Mem.*, ed. J. C. Moore. 166: 117–33

Lachenbruch, A. H., Sass, J. H. 1973. Thermo-mechanical aspects of the San Andreas fault system. *Proc. Conf. Tecton. Probl. of San Andreas Fault Syst.*, ed. R. L. Kovach, A. Nur, pp. 192–205. Stanford, Calif: Stanford Univ. Press

Lachenbruch, A. H., Sass, J. H. 1980. Heat flow and energetics of the San Andreas fault zone. *J. Geophys. Res.* 85: 6185–6223

Lachenbruch, A. H., Sass, J. H. 1988. The stress–heat flow paradox and thermal results from Cajon Pass. *Geophys. Res. Lett.* 15: 981–84

Laubscher, H. P. 1960. Role of fluid pressure in mechanics of overthrust faulting. *Geol. Soc. Am. Bull.* 71: 611–16

Lee, C. R., Cheng, W. T. 1986. *Preliminary heat flow measurements in Taiwan.* Presented at Circum-Pac. Energy and Miner. Resour. Conf., 4th, Singapore

Lehner, F. K. 1979. A derivation of the field equations for slow viscous flow through a porous medium. *Ind. Eng. Chem. Fundam.* 18: 41–45

Lehner, F. K. 1986. Comments on "Noncohesive critical Coulomb wedges: an exact solution" by F. A. Dahlen. *J. Geophys. Res.* 91: 793–96

Li, Y. H. 1976. Denudation of Taiwan island since the Pliocene epoch. *Geology* 4: 105–7

Logan, J. M., Rauenzahn, K. M. 1987. Frictional dependence of gouge mixtures of quartz and montmorillonite on velocity, composition and fabric. *Tectonophysics* 144: 87–108

Malvern, L. E. 1969. *Introduction to the Mechanics of a Continuous Medium*. Englewood Cliffs, NJ: Prentice-Hall

Minster, J. B., Jordan, T. H. 1978. Present-day plate motion. *J. Geophys. Res.* 83: 5331–54

Moore, J. C. 1989. Tectonics and hydrogeology of accretionary prisms: role of the décollement zone. *J. Struct. Geol.* 11: 95–106

Moore, J. C., Biju-Duval, B., Bergen, J. A., Blackington, B., Claypool, G. E., et al. 1982. Offscraping and underthrusting of sediment at the deformation front of the Barbados Ridge: Deep Sea Drilling Project Leg 78A. *Geol. Soc. Am. Bull.* 93: 1065–77

Moore, J. C., Mascle, A., Taylor, E., Andreieff, P., Alvarez, F., et al. 1988. Tectonics and hydrogeology of the northern Barbados Ridge: results from Leg 110 ODP. *Geol Soc. Am. Bull.* 100: 1578–93

Moore, W. L. 1961. Role of fluid pressure in overthrust faulting: a discussion. *Geol. Soc. Am. Bull.* 72: 1581–86

Morrow, C. A., Shi, L. Q., Byerlee, J. D. 1981. Permeability and strength of San Andreas fault gouge. *Geophys. Res. Lett.* 8: 325–28

Mulugeta, G. 1988. Modelling the geometry of Coulomb thrust wedges. *J. Struct. Geol.* 10: 847–59

Neumann, S. P. 1977. Theoretical derivation of Darcy's law. *Acta Mech.* 25: 153–70

Orowan, E. 1949. The flow of ice and other solids. *J. Glaciol.* 1: 231–40

Paterson, M. S. 1978. *Experimental Rock Deformation: The Brittle Field.* New York: Springer-Verlag

Peng, T. H., Li, Y. H., Wu, F. T. 1977. Tectonic uplift rates of the Taiwan island since the early Holocene. *Geol. Soc. China Mem.* 2: 57–69

Ranken, B., Cardwell, R. K., Karig, D. E. 1984. Kinematics of the Philippine Sea plate. *Tectonics* 3: 555–75

Roeder, D., Gilbert, O. E., Witherspoon, W. D. 1978. Evolution and macroscopic structure of Valley and Ridge thrust belt, Tennessee and Virginia. In *Studies in Geology 2*. Knoxville: Dept. Geol. Sci., Univ. Tenn. 25 pp.

Saffman, P. G. 1971. On the boundary condition at the surface of a porous medium. *Stud. Appl. Math.* 50: 93–101

Screaton, E. J., Wuthrich, D. R., Dreiss, S. J. 1989. Permeabilities, fluid pressures, and flow rates in the Barbados Ridge complex. *J. Geophys. Res.* In press

Seno, T. 1977. The instantaneous rotation vector of the Philippine Sea plate relative to the Eurasian plate. *Tectonophysics* 42: 209–26

Shi, Y., Wang, C. Y. 1988. Generation of high pore pressures in accretionary prisms: inferences from the Barbados subduction complex. *J. Geophys. Res.* 93: 8893–8909

Skempton, A. W. 1960. Terzaghi's discovery of effective stress. In *From Theory to Practice in Soil Mechanics: Selections from the Writings of Karl Terzaghi*, ed. L. Bjerrum, A. Casagrande, R. B. Peck, A. W. Skempton, pp. 42–53. New York: Wiley

Slattery, J. C. 1972. *Momentum, Energy, and Mass Transfer in Continua*. New York: McGraw-Hill

Stockmal, G. S. 1983. Modeling of large-scale accretionary wedge formation. *J. Geophys. Res.* 88: 8271–87

Suppe, J. 1980. A retrodeformable cross section of northern Taiwan. *Geol. Soc. China Proc.* 23: 46–55

Suppe, J. 1981. Mechanics of mountain building and metamorphism in Taiwan. *Geol. Soc. China Mem.* 4: 67–89

Suppe, J. 1984. Kinematics of arc-continent collision, flipping of subduction, and backarc spreading near Taiwan. *Geol. Soc. China Mem.* 6: 21–33

Suppe, J. 1987. The active Taiwan mountain belt. In *The Anatomy of Mountain Ranges*, ed. J. P. Schaer, J. Rodgers, pp. 277–93. Princeton, NJ: Princeton Univ. Press

Suppe, J., Wittke, J. H. 1977. Abnormal pore-fluid pressures in relation to stratigraphy and structure in the active fold-and-thrust belt of northwestern Taiwan. *Pet. Geol. Taiwan* 14: 11–24

Sykes, L. R., McCann, W. R., Kafka, A. L. 1982. Motion of Caribbean plate during last 7 million years and implications for earlier Cenozoic motions. *J. Geophys. Res.* 87: 10,656–76

Terzaghi, K. 1923. Die berechnung der durchlässigkeitsziffer des tones aus dem verlauf der hydrodynamischen spunnungserscheinungen. *Sitzungsber. Akad. Wiss. Wien* 132: 125–38

Westbrook, G. K. 1975. The structure of the crust and upper mantle in the region of Barbados and the Lesser Antilles. *Geophys. J. R. Astron. Soc.* 43: 201–42

Westbrook, G. K. 1982. The Barbados Ridge complex: tectonics of a mature forearc system. In *Trench-Forearc Geology: Sedimentation and Tectonics on Modern and Ancient Active Plate Margins*, ed. J. K. Leggett, pp. 275–90. London: Geol. Soc.

Westbrook, G. K., Ladd, J., Bangs, N. 1989. Structure of the northern Barbados accretionary prism. *Geology*. In press

Westbrook, G. K., Smith, M. J. 1983. Long décollements and mud volcanoes: evidence from the Barbados Ridge complex for the role of high pore-fluid pressure in the development of an accretionary complex. *Geology* 11: 279–83

Westbrook, G. K., Smith, M. J., Peacock, J. H., Poulter, M. J. 1982. Extensive underthrusting of undeformed sediment beneath the accretionary complex of the Lesser Antilles subduction zone. *Nature* 300: 625–28

Whitaker, S. 1969. Advances in theory of fluid motion in porous media. *Ind. Eng. Chem.* 61: 14–28

Zhao, W. L., Davis, D. M., Dahlen, F. A., Suppe, J. 1986. Origin of convex accretionary wedges: evidence from Barbados. *J. Geophys. Res.* 91: 10,246–58

Annu. Rev. Earth Planet. Sci. 1990. 18:101–22
Copyright © 1990 by Annual Reviews Inc. All rights reserved

LATE PRECAMBRIAN AND CAMBRIAN SOFT-BODIED FAUNAS

S. Conway Morris

Department of Earth Sciences, University of Cambridge, Downing Street, Cambridge CB2 3EQ, England

INTRODUCTION

No scientist ever has enough data, but to the outsider the lot of the paleontologist must seem bleak. A glance at any marine community will show that not only do the great majority of species lack sufficiently robust skeletal parts to survive fossilization in normal circumstances, but that biases also operate in terms of taxonomic grouping, trophic style, and position relative to the sediment-water interface. However, taphonomic processes provide a spectrum of possibilities. The normal expectation is a death assemblage composed of only well-skeletalized taxa, but such examples grade into a succession of more completely preserved biotas, which in rare instances even include soft parts.

What relevance do these observations have to the origin and early evolution of metazoans? In terms of the fossil record the earliest widely accepted occurrence is that of the Ediacaran faunas, of latest Precambrian age (ca. 570 Ma). Claims for pre-Ediacaran metazoan fossils are highly contentious, but data from molecular biology suggest that metazoans first appeared possibly in excess of 1000 Ma. The Ediacaran faunas are succeeded by an extraordinary evolutionary irruption near the base of the Cambrian, best known from the appearance of skeletal elements (Conway Morris 1987). However, concurrent diversification of trace fossils demonstrates that these adaptive radiations (antonomastically the "Cambrian explosion") also involved a wide variety of soft-bodied organisms. The most important insights in this category come from the Burgess Shale–type faunas, which provide unparalleled data into the nature of these evolutionary and ecological diversifications (Conway Morris 1989a).

0084–6597/90/0515–0101$02.00

PRE-EDIACARAN METAZOANS?

Study of molecular sequences in compounds such as hemoglobin (Runnegar 1982a), collagen (Runnegar 1985), and ribosomal RNA (Field et al 1988) suggests that if rates of substitution of component amino acids and nucleotides, respectively, are approximately constant, then times of metazoan origination might lie between ca. 700 and 1000 Ma. Claims for early metazoans, mostly in the form of putative trace fossils, remain almost entirely unsubstantiated. This need not imply, however, that paleontological and molecular data are in hopeless discrepancy. Rather, it may be that our "search image" has been too strongly governed by Phanerozoic expectations. If early metazoans were small, analogous to the Recent benthic meiofauna or zooplankton, then their detection might rely on subtle interpretation of bioturbation fabrics, minute burrows (Horodyski 1988), and fecal pellets (Robbins et al 1985). Nevertheless, a general absence of corresponding body fossils from exceptionally preserved biotas in shales (e.g. Butterfield et al 1988) and cherts is surprising. Regarding the latter, the tendency for cherts to form in hypersaline peritidal conditions, inhospitable to metazoan life, may provide a partial explanation.

The principal claimants for soft-bodied pre-Ediacaran metazoans are vermiform carbonaceous fossils from the Liulaobei and Jiuliqiao Formations of Anhui and Jiangsu provinces, northern China (Sun et al 1986, Chen 1988). From the former horizon, elongate sausagelike structures with prominent annulations are referred to *Sinosabellidites*. Apart from the annuli, the specimens approach closely in overall shape and dimensions cooccurring examples of *Tawuia*. A metazoan relationship has been entertained also for *Tawuia*, but the consensus places it as a macroscopic alga (Hofmann 1985). In any event, the lack of evidence for cephalization or internal organs in *Sinosabellidites* make its assignment to the metazoans questionable (Chen 1988). From the overlying Jiuliqiao Formation, carbonaceous structures, referred to as *Pararenicola* and *Protoarenicola*, are also interpreted as primitive worms. The former taxon is stout and annulated. Identification at one end of an aperture or proboscis, as well as internal anatomy, seems questionable. Comparable features in *Protoarenicola* seem to be equally open to taphonomic explanations of partial decay and disruption. The body of *Protoarenicola* may also bear prominent constrictions (Chen 1988), but the notion that these reflect muscular contractions seems tenuous. A metazoan affinity for these fossils, therefore, is still open to debate. Even if it does transpire that some are actual worms, the radiometric dates of 840 and ca. 740 Ma quoted for the Liulaobei and Jiuliqiao Formations, respectively (Sun et al 1986), may require revision, given that both are based on techniques that elsewhere have not always given reliable ages for sedimentation (Cloud 1986).

EDIACARAN FAUNAS

The first unequivocal fossil metazoans are represented by the relatively diverse Ediacaran faunas (Glaessner 1984, Conway Morris 1985a), the type locality of which is the Ediacara Hills in the Flinders Range, South Australia. It is clear that Ediacaran faunas have an effectively global distribution (Figure 1). Other important localities include those in the Soviet Union (Podolia area of Ukraine, White Sea area, and Olenek region of northern Siberia), Canada (Mackenzie and Wernecke Mountains of northwest Canada and southeast Newfoundland, respectively), Namibia, and England (Charnwood Forest). More restricted faunas have been reported from many other localities, with the authenticity of a few still requiring confirmation. Paleocontinental reconstructions allow tentative steps toward an Ediacaran paleobiogeography (Donovan 1987). Strong similarities between the faunas of southeast Newfoundland and England reflect their former proximity on one side of the Iapetus Ocean. However, the faunal similarity between South Australia and the USSR (especially the White Sea) is less easy to explain, given the postulated paleogeographic separation.

All are agreed that however and wherever the Precambrian-Cambrian boundary (Conway Morris 1987) is defined, the Ediacaran faunas are late Precambrian. Indeed, formal proposals exist to recognize an Ediacaran System based on type sections in South Australia (Jenkins 1981, Cloud & Glaessner 1982). This would be more or less equivalent to at least the upper parts of the Vendian and Sinian as used by Soviet and Chinese workers, respectively. The age range of Ediacaran faunas has not been well constrained, although values of ca. 620–580 Ma are widely accepted (Glaessner 1984). However, recent U-Pb determinations on zircons from an ash fall that entombed one of the assemblages from southeast Newfoundland (Mistaken Point fauna) give an apparently reliable date of 565 Ma (Benus 1988). Nevertheless, the overall age range of Ediacaran faunas remains speculative.

The majority of Ediacaran assemblages occur in shallow-water deposits and often owe their preservation to rapid burial by sands. Local facies may control whether individual specimens stand a high chance of fossilization (Wade 1968), but occurrences of Ediacaran faunas in carbonates from northern Siberia (Sokolov & Fedonkin 1984) show that some benthic species could tolerate a variety of substrates. Other faunas appear to have

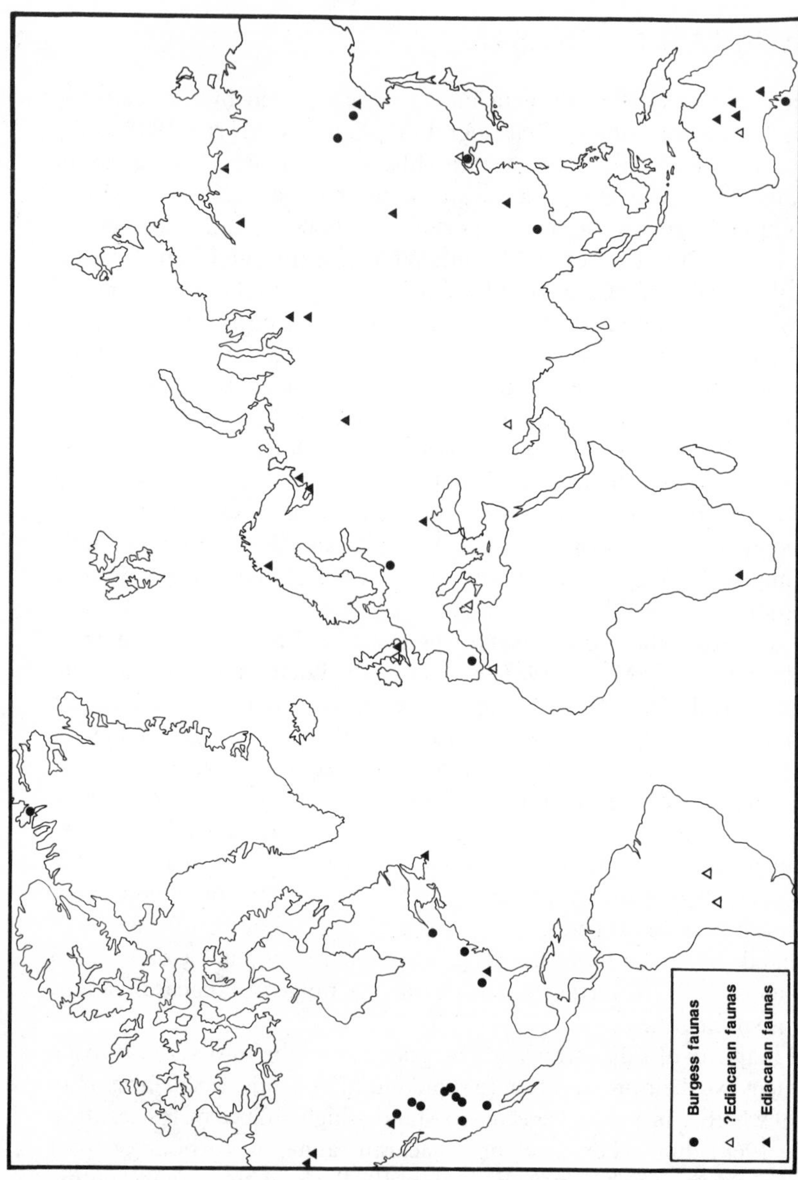

Figure 1 Distribution of soft-bodied faunas of Ediacaran age, including some questionable occurrences, and Lower–Middle Cambrian Burgess Shale–type faunas. Data on Ediacaran occurrences are largely from Glaessner (1984), and those on Burgess faunas are from Conway Morris (1989b).

occupied deeper waters, of which the most significant are the spectacular examples from southeast Newfoundland. Here large bedding planes strewn with fossils represent episodes of catastrophic burial by volcanic ash of communities living on and above turbiditic sands (Benus 1988) that accumulated on the flanks of an island-arc complex, probably well beneath the photic zone.

Faunal Diversity

The principal problems in understanding the Ediacaran faunas are fourfold: (*a*) What is their ancestry? (*b*) Do they show any evolutionary progression? (*c*) What are their affinities? (*d*) What links lie with the succeeding Cambrian faunas? From the discussion above on pre-Ediacaran metazoans, it is obvious that the first question is unresolved. Perhaps the appearance of Ediacaran fossils is a result of a taphonomic breakthrough whereby size increase, possibly mediated by changes in atmospheric levels of oxygen, ended a protracted period of cryptic metazoan evolution. The question of evolutionary change during Ediacaran times also remains speculative. Although some localities yield faunas over a considerable stratigraphic interval, few examples of evolutionary lineages are evident.

The third question, that of affinities, might seem more amenable to inspection, but radical divergences of opinion exist. Traditionally, the majority of Ediacaran faunas have been assigned to cnidarians (or a comparable grade), with a smaller number interpreted as triploblastic and ranging in complexity from platyhelminthes to echinoderms (Glaessner 1984, Fedonkin 1986, 1987). However, in a radical reappraisal, Seilacher (1984, 1989) has proposed that many, if not all, of the Ediacaran body fossils should be regarded as belonging to a major extinct group, with an architecture at variance with any metazoan. This proposal, which lacks detailed documentation, argues that the fundamental structure consisted of a series of baglike units without complex organ systems such as an alimentary canal. Such an interpretation was inspired in part by the contrasts between the prevalence of soft-part preservation in Ediacaran times as against the relative scarcity of preserved soft parts for the remainder of the Phanerozoic, perhaps indicating that the tissues of Ediacaran body fossils had a peculiar preservation potential. It implies also that the only genuine metazoans are the abundant, if simple, trace fossils (Crimes 1989). Seilacher's (1984) interpretation has not won wide support, but it focuses attention on neglected problems of taphonomic interpretation.

EDIACARAN CNIDARIANS If the consensus still regards Ediacaran fossils as metazoans, what then are their relationships? Among the purported cnidarians (Figure 2), examples of scyphozoans, cubozoans, hydrozoans (as chondrophores), and anthozoans have all been identified, as well as members of the extinct group Conulata, which are regarded as relatives of the scyphozoans (e.g. Sun 1985). Many of the medusiform elements are regarded as scyphozoans. It must be admitted, however, that some attempts to reconcile interpretations of anatomical structures, themselves equivocal, to expected scyphozoan anatomy appear somewhat forced (Glaessner 1984; see also Fedonkin 1986, 1987). This problem is compounded by a lack of diagnostic features in many specimens. The rarity in Ediacaran faunas of medusoids with tetraradial symmetry, one characteristic of scyphozoans, has also been a point of comment. Exceptions include *Conomedusites*, which shows a fourfold division and a fringe of tentacles, and possibly *Persimedusites*.

Cubozoans have also been identified from South Australia, where sandstone casts of *Kimberella* (Figure 2*e*) suggest a complex internal anatomy, while the characteristic bunches of trailing tentacles seem also to be identifiable (Jenkins 1984). Chondrophores are represented by floats, with annular traces of the original gas-filled chambers (Figure 2*d*), and sometimes the remains of the underlying zooids.

Even if assignments to the Scyphozoa, Cubozoa, and hydrozoan chondrophores are accepted, a substantial number of Ediacaran medusiforms remain of more enigmatic status. In some cases this is due to the absence of diagnostic features (Figure 2*c*), but a distinctive group with triradial symmetry may represent a major taxonomic group, the Trilobozoa (Fedonkin 1986, 1987). These include *Albumares* (Figure 2*b*) and *Tribrachidium* (Figure 2*a*). The arrangement of internal organs, which includes canallike structures, is quite variable, and the anatomical significance of these organs is largely unknown.

The principal candidates for anthozoans are stalked animals that more or less closely approach recent pennatulaceans (Glaessner 1984). In no case is there an exact correspondence, but forms such as *Charniodiscus*, with foliate extensions arising from an elongate rachis and a bulbous holdfast that anchored the organism in the substrate, provide fairly close analogies. However, such forms grade into baglike structures that are decidedly less pennatulidlike, e.g. *Pteridinium* and *Inkrylovia* (Figure 2*f*).

Figure 2 Typical Ediacaran taxa attributed either to cnidarians or to a comparable grade of organization: the trilobozoan medusiforms (*a*) *Tribrachidium heraldicum* and (*b*) *Albumares brunsae*, (*c*) the medusiform *Cyclomedusa plana*, (d) the chondrophore *Ovatoscutum concentricum*, (*e*) the cubozoan *Kimberella quadrata*, and (*f*) the baglike ?cnidarian *Inkrylovia lata*. Taxa (*a, c–e*) are from South Australia, and (*b, f*) are from the White Sea, USSR. The length of the scale bars is equivalent to either 5 mm (*b*) or 10 mm (*a, c–f*).

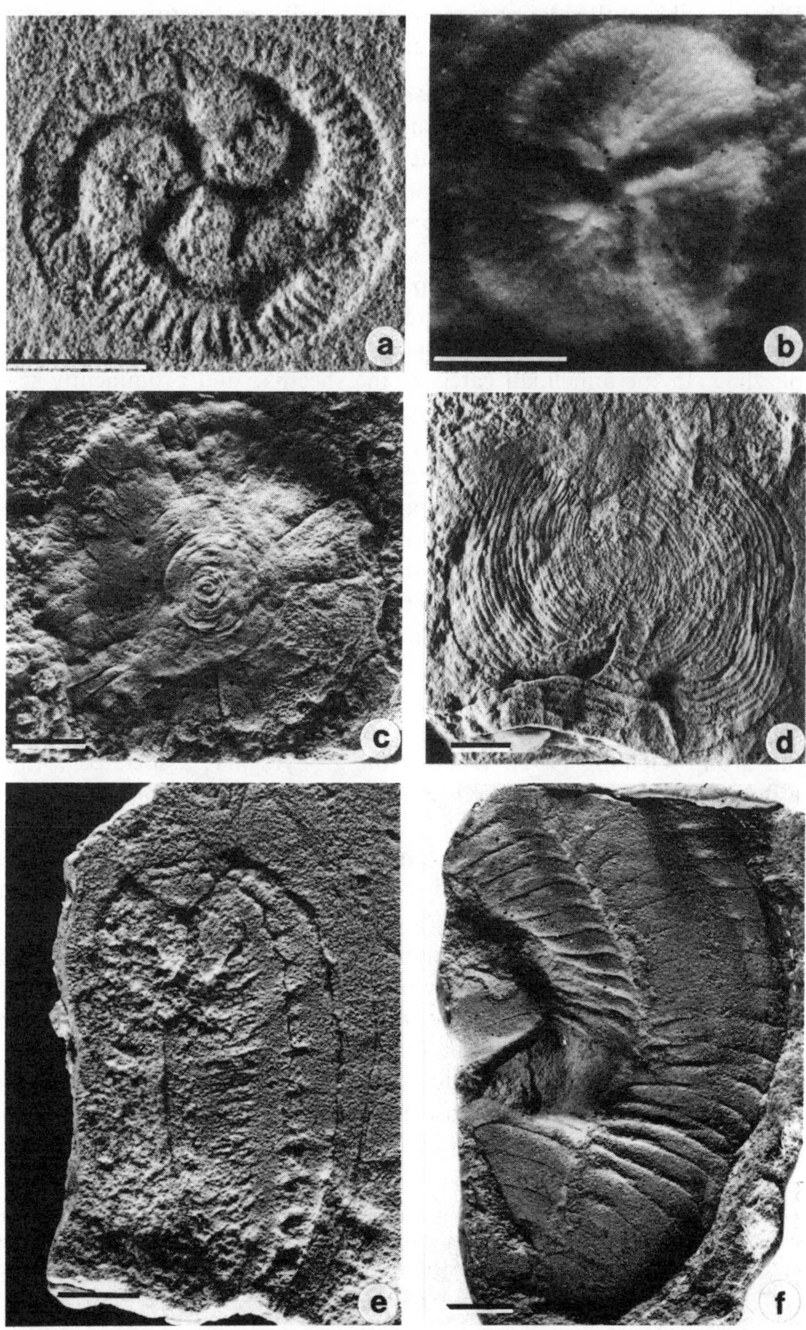
a
b
c
d
e
f

Other possible anthozoans are more or less bulbous structures interpreted as anemonelike structures (see, for example, Narbonne & Hofmann 1987).

EDIACARAN TRIPLOBLASTS Organisms of a higher grade include putative arthropods and annelids (Glaessner 1984). The former are relatively diverse and include types with a head-shield and an elongate segmented body [*Spriggina* (Figure 3*b*), *Marywadea*]. The nature of the appendages, however, is speculative. Others are smaller and bear what appears to be a broad carapace, sometimes with segmental divisions. In *Parvancorina* (Glaessner 1980) the putative carapace has an anterior and median thickening (Figure 3*a*). Elongate structures beneath the carapace are interpreted as appendages, although their disposition and angular relationships are puzzling. In the annelidan category the sheetlike and prominently segmented *Dickinsonia* (Figure 3*c*) has been compared to the modern ectoparasitic polychaete *Spinther*. This is probably a result of convergence. However, indirect arguments, based on (*a*) an inferred circulatory system that would be necessary given the body thickness and likely oxygen levels and (*b*) the style of segment accretion, suggest an annelid grade of organization (Runnegar 1982b). *Dickinsonia* has also been compared to the platyhelminthes, to which a number of other taxa have been assigned by Fedonkin (1987). These proposals must be regarded as tentative.

Although *Tribrachidium* (Figure 2*a*) has been removed from the echinoderms, discovery of minute (6 mm) discoidal fossils from South Australia with a pentaradial arrangement of grooves has prompted comparison to this group (Gehling 1987). Whether the grooves were associated with a water vascular system is conjectural, and no evidence exists for a calcareous skeleton.

Ediacaran faunas also contain a trace fossil assemblage (Crimes 1989), which appears to have been produced by animals other than those known as body fossils. In comparison with the number found in the Cambrian, the ichnological tally is low and the traces are two dimensional, showing little ability to penetrate the substrate. Particularly characteristic are simple sinuous trails (Figure 3*d*). Somewhat more complex traces have been described, but most appear to be very rare. The grade of organization of the animals responsible is difficult to judge. Peristaltic motion through sediment could argue for a molluscan-annelid organization, but the ability

Figure 3 Typical Ediacaran taxa attributed to metazoans of triploblastic grade: the possible arthropods (*a*) *Parvancorina minchami* and (*b*) *Spriggina floundersi*, (*c*) the possible annelid *Dickinsonia costata*, and (*d*) sinuous trace fossils (possibly the feeding burrows of a coelomate worm). All specimens are from South Australia. The length of the scale bars is equivalent to either 5 mm (*a*) or 10 mm (*b–d*).

a
b
c
d

of even nemerteans to burrow (Turbeville & Ruppert 1983) suggests a need for caution. There is, however, a general absence of scratch marks; such marks appear in abundance in the Cambrian and are reliably attributed to arthropods.

BURGESS SHALE–TYPE FAUNAS

Our knowledge of post-Ediacaran soft-bodied faunas has been critically enhanced by a continuing series of discoveries of Burgess Shale–type faunas from the Lower and Middle Cambrian (Conway Morris 1989b). The type occurrence is the Phyllopod bed fauna (Figure 4) of the Burgess Shale, located near the town of Field, British Columbia (Whittington 1985). The Burgess Shale is an informal stratigraphic unit within the "thick" Stephen Formation, a basinal sequence of shales and silts that accumulated beside a carbonate reef (the Cathedral escarpment), now dolomitized. Recently, however, the existence of this reef has been questioned by Ludvigsen (1989), who suggests that the Burgess Shale accumulated on a ramp.

Two lines of evidence demonstrate how the Burgess Shale faunas are an integral part of Cambrian life. First, assume that the extraordinary taphonomic conditions responsible for soft-part preservation had been held in abeyance. The resulting assemblage of shelly fossils, although very much impoverished, would be represented by trilobites, brachiopods, hyoliths, monoplacophorans, echinoderms, and sponges (the last two mostly occurring as ossicles and spicules owing to postmortem disaggregation). In broad outline such a taxonomic profile is indistinguishable from most other open-shelf Cambrian faunas (Conway Morris 1986). The second piece of evidence is the increasing number of localities yielding Burgess fossils (Figure 1). Taken together, these lines of evidence demonstrate both a widespread geographical distribution and a lengthy stratigraphic duration that encompasses most of the Lower and Middle Cambrian (Conway Morris 1989b). This latter property imparts a rather

Figure 4 Typical Burgess Shale–type taxa: (*a*) a possible Ediacaran survivor, *Mackenzia costalis*, (*b*) the problematic worm *Banffia constricta*, (*c*) the problematic organism *Dinomischus isolatus*, (*d*) the problematic pelagic organism *Nectocaris pteryx*, (*e*) the problematic medusiform *Eldonia ludwigi*, (*f*) the primitive chordate *Pikaia gracilens*, (*g*) the problematic organism *Hallucigenia sparsa*, (*h*) the problematic organism *Opabinia regalis*, (*i*) the problematic pelagic worm *Amiskwia sagittiformis*, and (*j*) the coeloscleritophoran *Wiwaxia corrugata*. All specimens are from the Burgess Shale, Phyllopod bed, Field, British Columbia. The length of the scale bars is equivalent to either 5 mm (*c*, *d*, *g*, *i*, *j*), 10 mm (*a*, *b*, *f*, *h*), or 30 mm (*e*).

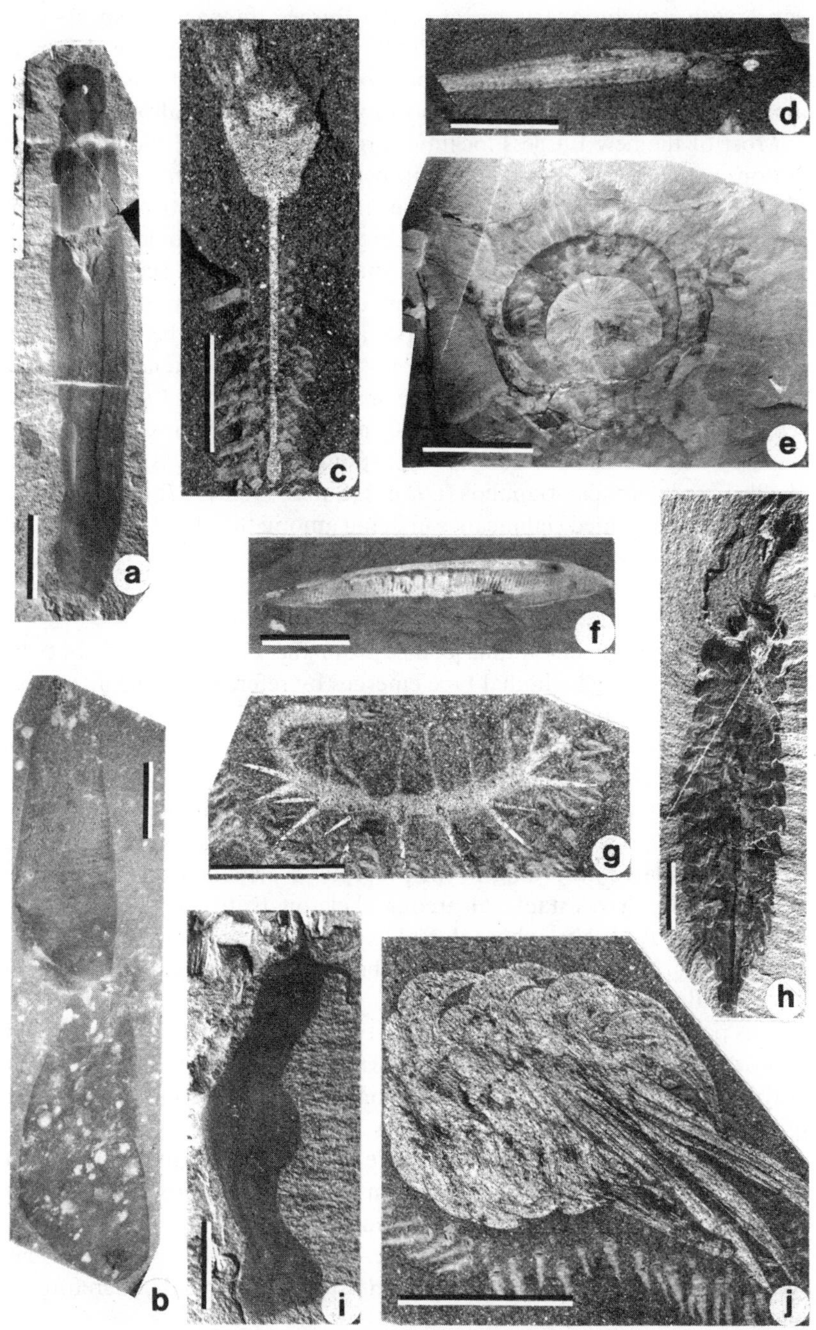
a
b
c
d
e
f
g
h
i
j

conservative evolutionary aspect to the Burgess faunas. It appears also that the unique importance of the Phyllopod bed is as a sampling horizon: Of the ca. 40 soft-bodied genera that cooccur in this and other deposits, about 90% are known prior to deposition of the type locality.

Most of the new Burgess localities are arrayed around the Laurentian craton, generally close to the margin between the median carbonate belt and outer detrital belt (Figure 1). The circumcratonal distribution is consistent with ease of migration, a feature also noted in some shelly taxa. Beyond the Laurentian craton, by far the most important occurrence is at Chengjiang, Yunnan province, South China (see, for example, Chen et al 1989a,b,c, Hou et al 1989). The diverse fauna there is still being described, but its overall similarity to the Burgess Shale fauna extends to generic identity, including arthropods [*Naraoia* (Zhang & Hou 1985)], incertae sedis [*Dinomischus* (Chen et al 1989c) and *Eldonia* (Sun & Hou 1987a, Conway Morris & Robison 1988)], sponges [*Leptomitus* (Chen et al 1989b)], and perhaps priapulids (*Ottoia*) (Sun & Hou 1987b). The Chengjiang fauna has added significance in being among the oldest known (most probably upper Atdabanian).

Faunal Composition

By far the most diverse fauna comes from the Phyllopod bed, but the identity of the Burgess faunal type emerges by reference to the numerous other localities (Conway Morris 1989b). The most important component is arthropods, of which the trilobites constitute only a small proportion (Conway Morris 1986). Although representatives of the three other principal groups (chelicerates, crustaceans, and uniramians) have all been identified, the greater part of this extraordinarily diverse fauna cannot be so accommodated according to the precepts of orthodox taxonomy. The trilobites have a resistant calcareous skeleton [with the exceptions of *Naraoia* and *Tegopelte*, whose dorsal exoskeletons are uncalcified (Whittington 1985, Zhang & Hou 1985)]. All the remaining Burgess Shale arthropods are also uncalcified.

This pattern of morphological diversity is apparent also in the priapulid (Conway Morris 1977a, Conway Morris & Robison 1986, 1988) and polychaete (Conway Morris 1979) worms. Unlike Recent seas, where priapulids are generally inconspicuous and polychaetes dominant, priapulids in the Cambrian are much more abundant. Although not known in the Phyllopod bed, palaeoscolecidans are quite widespread (Conway Morris & Robison 1986, Barskov & Zhuravlev 1988). This group is characterized by prominent rows of papillae, but whether they bore chaetae is conjectural (Kraft & Mergl 1989). Their status as annelids, therefore, is uncertain.

Other soft-bodied groups include what appear to be chordates and hemichordates; the latter include large numbers of ?enteropneusts in the Phyllopod bed. Both groups await detailed study, but forms such as *Pikaia* (Figure 4*f*) are expected to reveal much concerning the evolutionary lines leading to fish and higher vertebrates. Also identified are a number of ?cnidarians, including a putative actiniarian anemone (Figure 4*a*) and a pennatulaceanlike creature. While assignment to this phylum or a comparable organizational grade seems reasonable, they and related forms may be considered also in the context of Ediacaran survivors (see below).

Particular interest has devolved upon a remarkable assemblage whose wider phyletic relationships remain enigmatic. At least 20 distinctive body plans, perhaps equivalent to phyla in terms of orthodox taxonomy, have been recognized. Some are segmented, such as *Anomalocaris* (Whittington & Briggs 1985) and *Opabinia* (Whittington 1975). The former has a mouth composed of a circlet of spinose plates whose contraction would have held and punctured prey; the prey was manipulated by means of a pair of jointed appendages that flanked the mouth. *Opabinia* (Figure 4*h*) is equally remarkable, with an elongate and distally clawed proboscislike structure that presumably grasped prey and an elongate body terminating in a prominent tail fan.

Cataphract metazoans are represented by *Wiwaxia* (Conway Morris 1985b), with a body largely mantled by sclerites and bearing dorsolateral rows of spines (Figure 4*j*), both apparently conferring protection. Mollusclike features include evidence for a ventral foot and an anterior radulalike structure, but the sclerites are secreted in a manner quite distinct from the molluscan shell. Other members of the incertae sedis include *Banffia* (Figure 4*b*), a peculiar worm with a prominent median constriction separating a saclike posterior from an anterior annulated unit. *Dinomischus* (Figure 4*c*) has a calyxlike body supported by an elongate stalk. Some fossils may be fragments of yet larger unidentified creatures. *Hallucigenia* (Figure 4*g*) has seven paired spines arising from an elongate trunk, whereas on the opposite side a row of tentacles arises, the more elongate of which terminate in a snapperlike structure (Conway Morris 1977b). *Microdictyon* has superficial similarities to *Hallucigenia*, but at regular intervals the trunk (or ?appendage) bears paired phosphatic plates (Chen et al 1989a).

The above examples appear to have been either benthic or nektobenthic. Other problematical taxa were apparently pelagic. Some had gelatinous bodies, such as the medusiform *Eldonia* (Figure 4*e*) and swimming worm *Amiskwia* (Figure 4*i*). Others, such as *Nectocaris* (Figure 4*d*), had fins supported by fin rays on either side of the streamlined body.

Other groups within the Burgess fauna are more or less familiar components of the typical Cambrian shelly assemblages. They include monoplacophorans, hyoliths, inarticulate and articulate brachiopods, echinoderms, and sponges. Echinoderms are frequently articulated, and much of our knowledge of sponge diversity comes from specimens preserving original spicule arrangement (Rigby 1986, Chen et al 1989b). What appear to be monoplacophorans are represented by the breviconic *Scenella*. Attempts to reassign these fossils as the floats of chondrophore cnidarians (Babcock & Robison 1988) seem difficult to reconcile with evidence for the shells being mineralized, most probably as calcium carbonate.

Wider Significance of Burgess Faunas

PALEOECOLOGY Cambrian faunas, as represented by shelly assemblages, have been characterized as lacking ecological complexity, dominated by deposit and suspension feeders, with little evidence for resource partitioning. The Burgess faunas, however, demonstrate that this is a travesty of the original situation (Conway Morris 1986). Their study makes clear that the Phanerozoic ecological framework, in terms of feeding modes and position relative to the sediment-water interface, was established early in the Cambrian.

An abundant and diverse infauna is dominated by priapulids, some active burrowers, and others probably more sedentary. Additional infaunal elements include some polychaetes and ?hemichordate enteropneusts. The vagrant epifauna is composed largely of a remarkable diversity of arthropods, as well as monoplacophorans, hyoliths, and various members of the incertae sedis (e.g. *Hallucigenia*, *Wiwaxia*; Figures 4*g,j*). A sessile epifauna comprises numerous sponges, the spongelike *Chancelloria*, brachiopods, and some incertae sedis (e.g. *Dinomischus*; Figure 4*c*). Animals that swam a short distance above the seafloor are at high risk of burial by turbid suspensions. Such nektobenthos appear to include some polychaetes, arthropods, and chordates (Figure 4*f*), as well as more bizarre creatures that include *Anomalocaris* and *Opabinia* (Figure 4*h*). Much rarer animals, typically with well-developed fins and/or evidence for a gelatinous composition, provide a glimpse into the pelagic fauna (Figure 4*d,e,i*).

The trophic structure of the Burgess faunas is also more complex than coexisting shelly faunas would indicate (Conway Morris 1986). Suspension feeders are dominated by sponges and brachiopods and provide evidence for a rudimentary tiering of the water column. Deposit feeders are inferred to include many of the arthropods, *Wiwaxia* (Figure 4*j*), hyoliths, and monoplacophorans. In this trophic category there is evidence, from dominance diversity curves, that the epifaunal vagrants might have partitioned resources according to the model of niche preemption.

The most surprising aspect of the trophic makeup is an abundance of predators and scavengers (Conway Morris 1986). Not only had there been speculation that the absence of Cambrian predators was an artifact, but that the evolution of hard parts across the Precambrian-Cambrian boundary was a response to the rise of durophages (Conway Morris 1987). In the Burgess faunas, mouth parts and, more importantly, gut contents provide compelling evidence. Among the arthropods, spinose appendages (especially as gnathobases) are linked to predatory activity. Inferred methods of prey capture by the enigmatic taxa *Anomalocaris* and *Opabinia* (Figure 4*h*) were mentioned above. Gut contents include the comminuted debris of hyoliths and brachiopods in the hindgut of the arthropod *Sidneyia*. In the priapulid *Ottoia*, entire hyoliths were ingested alive, and this worm was also cannibalistic.

EDIACARAN SURVIVORS? The lack of evolutionary continuity between Ediacaran and Cambrian faunas is noteworthy. Given the taphonomic contrasts across the Precambrian-Cambrian boundary (that is, the predominantly soft-part preservation of Ediacaran forms and shelly preservation of Cambrian fossils), it might be sensible to ascribe this phyletic caesura to changes in fossilization potential. It is clear, however, that the Cambrian shelly assemblages are not simply armored equivalents of the preceding Ediacaran taxa. Inspection of the Burgess faunas confirms that in general there is little overall similarity between the respective soft-bodied faunas. Dissimilarity, however, is not total, because there appear to be a number of possible Ediacaran survivors. Their presence might be accounted for as taxa better adapted to withstand competitive interaction of the newly evolving Cambrian forms. Alternatively, if the hypothesized Ediacaran mass extinction has any reality (Conway Morris 1989a), then these might be simply relicts that filtered through this catastrophe, analogous to Paleocene dinosaurs that are conjectured to have crossed the Cretaceous-Tertiary boundary (Rigby et al 1987).

In shallow-water deposits from the Lower Cambrian of Kazakhstan, a possible example of *Dickinsonia* (Glaessner 1984) has been recorded, while specimens of medusiform organisms from Middle Cambrian sandstones of Poland (Stasinska 1960) and Upper Cambrian arenites in New Brunswick (Pickerill 1982) conceivably represent Ediacaran descendants. In the Burgess faunas, the roster of possible Ediacaran survivors include a chondrophore (*Rotadiscus*) from Chengjiang (Sun & Hou 1987a) and a possible example from the Phyllopod bed. Other records of medusiforms in this context, however, are more problematical. The supposed scyphozoan *Heliomedusa* from Chengjiang (Sun & Hou 1987a) is actually an inarticulate brachiopod with well-preserved mantle setae, while others (*Yunnanomedusa, Stellostomites*) appear similar to the medusiform *Eldonia* (Conway Morris & Robison 1988). The affinities of the latter animal (Figure

4*e*), which otherwise is known from the Middle Cambrian of British Columbia and Utah, are problematic.

In the Phyllopod bed there are also pennatulaceanlike organisms that resemble the Ediacaran forms. If a phyletic link can be established between them, then further study should help to resolve whether they are genuine pennatulaceans, especially given the contrasts in style of preservation between Ediacaran and Burgess faunas. Another possible candidate as an Ediacaran survivor is *Mackenzia* (Figure 4*a*), an elongate baglike animal that was benthic. Little is known of its internal anatomy, but *Mackenzia* recalls broadly a number of the saclike Ediacaran taxa.

EARLY SKELETAL FOSSILS The association of soft parts with Burgess taxa, known otherwise only from skeletal parts, is of considerable significance, with examples known from trilobites, brachiopods, and hyoliths. In the first group it remains uncertain as to why information on the appendages is almost entirely restricted to a few genera (especially *Olenoides*) when many other taxa occur as articulated material. In the brachiopods, preservation includes mantle setae and pedicle, while in the Chengjiang material tentacular traces (Sun & Hou 1987a) within the valves might represent the lophophore. Among hyoliths, articulated material with in situ operculum and helens is well known, but traces of soft parts are rarely found (Babcock & Robison 1988).

The Burgess faunas also throw light on the interrelationships of early skeletal forms of otherwise problematical status. In the Cambrian, many metazoans bore an armored coat of individual sclerites. Even though these sclerites often articulated closely, after death of the organism they disaggregated, so that reconstruction of the original scleritome is difficult. Coeloscleritophorans, for example, are united as a group by secreting sclerites with a prominent internal cavity that usually has a restricted opening to the exterior. The overall anatomy and relationship of the sclerites to the soft parts remains mostly speculative. Two exceptions are *Chancelloria* and *Wiwaxia*. In the former genus, which ranges through the Cambrian, the sclerites are fused into rosettes that in Burgess material are embedded separately in the wall of a vaselike organism. Articulated material of *Wiwaxia* (Figure 4*j*) demonstrates that it possessed a coating of scalelike sclerites over all but its ventral surface, and that it also bore two dorsolateral rows of elongate spines (Conway Morris 1985b). Convincing similarities can be drawn between these sclerites and those of the Lower Cambrian halkieriids. In the latter group the scleritome differed from *Wiwaxia* in being more tightly integrated and in having walls of calcium carbonate. However, sufficient similarities exist to allow *Wiwaxia* to be used as template for halkieriid reconstructions (Bengtson & Conway Morris 1984).

A particularly enigmatic skeletal fossil is *Microdictyon*. Typically the fossil occurs as convex phosphatic plates, with numerous perforations defined by a hexagonal network. The functional morphology and affinities of *Microdictyon* were both highly conjectural, but the discovery at Chengjiang of material in association with soft parts (Chen et al 1989a) has begun to unravel this conundrum. In these specimens the phosphatic plates form a paired series along an elongate structure bearing tentacular extensions.

THE ORIGIN OF PHYLA Of all the insights provided by the Burgess faunas, arguably none is of greater importance than those 20 or so genera, nearly all monospecific, with body plans so unusual that according to our frames of taxonomic reference they could represent extinct phyla. Their diversity is so remarkable that even a series of vignettes of the most conspicuous taxa can only hint at the morphological range. Several forms have already been mentioned: These include *Eldonia* (Figure 4*e*), *Microdictyon*, and *Wiwaxia* (Figure 4*j*).

A number of the incertae sedis are segmented, so it is natural to draw comparisons with members of the Articulata (arthropods and annelids). When the anterior appendages of *Anomalocaris* (Whittington & Briggs 1985) were found in isolation, prior to recognition of the remainder of the body, they were unequivocally assigned to the Arthropoda. As reconstructed in its entirety, however, such a relationship seems tenuous. In particular the diaphragmlike mouth finds no counterpart in other arthropods. *Opabinia* (Figure 4*h*) might be related to *Anomalocaris*, with similarities including the lobelike appendages arising from the trunk. Both *Anomalocaris* and *Opabinia* had eyes, but whereas the former had a pair located on either side of the head, in *Opabinia* there was a more unusual arrangement of five eyes clustered on the top of head.

Among the worms, one of the most puzzling is *Banffia* (Figure 4*b*). The prominent constriction may have acted to dampen fluctuations in hydrostatic pressure generated in either unit, but the mode of life of *Banffia* remains largely unexplained. In contrast, *Dinomischus* (Figure 4*c*) was sessile and has been compared to the entoprocts. The similarities are probably superficial, not least because in *Dinomischus* the crown of feeding structures does not consist of flexible tentacles but rather rigid plates. New material from Chengjiang is important because the specimens are about twice as large and may show new features of internal anatomy. However, identification (Chen et al 1989c) of an extraordinarily elongate anal funnel extending upward from the feeding crown seems in error. It is more likely that the stem has recurved beneath the rest of the body. The cynosure of oddities, however, is surely *Hallucigenia* (Figure 4*g*). Its appearance is so ludicrous (Conway Morris 1977b) that any discussion of its phyletic affinities seems fruitless. Perhaps it should be considered as part of some larger organism; one might imagine, for instance, a number of *Hallucigenias* radiating from a central body.

How do we tackle the classification of these extraordinary animals? Rather than answer the question directly, it is more instructive to look at the comparable taxonomic problems set by the most diverse of Burgess groups, the arthropods. Their diversity is expressed in number of segments, degree of tagmosis, number and type of appendages, and relative body proportions. Taken together, these properties would yield a potentially astronomical number of arthropod types. Indeed, the number of known species may represent only a fraction of the original diversity, and the continuing discoveries of new arthropods from localities such as Chengjiang (Hou et al 1989), Utah (Conway Morris & Robison 1988), and northern Greenland (Conway Morris et al 1987) suggest an almost inexhaustible reservoir of variants. However, nestling among this welter of forms are representatives of the four major classes (or phyla) of arthropods: the trilobites and three lightly skeletalized groups—namely, the chelicerates [as *Sanctacaris* (Briggs & Collins 1988)], crustaceans [as *Canadaspis* and perhaps related taxa, e.g. *Perspicaris* (Briggs 1983)], and uniramians [as *Aysheaia* (Whittington 1985) and *Luolishania* (Hou & Chen 1989)]. These forms are justifiably hailed as key ancestors, whose subsequent diversification endowed us with a planet dominated by arthropods. Consider, however, how a hypothetical Cambrian systematist might attempt to classify this radiation. Rather than using his taxonomy in hindsight (as we do), his would be effectively a "low-level" classification. Another important corollary is that our hypothetical observer would have no means of predicting which taxa, like *Canadaspis*, were destined for cladogenetic glory (i.e. as Crustacea), as against the myriads of similar forms doomed for extinction and failure to leave progeny.

Now consider the more problematic of the Burgess taxa. Suppose they had enjoyed cladogenetic success at the expense of other groups, e.g. *Pikaia* (Figure 4*f*) and its descendant chordates. Metazoan history might then have been very different. It might be objected that the bizarre appearance of the Burgess problematica means that they are "experiments," destined to extinction before the onslaught of superior competitors. However, the range of Cambrian metazoan morphologies is so enormous and the relative proportion of survivors so few that it seems more plausible to appeal to chance factors holding the balance rather than a specifically deterministic cause such as competition. Extirpation of species, perhaps under rates of extinction no different than those normally pertaining, allowed subsequent diversification of surviving clades into the eco- and morphospaces vacated by less fortunate groups.

Although best seen in the arthropods, this pattern of diversification is paralleled in other Cambrian groups. These include soft-bodied Burgess types (e.g. priapulids) and other shelly faunas (e.g. echinoderms). The latter group are particularly instructive because their cladistic classification shows how taxonomic order can be brought to the products of these remarkable diversifications (e.g. Smith 1988). We see also how macroevolutionary patterns emerge from microevolutionary processes by gradual acquisition of characters.

CONCLUSION

Soft-part preservation of Ediacaran and Cambrian fossils is more a matter of contrasts than similarities. Many Ediacaran taxa remain of problematical affinity. Even if their assignment to cnidarians, arthropods, and annelids is correct, they appear to have few connections with the succeeding Cambrian faunas. This biotic discontinuity across the Precambrian-Cambrian boundary in part reflects contrasts in taphonomic circumstances, but demise of much of the Ediacaran faunas during a mass extinction seems plausible and deserves study.

The Burgess faunas are an integral part of Cambrian marine life, showing both a wide geographical distribution and lengthy stratigraphic durations. The earliest examples (Chengjiang, Poland) may have lived in shallower water than many younger examples, which typically occur toward the outer-shelf edge in deeper (and cooler?) waters. Such a distribution lends itself to predictive searches, and many more Burgess faunas should be discovered in the next decade. Their diversity provides new insights into Cambrian paleoecology and the rates and style of early metazoan evolution. Nestling within the Burgess faunas appear to be both Ediacaran survivors and, more importantly, many of the earliest representatives of groups that were to dominate the history of Phanerozoic evolution.

ACKNOWLEDGMENTS

My knowledge of this area of research owes much to H. B. Whittington, S. Bengtson, and M. A. Anderson. I thank M. A. Fedonkin and N. Pledge for providing casts of Ediacaran specimens. Support by the Nuffield Foundation (One Year Science Research Fellowship) is appreciated, as is technical assistance by K. Harvey and S. Skinner and typing by S. J. Last. This paper is Cambridge Earth Sciences Publication 1456.

Literature Cited

Babcock, L. E., Robison, R. A. 1988. Taxonomy and paleobiology of some Middle Cambrian *Scenella* (Cnidaria) and hyolithids (Mollusca) from western North America. *Univ. Kans. Paleontol. Contrib. Pap.* 121: 1–22

Barskov, I. S., Zhuravlev, A. Yu. 1988. Soft-bodied organisms from the Cambrian of the Siberian platform. *Paleontol. Zh.* 1988(1): 3–9 (In Russian)

Bengtson, S., Conway Morris, S. 1984. A comparative study of Lower Cambrian *Halkieria* and Middle Cambrian *Wiwaxia*. *Lethaia* 17: 307–29

Benus, A. 1988. Sedimentological context of a deep-water Ediacaran fauna (Mistaken Point Formation, Avalon Zone, eastern Newfoundland). *Bull. NY State Mus.* 463: 8–9

Briggs, D. E. G. 1983. Affinities and early evolution of the Crustacea: the evidence of the Cambrian fossils. In *Crustacean Phylogeny*, ed. F. R. Schram, pp. 1–22. Rotterdam: Balkema

Briggs, D. E. G., Collins, D. 1988. A Middle Cambrian chelicerate from Mount Stephen, British Columbia. *Palaeontology* 31: 779–98

Butterfield, N. J., Knoll, A. K., Swett, K. 1988. Exceptional preservation of fossils in an upper Proterozoic shale. *Nature* 334: 424–27

Chen, J.-Y. 1988. Precambrian metazoans of the Huai River drainage area (Anhui, E. China): their taphonomic and ecological evidence. *Senckenb. Lethaea* 69: 189–215

Chen, J.-Y., Hou, X.-G., Lu, H.-Z. 1989a. Early Cambrian netted scale-bearing worm-like sea animal. *Acta. Palaeontol. Sin.* 28: 1–16

Chen, J.-Y., Hou, X.-G., Lu, H.-Z. 1989b. Lower Cambrian leptomitids (Demospongea), Chengjiang, Yunnan. *Acta Palaeontol. Sin.* 28: 17–30

Chen, J.-Y., Hou, X.-G., Lu, H.-Z. 1989c. Early Cambrian hock glass-like rare sea animal *Dinomischus* (Entoprocta) and its ecological features. *Acta Palaeontol. Sin.* 28: 58–71

Cloud, P. 1986. Reflections on the beginnings of metazoan evolution. *Precambrian Res.* 31: 405–8

Cloud, P., Glaessner, M. F. 1982. The Ediacaran Period and System: Metazoa inherit the Earth. *Science* 217: 783–92

Conway Morris, S. 1977a. Fossil priapulid worms. *Spec. Pap. Palaeontol.* 20: 1–95

Conway Morris, S. 1977b. A new metazoan from the Cambrian Burgess Shale of British Columbia. *Palaeontology* 20: 623–40

Conway Morris, S. 1979. Middle Cambrian polychaetes from the Burgess Shale of British Columbia. *Philos. Trans. R. Soc. London Ser. B* 285: 227–74

Conway Morris, S. 1985a. The Ediacaran biota and early metazoan evolution. *Geol. Mag.* 122: 77–81

Conway Morris, S. 1985b. The Middle Cambrian metazoan *Wiwaxia corrugata* (Matthew) from the Burgess Shale and *Ogygopsis* Shale, British Columbia, Canada. *Philos. Trans. R. Soc. London Ser. B* 307: 507–86

Conway Morris, S. 1986. The community structure of the Middle Cambrian Phyllopod bed (Burgess Shale). *Palaeontology* 29: 423–67

Conway Morris, S. 1987. The search for the Precambrian-Cambrian boundary. *Am. Sci.* 75: 156–67

Conway Morris, S. 1989a. Early metazoans. *Sci. Prog.* (*Oxford*) 73: 81–99

Conway Morris, S. 1989b. The persistence of Burgess Shale–type faunas: implications for the evolution of deeper-water faunas. *Trans. R. Soc. Edinburgh: Earth Sci.* 80: 271–83

Conway Morris, S., Peel, J. S., Higgins, A. K., Soper, N. J., Davis, N. C. 1987. A Burgess Shale–like fauna from the lower Cambrian of North Greenland. *Nature* 326: 181–88

Conway Morris, S., Robison, R. A. 1986. Middle Cambrian priapulids and other soft-bodied fossils from Utah and Spain. *Univ. Kans. Paleontol. Contrib. Pap.* 117: 1–22

Conway Morris, S., Robison, R. A. 1988. More soft-bodied animals and algae from the Middle Cambrian of Utah and British Columbia. *Univ. Kans. Paleontol. Contrib. Pap.* 122: 1–48

Crimes, T. P. 1989. Trace fossils. In *The Precambrian-Cambrian Boundary*, ed. J. W. Cowie, M. D. Brasier, pp. 166–85. Oxford: Clarendon

Donovan, S. E. 1987. The fit of the continents in the late Precambrian. *Nature* 327: 139–41

Fedonkin, M. A. 1986. Precambrian problematic animals: their body plan and phylogeny. In *Problematic Fossil Taxa*, ed. A. Hoffman, M. H. Nitecki, pp. 59–67. New York: Oxford Univ. Press

Fedonkin, M. A. 1987. Soft-bodied Vendian fauna and its place in metazoan evolution. *Tr. Paleontol. Inst. Akad. Nauk SSSR* 226: 1–176 (In Russian)

Field, K. G., Olsen, G. J., Lane, D. J., Giovannoni, S. J., Ghiselin, M. T. et al. 1988. Molecular phylogeny of the animal kingdom. *Science* 239: 748–53

Gehling, J. G. 1987. Earliest known echinoderm—a new Ediacaran fossil from the Pound Subgroup of South Australia. *Alcheringa* 11: 337–45

Glaessner, M. F. 1980. *Parvancorina*—an arthropod from the Late Precambrian (Ediacaran) of South Australia. *Ann. Naturhist. Mus. Wien* 83: 83–90

Glaessner, M. F. 1984. *The Dawn of Animal Life. A Biohistorical Study*. Cambridge: Univ. Press. 244 pp.

Hofmann, H. J. 1985. Precambrian carbonaceous megafossils. In *Paleoalgology: Contemporary Research and Applications*, ed. D. F. Toomey, M. H. Nitecki, pp. 20–33. Berlin: Springer-Verlag

Horodyski, R. J. 1988. Late Proterozoic fossils from the western U.S.: could they represent the oldest skeletal metaphytes and the oldest meiofaunal traces? *Geol. Soc. Am. Abstr. With Programs* 20: A256 (Abstr.)

Hou, X.-G., Chen, J.-Y. 1989. Early Cambrian arthropod-annelid intermediate sea animal, *Luolishania* gen. nov. from Chengjiang, Yunnan. *Acta Palaeontol. Sin.* 28: 207–13

Hou, X.-G., Chen, J.-Y., Lu, H.-Z. 1989. Early Cambrian new arthropods from Chengjiang, Yunnan. *Acta Palaeontol. Sin.* 28: 42–57

Jenkins, R. J. F. 1981. The concept of an "Ediacaran period" and its stratigraphic significance in Australia. *Trans. R. Soc. South Aust.* 105: 179–94

Jenkins, R. J. F. 1984. Interpreting the oldest fossil cnidarians. *Palaeontogr. Am.* 54: 95–104

Kraft, P., Mergl, M. 1989. Worm-like fossils (Palaeoscolecida; ?Chaetognatha) from the Lower Ordovician of Bohemia. *Sb. Geol. Ved. Paleontol.* 30: 9–36

Ludvigsen, R. 1989. The Burgess Shale: not in the shadow of Cathedral Escarpment. *Geosci. Can.* 16: 51–59

Narbonne, G. M., Hofmann, H. J. 1987. Ediacaran biota of the Wernecke Mountains, Yukon, Canada. *Palaeontology* 30: 647–76

Pickerill, R. K. 1982. Cambrian medusoids from the St. John Group, southern New Brunswick. *Curr. Res. B. Geol. Surv. Can.* 82–1B: 71–96

Rigby, J. K. 1986. Sponges of the Burgess Shale (Middle Cambrian), British Columbia. *Palaeontogr. Can.* 2: 1–105

Rigby, J. K., Newman, K. R., Smit, J., van der Kaars, S., Sloan, R. E., Rigby, J. K. 1987. Dinosaurs from the Paleocene part of the Hell Creek Formation, McCone County, Montana. *Palaios* 2: 296–302

Robbins, E. I., Porter, K. G., Haberyan, K. A. 1985. Pellet microfossils: possible evidence for metazoan life in early Proterozoic time. *Proc. Natl. Acad. Sci. USA* 82: 5809–13

Runnegar, B. 1982a. A molecular-clock date for the origin of the animal phyla. *Lethaia* 15: 199–205

Runnegar, B. 1982b. Oxygen requirements, biology and phylogenetic significance of the late Precambrian worm *Dickinsonia*, and the evolution of the burrowing habit. *Alcheringa* 6: 223–39

Runnegar, B. 1985. Collagen gene construction and evolution. *J. Mol. Evol.* 22: 141–49

Seilacher, A. 1984. Late Precambrian and Early Cambrian Metazoa: preservational or real extinctions? In *Patterns of Change in Earth Evolution*, ed. H. D. Holland, A. F. Trendall, pp. 159–68. Berlin: Springer-Verlag

Seilacher, A. 1989. Vendozoa: organismic construction in the Proterozoic biosphere. *Lethaia* 22: 229–39

Smith, A. B. 1988. Patterns of diversification and extinction in early Palaeozoic echinoderms. *Palaeontology* 31: 799–828

Sokolov, B. S., Fedonkin, M. A. 1984. The Vendian as the terminal system of the Precambrian. *Episodes* 7: 12–19

Stasinska, A. 1960. *Velumbrella czarnockii* n. gen., n. sp.—méduse du Cambrian inférieur des Monts de Sainte-Croix. *Acta Palaeontol. Pol.* 5: 337–44

Sun, W.-G. 1985. Late Precambrian scyphozoan medusa *Mawsonites randellensis* sp. nov. and its significance in the Ediacara metazoan assemblage, South Australia. *Alcheringa* 10: 169–81

Sun, W.-G., Hou, X.-G. 1987a. Early Cambrian medusae from Chengjiang, Yunnan, China. *Acta Palaeontol. Sin.* 26: 257–71

Sun, W.-G., Hou, X.-G. 1987b. Early Cambrian worms from Chengjiang, Yunnan, China: *Maotianshania* gen. nov. *Acta Palaeontol. Sin.* 26: 299–305

Sun, W.-G., Wang, G.-X., Zhou, B.-H. 1986. Macroscopic worm-like body fossils from the upper Precambrian (900–700 Ma), Huainan district, Anhui, China and their stratigraphic and evolutionary significance. *Precambrian Res.* 31: 377–403

Turbeville, J. M., Ruppert, E. E. 1983. Epidermal muscles and peristaltic burrowing in *Carinoma tremaphoros* (Nemertini): correlates of effective burrowing without segmentation. *Zoomorphology* 103: 103–20

Wade, M. 1968. Preservation of soft-bodied animals in Precambrian sandstones at Ediacarà, South Australia. *Lethaia* 1: 238–67

Whittington, H. B. 1975. The enigmatic animal *Opabinia regalis*, Middle Cambrian, Burgess Shale, British Columbia. *Philos. Trans. R. Soc. London Ser. B* 271: 1–43

Whittington, H. B. 1985. *The Burgess Shale*. New Haven, Conn: Yale Univ. Press. 151 pp.

Whittington, H. B., Briggs, D. E. G. 1985. The largest Cambrian animal, *Anomalocaris*, Burgess Shale, British Columbia. *Philos. Trans. R. Soc. London Ser. B* 309: 569–609

Zhang, W.-T., Hou, X.-G. 1985. Preliminary notes on the occurrence of the unusual trilobite *Naraoia* in Asia. *Acta Palaeontol. Sin*. 24: 591–95

Annu. Rev. Earth Planet. Sci. 1990. 18:123–71
Copyright © 1990 by Annual Reviews Inc. All rights reserved

GEOLOGICAL AND BIOLOGICAL CONSEQUENCES OF GIANT IMPACTS

Digby J. McLaren

248 Marilyn Avenue, Ottawa, Ontario K1V 7E5, Canada

Wayne D. Goodfellow

Geological Survey of Canada, 601 Booth Street, Ottawa, Ontario K1A 0E8, Canada

INTRODUCTION

It seems inescapable that large impacts have been an ongoing phenomenon throughout geological time. In the Phanerozoic the estimated frequency of impact by bodies of 5-km diameter or more is on the order of 10 every 100 Myr. The 5-km lower limit is chosen arbitrarily, but it is shown in this review that such a body is certainly large enough to cause very large terrestrial effects. A small decrease in this size limit would lead to a large increase in the frequency of occurrence. It may be claimed that bodies of a size likely to have drastic and sudden effects on many physical and biological features of the Earth have arrived with a frequency comparable to the average duration of a biostratigraphic stage. Nothing is said of the interval between arrivals, which may be random (i.e. stochastic). Currently there is not enough geological evidence to allow recognition of other than random spacing between a few positively identified events during the Phanerozoic.

Since impacts by extraterrestrial bodies have been clearly established, it is reasonable to postulate that they must leave a record in the stratigraphic column. Why, then, do we currently have so few horizons that have been claimed to contain evidence of a global impact event, since the effects we are considering are unquestionably global?

In this paper we bring together evidence from several disciplines that

0084–6597/90/0515–0123$02.00

bear upon the problem. We find that some events in the history of the Phanerozoic appear global, recognizable in several continents and/or ocean basins, and that these events transcend local differences and local postulates as to cause. The evidence may be summed globally and leads to an empirical conclusion based on biotic change, sedimentological events, and associated geochemical changes or signatures. Similar, coeval events observed at many independent stratigraphic occurrences suggest synchroneity and common cause. This might be seen as the application of the concept of simplicity in building a theory of common causation—sometimes referred to as Occam's razor (entities ought not to be multiplied except from necessity).

We find also that several successive global events at different geological times lead independently to similar conclusions, and consequently one may sum the probabilities of each to develop a unified hypothesis for a common cause. An impact or impacts by an extraterrestrial body or bodies, asteroid, or comet is the postulate that satisfies all observed conditions. Causes that have been suggested to explain specific events in time or in space are not excluded from the model, for secondary effects from an impact, as is demonstrated, are many and varied.

A final problem in discussing the effects of impacts is that if the impacts are large enough to cause major changes to the terrestrial environment, then their signatures should be found in the record at intervals approximating the mean duration of a stage. The geological record, however, has not yet provided a readily recognized measure of this frequency. It is suggested that some of the more easily detectable traces of impacts in the record are generated only when they reinforce crustal processes. When they strike at a time of stability or stasis, either physical or biological, they may be less likely to leave easily identifiable evidence.

In our survey, we find that nearly all major extinction events described in the literature have been reported as being due to some form of ongoing stress. Inadequate consideration of field observations on a global scale, combined with concern for taxonomic rather than biomass change, may lead to a forcing of the evidence toward a gradualistic or ongoing model for extinctions. The apparent contradiction between a gradualistic and a sudden catastrophic interpretation may be resolved by recognizing that climatic change, eustatic change, diversity rundown, and stepwise extinctions (commonly a local rather than global phenomenon) may indeed have been active over a time period during which a sudden and violent event occurred. The catastrophic effects of such an event would be superimposed on the evidence in the record for gradual and ongoing changes. The examples described in this paper show that in most cases the evidence for such sudden events is easily recognized and includes biomass disappearance at a bedding plane globally, a major sedimentary event, and primary and secondary elemental anomalies and stable isotope excursions.

The evidence at several horizons is summarized in this review, and we believe that a recognition of contrasting stratigraphic evidence differentiating between gradual and rapid events will lead to recognition of other horizons, resulting eventually in a well-defined stratigraphic time scale punctuated by global events that provide worldwide time horizons. Such recognition would allow theories of periodicity of extinction (i.e. stochastic versus deterministic) to be easily resolved. In addition, currently tentative suggestions on periodicity of other global phenomena could be evaluated.

The geological horizons chosen for examination in this paper are dealt with in order of increasing age. The first to be considered, however, is the Cretaceous-Tertiary boundary, which is taken out of order because it represents the best example of an abrupt event for which a bolide impact has been suggested as a cause and with which the other examples described herein may be compared.

GIANT IMPACTS AND THEIR GEOLOGICAL EFFECTS

Field Studies of Craters and Bolides

Since its formation over 4.5 Gyr ago, the Earth has been bombarded by extraterrestrial objects of variable size, composition, and frequency of arrival. The record of impacts is poorly preserved owing to crustal recycling by subduction and erosion, burial below sedimentary and volcanic cover, and masking by tectonic deformation. Grieve (1984), in an evaluation of the size-frequency distribution of craters, noted an increase in the rate for craters younger than about 120 Ma and ascribed this change to the lesser chance of preservation for older craters. Furthermore, this size-frequency distribution indicates that the Earth has retained a representative population of craters greater than 20 km in diameter, and only these structures may be used to calculate meaningful cratering rates (Grieve & Dence 1979). Based on both the Earth cratering record documented by Grieve & Robertson (1979) for craters less than 120 Myr old and diameters $D \geq 100$ km (i.e. bolide diameter $d \geq 5$ km, assuming $D/d = 20$) and the number of asteroids and comets in Earth-crossing orbits (Shoemaker et al 1988), M. R. Dence (personal communication) has estimated that between 44 and 84 bolides of this size have impacted the Earth during the Phanerozoic, or one every 7 to 14 Myr. Similarly, bolides with $d \geq 10$ km (and thus capable of forming a crater with $D \geq 200$ km) have struck the Earth on average one every 55 Myr.

Impactor compositions have been determined by mineralogical and chemical analyses of meteorites, melt rock at impact sites, and ejecta layers. Because most meteorites are small and, in the case of achondrites, difficult to detect by magnetic methods, most are never discovered. Of those recovered, over 90% consist of chondrites, irons, and stony-irons (Anderson 1983); achondrites constitute about 9% of the total.

Although the chemical composition of any meteorite type is highly variable, most siderophile group elements,[1] particularly Fe, Ni, Cr, Co, Ir, Pt, Au, and Os, display order-of-magnitude enrichment in chondrites, irons, and most stony-irons when compared with oceanic and continental crustal rocks (Mason 1971). Achondrites, however, have low siderophile group element contents. Chalcophile element abundances in meteorites generally fall within the range of most terrestrial rocks.

With the exception of the Brent crater, which was formed by a chondritic meteorite, all simple craters (less than 2 km in diameter) were caused by iron or stony-iron meteorites (Grieve 1982). At large impact structures (i.e. complex craters), bolide identification is based on the siderophile group element content of melt rock and ejecta. Craters with anomalously high siderophile group element contents include the Clearwater East and Brent craters in Canada (Palme et al 1978); the Lappajärvi and Sääksjärvi craters, Finland (Palme et al 1980, Palme 1980); the Rochechouart crater, France (Janssens et al 1977); the Mien and Dellen craters, Sweden (Palme et al 1980); and the Strangways crater, Australia (Morgan et al 1981). In addition, ejecta from the Bosumtwi, Ghana (Palme et al 1978), and Lake Acraman, Australia (Gostin et al 1989), craters and metallic spheres from soils in the Tunguska region of central Siberia (Ganapathy 1983) also contain anomalously high contents of Ir. Impact melt rocks from the Clearwater West, Manicouagan, and Mistastin craters in Canada (Palme et al 1978), the Popigay crater in the USSR (Masaytis & Raykhlin 1986), and the Ries crater in Germany (Morgan et al 1979) all show no detectable meteoritic component. The low abundances of siderophile group elements in melt rock may be due to high-velocity impact, bolides poor in siderophile group elements (i.e. achondrites, some stony-irons), or the contamination of only a fraction of the melt rock by meteoritic material (Palme et al 1978). The low siderophile group element content of melt rock from the Lake Acraman crater, despite the nearby occurrence of associated Ir-enriched ejecta (Gostin et al 1989), indicates that some melt rocks provide a poor measure of meteoritic composition.

[1] For simplicity of presentation, siderophile and associated noble metals are referred to as siderophile group elements in this paper. While the siderophile behavior of noble metals dominated during the formation of the Earth's core, many also have a chalcophile character, which has substantially influenced their distribution in the mantle and crust.

In summary, because about 90% of all meteorites that have impacted the Earth are enriched in siderophile group elements such as Ir, there is a similar chance that impact-generated mass extinction horizons will contain anomalously high Ir and other siderophile group elements. But this assumes total preservation of the ejecta layer and of the boundary interval in most extinction boundary sequences, and both of these are unlikely assumptions if additional factors (e.g. nondeposition and sedimentary reworking, particularly in high-energy depositional environments) are taken into account.

Computer Modeling of Impact Effects

Estimates of the energy released from a major impact indicate catastrophic consequences for the biosphere, hydrosphere, and atmosphere. The effects of a 10-km-diameter asteroid traveling at a velocity of 20 km s^{-1} that impacts into an ocean 5 km deep have been modeled by Roddy et al (1987, 1988). The initial kinetic energy of the asteroid was 2.6×10^{30} ergs (6.2×10^7 Mt TNT equivalent).

The effects of impact are the following:

1. A large, heated mass of low-density air with peak temperatures of about 20,000 K forms adjacent to the Earth's surface and moves out rapidly from the impact area (Hassig et al 1987); this can be expected to cause surface fires if combustible materials are available.
2. An enormous tsunami propagates outward with a wave height comparable to the ocean depth due to the collapse of the uplifted rim of the crater at an outward velocity of 0.5 km s^{-1}. This wave is of sufficient energy to scour the seafloor, rework sediments, and form tsunami deposits, particularly in shallow-water shelf environments. The effect of sedimentary reworking is to destroy or obscure the sedimentary record, including ejecta, mineralogical, and geochemical signatures.
3. Enough energy is released to produce an earthquake of magnitude 12.4 on the Richter scale. This energy is sufficient to release sedimentary prisms along continental slopes and to initiate or accelerate volcanism and hydrothermal activity.
4. Both 105 km^3 of excavated rock and 9×10^{13} t of vaporized asteroid, crust, and ocean are lofted to altitudes of up to 100 km (Grant et al 1987). The mass of vaporized asteroid is less than 1% of the mass of target material ejected (Roddy et al 1988). It is estimated that about 80–90% of the ejecta will fall back and form a continuous ejecta blanket surrounding the crater. The remaining 10–20% of ejecta mass, including vaporized asteroid, would be lofted to approximately 100 km, to fall back to Earth over a period of months.

The condensation of vaporized material will produce liquid droplets, which when quenched form microtektites (O'Keefe et al 1988). These congealed droplets and the fine dust ejected from the impact crater will remain suspended in the upper atmosphere long enough to circle the globe, block out sunlight, cool the Earth's surface, and disrupt the food chain by reducing or preventing photosynthesis. It has been estimated, using models of aerosol physics (Toon et al 1982, Pollack et al 1983), that dust from the Cretaceous-Tertiary impact settled from the atmosphere in less than six months and that minimum photosynthesis was restored in about four months. The oceans are cooled by about 3–4°C and the surface of continents by up to 40°C (Roddy et al 1988). These estimates, however, do not include the effects of soot particles, which are smaller and settle more slowly but are also better light absorbers than fine ejecta (Wolbach et al 1985). Even if we consider the added effects of soot, darkness is unlikely to have lasted much longer than six months.

Another environmental consequence of a giant impact is the shock heating of the proximal atmosphere and the formation of nitrous and nitric acid (Lewis et al 1982, Prinn & Fegley 1987). Nitrogen compounds inhibit photosynthesis by reducing solar radiation, and they asphyxiate fauna through exposure to NO_2. Furthermore, a nitric oxide cloud generated by the Tunguska meteorite is estimated to have caused a 45% ozone depletion in the Northern Hemisphere that persisted for four years (Turco et al 1981). Acid rain defoliates forests, mobilizes metals by accelerated weathering that creates toxicity problems for some organisms, and acidifies oceans sufficiently to destabilize calcareous-shelled organisms.

Seyfert & Sirkin (1979) first suggested that giant impacts may initiate mantle plumes and the breakup of plates. This is supported by a correspondence between tectonic episodes and impact events during the Cenozoic (Burek & Wanke 1988), by the accelerated outpouring of hot-spot flood basalts at times of mass extinction over the past 250 Myr (Rampino & Stothers 1988), and by the close temporal association of magnetic reversals with tektites of the Australian and Ivory Coast strewn fields (Glass et al 1979) and with sediments immediately postdating the Ries impact (Pohl 1977).

MASS EXTINCTIONS AND GIANT IMPACTS

Cretaceous-Tertiary (K-T)

INTRODUCTION In the marine realm, the Cretaceous-Tertiary boundary is marked essentially by a huge biomass loss of planktonic foraminifera, which coincides with the end of many invertebrate taxa ranging in rank from species to order. On land, there was an abrupt end of flowering plants, associated with an Ir anomaly at many localities, followed by a brief dominance of ferns. Many taxa survived, however, to reappear later. The last of the dinosaurs disappeared at or near this horizon.

There is a hiatus in 90% of the sequences that span the K-T boundary worldwide (Kauffman 1979). Despite this fact, it is still the best preserved major extinction event because of its relatively young age and the less than 30% subduction of older oceanic crust that has occurred over the past 66 Myr (A. G. Smith in Alvarez et al 1982). The K-T boundary is also the most extensively studied because of the Ir anomalies reported by Alvarez et al (1980) and the fact that boundary sediments in virtually every ocean basin have been cored by numerous Deep Sea Drilling Project (DSDP) and Ocean Drilling Program (ODP) drill holes. In less than a decade, following the initial discovery of high Ir at the K-T boundary, more than 60 marine and nonmarine sections spanning this extinction event have been shown to be anomalous in Ir (Alvarez et al 1980, 1982, Kyte et al 1980, 1985, Ganapathy 1980, Ganapathy et al 1981, Hsü et al 1982, Orth et al 1982, Brooks et al 1984, 1985, 1986, Nichols et al 1986, Preisinger et al 1986, Strong et al 1987, Bohor et al 1987a, Lerbekmo et al 1987, Izett 1988, Bhandari et al 1988). Most of these come from DSDP cores, marine sections exposed in the Apennines and Alpine mountain belts of Europe and Asia, and terrestrial sections from the interior of North America. A large majority of Ir-enriched marine sections, including those now exposed on land, consist of pelagic and hemipelagic sediments that were deposited off the continental margins in deep-water environments (Smit & Romein 1985). Exceptions are K-T sections of continental shelf sediments at Braggs, Alabama (Jones et al 1987), Brazos River, Texas (Bourgeois et al 1988), Stevns Klint, Denmark (Birkelund & Hakansson 1982), Woodside Creek, New Zealand (Strong 1977), and in New Jersey and Haiti (Hildebrand & Boynton 1988), and nonmarine sections from the Raton Basin, New Mexico (Orth et al 1982), Wyoming (Bohor et al 1987b), Montana (Smit & van der Kaars 1984), Alberta (Lerbekmo et al 1987), and Saskatchewan (Nichols et al 1986). With the exception of Stevns Klint and Woodside Creek, K-T boundary sections representing shallow-water environments display features of high-energy sedimentation.

BIOSTRATIGRAPHY (FIGURE 1) There remain, however, several major controversies in the interpretation of biological, physical, and chemical events at the boundary horizon. We are faced with a contrast between slow-acting changes in the physical environment and in the biota and a sudden, universal and synchronous biomass loss. There are many discussions on the changes and rates of change during the Late Cretaceous (particularly the Maastrichtan Stage), which involved either a gradual

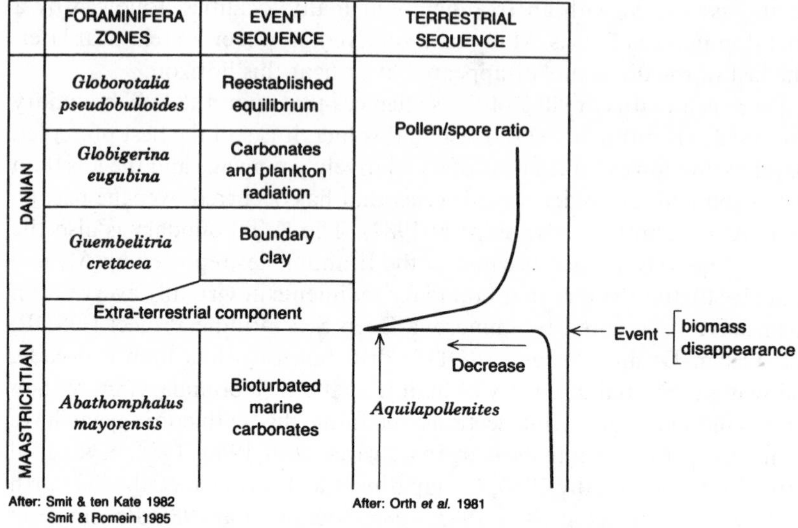

Figure 1 Cretaceous-Tertiary boundary—terminology and events.

decline in biotic diversity in the marine realm (e.g. Alvarez et al 1984, Wiedmann 1988) or a series of stepwise extinctions spread over some 2.5 Myr (Kauffman 1988). Close to the boundary, it is claimed that there were small stepwise extinctions before the main event. Keller (1988), for instance, claims stepwise extinctions of planktonic foraminifera over a thickness of some 32 cm of section at El Kef, Tunisia, but Smit et al (1988) find no significant progressive changes immediately before the horizon at which the foraminifera disappear, nor elsewhere in comparable sections in Spain. The disappearance of the planktonic foraminifera and the changes after the event have been well documented (Smit 1982, Smit & Romein 1985, Stradner & Rögel 1988). Among larger marine invertebrates, there appears to be a reduction in diversity. The slow decline in biodiversity is associated with a global cooling and eustatic changes (Stanley 1984).

Nevertheless, there was a sudden loss of biomass in planktonic foraminifera, brachiopods, ammonites, bryozoans, and many bivalves and gastropods (Surlyk & Johansen 1984, Alvarez et al 1984, Ward 1988). The calcareous nannoplankton biomass was also strongly affected, although the change from dominantly Cretaceous to dominantly Tertiary taxa takes place higher in the section (Thierstein 1981, Smit & Romein 1985). The killing event is sudden, and it cannot be claimed that there were premonitory signals in the biota. The duration of the killing was brief, with a scale of resolution down to years (Smit 1982, Alvarez et al 1984, Smit & Romein 1985, Preisinger et al 1986, 1988).

The detailed sedimentology across the K-T boundary has been described by Smit & Romein (1985) for marine sections and consists of the following units:

1. The bottom unit is a Late Cretaceous homogeneous sequence of calcareous oozes, limestones, and marls, with the top 10–20 cm commonly bioturbated.
2. This is overlain by an extraterrestrial layer with shocked quartz and graded spherules, high contents of siderophile group (including Ir) and chalcophile group elements, and anoxic conditions as indicated by the near absence of benthic organisms and a high sulfide content. This unit is commonly red in weathered sections owing to the oxidation of pyrite to goethite. In shallower marine sections, the sudden event is marked by a brief erosion episode (Jones et al 1987) or by an horizon with wave deposits or a huge tsunami event (Bourgeois et al 1988).
3. On top of this lies a dark to pale gray, organic-rich clay unit containing normal detrital clays and 20–40% carbonate from reworked foraminifera and a few survivors. This unit is characterized by a "Strangelove" environment (conditions of very low productivity following a biological crisis), which has been shown to follow the K-T extinction event (Hsü & Mackenzie 1985).
4. This is overlain by a unit defined by the first appearance of Paleocene planktonic foraminifera and by increased carbonate sedimentation resulting from the explosive development of planktonic organisms.
5. Finally, the uppermost unit shows a return to normal marine sediments similar to the Late Cretaceous carbonates, except for completely new planktonic biota.

In terrestrial successions the correlates of the marine K-T extinction horizon are associated with an abrupt end of flowering plants in all floral facies and an iridium anomaly (Orth et al 1981). Dominant angiosperm pollen is succeeded by a dominance of fern spores, which are replaced in turn by a return of pollen floras, reduced in diversity, in a permanently reorganized flora with few new group taxa (Nichols & Fleming 1988). In Montana, Alberta, and Wyoming, Sloan et al (1986) claim a slow rundown in biodiversity with lowering of the sea level toward the end of the Cretaceous, followed by the extinction of the dinosaurs. The horizon at which this occurred remains controversial. Retallack & Leahy (1986) suggest that taxonomic and taphonomic uncertainties exist concerning dinosaur disappearance and consider that diversity decline is difficult to prove. The timing of the end of the dinosaurs is, however, of little importance in terms of the total event at the end of the Cretaceous. The most important single indicator of this event is a huge biomass loss in marine and terrestrial successions.

It is argued that the killing was due to volcanic paroxysms over a relatively short time scale of about 500,000 yr. Hut et al (1987) suggest that the apparent stepwise extinctions at the close of the Cretaceous, for instance, may have been caused by multiple cometary impacts extending over an interval of 1–3 Myr. It is quite clear, however, that there was only one brief massive killing event at the K-T boundary, to be followed by an almost total loss of biomass, a Strangelove ocean (Zachos et al 1988), and the development of a new biota of microplankton.

MINERALOGY Spherules, which are common in unit 2 of Smit & Romein (1985), are highly variable in size (0.1–1 mm in diameter), shape (e.g. spheroid, teardrop, dumbbell, and other splash forms), and composition (e.g. goyazite, pyrite, clinopyroxene, feldspar, magnetite) (Montanari et al 1983, Smit & Kyte 1984, Brooks et al 1985, Montanari 1986, Izett 1987, Schmitz 1988). The origin of these spherules is a subject of debate. Schmitz (1988) and Izett (1988) have argued that they formed authigenically, whereas Bohor (1988) and Smit & Romein (1985) consider them to be microtektites modified during diagenesis.

Quartz grains displaying up to nine sets of shock lamellae have been described from five European marine K-T boundary sites, Woodside Creek, New Zealand, and core GPC-3 from the north-central Pacific (Bohor et al 1987a) and from numerous continental K-T boundary sections (Bohor et al 1984, Izett 1988). The global size distribution and abundance of shock-metamorphosed minerals suggest that impact occurred near North America (Izett 1987). In addition to shocked quartz, Izett (1987) found composite shocked grains and lithic fragments of quartz-quartz and quartz-feldspar. Owen & Anders (1988) noted that 40% of grains from the K-T boundary in the Raton Basin having one or more sets of continuous, planar, parallel shock-deformation lamellae were polycrystalline, with the shock lamella directions controlled by differing crystallographic orientations of the subcrystals. Furthermore, cathodoluminescence of shocked quartz showed that more than 97% of all shocked grains were of non-volcanic origin and probably derived from a metamorphic or intrusive basement. More recently, McHone et al (1989) confirmed the presence of stishovite (a very high pressure form of silica widely accepted as an indicator of terrestrial impact) in the K-T boundary clay.

GEOCHEMISTRY The Ir content of K-T boundary sediments is highly variable, ranging up to a peak value of 87 ppb at Stevns Klint. This wide range of values may be due to proximity to the impact site, variable rates of dilution by sediments deposited in different sedimentary environments, the level of sedimentary reworking (particularly in shallower marine settings), and geochemical redistribution during diagenesis and burial metamorphism. Iridium profiles across the boundary upsection typically display a sharp increase to a peak value at the extinction horizon, followed by a gradual decrease to normal values tens of centimeters above the horizon. The spread of anomalous Ir values above the K-T boundary is due to a number of secondary processes, including sedimentary reworking, bioturbation, diagenetic remobilization, and influx into ocean basins of Ir-rich ejecta eroded from the continents and continental margins. High-Ir values extending below the boundary (Crocket et al 1988) are probably due to bioturbation or diagenetic remobilization. The latter process may explain the large range of Ir to siderophile element ratios in sediments from many K-T extinction horizons (Kyte et al 1985; Noreen Evans, personal communication) and the Lake Acraman ejecta horizon, South Australia (Gostin et al 1989).

Despite several studies of the form of Ir in K-T boundary beds, the Ir carrier phase(s) is not known for most sites, and there is a lack of consensus arising from studies of sections reporting a carrier phase. For example, Preisinger et al (1986) have argued that the Ir in K-T boundary sediments at Gosau, Austria, is related to a Ti mineral, whereas Hansen et al (1987) have presented evidence that Ir is hosted in carbonaceous matter from Stevns Klint, Denmark. Margolis & Doehne (1988), after examining a series of samples from boundary sections at Zumaya and Sopelano, Spain, failed to detect Ir in specific mineral phases. The problem of low-Ir abundances is compounded by the possibility of redistribution during diagenesis. As a result, the present form of Ir in sediments may reflect secondary processes and therefore provide few clues on the primary carrier phase(s) of Ir at event horizons.

In addition to Ir, marine K-T boundary sections at Caravaca, Zumaya, Sopelano, Agost, and Barranco in Spain, at Stevns Klint in Denmark, at Gosau in Austria, and at Woodside Creek and Flaxbourne River in New Zealand are enriched both in siderophile group elements Au, Os, Pd, Pt, Re, Ni, Cr, Fe, and Co and in chalcophile elements As, Sb, Se, Zn, Cu, and Mo (Kyte et al 1980, 1985, Smit & Hertogen 1980, Ganapathy 1980, Brooks et al 1984, 1985, Schmitz 1985, Preisinger et al 1986, Strong et al 1987, Margolis & Doehne 1988). Nonmarine K-T boundary sections, however, have high contents of siderophile group elements Cr, Ni, Ir, Pt, and Au but generally low contents of most chalcophile elements (Orth et al 1982, Izett 1988).

Margolis & Doehne (1988) have shown that Pt in K-T boundary samples from Zumaya, Spain, occurs as native platinum grains; Ni and Cr in spinels and spherules; As and Ni in pyrite/arsenopyrite; and Pb, Cu, Zn, and Sb in sulfides. These results are in general agreement with the occurrence of Ni, Co, As, Sb, and Zn in pyrite from the K-T boundary at Stevns Klint (Schmitz 1985, Schmitz et al 1988). Luck & Turekian (1983) report $^{187}Os/^{186}Os$ values of 1.65 for the Ir-rich layer of the marine section at Stevns Klint, and 1.29 for the terrestrial section in the Raton Basin. These differ markedly from the crustal ratio of ~ 10 but are similar to the meteoritic and mantle value of ~ 1. The simplest explanation is that the data from the K-T sections represent a meteoritic origin.

The occurrence of most chalcophile elements as sulfides in K-T marine sediments, the low chalcophile element content of K-T continental sediments, and the low abundance of these elements in chondrites support the interpretation that they were precipitated during reaction with H_2S generated in organic-rich sediments by the bacterial reduction of seawater sulfate (Schmitz 1985). The organic-rich sediment that rained to the seafloor following the mass killing event generated reducing conditions locally in an otherwise oxidized sedimentary sequence and served therefore as a trap for elements that form insoluble sulfides. This mechanism may explain the high concentration of chalcophile elements in a green, reduced envelope surrounding the Lake Acraman ejecta horizon (Wallace et al 1990). In nonmarine K-T sections, the low sulfate content in groundwater limits the production of H_2S by sulfate-reducing bacteria, thereby diminishing the role of dissolved sulfide in fixing chalcophile elements.

ISOTOPES At the K-T boundary horizon in marine sections, there is a record of sudden decreases in $\delta^{13}C$ and $\delta^{18}O$ values in carbonates that persists into the Paleocene (Scholle & Arthur 1980, Romein & Smit 1981, Hsü et al 1982, Perch-Nielsen et al 1982, Shackleton et al 1983, Gerstel et al 1986, Margolis et al 1987, D'Hondt & Lindinger 1988, Preisinger et al 1988). Such decreases have been explained by a biomass reduction and a warming trend, respectively. Because ^{12}C is preferentially accumulated by marine organisms during photosynthesis, a decrease in biomass following an extinction event reduces the amount of ^{12}C fixed in organic matter, resulting in a decrease of $\delta^{13}C$ values in carbonates. Carbon isotopic trends for coexisting shallow-water and deep-water foraminifera converge at the K-T boundary because of the destruction of biologically mediated isotope gradients associated with the high productivity of shallow-water photosynthetic organisms, and they diverge 0.4 Myr after the extinction event as ocean conditions return to normal (D'Hondt & Lindinger 1988, Zachos & Arthur 1989).

Hsü et al (1982) estimated that the oxidation of all carbon fixed in the present biomass and organic matter of the upper 100 m of the water column would decrease $\delta^{13}C$ values in dissolved carbonates and carbonate precipitates by about 1.8‰. These calculations do not include the influence of accelerated carbonate dissolution resulting from the lowering of ocean water pH following a major impact. Because of steady-state equilibrium conditions established between hydrospheric and atmospheric CO_2, an increase of dissolved carbonate can be expected to increase atmospheric CO_2 levels, causing a warming trend due to a greenhouse effect.

Carbon and oxygen isotope trends must be treated with caution, however, because of possible effects of diagenetic processes on measured $\delta^{13}C$ values in carbonate precipitates. Because extinction horizons are commonly depleted in sedimentary carbonate and enriched in organic matter, generation of isotopically light diagenetic carbonate will occur as a result of coupling of sulfate reduction and carbon oxidation during diagenesis. Isotopic exchange between these two sources of carbonate may lead to an overall decrease in $\delta^{13}C$ values. High sulfide contents at the Stevns Klint K-T boundary horizon (Schmitz 1985), for example, provide evidence of extensive organic matter oxidation because of such coupling interactions.

The most meaningful isotope results come, therefore, from studies where measurements were made on particular organisms for which the ecology is understood. This approach avoids the overprinting of a normal water column isotope signature by diagenetic carbonate and allows for the evaluation of vertical isotope gradients in the ocean. Because of these limitations, reports by Mount et al (1986) of major negative $\delta^{13}C$ excursions measured in total carbonate and predating the K-T extinction are subject to question.

CAUSE An episode of major flood basalt volcanism represented by the Deccan Traps has been proposed as the cause of the K-T mass extinction (Officer & Drake 1983, 1985). The evidence for a causal link between catastrophic volcanism and mass extinction consists of a close temporal relationship between the Deccan basalts and the K-T event (Courtillot et al 1986, 1988a). The high Ir content in airborne particles from Kilauea volcano, Hawaii (Zoller et al 1983), and the report of shock mosaicism in plagioclase from the Toba crater, Sumatra (Carter et al 1986), have been presented as evidence that the Ir anomalies and shocked minerals at the K-T boundary are of volcanic origin. The felsic volcanism of Toba represents a style of volcanism considerably more explosive than flood basaltic volcanism.

There are a number of problems, however, with a volcanic extinction mechanism. First, the most recent $^{40}Ar/^{39}Ar$ age dates show that Deccan

Trap volcanism spans a time range of 69–65 Ma and thus was initiated prior to the extinction event (Courtillot et al 1988b). These dates are in general agreement with paleontological data that demonstrate that volcanism started in the Maastrichtian.

Second, in sections that have not undergone major reworking (i.e. resedimentation, bioturbation, and mixing during coring), peak Ir anomalies coincide remarkably well with the extinction horizon. If Deccan Trap volcanism was the source of Ir, there should be several Ir anomalies spanning the time interval represented by each volcanic episode. Interflow sediments associated with every episode of flood basalt volcanism should be anomalous in Ir. Furthermore, the amount of Ir in airborne particles is likely to be insufficient to account for the total flux of Ir estimated for the K-T event.

Third, de Silva & Sharpton (1988) present theoretical arguments, field evidence, and petrographic data that explosive volcanism cannot generate the shock-metamorphic features observed in quartz from the K-T boundary. Despite over 150 years of research in volcanology, there is no evidence of features common to impact structures, such as wall rocks with abundant planar features, shatter cones, pseudotachylite dikes, and high-pressure phases of quartz. Reports of "shocked minerals" associated with the Toba and Bishop Tuffs (Carter et al 1986) actually refer to a single set of deformation lamellae in feldspar and rare quartz clasts that bear little resemblance to those observed in shocked minerals from impact structures (Robertson et al 1968) and the K-T boundary (Bohor et al 1987a, Bohor 1988).

The Earth has been bombarded by giant bolides that have left a poorly preserved record of craters. These impact structures and associated ejecta deposits are characterized by the suite of shock effects and geochemical signatures previously referred to. The K-T event would appear to furnish well-documented examples of all the major effects that would be expected from an impact and, in fact, provide the "type specimen" with which other such events might be compared.

Eocene-Oligocene

Discussion of this reported extinction event and its association in time with microtektites belonging to the North American tektite strewn field is confused by the complexity and volume of data arising from biological, sedimentological, and geochemical studies globally (Pomerol & Premoli-Silva 1986). There continues to be uncertainty as to the horizon at which the base of the Oligocene should be defined that involves foraminiferal and nannoplankton biostratigraphy, chemostratigraphy, and sedimentological calibration. Consensus appears to favor a horizon that approximately corresponds to the boundary between planktonic foraminiferal zones P17/P18 of Blow (1969). This lies in magnetic polarity zone C13R (Lowrie 1986). The age of this horizon is still under discussion, with perhaps an upper limit established at 34.5 ± 0.4 Ma (Glass 1986).

The boundary represented by this horizon is a major evolutionary break in the Cenozoic. From the middle Eocene, changes have taken place successively, and this has continued into the lowest Oligocene; there is evidently not a single terminal event but an accelerated replacement rate among much of the biota within the marine realm (Pomerol & Premoli-Silva 1986). In the terrestrial realm, Russell & Tobien (1986) record a major change in mammalian faunas in North America and Asia during the late Eocene but see no sudden event. Rage (1986) finds that changes in amphibia and reptiles in Europe at the end of the Eocene have been more abrupt—"Grande Coupure" is still used as a term to describe them.

There appears, therefore, to be little evidence of rapid or catastrophic changes during the interval in question, although Keller (1986) and Hut et al (1987) have described four relatively sudden, stepwise taxonomic extinctions (which could be as brief as 1 Myr) that involve gastropods, bivalves, and planktonic foraminifera. The final of the four extinctions is the selected boundary horizon. From evidence compiled from deep-sea cores in the Gulf of Mexico, the Caribbean, and the western equatorial Atlantic, Glass (1986) has described a late Eocene microtektite horizon as part of the North American tektite strewn field. This has also been found in an outcrop at Bath Cliff, Barbados, where it occurs stratigraphically 26 m below the boundary (Saunders et al 1984). Two Ir peaks have been reported from approximately the same horizon as the microtektites and thus confirm an extraterrestrial origin (Montanari 1988). Glass (1986) points out that there is no evidence in the deep-sea record for extinctions or climatic changes associated with the tektite layer.

Although the extinction process appears to be long and drawn out in the latest Eocene, little information is available concerning biomass changes as opposed to taxonomic changes at any of the described extinction horizons. Hut et al (1987) have suggested that the end-Eocene events are an example of successive impacts arising from a cometary shower. Montanari (1988) points out that multidisciplinary stratigraphic information from disconnected and incomplete cores may constitute a major obstacle in attempts to resolve cause and effect relationships between extraterrestrial impacts and biological crises of the terminal Eocene.

Triassic-Jurassic (T_R-*J*)

BIOSTRATIGRAPHY The Triassic System may be considered distinct from both the Paleozoic and the rest of the Mesozoic in much of its faunal development (Boucot 1989). The top of the Triassic marks a major change—the disappearance of a distinctive and highly diverse fauna, replaced by (initially) a greatly reduced biota that developed into the rich molluscan and ammonite faunas of the Jurassic.

The terminology for the highest part of the Triassic is a subject of controversy (Tozer 1984, 1986a, 1988, Ager 1987, 1988), although there is currently little disagreement on where the top of the system should be defined. A major biomass reduction associated with taxonomic extinction occurs at the top of the highest Triassic ammonoid zone of *Choristoceras marshi* or *C. crickmayi* of the Norian Stage (or Rhaetian Substage). The event sees the end of the distinctive Triassic ammonoid fauna, conulariids, and conodonts and a reduction of brachiopod genera and many mollusks, although they were the least affected of benthic organisms (Hallam 1988, Boucot 1989). In most sections, the horizon of the event appears to represent a time-break, and the characteristic ammonoids of the latest Triassic and earliest Jurassic rarely occur in close juxtaposition. A continuous sequence may occur in the Gabbs Valley Range in Nevada, where Muller & Ferguson (1939) first reported the Jurassic ammonite genus *Psiloceras* overlying *Choristoceras*. Subsequently, Guex (1982) has closed the gap to within 1 m in the same section. In the northern Calcareous Alps in Austria, near Zinkenbach, a major biomass loss occurs between the bioclastic Kossen Formation (the typical Rhaetic facies) and the overlying dark gray calcareous shales characterized by Jurassic nannofossils of the "Preplanorbis" beds (Golebiowski & Braunstein 1988).

In the Late Triassic in eastern North America, extending from North Carolina to Nova Scotia, a sequence of lacustrine sediments exhibits sedimentary cycles controlled by orbitally induced climate change that allows accurate time discrimination (Olsen 1986). The sequence carries a rich vertebrate fauna throughout. A major and apparently sudden extinction of many forms of reptiles is reported at or near the top of the Triassic, and this is succeeded by a residual fauna of small survivors that occurs in the earliest Jurassic (Olsen et al 1987, Olsen & Cornet 1988; discussed in Padian 1988, Olsen et al 1988).

GEOCHEMISTRY Geochemical and isotopic profiles across the T_R-J boundary in two classical European sections (i.e. Kendelbach, Austria, and St. Audries Bay, England) display anomalous Ir, Fe, P, Y, Zr, La, Yb, Zn, As, and Sb values at the boundary and a marked and sympathetic decrease of $\delta^{13}C$ and $\delta^{18}O$ values in boundary sediments (Hallam et al 1989). The phosphatic nodules at the T_R-J boundary in the St. Audries section have a maximum Ir value of 0.40 ppb. The enrichment of Ir and of rare earth and chalcophile elements corresponds to the boundary unit containing phosphatic nodules and reworked limestone clasts in both sections. The facies changes across the boundary beds are consistent with a brief episode of regional sedimentary reworking and erosion at the time of extinction, followed by a major facies change to carbonaceous shales. This stage of reworking probably accounts for the concentration of relatively heavy apatite nodules. These features have been interpreted by Hallam (1981) as evidence for a marked regression, followed by a major transgression and a change to anoxic conditions. Hallam (1988) has suggested that the flooding of the platforms with anoxic waters was the most likely cause of the T_R-J extinction. The enrichment of elements at the extinction boundary would then be due to normal marine processes.

CAUSE It is possible that a brief erosional episode at the Triassic-Jurassic boundary was not caused by a sudden sea-level drop but by a tsunami that eroded and reworked sediments deposited in relatively shallow-water shelf environments. Such an impact-generated erosional event is consistent with the brief "regression" at the boundary. The facies change to carbonaceous shale at the end of the Triassic, therefore, may not be due to a transgression but rather to an environmental change resulting from conditions of low productivity and ocean anoxia [i.e. a "Strangelove" ocean (Hsü & Mackenzie 1985)] following a mass killing event. A shift to negative $\delta^{13}C$ values in oldest Jurassic calcareous shales of the Grenzmergel Formation, Kendelbach, Austria (Hallam et al 1989), is consistent with biomass reduction as a cause, although there are problems with the possible addition of isotopically light diagenetic carbonate. The chalcophile element and sulfide contents of the postextinction shale probably reflect both increased organic sedimentation and less biogenic recycling of organic matter following the mass killing.

Rampino & Stothers (1988) have shown a general correlation between mass extinctions and periods of major flood basalt volcanism. At the T_R-J boundary in the Newark Supergroup of eastern North America, episodes of mafic volcanism either straddle the Triassic boundary or are entirely Early Jurassic (Olsen et al 1987). As a result, the cause and effect relationship between this extinction and volcanism remains unclear.

Conditions at the time of the Late Triassic mass killing were not unfavorable climatically, and there was little orogenic or volcanic activity, although there was a regression at the close of the system. The killing represents a synchronous global event that may be found in any fossiliferous marine section. It occurs in many diverse facies, and the missing time interval in most sections is not great. The killing horizon is paralleled by a taxonomic extinction affecting a variety of different communities and including a major change in terrestrial vertebrates. An impact may provide a single unifying ultimate cause for an event that is highly variable in its development across the world.

Geochemical evidence for a bolide impact coincident with the T_R-J extinction is equivocal because of the low magnitude of the Ir anomaly and the lack of enrichment of associated siderophile group elements. The T_R-J boundary sections analyzed by Hallam et al (1989), however, represent a shallow-water environment and clearly have been reworked. By analogy with low-intensity Ir anomalies associated with reworked K-T boundary sediments deposited in shelf environments (Jones et al 1987, Bourgeois et al 1988), similar low-level Ir abundances for T_R-J boundary units should not therefore be used to exclude an extraterrestrial Ir source. The extinction horizon is of similar age to the large (70 km in diameter) Manicouagan impact structure in northern Quebec. Unfortunately, the radiometric ages of pseudotachylite from the crater and age estimates of the T_R-J boundary are not sufficiently precise to demonstrate temporal equivalence and therefore causal relationship. What is required is careful examination of continuous boundary sections for evidence of Manicouagan ejecta.

Permian-Triassic (P-T_R)

BIOSTRATIGRAPHY (FIGURE 2) The largest biomass and taxonomic extinction in the Phanerozoic occurred during the latest Permian. This represented a huge reduction in biomass involving marine organisms as well as terrestrial vertebrates and plants. The reduction was responsible for the loss of some 90–96% of marine invertebrate species (Sepkoski 1986), all but one of 90 genera of reptiles, and the Gondwana *Glossopteris* flora (Olson 1988). Although a rundown of diversity is evident in the latest Permian, there appears to have been a final, sudden disappearance of almost all the remaining biomass in at least the marine succession. In the

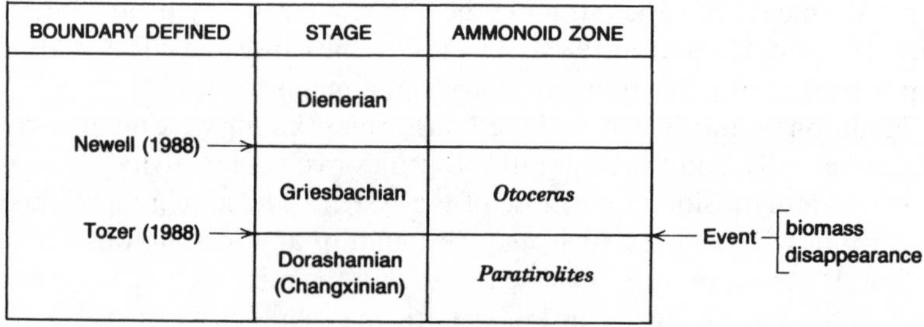

Figure 2 Permian-Triassic boundary—terminology.

rock record, this biomass loss is observable at a clearly defined horizon commonly associated with a change in lithology. The succeeding Lower Triassic is depauperate, with low diversity, and is marked by the gradual successive appearance of new forms during the first few stages. The faunal changes across the boundary have recently been summarized by Erwin (1989).

There is, nevertheless, disagreement on where to define the boundary between the eras. Tozer (1986a, 1988) has summarized the global development of boundary beds. In Arctic Canada, Greenland, Siberia, Iran, and several regions of the Himalayas, he considers that the boundary coincides with a horizon representing a period of nondeposition and doubts if there is a section preserved in the world where sedimentation was continuous from the Permian into the Triassic. At Meishan in Zhejian Province and elsewhere in South China, the top of the limestone Changxing Formation contains the last of many Permian invertebrates, which end abruptly with bioturbation and a sudden loss of biomass. It is succeeded by 30 cm of mudstone and limestone containing relict Permian brachiopod forms and some Triassic bivalves and brachiopods (Sheng et al 1987, Xu et al 1989). A huge biomass reduction is also noted on the Yangzi platform in Sichuan (Reinhardt 1988). Detailed study of the boundary at San Antonio in the Dolomites in Italy shows that it is marked by the disappearance of Paleozoic organisms, by cessation of bioturbation, and by shifts in carbon and sulfur isotope ratios. The cycle within which this change takes place is estimated to be a few $\times$ 10,000 yr (Brandner 1988). A richly fossiliferous section has been described in southern Tibet in which the Paleozoic fauna is overlain by beds bearing Griesbachian ammonoids (Wang et al 1988). The boundary is drawn at the base of beds with species of the genus *Otoceras*, which overlies the biomass extinction.

Newell (1988), a long-time worker on the problem of the P-T_R boundary, rejects Tozer's (1986b) choice of boundary horizon. He cites rare relict Permian brachiopod faunas in beds with the ceratite *Otoceras*, which he points out represents the last member of the Permian family. He would include the Griesbachian Stage in the Permian.

Irrespective of boundary definition, the main biomass change in global sections of the boundary interval occurs in beds with or equivalent to *Otoceras*. The universal unconformity below the *Otoceras* zone coupled with the quantitatively huge biomass loss must signify a single worldwide geological event. The biostratigraphic definition of the boundary in some sections by first appearances of Triassic taxa obscures the sedimentary event marked by loss of biomass (e.g. Brandner 1988). As is discussed below, the biomass loss is correlated with isotopic events that suggest a common cause.

GEOCHEMISTRY The Permian-Triassic boundary is characterized worldwide by shifts in the ratios of $^{13}C/^{12}C$, $^{34}S/^{32}S$, and $^{87}Sr/^{86}Sr$ in marine sections. Values of $\delta^{13}C$ for carbonates in boundary sections from South China (Xu et al 1986), the southern Alps, Italy (Magaritz et al 1988, Brandner 1988), the Carnic Alps, Austria (Holser et al 1989), India (Baud & Magaritz 1988), and the Transcaucasus of Iran (Holser & Magaritz 1987, and references therein) are positive in the Late Permian (+2‰ to +6‰) and decrease sharply at the boundary. The behavior of carbon isotopes before the boundary is variable. In carbonate sections from the Carnic Alps (Holser et al 1989) and South China (Xu et al 1986), for example, $\delta^{13}C$ values decrease gradually toward the extinction event, at which the rate of decrease accelerates. Values of $\delta^{13}C$ from sections in the Southern Alps (Magaritz et al 1988) and the Transcaucasus of Iran (Holser & Magaritz 1987) are more uniform below the extinction boundary.

Highly generalized secular profiles of $\delta^{34}S$ in evaporites (Claypoole et al 1980) and $^{87}Sr/^{86}Sr$ ratios in carbonates (Burke et al 1982) decrease upward in the Late Permian and then increase sharply above the Permian-Triassic boundary. Large sampling intervals and poor biostratigraphic control for samples used in these studies, however, preclude an evaluation of their significance with regard to events of short duration, such as a sudden mass extinction. A more recent $\delta^{34}S$ curve for evaporites spanning the P-T_R boundary in the Dolomites, southern Alps, Italy (Brandner 1988), confirm a marked shift to positive values above the boundary.

Holser et al (1989) have argued that the fixing of organic matter in swamps and forests of paralic basins and on shallow marine shelves probably accounts for the positive $\delta^{13}C$ values during the Late Permian. The oxidation of this organic matter during a major Late Permian regression would then cause a major decrease in $\delta^{13}C$ values in carbonates. An alternative interpretation involves the removal of isotopically light carbon from the ocean reservoir by sedimentation under anoxic conditions. Toxic bottom-water conditions would limit the recycling of carbon by bottom-dwelling organisms. A sudden shift to negative $\delta^{13}C$ values at the boundary could then result from the combined effects of vertical mixing of isotopically light anoxic waters with surface waters and a decrease in the rate of carbon removal by organisms due to a major biomass reduction.

In a marine anoxic environment, the sedimentation of isotopically light iron sulfides results in a shift to increasingly positive $\delta^{34}S$ values in residual sulfate (Goodfellow 1987). Vertical mixing of this isotopically heavy sulfur accumulated in a lower anoxic water column with surface waters during ocean turnover should result in an increase in $\delta^{34}S$ values. Near the P-T_R boundary, there is a marked increase in $^{34}S/^{32}S$ ratios, but this appears to postdate the extinction event (Holser & Magaritz 1987, Brandner 1988).

CAUSE Irrespective of boundary definition, the coincidence of isotope changes and biomass loss allows the following observations. The extinction of marine organisms in the Late Permian coincided with a major regression (Holser & Magaritz 1987), the termination of a major glacial period (Caputo & Crowell 1985), and a major shift to less positive $\delta^{13}C$ values. This event was then followed by transgression and the deposition of reduced sediments with high contents of chalcophile elements (Holser & Magaritz 1987, Holser et al 1989). Holser et al (1989) described weakly anomalous Ir and other metals in extinction boundary sediments from a drill-hole section from the Carnic Alps, Austria, marking a change from oxidized to reduced conditions. Nonchondritic Ir values and Co/Ir ratios have been used as evidence against an extraterrestrial Ir source. The authors have further argued that the metals were most likely precipitated as sulfides during reaction of dissolved metals with H_2S that was generated by sulfate-reducing bacteria feeding on decaying organic matter.

The unifying factor in discussing causes for the great killing is the fact that it is always found at a bedding-plane horizon that appears to be correlatable worldwide. Climatic changes combined with sea-level changes may well be responsible for the great variation in diversity during much of the Permian. The final killing, however, that occurred at the end of the era would appear to be sudden and drastic. It was also global. Thus, if one sums disappearances observed at each of the relatively few well-studied sections, it is legitimate to invoke a single cause that is independent of local facies, faunal province, or tectonic event. It evidently acted on land and in the sea, and it affected terrestrial animal and plant life. The coup de grace at the end of the Paleozoic was an unusually energetic event superimposed on a biota already deeply stressed environmentally. An extraterrestrial impact is very probably the ultimate cause of the Permian-Triassic killing event.

Devonian-Carboniferous (D-C)

BIOSTRATIGRAPHY The boundary between the Devonian and Carboniferous Systems is generally accepted to be the horizon that saw the disappearance of the abundant ammonoid faunas of the highest Famennian Stage, characterized by the genus *Wochlumeria* (Paproth & Streel 1985, Teichert 1988). Only one or two genera survived, and the earliest Carboniferous rocks contain few ammonoids, characterized by *Gattendorfia* and *Imitoceras* (Ramsbottom & Saunders 1985). There was a similar loss of diversity and biomass of trilobites and a reduction in the number of conodonts. The horizon of the biomass loss appears to be at the base of the *Siphonodella sulcata* or in the uppermost part of *S. praesulcata* conodont zones (Ziegler & Sandberg 1984). In Europe, this horizon is known as the Hangenberg event after a euxinic black shale formation of that name in the Rheinisches Schiefergebierge. The horizon is correlatable globally and is commonly associated with a black shale facies (Walliser 1984, House 1985, Streel 1986, Schönlaub et al 1988).

The sequence across the boundary in South China is similar to the Hangenberg Shale succession in Germany, both in lithology and fauna (Hou et al 1985, Bai & Ning 1988, Xu et al 1989). This may be due to the major eustatic fall that began in the middle *S. praesulcata* conodont zone and ended the Famennian (Johnson & Sandberg 1988), and to the fact that the Famennian is characterized by low diversity and an almost completely cosmopolitan fauna of both benthic and pelagic organisms (Boucot 1989). The biomass loss was large, and the taxonomic change, while almost total in the ammonoids, was less among benthic organisms, many of which disappeared at the killing horizon to reappear during the Early Carboniferous—i.e. the Lazarus effect (Jablonski 1986). A major change in spore assemblages appears to have been coincident with eustatic rise early in the *S. sulcata* conodont zone (Streel 1986).

GEOCHEMISTRY Four Ir anomalies in an extremely condensed sequence spanning three stages of Mississippian time in a 3-m interval of shale and carbonate deposited on the southern midcontinental cratonic region of the United States have been described by Orth et al (1988). The lowermost anomaly occurs about 50 cm above the first appearance of *Siphonodella sulcata*, near the top of the Woodford Shale formation. Younger anomalies extend several conodont zones above this extinction boundary. Associated with some Ir anomalies are elevated contents of Pt, Os, and Au. The lowermost anomaly in the Woodford Shale formation contains, in addition to Ir, high abundances of V, Ni, U, Cu, Zn, As, Se, Mo, Ag, Sb, and Hg. Overlying the Woodford Shale, the Welden Limestone contains two Ir anomalies, the upper enriched in Co, As, and Ni. The uppermost anomaly occurs within and immediately below a greenish-gray clay bed of the lower Caney Shale, also with elevated contents of Co, As, and Ni.

The origin of these elemental anomalies is not clearly understood. On the basis of nonchondritic platinum group element (PGE) ratios and the lack of shocked minerals and spheroids, Orth et al (1988) have suggested a number of possible mechanisms: direct precipitation from seawater in contact with organic- and sulfide-rich black shales for the lowermost anomaly, accumulation by bacteria for the two anomalies in the Welden Limestone, and erosion from an ultramafic source for the uppermost anomaly. Although the authors do not completely rule out an extraterrestrial Ir source, they consider that some undetermined terrestrial enrichment process is a more plausible alternative.

There are a number of concerns regarding the criteria used to distinguish extraterrestrial from terrestrial Ir. Siderophile group element ratios at the K-T boundary and ejecta layers associated with impact craters (e.g. Lake Acraman) are highly variable due to secondary remobilization and sedimentary reworking (Gostin et al 1989, Wallace et al 1990). This variability can be seen in anomalous samples from the Lower Mississippian, which display a large range of PGE/Ir ratios that overlap (with the exception of Pt/Ir) chondritic ratios (Orth et al 1988). The absence of shocked quartz and cosmic spheroids at older extinction boundaries must be interpreted with caution due to dilution by normal quartz during sedimentary reworking, particularly in relatively shallow-water shelf environments, and to overprinting effects of quartz dissolution and recrystallization during diagenesis and metamorphism.

CAUSE Several causes for the D-C event have been suggested. Most involve a sudden fall of sea level linked with a euxinic environment (e.g. Walliser 1984, Johnson & Sandberg 1988). Many kinds of organisms that disappeared, however, might be considered least likely to have been affected by loss of shelf area and development of anoxic basins. Shelly benthos, including brachiopods and the newly established Famennian coral fauna, survived as the ancestors of the Carboniferous benthos. Separating cause from effect may prove difficult, but an ultimate cause by an impact event must be considered and has been suggested (Bai & Ning 1988, Xu et al 1989).

The possibility remains that the siderophile group element anomalies are of meteoritic origin but have been extensively modified by sedimentary reworking and differential mobilization of PGE. Such processes would account for the wide range of siderophile group element ratios and highly concentrated chalcophile elements. The enrichment of chalcophile elements may be a secondary response to high-organic-matter contents and accelerated bacterial H_2S production following mass extinction. Multiple Ir anomalies over 3 m of section may therefore reflect differential diagenetic remobilization of PGEs or the prolonged input of Ir by streams and rivers, which have reworked unconsolidated continental ejecta deposits. The effects of reworking of Ir-rich ejecta are expected to be more pronounced in small restricted intracratonic basins than in open ocean settings.

Frasnian-Famennian (F-F)

BIOSTRATIGRAPHY (FIGURE 3)

Extinction event It has long been known that there was a massive biotic change at about the horizon of the boundary between the Frasnian and

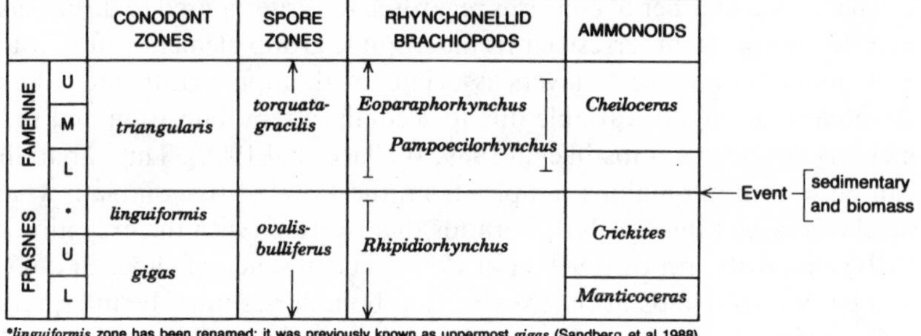

Figure 3 Frasnian-Famennian (Upper Devonian) boundary—terminology.

Famennian Stages of the Upper Devonian. McLaren (1970), after mapping the horizon over wide regions of Alberta and the Northwest Territories, suggested that the event was a sudden and huge killing recognizable in many facies on several continents, and that it was probably synchronous. An examination of the kinds of organisms that disappeared made it difficult to find an obvious common cause. Because of the global nature of the mass killing, McLaren suggested that it was the result of an oceanic impact by a large asteroid, citing an article by Dietz on the effects of large impacts on the planet (Dietz 1961, McLaren 1959, 1970). He also suggested that the horizon might be at the top or about the top of the *Manticoceras* ammonoid zone, or near the base of the *Palmatolepis triangularis* conodont zone, and that the event offered the possibility of being an accurate horizon correlatable globally and therefore an ideal criterion for boundary definition. McLaren (1982) subsequently suggested that the event was probably at or near the boundary between the uppermost *P. gigas* and lower *P. triangularis* conodont zones. The happening at the F-F boundary has been the subject of much discussion and research. Recent summaries may be found in McLaren (1983), Walliser (1986), Sandberg et al (1988), Boucot (1989), and the three volumes of *Devonian of the World*, edited by McMillan et al (1988).

During the Frasnian Stage, there were ongoing changes leading to loss in diversity among certain groups. Following the reduction of faunal provinces prominent during the Lower and Middle Devonian, there was no corresponding reduction in biomass of the more cosmopolitan biota that developed. Certainly, the warm-water shelly benthic faunas over much of the world flourished throughout the stage. Becker (1986) suggested a rundown over time of ammonoid taxa, Copper (1986) plotted ranges and changes in number of genera and families of brachiopods and other invertebrate groups and found a taxonomic reduction during the stage, and Stearn (1987) reported a similar decline in genera and families of stromatoporoids. Sorauf & Pedder (1986), on the other hand, documented an increasing coral diversity up to the top of the Frasnian Stage. Plotting taxa against intervals, however, does not lead to the easy identification of a sudden event (McLaren 1986, 1988, Teichert 1988).

The extinction event, wherever found, is a sudden happening. In the record—in basins, shelf margins, and midshelf and shallow-water environments—there is a sudden loss of biomass. This is commonly associated with a change in depositional regime, and in shallow-water environments there may be time missing in the rock record. Before the event, an abundant biota was frequently preserved in the sediments, including forms representing benthic and pelagic organisms. In the bed above the killing horizon there is an interval with little or no organic remains. Those that do occur may represent derived fossils as clasts from the earlier formation, survivors of preevent taxa that may persist or die out, and occasional bursts of new species forming an oligotaxic pulse that may disappear as other biota become established. The lower Famennian was extremely impoverished, in sharp contrast to the Frasnian (Boucot 1989).

Microfaunas were either less affected or recovered more quickly than larger forms. The conodonts underwent a sudden change in relative proportions of taxa that reflected a changing environment, but otherwise they continued unchanged and are important for time correlation across the killing horizon globally. The suddenness of the biomass loss is borne out by the sedimentary history of the latest Frasnian and early Famennian, traced in detail by Sandberg et al (1988) using conodonts.

The event horizon in any one stratigraphic section is frequently obscured by several factors: erosion or nondeposition of the underlying beds, nondeposition of the earliest postevent sediments, or failure to find diagnostic fossils in the lower postevent beds. Summing the events globally at many sedimentologically independent sections nevertheless allows the hiatus to be narrowed to considerably less than the time span of a conodont biozone (0.5 Myr; McLaren 1988). Recent work on conodont zonation of sequences that may be continuous has allowed accurate determination, consistent with the global evidence, that the killing event took place very near the end of the *P. linguiformis* conodont zone (renamed from the zone of uppermost *P. gigas*), shortly before the lower *P. triangularis* zone (Sandberg et al 1988). This horizon coincides with the upper Kellwasser Limestone of Germany, which has long been identified as the late Frasnian "Kellwasser event" and which corresponds to a black shale in a limestone sequence (House 1985, Walliser 1986).

Global development In Canada the boundary is well marked in the Alberta Rocky Mountains by erosion of carbonate banks and major slumping into cratonic and marginal basins (McLaren & Mountjoy 1962). In the Northwest Territories, the highest Frasnian westward from Great Slave Lake includes many reefs (the Kakisa Formation) with abundant stromatoporoids, corals, brachiopods, and trilobites. The killing event is at a bedding plane, and the overlying beds, across many facies, are depauperate (McLaren 1959, 1988, Belyea & McLaren 1962, Pedder 1982, Sorauf & Pedder 1986, Geldsetzer 1988, Goodfellow et al 1988). Terrestrial floras were also affected by the event, and a major abrupt change takes place between two spore zones within the Old Red Sandstone continent and in the adjoining seas in northern Canada, New York, and eastward to Europe and European USSR, coincident with the base of the lower *P. triangularis* zone (Richardson & McGregor 1986). The extinction horizon at Trout River, Northwest Territories, is enriched in chalcophile elements and records a sudden decrease of $\delta^{13}C$ values in carbonates (Goodfellow et al 1988).

Sandberg et al (1988) and Johnson & Sandberg (1988) describe an event in the western United States representing several different paleotectonic settings, from carbonate platform to deep basin. The sudden killing is accompanied by drastic eustatic and sedimentary events, including slumping, debris flows, and tsunami deposits. Sandberg et al (1988) describe a similar event in Germany, Belgium, and Morocco and construct a sedimentological model of the extinction involving eustatic changes that culminate in major storms and a sudden killing of most of the biomass. They suggest a bolide impact as the most probable ultimate cause. In sections at Steinbruch Schmidt in West Germany and Devil's Gate in Nevada, they show that the mass extinction took place in far less time than a few tens of thousands of years and was probably instantaneous. An Ir anomaly has not been found in any European section at the event horizon (McGhee et al 1986).

In South China, the event is well marked and closely similar to the patterns established in Canada, the USA, Belgium, and Germany. Hou et al (1988) have recently described sections representing basin, slope, platform, and nearshore environments in which the killing event is evident. Other sections in China confirm a correlation with the Kellwasser event (Jia et al 1988). The event is closely associated with a geochemical anomaly and carbon isotope shift between the *P. linguiformis* and lower *P. triangularis* conodont zones (H. Hou, personal communication, 1989). Geochemical anomalies reported from three horizons within the upper *P. triangularis* and *P. crepida* zone interval postdate the main biomass extinction (Wang & Bai 1988).

In the Canning Basin, Western Australia, Playford (1980) has described Middle and Upper Devonian reef complexes and basin sequences. The end-Frasnian killing event is well marked at a bedding plane throughout the reef front and platform, with an almost complete disappearance of abundant biomass, and it is succeeded by depauperate pelletoid and algal limestones. A section at McWhae Ridge in carbonates of lower marginal slope facies yielded a geochemical anomaly, including Ir associated with a cyanophyte bacterium (Playford et al 1984, McLaren 1985). This horizon was reported to be at the base of the upper *P. triangularis* conodont zone (Glenister & Klapper 1966, Druce 1976). The shelf extinction was believed to be equivalent to the horizon of the anomaly. This may be unlikely, however, in view of the additional evidence from China, which confirms yet again the Kellwasser horizon for the event globally. The McWhae Ridge anomaly may be independent of the platform killing event (cf. Wang & Bai 1988) or, alternatively, the conodont evidence may be open to other interpretations (McLaren 1988).

Recent work has called into question past differentiation of the lower and middle *P. triangularis* zones (Sandberg et al 1988). Further refinement of conodont work may adjust certain apparently anomalous occurrences of normal Frasnian faunas in beds reported as Famennian, e.g. in northwest Spain (van Loevezijn et al 1986) and in Eastern Europe and the USSR (Kalvoda 1986). It is evident, however, as with all extinctions, that there are some survivors of Frasnian-type faunas. In Moravia, Czechoslovakia, reefs of *P. crepida* zone age contain a low-diversity fauna of Frasnian type consisting of stromatoporoids, foraminifera, and tabulate corals, together with three forms of shallow-water corals (Friakova et al 1985, Galle et al 1988). Stearn (1987) described changes in stromatoporoid taxa across the boundary. He documented a decline in diversity during the Frasnian stage but recognized the huge biomass loss at the boundary. Many surviving forms died out in the lower Famennian, except for the labechiids; the latter group is reestablished in the latest part of the stage. The most striking feature of the early Famennian biota is its rarity and low diversity (Boucot 1989). Although much reduced, enough forms survived to establish gradually the late Famennian fauna, which is essentially Carboniferous in aspect, especially with regard to the corals and brachiopods. The ammonoids were essentially peculiar to the stage and might in fact suggest a separate system.

GEOCHEMISTRY The F-F extinction boundary displays many of the diagnostic characteristics of other mass killing events, including global synchroneity and short duration. A brief interval of sedimentary reworking and erosion is associated with a marked facies change in platformal sections to carbonaceous shales. These shales formed during a period of low biological productivity and anoxic conditions. Also associated with the interval are anomalous Ir and chalcophile elements and a decrease in $\delta^{13}C$ and $\delta^{18}O$ values due, respectively, to biomass loss and ocean warming (Goodfellow et al 1988). An ocean impact would rework sediments in shelf environments and form high-energy sedimentary deposits; tsunami deposits at the F-F boundary have been described at the Devil's Gate section (Sandberg et al 1988). However, some areas, such as the Jasper Basin, show only minor agitation of sediments, probably due to geographic isolation (Geldsetzer et al 1987). It is difficult to differentiate at any one locality between tsunami sediments generated by impact or terrestrial processes. Two important distinguishing characteristics of an impact event are its global extent and short duration, and these have been demonstrated for the F-F extinction in sections distributed worldwide (Goodfellow et al 1988).

The anomalous Ir, with siderophile group elements, referred to previously from the Canning Basin may not be correlated with the extinction boundary. The Ir content is low by comparison with peak anomalies measured for the K-T boundary and is concentrated in several beds containing the iron-secreting cyanobacterium "*Frutexites*" that span up to 60 cm of section (C. Orth, personal communication). The Ir may have been concentrated by cyanobacteria under unusual anoxic conditions. Its significance remains an open question.

CAUSE When reviewed globally, it is clear that there may be many immediate causes for the killing event. Much of the paleontological literature, based on plotting taxa against time, emphasizes the rundown in diversity during the Frasnian Stage. At least some of this rundown may be due to artificial range truncation—the Signor & Lipps (1982) effect, which is particularly marked when observations are made at few sections that may not be sedimentologically independent (McLaren 1988). The claim is made that the event took place over a time period of a million or more years, or even that there was no event at all (e.g. Copper 1986, Becker 1986, Farsan 1986, McGhee 1988, Becker et al 1989). Emphasis has been placed on extinction rates (i.e. disappearance of taxa over time), irrespective of biomass loss. McGhee (1988) plotted extinction rates against a variable estimated interval of 1 or 2 Myr with data from New York, the Appalachians, and the southern Urals. In fact, the killing event is manifest at a single bedding plane by a partial or total biomass loss in individual sections, irrespective of previous changes in diversity and taxa (McLaren 1988, Sandberg et al 1988). House (1985) considers the event to have been caused by a major transgression and black shale deposition, while others claim a major regression at the horizon (Johnson 1974, Johnson & Sandberg

1988). Stanley (1988) and others have claimed climatic cooling for the late Frasnian killing, and this has been discussed by Johnson (1974), Geldsetzer et al (1988), and Thompson & Newton (1988), who see no evidence for cooling in the early Famennian. There seems little evidence for sudden global cooling at the event horizon (Heckel & Witzke 1979, Boucot & Gray 1982, Boucot 1988). Warm-water pelletoid algal limestones succeed the event in many shallow shelf areas (e.g. Canning Basin, southeast China, Germany, Nevada, and Alberta). The much discussed Famennian glaciation in South America probably developed later in the stage (Geldsetzer et al 1988). Turnover of an anoxic ocean has been suggested as a general mechanism for black shale depositional episodes and extinction of marine organisms (Berry & Wilde 1978). An anoxic pulse in cratonic and marginal basins and carbonate shelves in western Canada at the end of the Frasnian provided a triggering mechanism for mass killing (Geldsetzer et al 1987; discussed in Wilde & Berry 1988, Geldsetzer et al 1988).

In summary, there is evidence from four continents for a globally synchronous F-F mass killing event that corresponds to a brief episode of sedimentary reworking and erosion (interpreted as a regression) and a facies change to carbonaceous shales (interpreted as a transgression) in platformal areas. Furthermore, the mass killing was followed in the earliest Famennian by a period of low productivity and reduced biological diversity, referred to as "Strangelove" ocean conditions (Hsü & Mackenzie 1985). Increased preservation of organic matter resulting from the termination of carbon-recycling organisms probably explains the enrichment of chalcophile elements at the extinction boundary in several sections (Goodfellow et al 1988) and a marked shift to positive $\delta^{34}S$ values in sedimentary pyrite (Geldsetzer et al 1987) and evaporites (Claypoole et al 1980). A similar isotopic shift has been described for the Late Ordovician extinction. The mass killing of marine organisms and the establishment of anoxic conditions facilitated the bacterial reduction of sulfate to sulfide. Because of the preferential reduction of ^{32}S by sulfate-reducing bacteria, $\delta^{34}S$ values for dissolved sulfate and sulfide would become progressively more positive as the fraction of reduced sulfate increased.

An examination of the many possible immediate causes for the global F-F event suggests either a large number of individual events that coincidentally occurred at about the same time or a single ultimate cause that was responsible for many of the suggested immediate causes. We return to the three prime possibilities: mantle-induced causes (e.g. volcanicity and plate movements), variations in the solar system (e.g. orbital and solar variations), or impact of an asteroid or comet. An impact explains the global observations of biomass loss and subsequent developments in every case.

Late Ordovician

BIOSTRATIGRAPHY (FIGURE 4) The base of the Silurian System has been defined by the Working Group on the Ordovician-Silurian boundary at a type section near Moffat in Scotland at an horizon corresponding to the base of the graptolite biozone of *Akidograptus acuminatus* (Bassett 1985), and a global analysis of boundary beds has been published (Cocks & Rickards 1988). The succession downward from the boundary horizon includes the graptolite biozone *Glyptograptus persculptus*, which in turn overlies a global event horizon, commonly coincident with the top of the Hirnantian Stage of the Ashgillian Series and its equivalents. There is a biomass loss at the top of the Hirnantian that is paralleled by a taxonomic extinction of variable proportions. This is commonly overlain by a diastem or by a poorly fossiliferous interval. Beds of the *G. persculptus* zone are transgressive and initiate a new sedimentary cycle.

Brenchley (1988) has summarized the changes that take place at the Hirnantian–*G. persculptus* horizon in Britain, Scandinavia, and China, where shallow or midshelf clastics or carbonates with shelly faunas are overlain abruptly by graptolitic shales. In the North America midcontinent, few graptolites are found, and beds with Hirnantian-type faunas are overlain by Llandovery fossils with a varying degree of time missing (Bergström & Boucot 1988). Where they have been identified, *G. persculptus* zone beds form the base of the Silurian transgression in the Americas and frequently rest on an erosion surface [e.g. in the northern Yukon (Lenz & McCracken 1988) and Argentina (Cuerda et al 1988)].

Although the Hirnantian commonly underlies the *G. persculptus* zone, there are evidently some exceptions. Apollonov et al (1988) reported Hir-

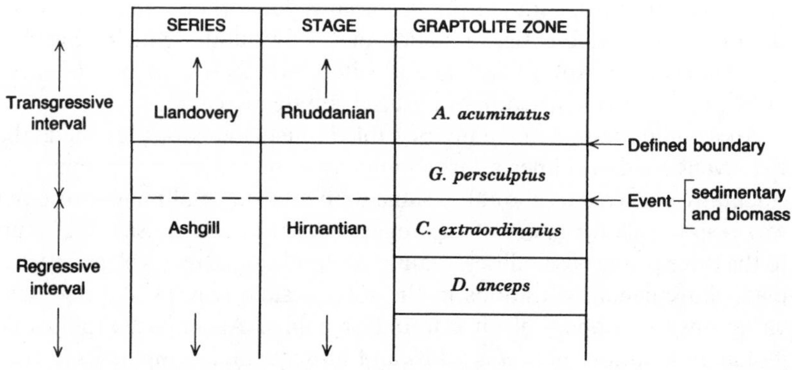

Figure 4 Ordovician-Silurian boundary—terminology.

nantian faunas in the *G. persculptus* zone in southern Kazakhstan, and Cocks (1988) cites other examples. Rong (1984) and Mu (1988) questioned the Kazakhstan interpretation and pointed out that there are no exceptions to the pre–*G. persculptus* age of the Hirnantian in all of China. On Anticosti Island in eastern Canada, Barnes (1988) reported a major faunal turnover in graptolites and conodonts in beds believed to be the equivalent of the upper *G. persculptus* in member 7 of the Ellis Bay Formation. The faunal change and biomass loss includes palynomorphs, aulacerid stromatoporoids, brachiopods, and trilobites and takes place near the horizon of an oncolite platform bed.

There are, thus, unresolved problems of correlation. The biomass loss, however, is sudden and over much of the world appears to be pre–*G. persculptus*. The occurrence of Hirnantian faunas in the *G. persculptus* zone may be due to one or more possibilities: (*a*) survivors from the killing event, (*b*) reworking of previously deposited material, (*c*) misdating of the associated fauna, and (*d*) miscorrelation between disconnected outcrops.

GEOCHEMISTRY Profiles of $\delta^{13}C$ and $\delta^{18}O$ values in carbonates from a section on Anticosti Island spanning the Late Ordovician extinction display a marked decrease at the extinction horizon, followed upward by an increase to values more positive than in preextinction carbonates (Orth et al 1986). A similar decrease in both carbon and oxygen isotope values has been observed in extinction horizon samples from sections in northwestern Canada (Nowlan et al 1988). The sharp negative deflection in $\delta^{13}C$ and $\delta^{18}O$ values at the time of the extinction event is consistent with a major biomass loss and a warming trend, respectively. An increase to more positive $\delta^{13}C$ values following the extinction probably reflects the mixing of isotopically heavier carbon from lower in the water column with surface water during ocean turnover. Similar trends to more positive $\delta^{13}C$ values following major biomass loss have been clearly documented for the K-T extinction (Zachos & Arthur 1989), although the increase of $\delta^{13}C$ values following the Ordovician extinction is greater, perhaps because of steeper vertical isotopic gradients in a previously stratified water column.

Six sections across the extinction horizon in northwestern Canada display a marked increase in organic carbon and a large suite of trace elements, including V, U, Ni, Se, Mo, As, Sb, Hg, Zn, Pb, and Cd (Nowlan et al 1988). The chalcophile elements are likely bound in pyrite, which shows an increase in $\delta^{34}S$ values immediately above the mass extinction (Goodfellow 1987). In the Anticosti Island section, Orth et al (1986) observed a weak Ir anomaly at the extinction horizon after normalizing the data to a carbonate-free basis, but they attributed this increase to sources other than bolides because of the lack of association with a chondritic suite of siderophile elements and the absence of detectable shocked quartz or spherules. In a similar study of the type Ordovician-Silurian boundary section at Dob's Linn, Scotland, Ir and Cr were shown to reach maximum values of about 0.25 ppb and 250 ppm, respectively, in the *Climacograptus extraordinarius* graptolite zone representing the extinction event (Wilde et al 1986). These anomalies near the extinction horizon were attributed to erosion of obducted deep crustal material, although it is not made clear why erosion of ophiolitic rocks should reach a maximum at this time.

An alternative interpretation is that the Ir is of extraterrestrial origin but has been redistributed and perhaps separated from other siderophile elements by sedimentary reworking and diagenetic remobilization. Sedimentary reworking at this extinction horizon has been observed at the Anticosti Island section (A. D. McCracken, personal communication) and may explain the low Ir content and the absence of detectable shocked quartz and spherules.

CAUSE Unlike the K-T boundary which is represented in most ocean basins, studies of older extinctions must rely on sections from shallow-water shelf environments, where the chances of preserving the critical evidence is considerably less. Sedimentary reworking results in a more restricted suite of anomalous siderophile group elements. A good deal has been written on the causes of the Late Ordovician biotic changes. A recent review of the current theories has been published by Sheehan (1988), who discusses the prevailing view that because the mass extinctions coincided with extensive glaciation in the African and South American parts of Gondwanaland, a combination of climatic deterioration and glacial-eustatic draining of epicontinental seas led to the eventual killing off of the biota. There was a reduction in diversity during the late Caradoc and early Ashgill Series. Many taxonomic extinctions are ascribed to plate movements and/or to glacial effects that took place before the Hirnantian, during which a series of regressions had little effect on diversity (Fortey 1984, Brenchley 1988). The mass killing event at the end of the Hirnantian marks the initiation of a transgression. This transgression was not a sudden event, and Spjeldnaes (1987) suggests that it was at least three orders of magnitude slower than most recent glacially induced eustatic transgressions.

Throughout the Late Ordovician and Early Silurian there were ongoing changes in diversity and taxonomic extinctions for which climatic and eustatic changes have been suggested as causes. The end-Hirnantian event appears to be global and sudden. It is marked by a facies change and a major biomass reduction. In any one section, it may not be possible to assess the duration of the event or estimate the missing time. Because it may be recognized in a large number of sedimentologically independent sections on five continents, it is reasonable to suggest a common cause and therefore synchroneity (McLaren 1988). It appears likely that the global event at the end of the Hirnantian may be independent of the suggested causes for the general changes in diversity and taxonomic extinctions during the Late Ordovician and Early Silurian.

The extinction was preceded by a glacial period during which time the oceans were stratified with anoxic bottom waters (Berry & Wilde 1978, Goodfellow & Jonasson 1984) and was followed by a major worldwide transgression. A possible cause of this extinction, which also accounts for a sudden and global regression followed by a transgression, is the impact of a major bolide into the ocean. A tsunami associated with an impact at the K-T boundary has resulted in a hiatus or erosion in over 90% of the K-T sections (Kauffman 1979). An impact in Late Ordovician time can be expected to have had similar effects. The destruction of biomass following a major impact would cause global warming resulting from an increase in oceanic and atmospheric CO_2 levels. Hunt (1984) has argued that the termination of a glacial episode would involve unusual mechanisms to warm the polar regions and suggested that a large increase of atmospheric CO_2 was the most likely cause. The warming of the oceans following the extinction would account for the rapid melting of polar ice sheets and transgression. A further consequence of an impact would be the vertical mixing of the water column with anoxic bottom waters and the establishment of Strangelove conditions throughout the entire water column. This might be achieved by the effects of a mass killing (cf. the K-T boundary) or by the combined effects of a mass killing and ocean turnover. The resulting high sedimentation rate of organic matter and the increased preservation of organic sediments might account for the high content of siderophile group and chalcophile elements. The fixation of chalcophile elements by bacterially generated H_2S following a mass killing was discussed earlier for the K-T event.

Precambrian-Cambrian (p€-€)

BIOSTRATIGRAPHY (FIGURE 5) Definition of the Precambrian-Cambrian boundary is still under discussion (Cowie et al 1989). Difficulties include our capacity to correlate intercontinentally, the lack of agreement on "period" or "epoch" terminologies, the unreliability of radiometric age determinations, and the uncertainty in interpretation of geochemical data, particularly PGE and carbon isotope changes.

Early metazoanlike fossils date back to 650–610 Ma (Pflug & Reitz 1986), but their affinities are obscure. They are assigned to the Vendian Series and earlier (Riphean). Small shelly fossils developed slowly during

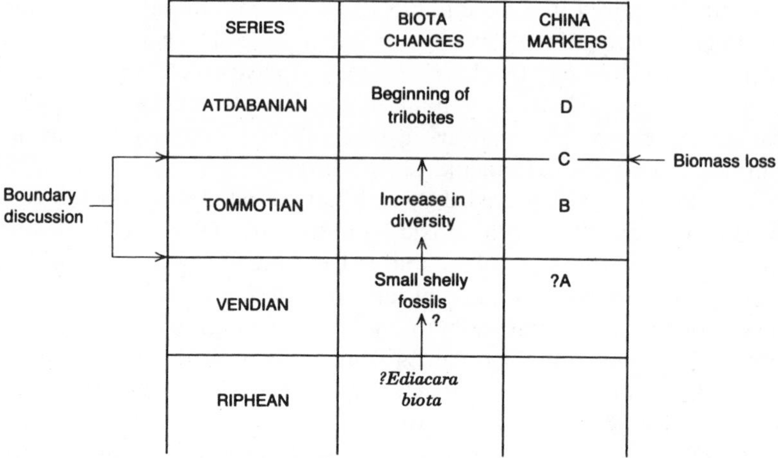

Figure 5 Precambrian-Cambrian boundary—terminology and biota.

an interval of time referred to as the Tommotian (Sokolov & Fedonkin 1986). Opinion appears to favor inclusion of the Tommotian (variously referred to as a series or stage) in that part of the Cambrian that predates the development of trilobites in the Atdabanian. The protometazoan faunas, loosely referred to as Ediacaran, appear to overlap in range with small shelly fossils (e.g. Xu et al 1989), and a boundary between biotas is difficult to define consistently.

Within the succession, however, these are events, although they are not necessarily sudden or easily correlatable. Xu et al (1989) have given an account of this interval in China and summarized events recognized on the subcontinent as China A, B, C, and D. Hsü et al (1985; discussed in Awramik 1986, Hsü 1986) describe a negative shift in δ^{13}C values and a marked increase in Ir, Os, and Au at the China C marker in Yangtze Gorge, southwest China. The China C marker is defined by an abrupt change from phosphatic-rich rock to black shale, above which trilobites appear. Hsü et al (1985) have interpreted this negative carbon isotope perturbation as a Strangelove ocean effect resulting from the catastrophic destruction of the biosphere that preceded the Cambrian radiation. Anomalous Ir at this boundary is interpreted as evidence for a major bolide impact, although it should be noted that reanalyses of the same bed have failed to detect an Ir enrichment (analyses by C. J. Orth reported in Awramik 1986).

GEOCHEMISTRY In a section across the Vendian-Tommotian boundary in Siberia, δ^{13}C values display an upward increasing trend to positive values in the Vendian, a marked decrease at the boundary, and a small increase at the base of the Tommotian before decreasing upward (Magaritz et al 1986). Highly positive values near the top of the Vendian, followed by a marked drop in values, have been interpreted as due to increased productivity accompanying the Cambrian explosive radiation, followed by a sharp biomass loss. The frequent occurrence of phosphate deposits is presented as evidence of a biological explosion. The lack, however, of a clear relationship between short-term carbon isotope changes and phosphogenic episodes indicates that phosphate deposits do not necessarily reflect (at least on a global scale) periods of increased productivity (Goodfellow 1986).

One of the difficulties in interpreting these isotopic trends is the uncertainty regarding when the major Cambrian radiation took place. The China C marker and Vendian-Tommotian boundary, for example, are just two of several p€-€ boundaries being considered by the Boundary Committee of the Commission of Stratigraphy. The question of global synchroneity of any observed event cannot therefore be answered. Without supporting evidence for global catastrophism, such as coincidental PGE anomalies, it is impossible to determine whether isotopic shifts reflect local or global processes.

Ejecta from what is probably one of the world's largest known impact structures, the Lake Acraman crater in South Australia (Gostin et al 1986, Williams 1986, Wallace et al 1990), is reported to underlie stratigraphically an Ediacaran assemblage and thus cannot be associated with the China C horizon. It should, however, be looked for as an influence on earlier biota and geochemistry in the Vendian. The age of the impact is not known precisely and has been bracketed between the Ediacaran "System" and a questionable Rb-Sr age of 593 ± 32 Ma for the Bunyeroo Formation (Compston et al 1987). Evidence for an impact origin includes shocked quartz, feldspar, and zircon; shatter cones; fragments in the ejecta layer up to 30 cm in diameter; and anomalous platinum group elements (Gostin et al 1986, 1989).

The Lake Acraman structure has an outer rim that is 140–160 km in diameter (Gostin et al 1989) and is associated with a 0–40 cm ejecta layer in red shales of the late Precambrian Bunyeroo Formation in the Adelaide geosyncline up to 300 km east of the Lake Acraman impact site (Gostin et al 1986). The ejecta horizon is surrounded by a reduced green envelope and has anomalous contents of Ir, Os, Pt, Au, Cr, Ni, Co, Cu, As, and Sb (Gostin et al 1989). Highly variable siderophile group element ratios (e.g. Ir/Pt, Ir/Au) indicate major remobilization of this elemental suite after deposition. Most of the chalcophile group elements, which reach values greater than 1% in the case of Cu, are not of meteoritic origin but were probably precipitated by reaction with bacterially generated H_2S during diagenesis.

The area of uncertainty in the global correlation of the events during the interval between the first appearance of the Ediacaran biota and the first trilobites is large. Time lines are few, and the signals from well-established events in one continent are not easily extended. The Lake Acraman event—the crater and its ejecta—offers at least the possibility of a globally defined time horizon and indicates the associated geochemical signals that might be expected in stratigraphic sections that span this long interval of time.

SEQUENCE OF EVENTS ACROSS ABRUPT EXTINCTION HORIZONS

A major impact by an extraterrestrial body will suddenly and catastrophically perturb the crustal environment and generate a state of disequilibrium in the biosphere, hydrosphere, and atmosphere. Biological, chemical, and physical reactions following this change of state represent a reestablishment of equilibrium under a new set of environmental conditions and at widely varying rates. In the early stages of chaos immediately following a major event, few reactions approach equilibrium conditions.

In the biosphere, the rate of change is important, since it determines what organisms are likely to survive. Under normal conditions, slow-acting processes (e.g. regression-transgression, glaciation, or tectonism) allow time for organisms to adapt to new environmental conditions. Fast-acting changes, by contrast, exceed the adaptive capacity of most and are therefore the major cause of mass extinction. The slow-acting non-catastrophic model for extinctions, identified by taxa changes, is well established and supported by a vast literature. The biological mechanisms appear to be well understood, although debate on causes includes many hypotheses—mantle movement, solar system changes, climate, sea-level changes, and ocean chemistry—with impacts being assigned a minor role, if any. Out of a very large number of syntheses, Jablonski (1986) and Boucot (1986, 1989) have given a balanced summary of the field. As examples of denials of impact effects, Hoffman (1989) and Hallam (1987) should be mentioned. The distinction between ongoing or background extinctions and mass killings and their evolutionary significance is discussed by Jablonski (1986) and Raup (1986). The former are detected by observations based on taxa changes through time; the latter, by biomass disappearance at a bedding plane.

Both slow-acting and abrupt or catastrophic extinction models are supported by observations and consequent hypotheses. The reality of both mechanisms must be recognized and an examination made of the evidence to decide which has been dominant at specific times in Earth history. For much of geological time, major biotic change has been an equilibrium process responding to gradual environmental changes. This state of biological equilibrium has been punctuated episodically (perhaps periodically) by catastrophic events of enormous energy and short duration. At several mass killing horizons, some of the disintegration products of an impacting body (e.g. anomalous Ir and other siderophile group elements, chondritic $^{187}Os/^{186}Os$ ratios, shocked minerals, spherules, and tsunami deposits) are preserved. At others, this direct evidence has been destroyed by sedimentary reworking, particularly in shallower shelf environments.

For extinction horizons lacking the direct products of an impact, it is possible to reconstruct a diagnostic sequence of events by summing probabilities from observations at an event horizon distributed worldwide. After establishing the microstratigraphy at event horizons, observed patterns may exhibit similarities or differences and lead to suggested causes. The microstratigraphy of extinction horizons records the magnitude and rate of environmental change following a major impact. Such a record is known for several boundaries at which the duration of initial events represents extremely short time periods, e.g. years or hundreds of years. Furthermore, the sequence of marine events is common to most sudden extinction horizons and consists of (*a*) mass killing, (*b*) extraterrestrially induced sedimentary effects, (*c*) depauperate Strangelove ocean, (*d*) biomass recovery with the slow or rapid radiation of new forms, and (*e*) return to preevent oceanic conditions but with a largely new biota.

The sedimentary record of these events, which may give evidence for a sudden and violent impact, includes the following:

1. A bedding plane surface is observed at the base in which a very large proportion of the biota disappears, or, if the biota is scarce in the bed below, disappears completely. The bedding plane commonly shows signs of erosion and may be cut by fissures or neptunian dikes filled with allochthonous sediments.
2. This is overlain by an extraterrestrial layer containing Ir and other siderophile element anomalies, chondritic $^{187}Os/^{186}Os$ ratios, radiogenic $^{87}Sr/^{86}Sr$ ratios, shocked minerals, soot and spheroids, and clastic deposits that may contain derived sediments, including fossils from previous horizons and tsunami and storm wave deposits.
3. A marked facies change to a carbonate-poor, organic-rich sediment (commonly a chalcophile-element-enriched anoxic black shale) follows; this layer is distinguished by a sudden decrease in $\delta^{13}C$ and $\delta^{18}O$ values

in carbonate and by a marked increase in $\delta^{34}S$ values in sulfate and sulfide. The overlying beds are commonly highly depauperate.

4. A recovery interval then ensues, characterized by the introduction of some new taxa and a gradual increase in $\delta^{13}C$ and $\delta^{18}O$ values.
5. Finally, there is a return to a normal sedimentary regime, probably different from the previous regime and displaying the radiation and development of a new biota.

The causes of the biomass killing observed at such horizons may be interpreted variously from independent local observations that suggest one or more terrestrial effects: climate change including glaciation and/or deglaciation, rapid transgression/regression, volcanicity, anoxic oceans, and even erosion and nondeposition. In summarizing the horizons considered in this paper, it proved impossible to select a single global cause from those listed above. One of the most common proximate causes invoked to account for extinctions has been anoxia and associated black shale deposition, but even here, at none of the horizons discussed was this present at every section globally.

The interpretation of a killing event as described here requires generalizing many local observations into a pattern that allows a global cause to be postulated. None of the causes listed above can meet this requirement. There may be a possible exception, however, with regard to a major volcanic paroxysm rather than an impact event causing the kinds of global effects described, although in the case of the K-T boundary, theories of volcanism have failed thus far to account for shocked minerals, siderophile group element anomalies, or chondritic $^{187}Os/^{186}Os$ ratios. While leaving the door open for a possible volcanic origin for some global events, it may be that periods of accelerated volcanism are more likely due to effects resulting from major impacts rather than the primary cause of mass killings. It is difficult to deny the reality of the bolide flux to which the Earth has been subjected throughout its history.

The individual extinction horizons discussed in this paper have been generalized around the world and found to be amenable to a single explanation in each case. It has been postulated, therefore, that because each appeared to represent a global series of correlatable events, they were synchronous and had a common cause. If we now take the individual extinction horizons and consider them through time by comparing them with each other, we find that we have similar probabilities for each globally. We may sum events through time in a manner similar to the summation of single events through space, and we find similar sequences displaying the characteristics listed above. We may sum these global events, each with a unique biota and sequence of change, and claim a common cause that produced a common killing and common chemical and physical effects.

The initial conditions of the biota before several of the events described were similar, and it has been claimed, at least locally, that there was a change in climate or sea-level or in "continentality," resulting in a lowering of diversity, and that consequently a slow extinction (spread over a period of a stage or more) was already taking place. Some of the change in diversity observed locally may be ascribed to a decline in taxa due to an artificial range truncation prior to the defined event horizon (Signor & Lipps 1982). But granting the changing conditions, some diversity reduction might be expected. Thus, near the end of the Ordovician, the biota was stressed owing to the acme of a glacial event, which ended approximately at the biomass killing horizon at the end of the Hirnantian. Diversity reduction was under way during the latest Permian, and this was presumably due to maximum continentality, minimal coastline length, very severe climatic conditions, and possibly a major regression. At the close of the Triassic, we have, again, maximum continentality, a poorly represented marine succession, and regression at or about the killing horizon. The Maastrichtian Stage at the end of the Cretaceous was evidently also a time of biotic rundown and climate cooling.

However, although there was a long, drawn-out change in biota, in the above four examples, there was in every case a single horizon, recognizable globally, at which the main biomass disappeared, bioturbation ceased, and (for a variable time) an almost total loss of life occurred. The worldwide killing event is easily found and represents a unique biologically, chemically, and physically anomalous horizon that appears to have formed instantaneously and to have been followed by the pattern of events described above.

The end-Frasnian event, on the other hand, differs in many ways from the others mentioned. There is little or no evidence for climatic change at the end of this stage, and in some parts of the world the warm shallow seas with widespread reefs were succeeded by further warm shallow-water limestones (although highly depauperate) of the Famennian. There are claims of lowered biodiversity in certain groups during the Frasnian, especially among brachiopods; however, corals, for instance, increased in diversity throughout the stage, and there is no evidence for a reduction in biomass up to the event itself. The killing was very great, perhaps the second largest after the Permian-Triassic event. The Famennian fauna established itself slowly during the early part of the stage.

We have smaller events at the end of the Devonian, and others could be named; however, time and space limitations have precluded their consideration in this paper. These, too, appear to be global but may represent the effect of a somewhat smaller bolide. Each, nevertheless, exhibits part of the "type" sequence associated with an impact event, although more weakly than the others considered above.

The Eocene-Oligocene event is unresolved. The evidence has not yet been sufficiently generalized globally, and although there is some evidence for impact, it is not necessarily conclusive at any particular horizon. Further analysis of sections in many facies may offer a globally acceptable solution.

Finally, the Lake Acraman impact event at about 600 Ma presents a challenge to stratigraphers to find a signal in that long unresolved interval currently being discussed in attempting to arrive at a definition of the Precambrian-Cambrian boundary and of the associated biological and chemical events.

Future research into the geological and biological consequences of impacts should focus on the use of biomass changes to define extinction horizons that may be tested as global events that are recognizable in a full range of sedimentary environments and to determine their relation to microstratigraphy and geochemistry. For events in which the evidence is poorly preserved owing to an incomplete depositional sequence, emphasis must be placed on deeper water sedimentary environments (e.g. pelagic and hemipelagic) whenever possible, for in such environments the chances of obtaining direct causal evidence is greater.

Further efforts should involve field studies of the effects of known impacts, focusing on ejecta horizons. These should involve careful sedimentological and geochemical study of continuously deposited sequences spanning impact events.

ACKNOWLEDGMENTS

We gratefully acknowledge advice and assistance received during the preparation of this manuscript from S. Bai, C. R. Barnes, A. J. Boucot, E. M. Cameron, M. R. Dence, N. Evans, H. H. J. Geldsetzer, H. Hou, K. J. Hsü, I. R. Jonasson, E. G. Kauffman, R. R. Keays, B. S. Norford, G. S. Nowlan, C. Orth, C. Sandberg, P. Sartenaer, J. M. Shaw, E. T. Tozer, K. Wang, D.-Y. Xu, and W. Ziegler. D. J. McLaren acknowledges support in funding and research facilities from the Geological Survey of Canada.

Literature Cited

Ager, D. V. 1987. A defence of the Rhaetian stage. *Albertiana* 6: 4–13

Ager, D. V. 1988. Dogmatism versus pragmatism in the Rhaetian stage. *Albertiana* 7: 16

Alvarez, L. W., Alvarez, W., Asaro, F., Michel, H. V. 1980. Extraterrestrial cause for the Cretaceous-Tertiary extinction. *Science* 208: 1095–1108

Alvarez, W., Alvarez, L. W., Asaro, F., Michel, H. V. 1982. Current status of the impact theory for the terminal Cretaceous extinction. See Silver & Schultz 1982, pp. 305–15

Alvarez, W., Alvarez, L. W., Asaro, F., Michel, H. V. 1984. The end of the Cretaceous: sharp boundary or gradual transition? *Science* 223: 1183–86

Anderson, D. L. 1983. Chemical composition of the mantle. *J. Geophys. Res.* 88: B41–B52 (Suppl.)

Apollonov, M. K., Koren, T. N., Nikitin, I. F., Paletz, L. M., Tzai, D. T. 1988. Nature of the Ordovician-Silurian boundary in south Kazakhstan, USSR. See Cocks & Rickards 1988, pp. 145–54

Awramik, S. M. 1986. The Precambrian-Cambrian boundary and geochemical perturbations: discussion. *Nature* 319: 696–97

Bai, S., Ning, Z. 1988. Faunal change and events across the Devonian-Carboniferous boundary of Huangmao section, Guangxi, South China. See McMillan et al 1988, pp. 147–58

Barnes, C. R. 1988. Stratigraphy and palaeontology of the Ordovician-Silurian boundary interval, Anticosti Island, Quebec, Canada. See Cocks & Rickards 1988, pp. 195–219

Bassett, M. G. 1985. Towards a "common language" in stratigraphy. *Episodes* 8: 87–92

Baud, A., Magaritz, M. 1988. Carbon isotope profile in the Permian-Triassic of the Central Tethys: the Kashmir sections (India). *Ber. Geol. Bundesanst.* 15: 2

Becker, R. T. 1986. Ammonoid evolution before, during and after the "Kellwasser-event"—review and preliminary new results. See Walliser 1986, pp. 181–88

Becker, R. T., Feist, R., Flajs, G., House, M. R., Klapper, G. 1989. Frasnian-Famennian extinction events in the Devonian at Coumiac, southern France. *C.R. Acad. Sci. Paris* 309(Sér. II): 259–66

Belyea, H. R., McLaren, D. J. 1962. Upper Devonian formations, southern part of Northwest Territories, northeastern British Columbia, and northwestern Alberta. *Geol. Surv. Can. Pap. 61-29.* 74 pp.

Bergström, S. M., Boucot, A. J. 1988. The Ordovician-Silurian boundary in the United States. See Cocks & Rickards 1988, pp. 273–84

Berry, W. B. N., Wilde, P. 1978. Progressive ventilation of the oceans—an explanation for the distribution of the Lower Paleozoic black shales. *Am. J. Sci.* 278: 257–75

Bhandari, N., Gupta, M., Pandey, J., Skukla, P. N. 1988. Geochemistry of the K/T boundaries in India and contributions of Deccan volcanism. See Lunar Planet. Inst./Natl. Acad. Sci. 1988, pp. 15–16 (Abstr.)

Birkelund, T., Hakansson, E. 1982. The terminal Cretaceous extinction in Boreal shelf areas—a multicausal event. See Silver & Schultz 1982, pp. 373–84

Blow, W. H. 1969. Late middle Eocene to Recent planktonic foraminiferal biostratigraphy. *Proc. Int. Conf. Planktonic Microfoss., 1st, Geneva, 1967*, ed. P. Bronnimann, H. H. Renz, 1: 199–422. Leiden: E. J. Brill

Bohor, B. F. 1988. Shocked quartz and more: impact signatures in K-T boundary clays. See Lunar Planet. Inst./Natl. Acad. Sci. 1988, pp. 17–18 (Abstr.)

Bohor, B. F., Foord, E. E., Modreski, P. J., Triplehorn, D. M. 1984. Mineralogic evidence for an impact event at the Cretaceous-Tertiary boundary. *Science* 224: 867–68

Bohor, B. F., Modreski, P. J., Foord, E. E. 1987a. Shocked quartz in the Cretaceous-Tertiary boundary clays: evidence for a global distribution. *Science* 236: 705–9

Bohor, B. F., Triplehorn, D. M., Nichols, D. J., Millard, H. T. 1987b. Dinosaurs, spherules, and the "magic" layer: a new K-T boundary clay site in Wyoming. *Geology* 15: 896–99

Boucot, A. J. 1986. Ecostratigraphic criteria for evaluating the magnitude, character and duration of bioevents. See Walliser 1986, pp. 25–46

Boucot, A. J. 1988. Devonian biogeography: an update. See McMillan et al 1988, pp. 211–17

Boucot, A. J. 1989. Phanerozoic extinctions: how similar are they to each other? In *Abrupt Changes in the Global Biota. Lect. Notes Earth Sci.*, ed. E. G. Kauffman, O. H. Walliser. Berlin: Springer-Verlag. In press

Boucot, A. J., Gray, J. 1982. Paleozoic data of climatological significance and their use for interpreting Silurian-Devonian climate. In *Climate in Earth History—Studies in Geophysics*, pp. 189–98. Washington, DC: Natl. Acad. Press

Bourgeois, J., Hansen, T. A., Wiberg, P. L., Kauffman, E. G. 1988. A tsunami deposit at the Cretaceous-Tertiary boundary in Texas. *Science* 241: 567–69

Brandner, R. 1988. The Permian-Triassic boundary section in the Dolomites (Southern Alps, Italy), San Antonio section. *Ber. Geol. Bundesanst.* 15: 49–56

Brenchley, P. J. 1988. Environmental changes close to the Ordovician-Silurian boundary. See Cocks & Rickards 1988, pp. 377–85

Brooks, R. R., Reeves, R. D., Yang, X.-H., Ryan, D. E., Holzbecher, J., et al. 1984. Elemental anomalies at the Cretaceous-

Tertiary boundary, Woodside Creek, New Zealand. *Science* 226: 539–42

Brooks, R. R., Hoek, P. L., Reeves, R. D., Wallace, R. C., Johnston, J. H., et al. 1985. Weathered spheroids in a Cretaceous/Tertiary boundary shale at Woodside Creek, New Zealand. *Geology* 13: 738–40

Brooks, R. R., Strong, C. P., Lee, J., Orth, C. J., Gilmore, J. S., et al. 1986. Stratigraphic occurrences of iridium anomalies at four Cretaceous/Tertiary boundary sites in New Zealand. *Geology* 14: 727–29

Burek, P. J., Wanke, H. 1988. Impacts and glacio-eustasy, plate-tectonic episodes, geomagnetic reversals: a concept to facilitate detection of impact events. *Phys. Earth Planet. Inter.* 50: 183–94

Burke, W. H., Denison, R. E., Hetherington, E. A., Koepnick, R. B., Nelson, H. F., Otto, J. B. 1982. Variation of seawater $^{87}Sr/^{86}Sr$ throughout Phanerozoic time. *Geology* 10: 516–19

Caputo, M. V., Crowell, J. C. 1985. Migration of glacial centres across Gondwana during Paleozoic era. *Geol. Soc. Am. Bull.* 96: 1020–36

Carter, N. L., Officer, C. B., Chesner, C. A., Rose, W. I. 1986. Dynamic deformation of volcanic ejecta from the Toba caldera: possible relevance tò Cretaceous/Tertiary boundary phenomena. *Geology* 14: 380–83

Claypoole, C. E., Holser, W. T., Saki, I. R., Zak, I. 1980. The age curves for sulfur and oxygen isotopes in marine sulfate and their mutual interpretations. *Chem. Geol.* 28: 199–260

Cocks, L. R. M. 1988. Brachiopods across the Ordovician-Silurian boundary. See Cocks & Rickards 1988, pp. 311–15

Cocks, L. R. M., Rickards, R. B., eds. 1988. A global analysis of the Ordovician-Silurian boundary. *Bull. Br. Mus.* (*Nat. Hist.*) *Geol. No. 43*. 394 pp.

Compston, W., Williams, I. S., Jenkins, R. J. F., Gostin, V. A., Haines, P. W. 1987. Zircon age evidence for the late Precambrian Acraman ejecta blanket. *Aust. J. Earth Sci.* 34: 435–45

Copper, P. 1986. Frasnian/Famennian mass extinction and cold-water oceans. *Geology* 14: 835–39

Courtillot, V., Besse, J., Vandamme, D., Montigny, R., Jaeger, J.-J., Cappetta, H. 1986. Deccan flood basalts at the Cretaceous/Tertiary boundary. *Earth Planet. Sci. Lett.* 80: 361–74

Courtillot, V., Feraud, G., Maluski, H., Vandamme, D., Moreau, M. G., Besse, J. 1988a. Deccan flood basalts and the Cretaceous-Tertiary boundary. *Nature* 333: 843–46

Courtillot, V., Vandamme, D., Besse, J., Jaeger, J.-J. 1988b. Deccan volcanism at the Cretaceous-Tertiary boundary. See Lunar Planet. Inst./Natl. Acad. Sci. 1988, pp. 31–32 (Abstr.)

Cowie, J. W., Ziegler, W., Remane, J. 1989. Stratigraphic Commission accelerates progress: 1984 to 1989. *Episodes* 12: 79–83

Crocket, J. H., Officer, C. B., Wezel, F. C., Johnson, G. D. 1988. Distribution of noble metals across the Cretaceous/Tertiary boundary at Gubbio, Italy: iridium variation as a constraint on the duration and nature of Cretaceous/Tertiary boundary events. *Geology* 16: 77–80

Cuerda, A., Rickards, R. B., Cingolani, C. 1988. The Ordovician-Silurian boundary in Bolivia and Argentina. See Cocks & Rickards 1988, pp. 291–94

de Silva, S. L., Sharpton, V. L. 1988. Explosive volcanism, shock metamorphism and the K-T boundary. See Lunar Planet. Inst./Natl. Acad. Sci. 1988, pp. 38–39 (Abstr.)

D'Hondt, S., Lindinger, M. 1988. An extended Cretaceous-Tertiary (K/T) stable isotope record: implications for paleoclimate and the nature of the K/T boundary event. See Lunar Planet. Inst./Natl. Acad. Sci. 1988, pp. 40–41 (Abstr.)

Dietz, R. S. 1961. Astroblemes. *Sci. Am.* 1961(Aug.): 50–58

Druce, E. C. 1976. Conodont biostratigraphy of the Upper Devonian reef complexes of the Canning Basin, Western Australia. *Aust. Bur. Miner. Resour. Geol. Geophys. Bull. 158*. 303 pp.

Elliott, D. K., ed. 1986. *Dynamics of Extinction*. New York: Wiley. 294 pp.

Erwin, D. H. 1989. The end-Permian extinction event. In *Encyclopedia of Palaeontology*, ed. D. E. G. Briggs, P. Crowther. London: Blackwell. In press

Farsan, N. M. 1986. Frasnian mass extinction—a single catastrophic event or cumulative? See Walliser 1986, pp. 189–98

Fortey, R. A. 1984. Global earlier Ordovician transgressions and their biological implications. In *Aspects of the Ordovician System. Paleontol. Contrib. Univ. of Oslo*, ed. D. L. Bruton, 295: 37–50

Friakova, O., Galle, A., Hladil, J., Kalvoda, J. 1985. A lower Famennian fauna from the top of reefoid limestones at Mokra (Moravia, Czechoslovakia). *Newsl. Stratigr.* 15: 43–56

Galle, A., Friakova, O., Hladil, J., Kalvoda, J., Krejci, Z., et al. 1988. Biostratigraphy of middle Upper Devonian carbonates of Moravia, Czechoslovakia. See McMillan et al 1988, pp. 633–45

Ganapathy, R. 1980. A major meteorite impact on the Earth 65 million years ago:

evidence from the Cretaceous-Tertiary boundary clay. *Science* 209: 921–23

Ganapathy, R. 1983. The Tunguska explosion of 1908: discovery of meteoritic debris near the explosion site and at the South Pole. *Science* 220: 1158–61

Ganapathy, R., Gartner, S., Jiang, M.-J. 1981. Iridium anomaly at the Cretaceous-Tertiary boundary in Texas. *Earth Planet. Sci. Lett.* 54: 393–96

Geldsetzer, H. H. J. 1988. The Frasnian-Famennian extinction event(s) and anoxic sedimentation: are they related? *Ber. Geol. Bundesanst.* 15: 7

Geldsetzer, H. H. J., Goodfellow, W. D., McLaren, D. J., Orchard, M. J. 1987. Sulfur-isotope anomaly associated with the Frasnian-Famennian extinction, Medicine Lake, Alberta, Canada. *Geology* 15: 393–96

Geldsetzer, H. H. J., Goodfellow, W. D., McLaren, D. J., Orchard, M. J. 1988. Sulfur-isotope anomaly associated with the Frasnian-Famennian extinction, Medicine Lake, Alberta, Canada: reply. *Geology* 16: 86–88

Gerstel, J., Thunell, R. C., Zachos, J. C., Arthur, M. A. 1986. The Cretaceous/Tertiary boundary event in the North Pacific: planktonic foraminiferal results from Deep Sea Drilling Project site 577, Shatsky Rise. *Paleoceanography* 1: 97–117

Glass, B. P. 1986. Late Eocene microtektites and clinopyroxene-bearing spherules. See Pomerol & Premoli-Silva 1986, pp. 395–401

Glass, B. P., Swincki, M. B., Zwart, P. A. 1979. Australasian, Ivory Coast and North American tektite strewn fields: size, mass and correlation with geomagnetic reversals and other Earth events. *Proc. Lunar Planet. Sci. Conf., 10th*, pp. 25–37. New York: Pergamon

Glenister, B. F., Klapper, G. 1966. Upper Devonian conodonts from the Canning Basin, Western Australia. *J. Paleontol.* 40: 777–842

Golebiowski, R., Braunstein, R. E. 1988. A Triassic/Jurassic boundary section in the northern Calcareous Alps (Austria). *Ber. Geol. Bundesanst.* 15: 39–46

Goodfellow, W. D. 1986. Anoxic oceans and short-term carbon isotope trends. *Nature* 322: 116–17

Goodfellow, W. D. 1987. Anoxic stratified oceans as a source of sulphur in sediment-hosted stratiform Zn-Pb deposits (Selwyn Basin, Yukon, Canada). *Chem. Geol.* 65: 359–82

Goodfellow, W. D., Jonasson, I. R. 1984. Ocean stagnation and ventilation defined by δ^{34}S secular trends in pyrite and barite, Selwyn Basin, Yukon. *Geology* 12: 583–86

Goodfellow, W. D., Geldsetzer, H. H., McLaren, D. J., Orchard, M. J., Klapper, G. 1988. The Frasnian-Famennian extinction: current results and possible causes. See McMillan et al 1988, pp. 9–22

Gostin, V. A., Haines, P. W., Jenkins, R. J. F., Compston, W., Williams, I. S. 1986. Impact ejecta horizon within late Precambrian shales, Adelaide geosyncline, South Australia. *Science* 233: 198–203

Gostin, V. A., Keays, R. R., Wallace, M. W. 1989. Iridium anomaly from the Acraman impact ejecta horizon: impacts can produce sedimentary iridium peaks. *Nature* 340: 542–44

Grant, L. B., Schuster, S., Roddy, D. J. 1987. Analytical simulation of a 10-km-diameter asteroid impact into a terrestrial ocean: part 3—cratering mechanics. *Lunar Planet. Sci. XVII*, pp. 281–82. Houston: Lunar Planet. Inst. (Abstr.)

Grieve, R. A. F. 1982. The record of impact on Earth: implications for a major Cretaceous/Tertiary impact event. See Silver & Schultz 1982, pp. 25–37

Grieve, R. A. F. 1984. The impact cratering rate in Recent time. *Proc. Lunar Planet. Sci. Conf., 14th. J. Geophys. Res.* 89: B403–8 (Suppl.)

Grieve, R. A. F., Dence, M. R. 1979. The terrestrial cratering record II. The crater production rate. *Icarus* 38: 230–42

Grieve, R. A. F., Robertson, P. B. 1979. The terrestrial cratering record I. Current status of observations. *Icarus* 38: 212–29

Guex, J. 1982. Rélations entre le genre Psiloceras et les Phylloceratida au voisinage de la limite Trias-Jurassique. *Bull. Geol. Lausanne* 1982: 47–51

Hallam, A. 1981. The end-Triassic bivalve extinction event. *Palaeogeogr. Palaeoclimatol. Palaeoecol.* 35: 1–44

Hallam, A. 1987. End-Cretaceous mass extinction event: argument for terrestrial causation. *Science* 238: 1237–42

Hallam, A. 1988. The end-Triassic mass extinction event. See Lunar Planet. Inst./Natl. Acad. Sci. 1988, pp. 66–67 (Abstr.)

Hallam, A., Asaro, F., Goodfellow, W. D. 1989. Sedimentary and geochemical profiles of two classic sections across the Triassic-Jurassic boundary. In preparation

Hansen, H. J., Rasmussen, K. L., Gwozdz, R., Kunzendorf, H. 1987. Iridium-bearing Carbon Black at the Cretaceous-Tertiary boundary. *Bull. Geol. Soc. Den.* 36: 305–14

Hassig, P. J., Rosenblatt, M., Roddy, D. J. 1987. Analytical simulation of a 10-km-diameter asteroid impact into a terrestrial ocean: Part 2—atmospheric response. *Lunar Planet. Sci. XVII*, pp. 321–22. Houston: Lunar Planet. Inst. (Abstr.)

Heckel, P. H., Witzke, B. J. 1979. Devonian world palaeogeography determined from distribution of carbonates and related lithic palaeoclimatic indicators. *Palaeontol. Assoc. Spec. Pap. 23*, pp. 99–123

Hildebrand, A. R., Boynton, W. V. 1988. Impact wave deposits provide new constraints on the location of the (K/T) boundary impact. See Lunar Planet. Inst./ Natl. Acad. Sci. 1988, pp. 76–77 (Abstr.)

Hoffman, A. 1989. Mass extinctions: the view of a sceptic. *J. Geol. Soc. London* 146: 21–35

Holser, W. T., Magaritz, M. 1987. Events near the Permian-Triassic boundary. *Mod. Geol.* 11: 155–80

Holser, W. T., Schönlaub, H.-P., Attrep, M. Jr., Boeckelmann, K., Klein, P., et al. 1989. A unique geochemical record at the Permian-Triassic boundary. *Nature* 337: 39–44

Hou, H., Ji, Q., Wu, X., Xiong, J., Wang, S., et al. 1985. *Muhua Sections of Devonian-Carboniferous Boundary Beds*. Beijing: Geol. Publ. House. 226 pp.

Hou, H., Ji, Q., Wang, J. 1988. Preliminary report on Frasnian-Famennian events in South China. See McMillan et al 1988, pp. 63–70

House, M. R. 1985. Correlation of mid-Paleozoic ammonoid evolutionary events with global sedimentary perturbations. *Nature* 313: 17–22

Hsü, K. J. 1986. The Precambrian-Cambrian boundary and geochemical perturbations: reply. *Nature* 319: 697

Hsü, K. J., Mackenzie, J. A. 1985. A "Strangelove" ocean in earliest Tertiary. In *The Carbon Cycle and Atmospheric CO_2: Natural Variations, Archean to Present. Geophys. Monogr.*, ed. E. T. Sunquist, W. Broecker, 32: 487–92. Washington, DC: Am. Geophys. Union

Hsü, K. J., He, Q., McKenzie, J. A., Weissert, H., Perch-Nielsen, K., et al. 1982. Mass mortality and its environmental and evolutionary consequences. *Science* 216: 249–56

Hsü, K. J., Oberhansli, H., Gao, J. Y., Shu, S., Haihong, C., Krahenbuhl, V. 1985. "Strangelove ocean" before the Cambrian explosion. *Nature* 316: 809–11

Hunt, B. G. 1984. Polar glaciation and the genesis of ice ages. *Nature* 308: 48–51

Hut, P., Alvarez, W., Elder, W. P., Hansen, T., Kauffman, E. G., et al. 1987. Comet showers as a cause of mass extinctions. *Nature* 329: 118–26

Izett, G. A. 1987. Authigenic "spherules" in K-T boundary sediments at Caravaca, Spain, and Raton Basin, Colorado and New Mexico, may not be impact derived. *Geol. Soc. Am. Bull.* 99: 78–86

Izett, G. A. 1988. The Cretaceous-Tertiary (K-T) boundary interval, Raton Basin, Colorado and New Mexico, and its content of shock-metamorphosed minerals: implications concerning the K-T boundary impact extinction theory. *U.S. Geol. Surv. Open-File Rep. 87-606*. 125 pp.

Jablonski, D. 1986. Causes and consequences of mass extinctions: a comparative approach. See Elliott 1986, pp. 183–229

Janssens, M.-J., Hertogen, J., Takahashi, H., Anders, E., Lambert, P. 1977. Rochechouart meteorite crater: identification of projectile. *J. Geophys. Res.* 82: 750–58

Jia, H.-C., Xian, S.-Y., Yang, D.-L., Zhou, H.-L., Han, Y.-J., et al. 1988. An ideal Frasnian/Famennian boundary in Ma-anshan, Zhongping, Xiangzhou, Guangxi, South China. See McMillan et al 1988, pp. 79–92

Johnson, J. G. 1974. Extinction of perched faunas. *Geology* 2: 479–82

Johnson, J. G., Sandberg, C. A. 1988. Devonian eustatic events in the western United States and their biostratigraphic responses. See McMillan et al 1988, pp. 171–78

Jones, D. S., Mueller, P. A., Bryan, J. R., Dobson, J. P., Channell, J. E. T., et al. 1987. Biotic, geochemical, and paleomagnetic changes across the Cretaceous/ Tertiary boundary at Braggs, Alabama. *Geology* 15: 311–15

Kalvoda, J. 1986. Upper Frasnian and Lower Tournaisian events and evolution of calcareous foraminifera—close links to climatic changes. See Walliser 1986, pp. 225–36

Kauffman, E. G. 1979. The ecology and biostratigraphy of the Cretaceous-Tertiary extinction event. In *Proc. Cretaceous and Tertiary Boundary Events II*, ed. W. K. Christensen, T. B. Birkland, pp. 29–37. Copenhagen: Univ. Copenhagen

Kauffman, E. G. 1988. Oceanic impact and marine mass extinctions: Cretaceous examples. See Lunar Planet Inst./Natl. Acad. Sci. 1988, unpaginated (Abstr.)

Keller, G. 1986. Late Eocene impact events and stepwise mass extinctions. See Pomerol & Premoli-Silva 1986, pp. 403–12

Keller, G. 1988. Extended period of K/T boundry mass extinction in the marine realm. See Lunar Planet. Inst./Natl. Acad. Sci. 1988, pp. 88–89 (Abstr.)

Kyte, F. T., Zhou, Z., Wasson, J. T. 1980. Siderophile-enriched sediments from the Cretaceous-Tertiary boundary. *Nature* 288: 651–56

Kyte, F. T., Smit, J., Wasson, J. T. 1985. Siderophile interelement variations in the Cretaceous-Tertiary boundary sediments from Caravaca, Spain. *Earth Planet. Sci. Lett.* 73: 183–95

Lamolda, M. A., Kauffman, E. G., Walliser, O. H., eds. 1988. *Paleontology and Evolution: Extinction Events. Int. Conf. Global Bioevents, 2nd. Rev. Esp. Paleontol., Numero Extraordinario.* 155 pp.

Lenz, A. C., McCracken, A. D. 1988. Ordovician-Silurian boundary, northern Yukon, Canada. See Cocks & Rickards 1988, pp. 265–71

Lerbekmo, J. F., Sweet, A. R., St. Louis, R. M. 1987. The relationship between the iridium anomaly and palynological floral events at three Cretaceous-Tertiary boundary localities in western Canada. *Geol. Soc. Am. Bull.* 99: 325–30

Lewis, J. S., Watkins, G. H., Hartman, H., Prinn, R. G. 1982. Chemical consequences of major impact events on Earth. See Silver & Schultz 1982, pp. 215–21

Lowrie, W. 1986. Magnetic stratigraphy of the Eocene/Oligocene boundary. See Pomerol & Premoli-Silva 1986, pp. 357–62

Luck, J. M., Turekian, K. K. 1983. Osmium-187/osmium-186 in manganese nodules and the Cretaceous-Tertiary boundary. *Science* 222: 613–15

Lunar and Planetary Institute/National Academy of Sciences. 1988. *Conference on Impacts, Volcanism and Mass Mortality: Global Catastrophes, 2nd, Snowbird, Utah, Abstr. Vol. Lunar Planet. Inst. Contrib. 673*

Magaritz, M., Holser, W. T., Kirschvink, J. L. 1986. Carbon-isotope events across the Precambrian/Cambrian boundary in the Siberian platform. *Nature* 320: 258–59

Magaritz, M., Bar, R., Baud, A., Holser, W. T. 1988. The carbon isotope shift at the Permian/Triassic boundary in the southern Alps is gradual. *Nature* 331: 337–39

Margolis, S. V., Doehne, E. F. 1988. Trace element and isotope geochemistry of Cretaceous/Tertiary boundary sediments. See Lunar Planet. Inst./Natl. Acad. Sci. 1988, pp. 113–14 (Abstr.)

Margolis, S. V., Mount, J. F., Doehne, E., Showers, W., Ward, P. 1987. The Cretaceous/Tertiary boundary carbon and oxygen isotope stratigraphy, diagenesis, and paleoceanography at Zumaya, Spain. *Paleoceanography* 2: 361–77

Masaytis, V. L., Raykhlin, A. I. 1986. The Popigay crater formed by the impact of an ordinary chondrite. *Trans. USSR Acad. Sci.* 286: 121–22

Mason, B., ed. 1971. *Handbook of Elemental Abundances in Meteorites.* London/New York: Gordon & Breach. 600 pp.

McGhee, G. R. Jr. 1988. Evolutionary dynamics of the Frasnian-Famennian boundary interval in western Canada. See McMillan et al 1988, pp. 23–28

McGhee, G. R. Jr., Orth, C. J., Quintana, L. R., Gilmore, J. S., Olsen, E. J. 1986. Late Devonian "Kellwasser event" mass extinction horizon in Germany: no geochemical evidence for a large-body impact. *Geology* 14: 776–79

McHone, J. F., Niema, R. A., Lewis, C. F., Yates, A. M. 1989. Stishovite at the Cretaceous-Tertiary boundary, Raton, New Mexico. *Science* 243: 1182–84

McLaren, D. J. 1959. The role of fossils in defining rock units with examples from the Devonian of western and arctic Canada. *Am. J. Sci.* 257: 734–51

McLaren, D. J. 1970. Presidential address: Time, life and boundaries. *J. Paleontol.* 44: 801–15

McLaren, D. J. 1982. Frasnian-Famennian extinctions. See Silver & Schultz 1982, pp. 477–84

McLaren, D. J. 1983. Bolides and biostratigraphy. *Geol. Soc. Am. Bull.* 94: 313–24

McLaren, D. J. 1985. Mass extinction and iridium anomaly in the Upper Devonian of Western Australia: a commentary. *Geology* 13: 170–72

McLaren, D. J. 1986. Abrupt extinctions. See Elliott 1986, pp. 37–46

McLaren, D. J. 1988. Detection and significance of mass killings. See McMillan et al 1988, pp. 1–8

McLaren, D. J., Mountjoy, E. W. 1962. Alexo equivalents in the Japser region, Alberta. *Geol. Surv. Can. Pap. 62-23.* 36 pp.

McMillan, N. J., Embry, A. F., Glass, D. J., eds. 1988. *Devonian of the World. Proc. Can. Soc. Pet. Geol. Int. Symp. Devonian System III.* 714 pp.

Montanari, A. 1986. Spherules from the Cretaceous/Tertiary boundary clay at Gubbio, Italy: the problem of outcrop contamination. *Geology* 14: 1024–26

Montanari, A. 1988. Terminal Eocene impacts and mass extinctions: cause and effect relationship as a problem of timing. See Lunar Planet. Inst./Natl. Acad. Sci. 1988, unpaginated (Abstr.)

Montanari, A., Hay, R. L., Alvarez, W., Asaro, F., Michel, H. V., et al. 1983. Spheroids at the Cretaceous/Tertiary boundary are altered impact droplets of basaltic composition. *Geology* 11: 668–72

Morgan, J. W., Janssens, M.-J., Hertogen, J., Gros, J., Takahashi, H. 1979. Ries impact crater, southern Germany: search for meteoritic material. *Geochim. Cosmochim. Acta* 43: 803–15

Morgan, J. W., Wandless, G. A., Petrie, P. K. 1981. Strangeways crater: trace elements in melt rocks. *Lunar Planet. Sci. XII*, pp. 714–16. Houston: Lunar Planet. Inst. (Abstr.)

Mount, J. F., Margolis, S. V., Showers, W., Ward, P., Doehne, E. 1986. Carbon and oxygen isotope stratigraphy of the upper Maastrichtian, Zumaya, Spain: a record of oceanographic and biologic changes at the end of the Cretaceous period. *Palaios: Soc. Econ. Paleontol. Mineral. Res. Lett.* 1: 87–92

Mu, E.-Z. 1988. The Ordovician-Silurian boundary in China. See Cocks & Rickards 1988, pp. 117–31

Muller, S. W., Ferguson, H. G. 1939. Mesozoic stratigraphy of the Hawthorne and Tonopah quadrangles, Nevada. *Geol. Soc. Am. Bull.* 47: 1573–1624

Newell, N. D. 1988. The Paleozoic/Mesozoic erathem boundary. *Mem. Soc. Geol. Ital.* 34: 303–11

Nichols, D. J., Fleming, R. F. 1988. Plant microfossil record of the terminal Cretaceous event in the western United States and Canada. See Lunar Planet. Inst./Natl. Acad. Sci. 1988, pp. 130–31 (Abstr.)

Nichols, D. J., Jarzen, D. M., Orth, C. J., Oliver, P. Q. 1986. Palynological and iridium anomalies at Cretaceous-Tertiary boundary, south-central Saskatchewan. *Science* 231: 714–17

Nowlan, G. S., Goodfellow, W. D., McCracken, A. D., Lenz, A. C. 1988. Geochemical evidence for sudden biomass reduction and anoxic basins near the Ordovician-Silurian boundary in northwestern Canada. *Int. Symp. Ordovician Syst., 5th, St. John's, Can.*, p. 66 (Abstr.)

Officer, C. B., Drake, C. L. 1983. The Cretaceous-Tertiary transition. *Science* 219: 1383–90

Officer, C. B., Drake, C. L. 1985. Terminal Cretaceous environmental events. *Science* 227: 1161–67

O'Keefe, J. D., Ahrens, T. J., Koschny, D. 1988. Environmental effects of large impacts on the Earth . . . relation to extinction mechanisms. See Lunar Planet. Sci./Natl. Acad. Sci. 1988, pp. 133–34 (Abstr.)

Olsen, P. E. 1986. A 40-million-year lake record of early Mesozoic orbital climatic forcing. *Science* 234: 842–48

Olsen, P. E., Cornet, B. 1988. The Triassic-Jurassic boundary in eastern North America. See Lunar Planet. Sci./Natl. Acad. Sci. 1988, pp. 135–36 (Abstr.)

Olsen, P. E., Shubin, N. H., Anders, M. H. 1987. New Early Jurassic tetrapod assemblages constrain Triassic-Jurassic tetrapod extinction event. *Science* 237: 1025–29

Olsen, P. E., Shubin, N. H., Anders, M. H. 1988. Triassic-Jurassic extinctions: reply. *Science* 241: 1359–60

Olson, E. C. 1988. Permo-Triassic vertebrate extinctions: a program. See Lunar Planet. Inst./Natl. Acad. Sci. 1988, pp. 137–38 (Abstr.)

Orth, C. J., Gilmore, J. S., Knight, J. D., Pillmore, C. L., Tschudy, R. H., Fassett, J. E. 1981. An iridium abundance anomaly at the palynological Cretaceous-Tertiary boundary in northern New Mexico. *Science* 214: 1341–43

Orth, C. J., Gilmore, J. S., Knight, J. D., Pillmore, C. L., Tschudy, R. H., et al. 1982. Iridium abundance measurements across the Cretaceous/Tertiary boundary in the San Juan and Raton Basins of northern New Mexico. See Silver & Schultz 1982, pp. 423–33

Orth, C. J., Gilmore, J. S., Quintana, L. R., Sheehan, P. M. 1986. Terminal Ordovician extinction: geochemical analysis of the Ordovician/Silurian boundary, Anticosti Island, Quebec. *Geology* 14: 433–36

Orth, C. J., Quintana, L. R., Gilmore, J. S., Barrick, J. E., Haywa, J. N., et al. 1988. Pt-group metal anomalies in lower Mississippian of southern Oklahoma. *Geology* 16: 627–30

Owen, M. R., Anders, M. H. 1988. Cathodoluminescence of shocked quartz at the Cretaceous-Tertiary boundary. See Lunar Planet. Inst./Natl. Acad. Sci., pp. 141–42 (Abstr.)

Padian, K. 1988. Triassic-Jurassic extinctions: comment. *Science* 241: 1358–59

Palme, H. 1980. The meteoritic contamination of terrestrial and lunar impact melts and the problem of indigenous siderophiles in the lunar highland. *Proc. Lunar Sci. Conf., 11th*, pp. 481–506. New York: Pergamon

Palme, H., Janssens, M.-J., Takahashi, H., Anders, E., Hertogen, J. 1978. Meteoritic material at five large impact craters. *Geochim. Cosmochim. Acta* 42: 313–23

Palme, H., Rammensee, W., Reinhold, U. 1980. The meteoric component of impact melts from European impact craters. *Lunar Planet. Sci. XI*, pp. 848–51

Paproth, E., Streel, M. 1985. In search of a Devonian-Carboniferous boundary. *Episodes* 8: 110–11

Pedder, A. E. H. 1982. The rugose coral record across the Frasnian-Famennian boundary. See Silver & Schultz 1982, pp. 485–90

Perch-Nielsen, K., McKenzie, J., He, Q. 1982. Biostratigraphy and isotope stratigraphy and the "catastrophic" extinction of calcareous nannoplankton at the Cretaceous/Tertiary boundary. See Silver & Schultz 1982, pp. 353–71

Pflug, H. D., Reitz, E. 1986. Evolutionary

changes in the Proterozoic. See Walliser 1986, pp. 95–104

Playford, P. E. 1980. Devonian "Great Barrier Reef" of Canning Basin, Western Australia. *Am. Assoc. Pet. Geol. Bull.* 64: 814–40

Playford, P. E., McLaren, D. J., Orth, C. J., Gilmore, J. S., Goodfellow, W. D. 1984. Iridium anomaly in the Upper Devonian of the Canning Basin, Western Australia. *Science* 226: 437–39

Pohl, J. 1977. Paleomagnetische und gesteinsmagnetische untersuchungen an den kermen der Forschungsbohrung Noerdingen 1973. *Geol. Bavarica* 75: 329–48

Pollack, J. B., Toon, O. B., Ackerman, T. P., McKay, C. P., Turco, R. P. 1983. Environmental effects of an impact-generated dust cloud: implications for the Cretaceous-Tertiary extinctions. *Science* 219: 287–89

Pomerol, Ch., Premoli-Silva, I., eds. 1986. *Terminal Eocene Events.* Amsterdam: Elsevier. 414 pp.

Preisinger, A., Zobetz, E., Gratz, A. J., Lahodynsky, R., Becke, M., et al. 1986. The Cretaceous/Tertiary boundary in the Gosau Basin, Austria. *Nature* 322: 794–99

Preisinger, A., Stradner, H., Mauritsch, H. J. 1988. Bio-, magneto-, and event-stratigraphy across the K/T boundary. See Lunar Planet. Inst./Natl. Acad. Sci. 1988, pp. 143–44 (Abstr.)

Prinn, R. G., Fegley, B. 1987. Bolide impacts, acid rain, and biospheric traumas at the Cretaceous-Tertiary boundary. *Earth Planet. Sci. Lett.* 83: 1–15

Rage, J.-C. 1986. The amphibians and reptiles at the Eocene-Oligocene transition in western Europe: an outline of the faunal alterations. See Pomerol & Premoli-Silva 1986, pp. 309–10

Rampino, M. R., Stothers, R. B. 1988. Flood basalt volcanism during the last 250 million years. *Science* 241: 663–68

Ramsbottom, W. H. C., Saunders, W. B. 1985. Evolution and evolutionary biostratigraphy of Carboniferous ammonoids. *J. Paleontol.* 59: 123–39

Raup, D. M. 1986. Biological extinction in Earth history. *Science* 231: 1528–33

Reinhardt, J. W. 1988. Uppermost Permian reefs and Permo-Triassic sedimentary facies from the southeastern margin of Sichuan Basin, China. *Facies* 18: 231–88

Retallack, G., Leahy, G. D. 1986. Cretaceous-Tertiary dinosaur extinction. *Science* 234: 1170–71

Richardson, J. B., McGregor, D. C. 1986. Silurian and Devonian spore zones of the Old Red Sandstone continent and adjacent regions. *Geol. Surv. Can. Bull. 364.* 79 pp.

Robertson, P. B., Dence, M. R., Vos, M. A. 1968. Deformation of rock-forming minerals from Canadian craters. In *Shock Metamorphism of Natural Materials*, ed. B. M. French, N. M. Short, pp. 433–52. Baltimore: Mono Book Corp.

Roddy, D. J., Schuster, S., Rosenblatt, M., Grant, L. B., Hassig, P. J., Kreyenhagen, K. N. 1987. Analytical simulation of a 10-km-diameter asteroid impact into a terrestrial ocean: part 1—summary. *Lunar Planet. Sci. Conf. XVII*, pp. 720–21. Houston: Lunar Planet. Inst. (Abstr.)

Roddy, D. J., Schuster, S. H., Rosenblatt, M., Grant, L. B., Hassig, P. J., Kreyenhagen, K. N. 1988. Computer modeling of large asteroid impacts into continental and oceanic sites: atmospheric, cratering and ejecta dynamics. See Lunar Planet. Inst./Natl. Acad. Sci. 1988, pp. 158–59 (Abstr.)

Romein, A. J. T., Smit, J. 1981. The Cretaceous-Tertiary boundary: calcareous nannofossils and stable isotopes. *Proc. Kon. Ned. Akad. Wet. B* 84: 295–314

Rong, J.-Y. 1984. Brachiopods of the latest Ordovician in the Yichang district, western Hubei, central China. In *Stratigraphy and Palaeontology of Systemic Boundaries in China, Ordovician-Silurian Boundary. Nanjing Inst. Geol. Palaeont. Acad. Sin.*, 1: 111–90. Anhui Sci. Tech. Publ.

Russell, D. E., Tobien, H. 1986. Mammalian evidence concerning the Eocene-Oligocene transition in Europe, North America and Asia. See Pomerol & Premoli-Silva 1986, pp. 299–307

Sandberg, C. A., Ziegler, W., Dreesen, R., Butler, J. L. 1988. Part 3: Late Frasnian mass extinction: conodont event stratigraphy, global changes, and possible causes. *Cour. Forsch.-Inst. Senckenb.* 102: 263–307

Saunders, J. B., Bernoulli, D., Muller-Merz, E., Oberhansli, H., Perch-Nielsen, K., et al. 1984. Stratigraphy of the late Middle Eocene to Early Oligocene in the Bath Cliff section, Barbados, West Indies. *Micropaleontology* 30: 390–425

Schmitz, B. 1985. Metal precipitation in the Cretaceous-Tertiary boundary clay at Stevns Klint, Denmark. *Geochim. Cosmochim. Acta* 49: 2361–70

Schmitz, B. 1988. Marine and continental K/T boundary clays compared. See Lunar Planet. Inst./Natl. Acad. Sci. 1988, pp. 164–65 (Abstr.)

Schmitz, B., Andersson, P., Dahl, J. 1988. Iridium, sulfur isotopes and rare elements in the Cretaceous-Tertiary boundary clays at Stevns Klint, Denmark. *Geochim. Cosmochim. Acta* 52: 229–36

Scholle, P. A., Arthur, M. A. 1980. Carbon isotope fluctuations in Cretaceous pelagic limestones: potential stratigraphic and petroleum exploration tool. *Am. Assoc. Pet. Geol. Bull.* 64: 67–87

Schönlaub, H. P., Klein, P., Magaritz, M., Orth, C., Attrep, M. 1988. The D-C boundary event in the Carnic Alps (Austria). *Ber. Geol. Bundesanst.* 15: 24–24a

Sepkoski, J. J. 1986. Phanerozoic overview of mass extinctions. In *Patterns and Processes in the History of Life*, ed. D. M. Raup, D. Jablonski, pp. 277–96. Berlin: Springer-Verlag

Seyfert, C. K., Sirkin, L. A. 1979. *Earth History and Plate Tectonics.* New York: Harper & Row

Shackleton, N. J., Imbrie, J., Hall, M. A. 1983. Oxygen and carbon isotope record of East Pacific core V19-30: implications for the formation of deep water in the late Pleistocene North Atlantic. *Earth Planet. Sci. Lett.* 65: 233–44

Sheehan, P. M. 1988. Late Ordovician events and the terminal Ordovician extinction. *N. M. Bur. Mines Miner. Resour. Mem.* 44: 405–15

Sheng, J., Chen, C., Wang, Y., Rui, L., Liao, Z., et al. 1987. New advances on the Permian and Triassic boundary of Jiangsu, Zhejiang and Anhui. In *Stratigraphy and Palaeontology of Systemic Boundaries in China, Permian-Triassic Boundary*, 1: 1–21 (In Chinese, with Engl. summary)

Shoemaker, E. M., Shoemaker, C. S., Wolfe, R. F. 1988. Asteroid and comet flux in the neighborhood of the Earth. See Lunar Planet. Inst./Natl. Acad. Sci. 1988, pp. 174–75 (Abstr.)

Signor, P. W., Lipps, J. H. 1982. Sampling bias, gradual extinction patterns and catastrophes in the fossil record. See Silver & Schultz 1982, pp. 291–96

Silver, L. T., Schultz, P. H., eds. 1982. *Geological Implications of Impacts of Large Asteroids and Comets on the Earth. Geol. Soc. Am. Spec. Pap. 190.* 528 pp. Boulder, Colo: Geol. Soc. Am.

Sloan, R. E., Rigby, J. K., Van Valen, L. M., Gabriel, D. 1986. Gradual dinosaur extinction and simultaneous ungulate radiation in the Hell Creek formation. *Science* 232: 629–33

Smit, J. 1982. Extinction and evolution of planktonic foraminifera after a major impact at the Cretaceous/Tertiary boundary. See Silver & Schultz 1982, pp. 329–52

Smit, J., Hertogen, J. 1980. An extraterrestrial event at the Cretaceous-Tertiary boundary. *Nature* 285: 198–200

Smit, J., Kyte, F. T. 1984. Siderophile-rich magnetic spheroids from the Cretaceous-Tertiary boundary in Umbria, Italy. *Nature* 310: 403–5

Smit, J., Romein, A. J. T. 1985. A sequence of events across the Cretaceous-Tertiary boundary. *Earth Planet. Sci. Lett.* 75: 155–70

Smit, J., van der Kaars, S. 1984. Terminal Cretaceous extinctions in the Hell Creek area, Montana: compatible with catastrophic extinction. *Science* 223: 1177–79

Smit, J., ten Kate, W. G. H. Z. 1982. Trace-element patterns at the Cretaceous-Tertiary boundary—consequences of a large impact. *Cretac. Res.* 3: 307–32

Smit, J., Groot, H., de Jonge, R., Smit, P. 1988. Impact and extinction signatures in complete Cretaceous Tertiary (KT) boundary sections. See Lunar Planet. Inst./Natl. Acad. Sci. 1988, pp. 182–83 (Abstr.)

Sokolov, B. S., Fedonkin, M. A. 1986. Global biological events in the late Precambrian. See Walliser 1986, pp. 105–8

Sorauf, J. E., Pedder, A. E. H. 1986. Late Devonian rugose corals and the Frasnian-Famennian crisis. *Can. J. Earth Sci.* 23: 1265–87

Spjeldnaes, N. 1987. The events at the Ordovician/Silurian boundary. *Terra Cognita* 1987: 209 (Abstr.)

Stanley, S. M. 1984. Temperature and biotic crisis in the marine realm. *Geology* 12: 205–8

Stanley, S. M. 1988. Paleozoic mass extinctions: shared patterns suggest global cooling as a common cause. *Am. J. Sci.* 288: 334–52

Stearn, C. W. 1987. Effect of the Frasnian-Famennian extinction event on the stromatoporoids. *Geology* 15: 677–79

Stradner, H., Rögel, F. 1988. Microfauna and nannofauna of the Knappengraben section (Austria) across the Cretaceous/Tertiary boundary. *Ber. Geol. Bundesanst.* 15: 25–25a

Streel, M. 1986. Miospore correlation between North American, German and Uralian (Udmurtia) deep facies through Appalachian, Irish and Belgian platform and continental facies near the Devonian/Carboniferous boundary. See Walliser 1986, pp. 237–40

Strong, C. P. 1977. Cretaceous-Tertiary boundary at Woodside Creek, north-eastern Marlborough. *N.Z. Geol. Surv. Rec.* 3: 47–51

Strong, C. P., Brooks, R. R., Wilson, S. M., Reeves, R. D., Orth, C. J., et al. 1987. A new Cretaceous-Tertiary boundary site at Flaxbourne River, New Zealand: biostratigraphy and geochemistry. *Geochim. Cosmochim. Acta* 51: 2769–77

Surlyk, F., Johansen, M. B. 1984. End-Cretaceous brachiopod extinctions in the Chalk of Denmark. *Science* 223: 1174–77

Teichert, C. 1988. Crises in cephalopod evolution. *Int. Colloq. L'Evolution dans sa Réalité et ses Diverses Modalités, 1985*, 3: 7–34. Paris: Masson

Thierstein, H. R. 1981. Late Cretaceous nannoplankton and the change at the Cretaceous-Tertiary boundary. In *The Deep Sea Drilling Project: A Decade of Progress. Soc. Econ. Paleontol. Mineral. Spec. Publ.*, ed. J. Warme et al, 32: 355–94

Thompson, J. B., Newton, C. R. 1988. Late Devonian mass extinction: episodic climatic cooling or warming? See McMillan et al 1988, pp. 29–34

Toon, O. B., Pollack, J. B., Ackerman, T. P., Turco, R. P., McKay, C. P., et al. 1982. Evolution of an impact-generated dust cloud and its effects on the atmosphere. See Silver & Schultz 1982, pp. 187–200

Tozer, E. T. 1984. The Trias and its ammonoids: the evolution of a time scale. *Geol. Surv. Can. Misc. Rep. 35*. 171 pp.

Tozer, E. T. 1986a. Triassic stage terminology. *Albertiana* 5: 10–14

Tozer, E. T. 1986b. Definition of the Permian-Triassic (P-T) boundary: the question of the age of the *Otoceras* beds. *Mem. Soc. Geol. Ital.* 34: 291–301

Tozer, E. T. 1988. Towards a definition of the Permian-Triassic boundary. *Episodes* 11: 251–55

Turco, R. P., Toon, O. B., Park, C., Whitten, R. C., Pollack, J. B., et al. 1981. Tunguska meteor fall of 1908: effects on stratospheric ozone. *Science* 214: 19–23

van Loevezijn, G. B. S., Raven, J. G. M., van der Pol, W. 1986. The Cremenes Limestone, a late Frasnian biostrome in the Cantabrian Mountains (northwestern Spain). *Jahrb. Geol. Paläontol. Mineral.* 10: 599–612

Wallace, M. W., Gostin, V. A., Keays, R. R. 1990. Acraman impact ejecta and whole shales: evidence for low-temperature mobilization of iridium and other platinoids. *Geology* 18: 132–35

Walliser, O. H. 1984. Geologic processes and global events. *Terra Cognita* 4: 17–20

Walliser, O. H., ed. 1986. *Global Bio-Events. Lect. Notes Earth Sci.*, Vol. 8. Berlin: Springer-Verlag. 435 pp.

Wang, K., Bai, S. 1988. Faunal changes and events near the Frasnian-Famennian boundary of South China. See McMillan et al 1988, pp. 71–78

Wang, Y., Chen, C., Rui, L. 1988. A potential global stratotype of the Permian-Triassic boundary. *Permophiles, Newsl. Subcomm. Permian Stratigr.* 13: 3–7

Ward, P. D. 1988. Maastrichtian Ammonite and Inoceramid ranges from Bay of Biscay Cretaceous-Tertiary boundary sections. See Lamolda et al 1988, pp. 119–26

Wiedmann, J. 1988. The Basque coastal section of the K/T boundary. A key to understanding "mass extinction" in the fossil record. See Lamolda et al 1988, pp. 127–40

Wilde, P., Berry, W. B. N. 1988. Sulfur-isotope anomaly associated with the Frasnian-Famennian extinction, Medicine Lake, Alberta, Canada: comment. *Geology* 16: 86–88

Wilde, P., Berry, W. B. N., Quinby-Hunt, M. S., Orth, C. J., Quintana, L. R., et al. 1986. Iridium abundances across the Ordovician-Silurian stratotype. *Science* 233: 339–41

Williams, G. E. 1986. The Acraman impact structure: source of ejecta in Late Precambrian shales, South Australia. *Science* 233: 200–3

Wolbach, W. S., Lewis, R. S., Anders, E. 1985. Cretaceous extinctions: evidence for wildfires and search for meteoric material. *Science* 230: 167–70

Xu, D.-Y., Zheng, Y., Zhang, Q.-W., Sun, Y.-Y. 1986. Three main mass extinctions—significant indicators of major natural divisions of geological history in the Phanerozoic. *Mod. Geol.* 10: 365–75

Xu, D.-Y., Zhang, Q.-W., Sun, Y.-Y., Yan, Z., Chaj, Z.-F., He, J.-W. 1989. *Astrogeological Events in China*. Edinburgh: Scot. Acad. Press. 264 pp.

Zachos, J. C., Arthur, M. A. 1989. Geochemical evidence for suppression of pelagic marine productivity at the Cretaceous/Tertiary boundary. *Nature* 337: 61–64

Zachos, J. C., Arthur, M. A., Dean, W. E. 1988. The Cretaceous-Tertiary boundary marine extinction and global primary productivity collapse. See Lunar Planet. Inst./Natl. Acad. Sci., pp. 221–22 (Abstr.)

Ziegler, W., Sandberg, C. A. 1984. *Palmatolepis*-based revision of upper part of standard Late Devonian conodont zonation. *Geol. Soc. Am. Spec. Pap.* 196: 179–89

Zoller, W. H., Parrington, J. R., Kotra, J. M. P. 1983. Iridium enrichment in airborne particles from Kilauea volcano. *Science* 222: 1118–21

Annu. Rev. Earth Planet. Sci. 1990. 18:173–204

SEAFLOOR HYDROTHERMAL ACTIVITY: BLACK SMOKER CHEMISTRY AND CHIMNEYS[1]

K. L. Von Damm

Environmental Sciences Division, Oak Ridge National Laboratory, P.O. Box 2008, Oak Ridge, Tennessee 37831-6036

INTRODUCTION

Venting of hydrothermal solutions on the seafloor, first discovered in 1977 at the Galapagos spreading center (GSC) (Figure 1), has now been shown to be a relatively common phenomenon on the world ridge-crest system. It is a consequence of the emplacement of hot rock at divergent plate boundaries and the accompanying circulation of seawater through the oceanic crust. This mechanism for convective cooling of the crust was first suggested based on the deficiency of conductive heat loss extending from the ridge axes (zero crustal age) to several thousand kilometers away on the ridge flanks (crustal ages of several million years) (Anderson et al 1977). Although relatively little of the 55,000-km length of the world ridge-crest system has been explored, at least 10 hydrothermally active sites have been visited and their fluids sampled by submersible. Photographic evidence from camera tows and dredging of hydrothermally precipitated or altered minerals and rocks suggest past or present hydrothermal activity at numerous other sites. The modes of occurrence of these seafloor hot springs differ widely; they are found over a wide range of spreading rates, at sediment-covered and sediment-starved ridge axes, and at temperatures ranging from just a few degrees above that of the ambient seawater (2–4°C) to ≥350°C. Table 1 summarizes some of the physical properties of the areas that have had vent fluid sampled by submersible. Where the measured exit temperature and the inferred temperature of reaction within the hydrothermal system differ significantly, both values are given. Although the fluids have certain common characteristics to their chemistry,

[1] The US Government has the right to retain a nonexclusive, royalty-free license in and to any copyright covering this paper.

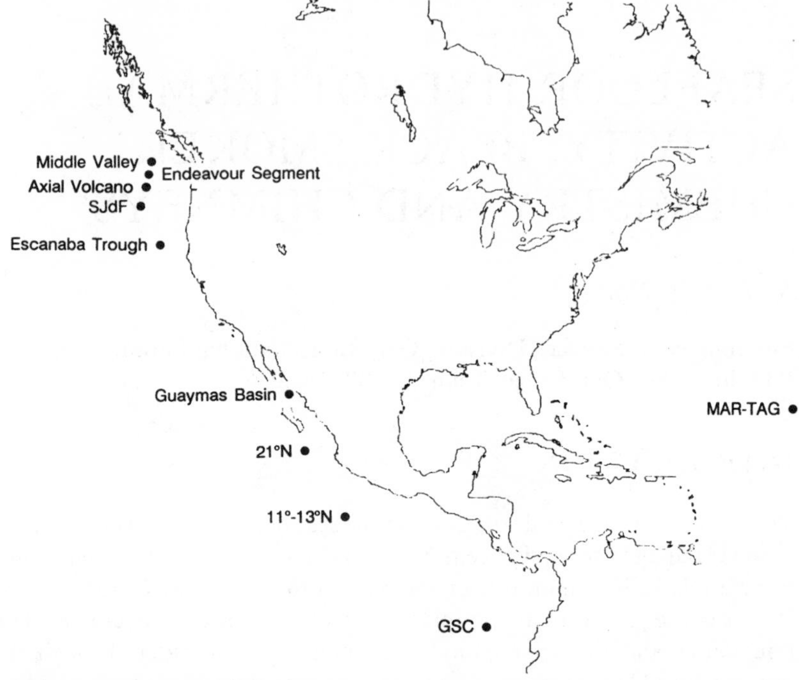

Figure 1 Map of sampled (except Middle Valley) seafloor hydrothermal systems. Not shown are the Red Sea (21°24′N, 38°03′E) and the Mariana Trough (18°13′N, 144°41′E). See Table 1 for explanation of abbreviations used in this figure.

in general each sampling of new vents has expanded our range of known compositional diversity.

CHEMISTRY OF THE FLUIDS

The chemistry of the vent fluids is the result of the interaction of seawater with basalt; and in some cases sediment, at elevated temperatures ($\geq 350°C$ in most cases). Most sampling to date has concentrated on the high-temperature fluids exiting from sulfide chimneys. Many chimneys are surrounded by diffuse flow of presumed lower temperature. The temperatures and the compositions of these diffuse flows remain largely unknown. All of the fluid samples have been obtained using submersibles. Sampling is often a difficult procedure, requiring the pilot to hold the submersible in one position while a sampling device is inserted into the chimney orifice, which is often only a few centimeters in diameter. When the hydrothermal fluids are sampled, some variable amount of seawater is usually entrained in the sample (Figure 2). Experimental work on basalt-

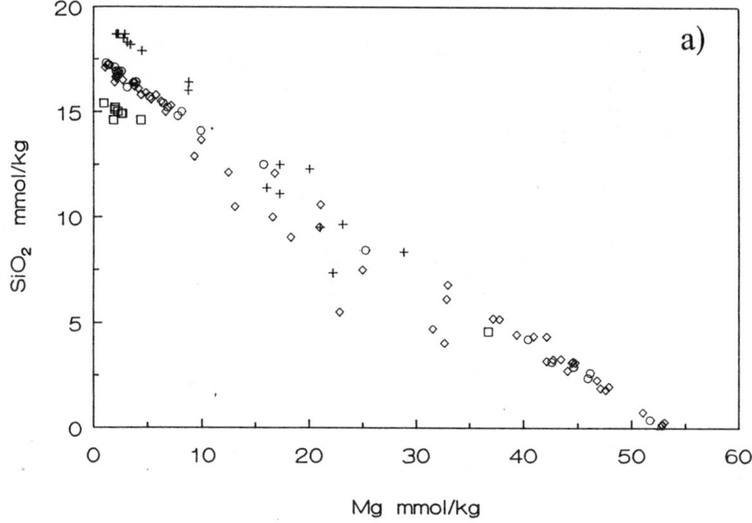

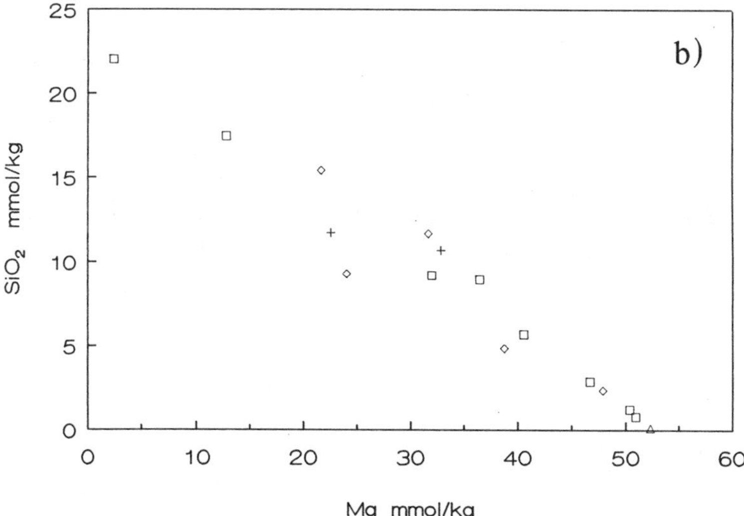

Figure 2 Graph of SiO_2 versus Mg for samples collected from (*a*) 21°N East Pacific Rise (EPR) in 1981 and (*b*) the southern Juan de Fuca Ridge (SJdF) in 1984, demonstrating the density of sample coverage and the amount of mixing occurring with seawater (Mg = 53 mmol kg^{-1}) during sampling. Different symbols designate different vent fields (see Table 1 for explanation of abbreviations). For (*a*), + = NGS, ○ = OBS, ◇ = SW, and □ = HG; for (*b*), □ = Plume, ◇ = Vent 1, and + = Vent 3. While values for 21°N EPR vents are well defined, those for Vent 3 of the SJdF are poorly defined owing to the collection of only two samples containing large amounts of admixed seawater from this site.

Table 1 Summary of physical characteristics of discovered seafloor hydrothermal areas

Vent area	Measured (inferred) temperature (°C)	Water depth (m)	Spreading rate (whole) ($cm\ yr^{-1}$)	Sediment cover (m)	Years sampled	Other comments
Galapagos spreading center (GSC)		2450	7	None	1977, 1979	Low exit temperatures are result of subsurface mixing with cool seawater
Clambake (CB)	< 13(350)					
Garden of Eden (GE)	< 13(350)					
Dandelions (DL)	< 13(350)					
Oyster Beds (OB)	< 13(350)					
21°N East Pacific Rise (EPR)		2600	6	None	1979, 1981, 1985	
National Geographic Society (NGS)	273					
Ocean Bottom Seismograph (OBS)	350					
Southwest (SW)	355					
Hanging Gardens (HG)	351					
Guaymas Basin		2000	6	~500	1982, 1985, 1988	Sediment cover very organic- and carbonate rich
1	291					
2	291					
3	285					
4	315					
5	287					
6	264					
7	300					
9	100					
Southern Juan de Fuca Ridge (SJdF)		2300	6	None	1984, 1987	
Plume	224(> 340)					
Vent 1	285(> 340)					
Vent 3	(> 340)					
11–13°N EPR		2600	12	None	1982, 1984, 1985	

N & S–13°N	317					
1–13°N						
2–13°N	354					
3–13°N	380?					
4–11°N	347					
5–11°N						
6–11°N						
Mid-Atlantic Ridge (MAR)		3600	2.6	None	1986	
TAG	290, 321					
MARK-1	350					
MARK-2	335					
Axial Volcano, Juan de Fuca Ridge		1542		None	1986, 1987, 1988	Shallowest system sampled
Hell, Hillock & Mushroom (HE, HI, MR)	136–323					
Inferno	149–328					
Virgin Mound (VM), Crack	5–299					
Escanaba Trough, Gorda Ridge	18–220	3200	2.3	≤500	1987	
Middle Valley, Juan de Fuca Ridge		2500	6	~300–≥1500	—	Black smokers, but not yet sampled. Sediment is mostly terrigenous
Endeavour segment, Juan de Fuca Ridge	>400	2200	6	None	1984	
Mariana Trough	285–287	3650	6		1987	Back-arc spreading center
Red Sea, Atlantis II Deep	62.3(>250)	2000	1.5	10–30	Since mid-1960s	Fluids form brine pools; direct venting not observed. Fluids react with Miocene evaporites and basalt

seawater interaction under hydrothermal conditions representative of the seafloor systems has shown that Mg is close to being quantitatively removed from the solutions at temperatures $\geq 150°C$ (e.g. Bischoff & Dickson 1975, Seyfried & Bischoff 1979). In order to compare all of the sampled fluids from different locations on the same basis, it is assumed that any Mg present is due to the entrainment of ambient seawater, and all fluid compositions are extrapolated to Mg = 0 mmol kg^{-1}. For some samples, the extrapolated value is within the experimental error of the analytical data, whereas in other cases the extrapolation is considerably larger. All data in the subsequent tables and discussion are for the extrapolated end-member (Mg = 0 mmol kg^{-1}) values. This approach may not be valid for lower temperature solutions, from which all the Mg may not have been removed by rock-water interaction, and may obfuscate evidence related to mixing of higher and lower temperature solutions. It remains, however, the best means of correcting for seawater entrainment during sampling.

Table 2 presents a compendium of data for most of the seafloor hydrothermal fluids sampled to date; also listed for purposes of comparison is the composition of seawater. Each vent area normally consists of a number of isolated vent fields, which may be separated by a few kilometers. These vent fields contain individual vents that may have subtle or even large differences in chemistry over very short (e.g. 10 m) geographical distances. While the 21°N East Pacific Rise (EPR), 11–13°N EPR, southern Juan de Fuca Ridge (SJdF), Galapagos spreading center (GSC), Mid-Atlantic Ridge (MAR), and Axial Volcano fluids have reacted only with basalt, the Guaymas Basin fluids have also reacted with an organic- and carbonate-rich sediment cover. The sediment covers in Escanaba Trough of the Gorda Ridge and Middle Valley of the Juan de Fuca Ridge are more terrigenous and contain less organic matter and carbonate. The Red Sea remains a unique example of an area where hydrothermal fluids form brine pools on the seafloor owing to their high salinity, gained through reaction with Miocene evaporites. The GSC fluids have mixed below the seafloor with cool seawater, and the Axial Volcano and SJdF fluids have probably been affected by phase separation.

In the remainder of this section, I briefly summarize the results found in Table 2.

The alkalis and ammonium The elements Li, K, Rb, and Cs are variably enriched with respect to the initial seawater value, whereas sodium exhibits both enrichments and depletions with respect to initial seawater. As Na is the predominant cation in these fluids, its behavior is closely tied to that of chloride, the predominant anion. Although the sediments at Guaymas

Basin provide an additional, more enriched source of the alkali elements than does the basalt alone at the other sites, Guaymas Basin fluids do not contain the largest enrichments of all of these elements. The SJdF fluids are in many cases the most enriched in the alkalis, probably as a result of the higher chloride content of these fluids, which in turn complexes these metals and permits higher levels to be maintained in solution. Although small amounts of ammonium have been found in other vents (M. D. Lilley, personal communication, 1988), only the Guaymas Basin fluids contain substantial amounts of this compound. The ammonium is produced as a consequence of the decomposition of organic matter in the sediment column.

Alkaline earths Magnesium is assumed to be zero in the pure hydrothermal fluid, and samples with less than 1 mmol kg^{-1} Mg have been collected from a number of vents. The elements Be and Ca are both variably enriched in all the vent areas, whereas strontium is one of the few elements that displays both enrichments and depletions relative to the seawater value. Calcium varies by almost an order of magnitude, and in some vents it is essentially identical to the seawater value. Although the carbonate-rich sediments in Guaymas Basin provide an additional source of both Ca and Sr, these elements reach their highest concentration in the SJdF solutions, again presumably as a result of enhanced chloride complexing. The Sr isotopes suggest a predominantly basaltic source for the Sr present in the hydrothermal solutions, with a variable (but small) component of seawater Sr. The Guaymas Basin fluids are significantly more radiogenic, which may reflect reaction with sedimentary carbonates or the admixture of additional seawater Sr. Barium forms the very insoluble sulfate mineral barite. As some sulfate is always present owing to the entrainment of seawater during sampling, only a lower limit can be given for the barium dissolved in the hydrothermal solutions. Radium-226 and its daughters (radon-222 and lead-210) are also enriched (Krishnaswami & Turekian 1982, Dymond et al 1983, Kadko 1988), although radium enrichments are undoubtedly also affected by the precipitation of sulfate minerals.

Silica The silica contents of all the measured vent solutions vary less than many of the other chemical parameters. This relative lack of variation is believed to be due to the control of dissolved silica by equilibrium with quartz in these systems (Von Damm et al 1985a). A possible exception is the Guaymas Basin fluids, where the presence of other phases controlling the activity of silica cannot be ignored owing to both the lower temperatures (favoring amorphous silica) and authigenic aluminosilicates found in the sediment cover.

Table 2a Summary of chemical data for seafloor hydrothermal solutions. Alkali and alkaline earth metals, ammonium, and silica[a]

Vent	Temp. (°C)	Li (μmol kg^{-1})	Na (mmol kg^{-1})	K (mmol kg^{-1})	Rb (μmol kg^{-1})	Cs (nmol kg^{-1})	NH_4 (mmol kg^{-1})	Be (nmol kg^{-1})	Mg (mmol kg^{-1})	Ca (mmol kg^{-1})	Sr (μmol kg^{-1})	$^{87}Sr/^{86}Sr$	Ba (μmol kg^{-1})	SiO_2 (mmol kg^{-1})
Galapagos spreading center								11–37						
CB	<13	1142	487	18.7	20.3				0	40.2	87		>42.6	21.9
GE	<13	1142	451	18.8	21.2				0	34.3	87		>17.2	21.9
DL	<13	1142	313	18.8	17.3				0	34.3	87		>17.2	21.9
OB	<13	689	259	18.8	13.4				0	24.6	87		>17.2	21.9
21°N EPR														
NGS	273	1033	510	25.8	31.0		<0.01	37	0	20.8	97	0.7030	>16	19.5
OBS	350	891	432	23.2	28.0	202	<0.01	15	0	15.6	81	0.7031	>8	17.6
SW	355	899	439	23.2	27.0		<0.01	10	0	16.6	83	0.7033	>10	17.3
HG	351	1322	443	23.9	33.0		<0.01	13	0	11.7	65	0.7030	>11	15.6
Guaymas Basin														
1	291	1054	489	48.5	85.0		15.6	12	0	29.0	202		>12	12.9
2	291	954	478	46.3	77.0		15.3	18	0	28.7	184		>20	12.5
3	285	720	513	37.1	57.0		10.3	42	0	41.5	253	0.7052	>15	13.5
4	315	873	485	40.1	66.0		12.9	29	0	34.0	226	0.7052	>54	13.8
5	287	933	488	43.1	74.0		14.5	29	0	30.9	211		>13	12.4
6	264	896	475	45.1	74.0		14.5	60	0	26.6	172	0.7059	>16	10.8
7	300	1076	490	49.2	86.0		15.2	17	0	29.5	212		>24	12.8
9	100	630	480	32.5	57.0		10.7	91	0	30.2	160			9.3

Southern Juan de Fuca												0.7034		
Plume	224	1718	796	51.6	37.0			95	0	96.4	312			23.3
Vent 1	285	1108	661	37.3	28.0			150	0	84.7	230			22.8
Vent 3		1808	784	45.6	32.0			150	0	77.3	267			22.7
11–13°N EPR														
N & S–13°N	317	688	560	29.6	14.1				0	55.0	175	0.7041		22.0
1–13°N		614	587	29.8	18.0				0	44.6	171			21.9
2–13°N	354	592	551	27.5	19.0				0	53.7	182			19.4
3–13°N	(380)	591	596	28.8	20.0				0	54.8	168			17.9
4–11°N	347	884	472	32.0	24.0				0	22.5	80			18.8
5–11°N		623	577	32.9	25.0				0	35.2	135			20.6
6–11°N		484	290	18.7	15.0				0	10.6	38			14.3
Mid-Atlantic Ridge														
TAG	290, 321	411	584	17.0	10.0	100			0	26.0	99	0.7029		22.0
MARK-1	350	843	510	23.6	10.5	177		38.5	0	9.9	50	0.7028		18.2
MARK-2	335	849	509	23.9	10.8	181		38.0	0	10.5	51	0.7028		18.3
Axial Volcano														
HE, HI, MR	136–323	512	415	22.0					0	37.3				15.1
Inferno	149–328	637	500	27.5					0	46.8				15.1
VM, Crack	5–299	204	159	7.6					0	10.2				13.5
Seawater	2	26	464	9.8	1.3	2.0	<0.01	0.0	52.7	10.2	87	0.7090	0.14	0.16

[a] Data references listed at end of Table 2*c*.

Table 2b Summary of chemical data for seafloor hydrothermal solutions. pH, carbon and sulfur systems, halogens, boron, aluminum, and water isotopes[a]

Vent	pH	Alk_T (meq kg^{-1})	Total CO_2 (mmol kg^{-1})	SO_4 (mmol kg^{-1})	H_2S (mmol kg^{-1})	$\delta^{34}S$	As (nmol kg^{-1})	Se (nmol kg^{-1})	Cl (mmol kg^{-1})	Br (μmol kg^{-1})	B (μmol kg^{-1})	$\delta^{11}B$	Al (μmol kg^{-1})	$\delta^{18}O$	δD
Galapagos spreading center			9.3–11.3							832–835					
CB	–	<0		0	+				595						
GE	–	<0		0	+				543						
DL	–	<0		0	+				395						
OB	–	<0		0	+				322						
21°N EPR			5.72											1.6–2.0	2.5
NGS	3.8	−0.19		0.00	6.6	3.4	30	<0.6	579	929	507	32.7	4.0		
OBS	3.4	−0.40		0.50	7.3	1.3–1.5	247	72	489	802	505	32.2	5.2		
SW	3.6	−0.30		0.60	7.5	2.7–5.5	214	70	496	877	500	31.5	4.7		
HG	3.3	−0.50		0.40	8.4	2.3–3.2	452	61	496	855	548	30.0	4.5		
Guaymas Basin															
1	5.9	10.60		−0.15	5.8		283	82	601	1054–1117			0.9		
2	5.9	9.60		−0.09	4.0		732	87	589		1630	17.4	1.2		
3	5.9	6.50		−0.34	5.2		1071	38	637				6.7		
4	5.9	8.10		0.06	4.8		1074	103	599	1063	1570	23.2	3.7		
5	5.9	9.70		−0.07	4.1		516		599				3.0		
6	5.9	7.30		−0.32	3.8		669	49	582				3.9		
7	5.9	10.50		−0.06	6.0		711	92	606	1054–1117	1730	19.6	1.0		
9	5.9	2.80		−4.20	4.6		577		581				7.9		

Southern Juan de Fuca			3.92–4.46												−2.5–+0.5
Plume	3.2			−0.50	3.5	4.2–7.3		<1	1087	1832	496	34.2		0.65	
Vent 1	3.2			−1.30	(3.0)	4.0–6.4		<1	896	1580				0.60	
Vent 3	3.2			−1.70	(4.4)	7.2–7.4		<1	951	1422				0.80	
11–13°N EPR															
N & S–13°N	3.2		10.8–16.7	0.00					740	1163				0.39–0.69	0.62–1.49
1–13°N	3.2	−0.64			2.9				718	1131			13.3		
2–13°N	3.1	−0.74			8.2	4.7			712	1158	467	34.9	19.8		
3–13°N	3.3	−0.40			4.5	2.3–3.5			760	1242			20.0		
4–11°N	3.1	−1.02			8.0	4.6			563	940			12.9		
5–11°N	3.7	−0.28			4.4	4.7–4.9			686	1105	451	36.8	13.5		
6–11°N	3.1	−0.88			12.2	4.1–5.2			338	533	493	31.5	13.6		
Mid-Atlantic Ridge															
TAG									659						
MARK-1	3.9	−0.06			5.9	4.9			559	847	518	26.8	5.3	2.37	
MARK-2	3.7	−0.24			5.9	5.0			559	847	530	26.5	5.0	2.37	
Axial Volcano											476	34.7			
HE, HI, MR	3.5	−0.52			8.1				515	760	565			0.8–1.1	
Inferno	3.5	−0.45			7.0				625	950	565			0.8–1.1	
VM, Crack	4.4	0.58	150–170		(19.5)				188	240	503			0.7–0.9	
Seawater	7.8	2.3	2.3	27.9	0.0		27	2.5	541	840	416	39.5	0.020	0.0	0.0

[a] Data references listed at end of Table 2*c*.

Table 2c Summary of chemical data for seafloor hydrothermal solutions. Trace metals[a]

Vent	Mn (μmol kg^{-1})	Fe (μmol kg^{-1})	Co (nmol kg^{-1})	Cu (μmol kg^{-1})	Zn (μmol kg^{-1})	Ag (nmol kg^{-1})	Cd (nmol kg^{-1})	Pb (nmol kg^{-1})
Galapagos spreading center								
CB	1140	+		0			0	
GE	390	+		0			0	
DL	480	+		0			0	
OB	360	+		0			0	
21°N EPR								
NGS	1002	871	22	<0.02	40	<1	17	183
OBS	960	1664	213	35.00	106	38	155	308
SW	699	750	66	9.70	89	26	144	194
HG	878	2429	227	44.00	104	37	180	359
Guaymas Basin								
1	139	56	<5	<0.02	4.2	230	<10	265
2	222	49	<5	<0.02	1.8	<1	<10	304
3	236	180	<5	<0.02	40.0	24	46	652
4	139	77	<5	1.10	19.0	2	27	230
5	128	33	<5	0.10	2.2	<1	<10	<20
6	148	17	<5	<0.02	0.1	<1	<10	<20
7	139	37	<5	<0.02	2.2	<1	<10	<20
9	132	83	<5	<0.02	21.0	<1	<10	<20
Southern Juan de Fuca								
Plume	3585	18,739		<2	900			900
Vent 1	2611	10,349		<2	<600			
Vent 3	4480	17,770		<2				

11–13°N EPR								
N&S–13°N	1000	1450						
1–13°N	1689	3980			102.0		55	135
2–13°N	2932	10,370			5.0		70	27
3–13°N	2035	10,760			2.0		65	14
4–11°N	766	6470			105.0		30	50
5–11°N	742	1640			73.0		43	270
6–11°N	925	2640			44.0		1	9
Mid-Atlantic Ridge								
TAG	1000	1640						
MARK-1	491	2180		17.0	50.0			
MARK-2	493	1832		10.0	47.0			
Axial Volcano								
HE, HI, MR	1081	1006		12.0	113.0			302
Inferno	1081	1006		12.0	113.0			302
VM, Crack	162	9		0.70	2.3			101
Seawater	<0.001	<0.001	0.03	0.007	0.01	0.02	1.0	0.0100

[a] Data from following references:
Galapagos: Edmond et al 1979a,b, Welhan 1981.
21°N EPR: Craig et al 1980, Welhan 1981, Von Damm et al 1985a, Spivack & Edmond 1987, Woodruff & Shanks 1988, Campbell & Edmond 1989.
Guaymas: Von Damm et al 1985b, Spivack et al 1987, Campbell & Edmond 1989.
Juan de Fuca: Von Damm & Bischoff 1987, Shanks & Seyfried 1987, Hinkley & Tatsumoto 1987, Evans et al 1988, Campbell & Edmond 1989.
11–13°N EPR: Michard et al 1984, Merlivat et al 1987, Bowers et al 1988, Bluth & Ohmoto 1988, Campbell & Edmond 1989.
Mid-Atlantic Ridge: Campbell et al 1988b, Campbell & Edmond 1989.
Axial: Butterfield et al 1988, Massoth et al 1989.
( ) denotes large uncertainty in value.

pH The 25°C, 1 atm pH is quite acid for all of the vent fluids with the exception of the Guaymas Basin fluids. The calculation of an in situ pH requires a good knowledge of all the aqueous complexes. The original estimates of Bowers et al (1985) of values just a few tenths of a pH unit higher than the 25°C value have been superseded by values suggesting an in situ pH closer to 4.5 (Bowers et al 1988). The calculation of an in situ pH for the Guaymas Basin solutions is more problematic still, owing to the presence of unknown organic ligands. Best estimates suggest values close to neutrality.

Carbon system At the acid pHs observed for most of these fluids, zero, or even a slightly negative, total alkalinity has been measured. The exception is again the Guaymas Basin fluids, which have alkalinities significantly higher than seawater. The most abundant carbon species present in these solutions is dissolved CO_2. Carbon dioxide is also the most abundant gas in the vent fluids. Hence, measurements of the total condensable gases are primarily of the CO_2. Small amounts of CH_4 are also present, as are lesser amounts of short-chain hydrocarbons (Evans et al 1988). Measurements of the total CO_2 and other gases in Guaymas Basin fluids have been hindered by the extremely high gas content of these fluids, which tend to degas after sampling as the pressure decreases during the return of the submersible and samples to the surface. Gas-tight samples have shown the levels of total condensable gases to be very high, but the individual gases remain to be determined (J. Lupton, personal communication, 1989). Carbon isotopic measurements suggest that most of the carbon at GSC and 21°N EPR has a "mantle" signature, with $\delta^{13}C = -7.00‰$ vs. the PDB carbon standard at 21°N EPR (Welhan 1981).

Sulfur system Essentially all of the seawater sulfate is removed from the fluids, and variable amounts of sulfide are present. There is a net loss of sulfur from the solutions compared with the initial seawater concentration. The sulfur isotope systematics suggests that the sulfide present in the vent fluids is primarily from a basaltic source, with variable amounts of reduced-seawater sulfate being present as well. The majority of the seawater sulfate is probably being removed by the precipitation of anhydrite ($CaSO_4$) during the heating of cool seawater in the hydrothermal downflow zone. Several authors (Woodruff & Shanks 1988, Bluth & Ohmoto 1988) have suggested that the reduced-seawater sulfur is added within the upflow zone, either deep in the system or perhaps very close to the seafloor. Arsenic and selenium, which under certain conditions behave similarly to sulfur, are also enriched in most of the fluids compared with their initial seawater concentrations. The sediment cover in Guaymas Basin and Escanaba Trough provides an additional source for these two elements.

Halogens Fluoride has only been measured in a few of the hydrothermal fluids, and available evidence suggests that it is completely removed during the hydrothermal circulation (Edmond et al 1979a, Von Damm et al 1985a). Chloride is variably enriched and depleted in the hydrothermal solutions, with many of the sampled fluids having concentrations relatively close to that of seawater. The chloride concentration is important in controlling the concentration of other species, as it is the dominant anion in the fluids. For this reason its cycle is closely tied to that of sodium. The Na/chloride molar ratio (Table 3) suggests that in all vents except at 21°N EPR and MAR, there is a net loss of Na with respect to chlorides, regardless of whether chloride is enriched or depleted relative to the seawater concentration. The relative Na loss probably results from its uptake by the rock during albitization. The reasons for the variability in chloride concentrations remain open to speculation and are discussed in detail in a later section. Campbell & Edmond (1989) have recently shown that bromide appears to maintain a bromide/chloride ratio that is almost constant and close to the seawater value in the basaltic systems. They have also noted an input of Br and I to the fluids in sedimented systems from the decomposition of organic matter.

Boron The concentration of boron in seawater is relatively high, and only a relatively small addition is made to the hydrothermal fluids, with the exception of those in the Guaymas Basin. The boron isotope systematics confirms this additional sedimentary source (Spivack & Edmond 1987, Spivack et al 1987).

Aluminum Aluminum is significantly enriched in the vent fluids in all of these systems, although the amount added to the water is relatively small compared with the amount available from the rock. The 11–13°N EPR vent fluids contain the largest enrichments.

Table 3 Molar ratio of Na/Cl in vent fluids

Vent	Na/Cl
GSC	0.79–0.83
21°N EPR	0.88–0.89
Guaymas Basin	0.81–0.83
SJdF	0.73–0.82
11°–13°N EPR	0.76–0.86
MAR	0.89–0.91
Axial Volcano	0.80–0.85
Seawater	0.86

Water isotopes The oxygen and hydrogen isotopic contents of the vent fluids have been determined for a limited number of samples. Although most of the observed values differ from initial seawater, a systematic variation is not observed.

Trace metals Several difficulties are encountered in obtaining good trace-metal data for the vent fluids. At many vent areas, the fluids are clear as they exit from the sulfide chimneys and then rapidly begin to precipitate the characteristic black smoke of metal sulfides and sulfates as they mix with seawater. At other vents this mixing appears to begin within or below the chimney itself, as the fluids already appear to contain particles when they reach the chimney orifice. In some cases, chimney particles are also introduced into the fluid samples during the sampling procedure. The metals from particles precipitated during the sampling need to be incorporated into the metal content of the fluids, while chimney particles need to be excluded. It is sometimes possible for fluid samples containing very small amounts of admixed seawater to find essentially no particles, as is the case for some of the samples from 21°N EPR. Often, samples that were difficult to obtain suffer from the multiple problems of containing a large proportion of seawater, which results in an increased pH and hence in their having precipitated a large amount of their metal content, and of containing small pieces of chimney material. A consistent trace-metal concentration versus Mg concentration (e.g. the mixing line as shown in Figure 2) or several samples of similar Mg concentration with similar metal contents provide the best check on the accuracy of the trace-metal data. These difficulties are part of the reason that more comprehensive trace-metal data are not given for many of the vent areas (Table 2). Metals that are sulfide formers suffer from the largest uncertainties. Iron and manganese are extremely enriched in all of these fluids compared with the seawater value, although the enrichment is less so in the Guaymas Basin fluids. The enrichment of copper is also variable and appears to be related to temperature, with only those vents with exit temperatures $\geq 350°C$ containing substantial copper. Zinc is also quite enriched with respect to seawater in the hydrothermal solutions, especially in the high-chloride and $<300°C$ SJdF solutions. Cadmium, cobalt, silver, and lead are variably enriched in the hydrothermal fluids. The higher lead concentrations in the Guaymas Basin fluids reflect the additional sedimentary input of this element, and this conclusion is confirmed by the isotopic composition (Chen et al 1986).

Gases In addition to the enrichments in CO_2, CH_4, and H_2S noted above, the fluids are significantly enriched in ^3He, ^4He, and hydrogen (Craig et al 1980, Welhan 1981, Merlivat et al 1987, Evans et al 1988, Kennedy 1988).

Controls on the Composition of the Solutions

As we cannot at present see below the seafloor into an active seafloor hydrothermal system, we must infer the subsurface processes that are occurring either from the solution chemistry alone or by analogy with ophiolites or the rocks found within Deep Sea Drilling Project (DSDP) Hole 504B (Leinen, Rea et al 1986). Some of the parameters that influence solution chemistry and hence the chemistry of the chimney deposits include the following: (*a*) temperature, both in the subsurface and at the seafloor; (*b*) pressure, which is related to the water depth; (*c*) rock composition, which is relatively constant; (*d*) at some locations, sediment compositions, which can be highly variable; (*e*) whether the rock is fresh or has been previously altered; (*f*) whether the fluids have reached equilibrium with the rocks or, alternatively, whether kinetics is controlling the observed fluid compositions and hence placing importance on the residence time of the fluid in the system; (*g*) whether the fluids have undergone, or are presently undergoing phase separation; and (*h*) whether subsurface mixing of different fluids has occurred.

Bowers et al (1988), based on results from the geochemical modeling code EQ3/6 (Wolery 1983), have recently suggested that the fluids are in equilibrium with a mineral assemblage that is similar to a greenschist facies assemblage (Table 4). It should be noted that many of the needed thermodynamic constants are poorly known at the pressure and temperature conditions of interest. Fluids with salinities significantly higher than seawater or temperatures much above 350°C are also outside the range of available data and models, which limits the applicability of this approach to solutions with higher ionic strengths. Many elemental ratios

Table 4 Proposed equilibrium mineral assemblage based on observed fluid chemistry and EQ3/6 modeling results of Bowers et al (1988)

Mineral	Chemical formula
Quartz	SiO_2
K-feldspar	$KAlSi_3O_8$
Albite	$NaAlSi_3O_8$
Anorthite	$CaAl_2Si_2O_8$
Muscovite	$KAl_2(AlSi_3O_{10})(OH)_2$
Paragonite	$NaAl_2(AlSi_3O_{10})(OH)_2$
Ca-beidellite	$Ca_{0.165}Al_2(Al_{0.33}Si_{3.67}O_{10})(OH)_2$
Epidote	$Ca_2FeAl_2Si_3O_{12}(OH)$
Daphnite	$Fe_5Al(AlSi_3O_{10})(OH)_8$
Pyrrhotite (pyrite)	$FeS(FeS_2)$

and products and chemical geothermometers give similar results for all of the fluids (with the exception of Guaymas Basin), which also suggests an equilibrium or steady-state control (rather than kinetic) on the fluid chemistry (Von Damm 1988). The similarity between the model results and the more empirical elemental ratios for the areas examined also suggests that the conditions at depth within these hydrothermal systems are quite similar. Additional evidence bearing on the mineralogy of the subsurface alteration that is occurring in these hydrothermal systems comes from DSDP Hole 504B (Table 5), which penetrated over a kilometer of the oceanic crust. This drill hole cored through the sediments and pillow lavas into a sheeted dike section and intersected an area of hydrothermal alteration (Alt et al 1984). The mineralogy observed in the 504B core section is, with few exceptions, quite similar to that proposed by Bowers et al (1988). Studies of ophiolites are also providing insight into the subsurface behavior of the fluids, including flow paths and alteration processes (Haymon et al 1989, and references therein).

Mixing of hydrothermal fluids in the subsurface may occur more commonly than previously thought. Fluids from the GSC display clear chemical evidence for mixing between high-temperature hydrothermal fluids and cold seawater below the seafloor (Edmond et al 1979a,b). At SJdF, the chemistry of the fluids suggests that subsurface mixing of a brine with a hydrothermal solution having a chlorinity close to that of seawater has occurred (Von Damm & Bischoff 1987). Butterfield et al (1989) have also invoked subsurface mixing of hydrothermal fluids with differing compositions to explain the observed fluid chemistries at Axial Volcano. Mixing of a high-temperature fluid with a small amount of a low-temperature fluid may also explain the chemistry and temperature of the NGS vent at 21°N EPR, at which the best-sampled fluids appear to contain slightly more magnesium than is contained in other fluids from this location. The

Table 5 Observed mineral phases and potential elemental sinks in DSDP Hole 504B[a]

Element	Mineral phase(s)
Na	Albite
K	Orthoclase found in albite
Ca	Actinolite, epidote, sphene, chlorite-clay mixture
Fe	Sulfides, etc
Mn	Chlorite-clay mixture (actinolite, albite, epidote, sphene)
Si	Quartz
S	Sulfides

[a] Reference: Alt et al (1984).

concentrations of various elements (especially silica) in these solutions, which are interpreted to have mixed, suggest that the fluids are equilibrated with the same mineral assemblage as are the other vent fluids. This suggests in turn that the fluids have reequilibrated with the rock after mixing has occurred (Von Damm 1988, Butterfield et al 1989). These authors favor a model in which this observed composition of the hydrothermal fluids is explained as a mixture between a gas-depleted high-chlorinity fluid (brine), a gas-enriched low-chlorinity fluid (vapor), and a gas-containing fluid having a chlorinity close to that of seawater.

Whichever models are invoked to explain the chemistries of the vent fluids, they must account for several surprising observations. One of these is the time constancy of the chemistry of the vent solutions. Given the high heat output and the dynamic nature of these systems, it was initially thought that vents would rapidly come and go and that the chemistry of the fluids would reflect this instability on a time scale of years. Most observations to date, however, suggest that these vents, including the water chemistry, are stable on longer time scales. Repeated samplings of vent fluids at 21°N EPR in 1979, 1981, and 1985 (Campbell et al 1988a), the Guaymas Basin in 1982, 1985, and 1988 (Campbell et al 1988a, Gieskes et al 1988), the SJdF in 1984 and 1987 (Von Damm & Bischoff 1987, Massoth et al 1988), and the ASHES vent field on Axial Volcano in 1986, 1987, and 1988 (Massoth et al 1989) show very little change in fluid chemistries or temperatures. These observations support an equilibrium or steady-state model and also suggest that the at-depth hydrologic systems are very stable. This constancy provides a striking dichotomy to the episodicity of the two "megaplume" events observed on the Juan de Fuca Ridge (Baker et al 1987), in which large amounts of hydrothermal fluids and particles, and therefore heat, were released to the water column over a brief period of time, as evidenced by the presence of two large, particle-rich water column plumes. It remains an open question as to how common this type of event is on the world ridge-crest system. Evidence of changing fluid compositions also comes from the chimney deposits themselves. How these seemingly contradictory observations relate to the life cycle of an individual vent or vent field and to their relative input of heat and solutes to seawater remains unresolved.

Chloride Variability—Possible Explanations

The observed variability in the chloride content of vent solutions ranges from approximately one third to twice the normal seawater concentration. As chloride is by far the dominant anion in these solutions, its concentration plays a major role in their metal-carrying capacities, as well as in their physical and chemical properties, such as density and critical point. Several mechanisms that could significantly affect the chloride content of the solutions are available: (*a*) Although flow through evaporites has been invoked as a source of increased chlorinity for the Red Sea brines, it is unlikely for most of these other solutions, which occur on unsedimented, open ocean-ridge axes. (*b*) Although direct input of volatiles, such as water and HCl from magma degassing, cannot be ruled out, at present no evidence exists to suggest that this process is an important contributor to the chemistry of the hydrothermal fluids. (*c*) Small amounts of water can be lost from a solution as a result of rock hydration and could potentially be re-released at a later time owing to changes in the conditions in the hydrothermal system (e.g. a later thermal pulse, with resulting higher temperatures). (*d*) Not only water but also chloride could occupy sites in hydrous minerals that may serve as sources or sinks, depending on changing conditions. (*e*) Precipitation or dissolution of an unidentified chloride-rich phase with possible retrograde solubility has been suggested as a cause of the observed chloride anomalies. (*f*) Phase separation could result in the formation of low- and high-salinity fluids and may even result in the formation of halite, which could later redissolve. The high-salinity fluids, given their higher densities and lower buoyancies, may be stored in the oceanic crust between thermal pulses. Figure 3 shows the phase relation-

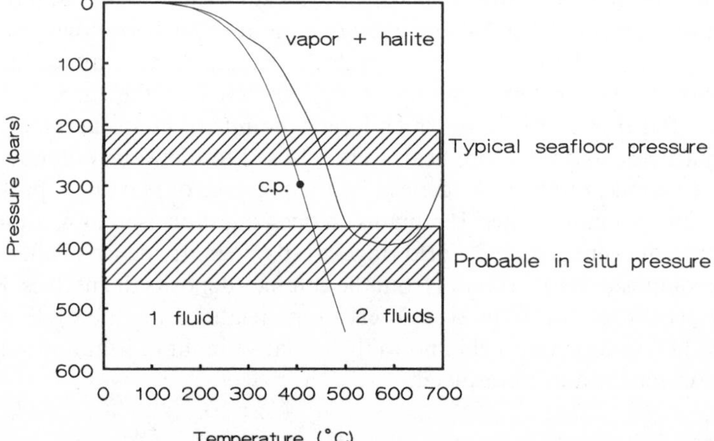

Figure 3 A pressure-temperature diagram applicable to seafloor hydrothermal systems. The critical point (c.p.) of seawater is shown (Bischoff & Rosenbauer 1988) along the two-phase boundary. The halite saturation curve is based on data from Sourirajan & Kennedy (1962). Also shown is a range of pressures typical of conditions at the seafloor for many of the sampled hydrothermal systems, as well as a range of inferred pressures occurring at depth within the hydrothermal systems.

ships that apply to a seawater hydrothermal system. In a seawater system, crossing the two-phase boundary below the critical point (i.e. boiling) will result in the formation of a small amount of a low-salinity "vapor," while crossing the two-phase boundary above the critical point (i.e. condensation) will result in the formation of a small amount of high-salinity "brine."

In addition to the sampled hydrothermal fluids, other evidence suggests that high-salinity fluids exist with the oceanic crust. In fact, the other evidence suggests that fluids of much higher salinities exist than have been sampled from vents. Table 6 summarizes the results from several fluid inclusion studies. In several of these studies (e.g. Cowan & Cann 1988) both high- and low-salinity inclusions were found, providing evidence consistent with phase separation as the cause. Although the data are limited, there is some evidence to suggest that these very high salinity fluids occur at greater depths in the oceanic crust than those depths at which axial hot springs are currently believed to circulate (Vanko 1988). This is one reason why Bischoff & Rosenbauer (1989) have recently suggested that two stacked hydrothermal circulation cells exist within the oceanic crust, with the deeper one being very saline and primarily responsible for the albitization of the oceanic crust and for transferring heat and some salt to the shallower, less saline cell. Several authors have also reported finding chloride-rich minerals within the oceanic crust. For example, Vanko (1986) and Stakes & Vanko (1986) found amphiboles containing up to 4.0 wt% chloride, compared with usual values of ≤ 0.05 wt%. They have inferred that these phases could only have formed in the presence of

Table 6 Summary of fluid inclusion evidence for the existence of high- and low-salinity fluids within the oceanic crust and critical point (C_p) data

Location	NaCl (wt%)	Inferred temperature, pressure (°C, bars)	Reference
Mathematician Ridge	58	600–>700, 600–1000	Vanko 1988
Kane Fracture	10	>407, >298.5	Delaney et al 1987
Zone	50	700, 1000–1200	Kelley & Delaney 1987
	1–2	400, 1000–1200	
Troodos	48(57)	500–525, 350–400	Cowan & Cann 1988
Ophiolite	0–2		
Seawater	3.2	$C_p = 407, 298.5$	Bischoff & Rosenbauer 1988
Distilled water	0	$C_p = 374.1, 220.4$	

fluids significantly more saline than seawater. Brett et al (1987) have reported finding lizardite [$Mg_3Si_2O_5(OH)_4$] containing up to 1.2 wt% chloride and a zinc-hydroxychlorosulfate [$Zn_{12}(OH)_{15}(SO_4)_3Cl_3 \cdot 5H_2O$] containing up to 7.0 wt% chloride. It should be noted that these phases are all relatively rare and could not provide the chloride source or sink needed to sustain the depletions and enrichments in chloride observed in a number of the hot springs. Edmond et al (1979a) first suggested a metal-hydroxycloride mineral found in altered ultrabasic rocks (Kohls & Rodda 1967, Ricklidge & Patterson 1977) as a possible source or sink for the chloride variations observed in the GSC hot springs. Seyfried et al (1986) were able to replicate the chloride depletions in an experimental study and suggested that the mineral phase involved was $Fe_2(OH)_3Cl$, although it was never identified as a run product. This phase has also yet to be identified from the seafloor. At the concentrations of chloride observed in the SJdF vents, which have been stable for at least three years, a very large amount of this phase would need to have formed and be redissolving, creating additional problems for the elemental cycles of Na and H^+ (Von Damm & Bischoff 1987). It has been suggested, as a test of the chloride-rich mineral hypothesis, that if a chloride-rich phase is forming it would discriminate against bromide, and hence that if high salinities in the fluids are the result of dissolution of a mineral phase, these same fluids should have a lowered bromide/chloride ratio compared with seawater. Conversely, low-salinity fluids formed by the precipitation of such phases should have elevated Br/Cl ratios. Campbell & Edmond (1989) have recently reported data on the bromide/chloride ratio for several of the high- and low-salinity vent fluids. They find that, with the exception of vents from sedimented systems where organic matter provides a significant additional source of bromide, the ratio is close to that of seawater. These results argue against the formation and dissolution of a chloride-rich phase having a significant influence on the chemistry of the vent fluids.

Von Damm & Bischoff (1987) and Von Damm (1988) have discussed the evidence for phase separation in the twice-seawater-chlorinity solutions found in the SJdF fluids. They have suggested, primarily on the basis of the gas data (Kennedy 1988, Evans et al 1988), that a fluid with approximately six times the chlorinity of seawater exists at depth within this system. Massoth et al (1989) have discussed the evidence for phase separation at Axial Volcano, where both high- and low-salinity fluids are found. The evidence is quite compelling that active phase separation of seawater is occurring at this site. Massoth et al (1988), in a revisit to the SJdF site, also found that the high chlorinity of the solutions had been maintained for three years. In addition, based on the heat and chemical balance between the SJdF vents and the overlying water column plume, they suggested that the high-temperature, high-salinity fluids were not representative of the vent field as a whole, because a heat and mass balance could not be made. Analysis of a sample of a diffuse water flow from the Plume site at SJdF has shown it to have a chlorinity below that of seawater (K. L. Von Damm, unpublished data). Contrary to the suggestion that phase separation had occurred at some unknown time in the past, resulting in the storage of brine within the oceanic crust at the SJdF site (Von Damm & Bischoff 1987), this finding suggests that phase separation may be an ongoing process at the site. Actively occurring phase separation could resolve the difficulties inherent in storing a three-year supply of brine within the crust, as well as the inconsistencies between the water column plume and vent chemistry data.

If brines (and vapors) exist within the oceanic crust, as suggested by fluid inclusion and other evidence described above, why do most of the hot springs that have been sampled fall within a relatively small compositional range close to seawater chlorinity? One problem relates to the segregation of the two phases from each other. Goldfarb & Delaney (1988) and Fox (1989) have recently discussed mechanisms related to this process, and although both agree that segregation can occur, their models give contrasting results. The model of Fox (1989) suggests that the low-salinity phase should form the diffuse flow, whereas that of Goldfarb & Delaney (1988) suggests that the brines will be the diffuse flow. A second problem relates to the sampling emphasis on "black smokers," which has resulted in an unrepresentative low sampling of the diffuse flows. Since it is the density of the fluids that is primarily responsible for their buoyancy, black smokers may, owing to the necessity to maintain rapid flow rates in order to build chimneys, be of a limited density and hence a limited compositional and temperature range. It is the density difference between the hydrothermal fluid and the overlying seawater that ultimately determines the hydrothermal fluid's buoyancy. McDuff (1988) has pointed out that the density of the hydrothermal fluid will affect the structure of the water column plume it forms. Density may, however, have a more primary control and affect which fluids exit from the seafloor and whether they exist as a diffuse flow. Table 7 gives the density of several different NaCl solutions at conditions representative of those at the seafloor (350°C, 200 bars) and at the higher temperature and pressure conditions (375°C, 400 bars) inferred to occur within the hydrothermal systems. With increasing salinity the density increases markedly, and even with larger increases in temperature (e.g. 450°C) it does not cause a large decrease in density and therefore does not result in a large increase in the density difference from seawater. High-salinity fluids, which have less of a density difference from bottom seawater, will also have less buoyancy, and this may result in lower

Table 7 Comparison of the densities of NaCl solutions similar to known or inferred hydrothermal solution chemistries at conditions representative of the seafloor (350°C, 200 bars) and within the hydrothermal system (375°C, 400 bars)[a]

Fluid	Molality NaCl	T (°C)	P (bars)	ρ (g cm^{-3})
Seawater	0.5	2	200	1.03
"Normal"	0.5	350	200	0.68
hydrothermal		375	400	0.66
Southern Juan	1.0	350	200	0.71
de Fuca		375	400	0.70
SJdF	3.0	350	200	—
brine		375	400	0.81
Brine	6.0	350	200	0.95
		375	400	0.94
		450	400	0.85
Distilled	0	350	200	0.59
water		375	400	0.61

[a] Data from Burnham et al (1969), Potter & Brown (1977) and Rogers & Pitzer (1982).

exit velocities on the seafloor. This lower velocity may also provide a longer transit time for the fluids to reach the seafloor, which may result in conductive cooling. The net effect of these processes may be that the highest salinity fluids have not been sampled because they appear as the diffuse flows, which are primarily identified by their "shimmering" (caused by their density differences with seawater). Similarly, in addition to the reasons given by Fox (1989) for the low-salinity phases to be diffuse flows, these solutions, although having the needed buoyancy to rise to the seafloor, will contain low metal and silica contents. Consequently, they cannot seal their channels by precipitation of metal sulfide and silica minerals and hence are more likely to entrain cold seawater "groundwater" during their passage through the crust to the seafloor. This entrainment may result in their cooling and hence a lowering of their velocity, with the result that they appear at the seafloor primarily as diffuse flows. An anhydrite chimney that is being formed as low-salinity and metal-poor fluids debouch from the seafloor at Axial Volcano (Massoth et al 1989) may result from an unusually tight "plumbing system" below the seafloor, since no other anhydrite chimneys have been observed.

While invoking phase separation as the primary mechanism to create high- and low-salinity seafloor hydrothermal fluids can explain many of the observations, it is not without its own problems. The concentrations of many of the dissolved species cannot be definitively used to determine if phase separation has occurred. In general, the metals should be concentrated in the high-salinity phase and the gases in the low-salinity phase. A specific example is the relationship between Fe, chloride, and H_2S. The higher chlorinity fluids appear to have $Fe/H_2S > 1$, while the inverse is true for the lower chlorinity fluids. In general, Fe content correlates with chloride content and is anticorrelated with H_2S content, but whether this is the result of Fe being primarily controlled by chloride-complexing or the solubility of Fe-sulfide minerals (or even some other factor) cannot be uniquely determined. The gas and isotopic data, which are of more limited availability for many of the vent fluids, in many cases do not necessitate phase separation and may even appear to argue against it. Welhan & Craig (1979) first addressed this issue with respect to the 21°N EPR fluids, most of which have chlorinities close to seawater. If phase separation were the primary mechanism responsible for chloride variability, one would expect to see the highest gas contents in the lowest chlorinity fluids, and vice versa. Similarly, for the species with more than one isotope, the heavy isotope would be expected to be enriched in the high-chlorinity phase. Both of these simple statements are complicated by the ongoing water-rock reactions. An example is the case of sulfur, whose cycle appears to be strongly influenced by additional reduction of heavy-seawater sulfate in the hydrothermal upflow zone (Woodruff & Shanks 1988). The oxygen isotopic data for the SJdF fluids are at least internally consistent, with the heavier solutions being the ones of highest chlorinity. These fluids are, however, lighter than lower chlorinity fluids found at other areas, such as at 21°N EPR. The reasons may be attributable to the extent and specific conditions of alteration occurring in the subsurface, as these isotopes are influenced by rock-water interactions in addition to phase separation. Cathles (1983) has demonstrated how the oxygen isotopic composition of the fluid and rock in a hydrothermal system can evolve with time. It may be that the isotopic data and chlorinities are not consistent from hydrothermal system to hydrothermal system because of differences between the evolutionary state of these systems, but that the data should be consistent within a single system.

MINERALOGY OF THE CHIMNEYS

Just as the fluid compositions retain certain common characteristics among all the vent areas studied despite individual differences, so too are there similarities and yet differences in the mineralogy of the chimneys. The iron sulfides (pyrite and pyrrhotite) are the most commonly observed minerals, followed by zinc sulfides (sphalerite and wurtzite) and variable amounts of copper sulfides (primarily isocubanite, bornite, and chalcopyrite). Lesser amounts of the lead sulfide galena have also been described from some areas. Several authors have presented models of chimney growth (Haymon 1983, Goldfarb et al 1983, Tivey & Delaney 1986, Graham et al 1988, Koski et al 1988), and there is a consensus that the growth of a chimney begins with an anhydrite ($CaSO_4$) shell. Anhydrite will form when seawater is heated to temperatures greater than $\sim 130°C$. It is commonly found in active chimneys but is absent from most extinct chimneys, presumably owing to its higher solubility under cool, seafloor conditions. Lesser amounts of barite are also found in some chimneys. Magnesium silicates and silica (as either amorphous silica or quartz) have also been described from a number of sites. Unusual or new minerals that have been described from one or more sites include caminite (a Mg-hydroxide-sulfate-hydrate) at 21°N EPR and other sites (Haymon & Kastner 1986); and a hydrated Zn, Fe-hydroxychlorosulfate, a (Mn, Mg, Fe)-hydroxide or hydroxy-hydrate, lizardite ($Mg_3Si_2O_5(OH)_4$), starkeyite ($MgSO_4 \cdot 4H_2O$), and anastase (TiO_2) at SJdF (Brett et al 1987). Aragonite ($CaCO_3$) has only been found in MAR samples (Thompson et al 1988). Chimneys on sedimented ridge crests, of which the Guaymas Basin is the best described, in general contain the same minerals but in quite different abundances from those on unsedimented ridges. Carbonate (primarily calcite) is much more abundant, as are galena and the sulfate minerals anhydrite and barite (Peter & Scott 1988). Organic matter also forms a major constituent of the chimneys at both Guaymas Basin and the Escanaba Trough. Escanaba Trough sulfides are also unusual for the amounts of arsenic, antimony, and silver that they contain (Koski et al 1988).

The general model of chimney growth includes the formation of an anhydrite shell due to the rapid mixing of the calcium-rich vent water and the sulfate-rich seawater. The building of the anhydrite shell occurs quite rapidly: Several observations from submersibles have noted growth rates on the order of centimeters to tens of centimeters per day. The deposition of metal sulfides then commences, and as the chimney walls thicken, so too does their isolation and insulation from seawater, which results in decreased mixing, a rise in temperature, and a progressive change in the mineral assemblage deposited. Owing to the temperature gradient through the wall, a "zone refining" occurs as previously deposited minerals are transformed to other mineral phases. In many cases, concentric mineral banding becomes well developed within the chimney walls.

In some cases the relationship of the chimney mineralogy to the fluid composition is quite clear. Some fluids that are zinc rich and copper poor appear to issue primarily from zinc-rich chimneys, and the alkalinity- and relatively lead-rich fluids at Guaymas Basin deposit chimneys containing abundant calcite and more galena. In other cases the relationship is less clear. Peter & Scott (1988) have described fluid inclusions from the Guaymas Basin chimneys that are much more saline than any fluids presently found in this area. Woodruff & Shanks (1988) and Bluth & Ohmoto (1988) have found substantial differences between the sulfur isotopic content of the vent fluids and that of the sulfide minerals of the chimneys. Several other authors (e.g. Marchig et al 1988) have noted other changes within the chimneys that lead them to suggest that the composition of the vent fluids has evolved through time.

In addition to the classic chimney shape, which narrows toward the top and may consist of multiple spires, several other morphologies have been noted. These include pagoda-shaped chimneys at Guaymas Basin, asparagus-shaped chimneys at SJdF, and onion-shaped chimneys at TAG, in addition to the flanges noted on Endeavour segment (Goldfarb 1988). Sizes vary from the tiny "fairy castles" on SJdF (on a scale of centimeters) to chimneys several tens of meters in height. The chimneys are unstable, and it is often possible to see collapsed chimneys being incorporated into sulfide mounds. Our understanding of the process of sulfide deposition on the seafloor is limited by our ability to sample only the uppermost, rather spindly chimneys that a submersible can most easily break off. Knowledge of the interior of these mounds is inferred from onland analogues. Similarly, the subsurface extent of these mounds is poorly known. Many hydrothermally active areas contain extensive lava drain-back features, and presumably a fair amount of void space exists below a thin lava skin. These voids may provide trapping areas for sulfide that would otherwise be dispersed as particles of smoke. It is currently thought that the sediment-hosted systems may contain larger deposits owing to higher trapping efficiencies. To date, the largest sulfide deposits on unsedimented ridges appear to be those at the Galapagos spreading center (Embley et al 1988), the Endeavour segment (Tivey & Delaney 1986), and the TAG area (Thompson et al 1988).

FLUXES

Several key questions related to seafloor hydrothermal activity remain: What is the net flux of these solutions to the ocean? What is their present influence on seawater chemistry? What has this influence been throughout geologic time? With the discovery of the GSC vents, Jenkins et al (1978) tied the flux to the global ^{3}He and heat fluxes to arrive at global hydrothermal fluxes. Subsequent authors challenged these values (Hart & Staudigel 1982, Mottl 1983), partly on the basis that they required more K to be input than was present in the amount of new crust generated each year. The issue has gotten more complicated with the realization that the

^3He/heat ratio varies between vent areas (Lupton et al 1989); with the continued debate over how much of the conductive heat flow anomaly at spreading centers to ascribe to convection of high-temperature axial hot springs; and, as additional hot springs have been sampled, with the realization that their chemistry is not uniform. Sampling of hydrothermal fluids at the ridge axis has concentrated on the highest temperature "black smokers," and the chemistry, as well as the relative importance in transporting heat, of the ubiquitous diffuse flows remains to be defined. Based on the conductive heat flow anomaly, convection of fluids through the oceanic crust should extend over a large area on the ridge flanks to crust millions of years old. Very few observations of these fluids have been made (e.g. Maris et al 1984, Leinen et al 1987), and their chemistry and average temperature remain poorly known, although their volumetric input to the ocean may be much larger than that from the axial springs (COSOD II 1987). Additional complications arise from events such as the megaplumes, which added the annual heat equivalent of an estimated 200–2000 high-temperature vents (Baker et al 1987), and the unknown importance of megaplume-type events on a global scale. Although the influence of hydrothermal activity is evaluated primarily on the basis of the chemistry of the solutions themselves, recent work has shown that the particles in the hydrothermal plumes may actually scavenge a significant proportion of certain dissolved species (e.g. P, As, V, rare earth elements) from seawater (Feely et al 1988, Olivares & Owen 1988, Metz & Trefry 1988). The river fluxes are also not well known for many trace elements, which makes it difficult to evaluate the relative importance of the river fluxes versus the hydrothermal fluxes. Palmer & Edmond (1989) have recently suggested the use of strontium isotopes to constrain the relative importance of river and hydrothermal fluxes. They favor an even higher axial flux than previously suggested, based on the average strontium concentrations and isotopic signatures in rivers and in high-temperature hydrothermal fluids. The discrepancy in the alkali metal fluxes may result from the reaction of the hydrothermal fluids with previously altered, and hence alkali-enriched, crust.

The importance of the flux of hydrothermal fluids to the ocean will continue to be debated with respect to the river flux and to the sink in low-temperature alteration of the oceanic crust, e.g. weathering of rocks on the seafloor. Hydrothermal alteration is probably the main source of iron, manganese, and several of the alkali metals to the ocean, but as a result of the low-temperature weathering-type reactions, the oceanic crust is probably a net sink for the alkalis (Thompson 1983). The relative importance of these various sources and sinks remains unclear for most of the other elements.

SUMMARY

Seafloor hydrothermal activity, often manifested as "black smokers," occurs in a variety of locations and geologic settings. Chlorinities range from approximately one third to twice the seawater value, with other dissolved constituents varying in concert. All of the fluids and chimneys show a certain consistency, although individual differences occur between different vent areas as well as between individual vents separated by only a few tens of meters. Although much progress has been made in understanding the controls on the fluids and chimneys in individual areas, extension to a global context remains difficult. High-salinity fluids occur within the oceanic crust, but phase separation is not the only mechanism to create them. These mechanisms have not yet been unequivocally identified. Uncertainty remains in quantifying the hydrothermal fluxes and in evaluating their overall influence on oceanic chemistry due to unknown compositional and geographic variabilities.

ACKNOWLEDGMENTS

This article is publication No. 3407 of the Environmental Sciences Division, Oak Ridge National Laboratory, operated by Martin Marietta Energy Systems, Inc. under contract DE-AC05-84OR21400 for the US Department of Energy.

Literature Cited

Alt, J. C., Laverne, C., Muehlenbachs, K. 1984. Alteration of the upper oceanic crust: mineralogy and processes in Deep Sea Drilling Project Hole 504B, Leg 83. In *Initial Reports of the Deep Sea Drilling Project*, 83: 217–247. Washington, DC: US Govt. Print. Off.

Anderson, R. N., Langseth, M. G., Sclater, J. G. 1977. The mechanisms of heat transfer through the floor of the Indian Ocean. *J. Geophys. Res.* 82: 3391–3409

Baker, E. T., Massoth, G. J., Feely, R. A. 1987. Cataclysmic hydrothermal venting on the Juan de Fuca Ridge. *Nature* 329: 149–51

Bischoff, J. L., Dickson, F. W. 1975. Seawater-basalt interaction at 200°C and 500 bars: implications for origin of seafloor heavy metal deposits and regulation of seawater chemistry. *Earth Planet. Sci. Lett.* 25: 385–97

Bischoff, J. L., Rosenbauer, R. J. 1988. Liquid-vapor relations in the critical region of the system NaCl-H_2O from 380 to 415°C: a refined determination of the critical point and two-phase boundary of seawater. *Geochim. Cosmochim. Acta* 52: 2121–26

Bischoff, J. L., Rosenbauer, R. J. 1989. Salinity variations in submarine hydrothermal systems by layered double-diffusive convection. *J. Geol.* 97: 613–23

Bluth, G. J., Ohmoto, H. 1988. Sulfide-sulfate chimneys on the East Pacific Rise, 11° and 13°N latitudes. Part II: sulfur isotopes. *Can. Mineral.* 26: 505–15

Bowers, T. S., Von Damm, K. L., Edmond, J. M. 1985. Chemical evolution of mid-ocean ridge hot springs. *Geochim. Cosmochim. Acta* 49: 2239–52

Bowers, T. S., Campbell, A. C., Measures, C. I., Spivack, A. J., Khadem, M., Edmond, J. M. 1988. Chemical controls on the composition of vent fluids at 13°–11°N and 21°N, East Pacific Rise. *J. Geophys. Res.* 93: 4522–36

Brett, R., Evans, H. T. Jr., Gibson, E. K. Jr., Hedenquist, J. W., Wandless, M.-V., Sommer, M. A. 1987. Mineralogical studies of sulfide samples and volatile concentrations of basalt glasses from the southern Juan de Fuca Ridge. *J. Geophys. Res.* 92: 11,373–79

Burnham, C. W., Holloway, J. R., Davis, N. F. 1969. Thermodynamic properties of water to 1000°C and 10,000 bars. *Geol. Soc. Am. Spec. Pap. 132*. 96 pp.

Butterfield, D. A., McDuff, R. E., Lilley, M. D., Massoth, G. J., Lupton, J. E. 1988. Chemistry of hydrothermal fluids from the ASHES vent field: evidence for phase separation. *Eos, Trans. Am. Geophys. Union* 69: 1468 (Abstr.)

Butterfield, D. A., Massoth, G. J., McDuff, R. E., Lupton, J. E., Lilley, M. D. 1989. The chemistry of phase-separated hydrothermal fluids from ASHES vent field, Juan de Fuca Ridge. *J. Geophys. Res.* In press

Campbell, A. C., Edmond, J. M. 1989. Halide systematics of submarine hydrothermal vents. *Nature*. In press

Campbell, A. C., Bowers, T. S., Measures, C. I., Falkner, K. K., Khadem, M., Edmond, J. M. 1988a. A time series of vent fluid compositions from 21°N, East Pacific Rise (1979, 1981, 1985) and the Guaymas Basin, Gulf of California (1982, 1985). *J. Geophys. Res.* 93: 4537–49

Campbell, A. C., Palmer, M. R., Klinkhammer, G. P., Bowers, T. S., Edmond, J. M., et al. 1988b. Chemistry of hot springs on the Mid-Atlantic Ridge. *Nature* 335: 514–19

Cathles, L. M. 1983. An analysis of the hydrothermal system responsible for massive sulfide deposition in the Hokuroko Basin of Japan. In *The Kuroko and Related Volcanogenic Massive Sulfide Deposits. Econ. Geol. Monogr.*, 5: 439–87

Chen, J. H., Wasserburg, G. J., Von Damm, K. L., Edmond, J. M. 1986. The U-Th-Pb systematics in hot springs on the East Pacific Rise at 21°N and Guaymas Basin. *Geochim. Cosmochim. Acta* 50: 2467–79

COSOD II. 1987. *Report of the Second Conference on Scientific Ocean Drilling, Strasbourg*. Strasbourg, Fr: Eur. Sci. Found./ Joint Oceanogr. Inst. Deep Earth Sampling

Cowan, J., Cann, J. 1988. Supercritical two-phase separation of hydrothermal fluids in the Troodos ophiolite. *Nature* 333: 259–61

Craig, H., Welhan, J. A., Kim, K., Poreda, R., Lupton, J. E. 1980. Geochemical studies of the 21°N EPR hydrothermal fluids. *Eos, Trans. Am. Geophys. Union* 61: 992 (Abstr.)

Delaney, J. R., Mogk, D. W., Mottl, M. J. 1987. Quartz-cemented breccias from the Mid-Atlantic Ridge: samples of a high salinity hydrothermal upflow zone. *J. Geophys. Res.* 92: 9175–92

Dymond, J., Cobler, R., Gordon, L., Biscaye, P., Mathieu, G. 1983. ^{226}Ra and ^{222}Rn contents of Galapagos Rift hydrothermal waters—the importance of low-temperature interactions with crustal rocks. *Earth Planet. Sci. Lett.* 64: 417–29

Edmond, J. M., Measures, C., McDuff, R. E., Chan, L. H., Collier, R., et al. 1979a. Ridge crest hydrothermal activity and the balances of the major and minor elements in the ocean: the Galapagos data. *Earth Planet. Sci. Lett.* 46: 1–18

Edmond, J. M., Measures, C., Magnum, B., Grant, B., Sclater, F. R., et al. 1979b. On the formation of metal-rich deposits at ridge crests. *Earth Planet. Sci. Lett.* 46: 19–30

Embley, R. W., Jonasson, I. R., Perfit, M. R., Franklin, J. M., Tivey, M. A., et al. 1988. Submersible investigation of an extinct hydrothermal system on the Galapagos Ridge: sulfide mounds, stockwork zone and differentiated lavas. *Can. Mineral.* 26: 517–39

Evans, W. C., White, L. D., Rapp, J. B. 1988. Geochemistry of some gases in hydrothermal fluids from the southern Juan de Fuca Ridge. *J. Geophys. Res.* 93: 15,305–13

Feely, R. A., Massoth, G. J., Baker, E. T., Cowen, J. P., Lamb, M. F. 1988. The effect of hydrothermal processes on phosphorous distributions in the northeast Pacific. *Eos, Trans. Am. Geophys. Union* 69: 1497 (Abstr.)

Fox, C. G. 1989. The consequences of phase separation on the distribution of hydrothermal fluids at ASHES vent field, Axial Volcano, Juan de Fuca Ridge. *J. Geophys. Res.* In press

Gieskes, J. M., Simoneit, B. R. T., Brown, T., Shaw, T., Wang, Y.-C., Magenheim, A. 1988. Hydrothermal fluids and petroleum in surface sediments of Guaymas Basin, Gulf of California: a case study. *Can. Mineral.* 26: 589–602

Goldfarb, M. S. 1988. Flanges and the formation of hydrothermal edifices, Endeavour segment, Juan de Fuca Ridge. *Eos, Trans. Am. Geophys. Union* 69: 1498 (Abstr.)

Goldfarb, M. S., Delaney, J. R. 1988. Response of two-phase fluids to fracture configurations within submarine hydrothermal systems. *J. Geophys. Res.* 93: 4585–94

Goldfarb, M. S., Converse, D. R., Holland, H. D., Edmond, J. M. 1983. The genesis of hot spring deposits on the East Pacific Rise, 21°N. In *The Kuroko and Related Volcanogenic Massive Sulfide Deposits. Econ. Geol. Monogr.*, 5: 184–97

Graham, U. M., Bluth, G. J., Ohmoto, H.

1988. Sulfide-sulfate chimneys on the East Pacific Rise, 11° and 13°N latitudes. Part 1: mineralogy and paragenesis. *Can. Mineral.* 26: 487–504

Hart, S. R., Staudigel, H. 1982. The control of alkalis and uranium in seawater by ocean crust alteration. *Earth Planet. Sci. Lett.* 58: 202–12

Haymon, R. M. 1983. The growth history of hydrothermal black smoker chimneys. *Nature* 301: 695–98

Haymon, R. M., Kastner, M. 1986. Caminite: a new magnesium-hydroxide-sulfate-hydrate mineral found in a submarine hydrothermal deposit, East Pacific Rise, 21°N. *Am. Mineral.* 71: 819–25

Haymon, R. M., Koski, R. A., Abrams, M. J. 1989. Hydrothermal discharge zones beneath massive sulfide deposits mapped in the Oman ophiolite. *Geology* 17: 531–35

Hinkley, T. K., Tatsumoto, M. 1987. Metals and isotopes in Juan de Fuca Ridge hydrothermal fluids and their associated solid materials. *J. Geophys. Res.* 92: 11,400–10

Jenkins, W. J., Edmond, J. M., Corliss, J. B. 1978. Excess ^{3}He and ^{4}He in Galapagos submarine hydrothermal waters. *Nature* 272: 156–58

Kadko, D. 1988. Radiochemistry of vent fluids from Axial volcano and the southern Cleft segment of the Juan de Fuca Ridge. *Eos, Trans. Am. Geophys. Union* 69: 1497 (Abstr.)

Kelley, D. S., Delaney, J. R. 1987. Two-phase separation and fracturing in mid-ocean ridge gabbros at temperatures greater than 700°C. *Earth Planet. Sci. Lett.* 83: 53–66

Kennedy, B. M. 1988. Noble gases in vent water from the Juan de Fuca Ridge. *Geochim. Cosmochim. Acta* 52: 1929–35

Kohls, D. W., Rodda, J. L. 1967. Iowaite, a new hydrous magnesium hydroxide–ferric oxychloride from the Precambrian of Iowa. *Am. Mineral.* 52: 1261–71

Koski, R. A., Shanks, W. C. III, Bohrson, W. A., Oscarson, R. L. 1988. The composition of massive sulfide deposits from the sediment-covered floor of the Escanaba trough, Gorda Ridge: implications for depositional processes. *Can. Mineral.* 26: 655–73

Krishnaswami, S., Turekian, K. K. 1982. ^{238}U, ^{226}Ra, and ^{210}Pb in some vent waters of the Galapagos spreading center. *Geophys. Res. Lett.* 9: 827–30

Leinen, M., Rea, D. K., et al. 1986. *Initial Reports of the Deep Sea Drilling Project*, Vol. 92: Washington, DC: US Govt. Print. Off.

Leinen, M., McDuff, R., Delaney, J., Becker, K., Schultheiss, P. 1987. Off-axis hydrothermal activity in the Mariana Mounds field. *Eos, Trans. Am. Geophys. Union* 68: 1531 (Abstr.)

Lupton, J. E., Baker, E. T., Massoth, G. J. 1989. Variable ^{3}He/heat ratios in submarine hydrothermal systems: evidence from two plumes over the Juan de Fuca Ridge. *Nature* 337: 161–64

Marchig, V., Rosch, H., Lalou, C., Brichet, E., Oudin, E. 1988. Mineralogical zonation and radiochronological relations in a large sulfide chimney from the East Pacific Rise at 18°25′S. *Can. Mineral.* 26: 541–54

Maris, C. R. P., Bender, M. L., Froehlich, P. N., Barnes, R., Leudtke, N. A. 1984. Chemical evidence for advection of hydrothermal solutions in the sediments of the Galapagos Mounds hydrothermal field. *Geochim. Cosmochim. Acta* 48: 2331–46

Massoth, G. J., Baker, E. T., Feely, R. A., Lupton, J. E., Butterfield, D. A., McDuff, R. E. 1988. Hydrothermal fluids and plumes of Cleft segment, Juan de Fuca Ridge. *Eos, Trans. Am. Geophys. Union* 69: 1497 (Abstr.)

Massoth, G. J., Butterfield, D. A., Lupton, J. E., McDuff, R. E., Lilley, M. D., Jonasson, I. R. 1989. Submarine venting of phase-separated hydrothermal fluids at Axial volcano, Juan de Fuca Ridge. *Nature*. In press

McDuff, R. E. 1988. Effects of vent fluid properties on the hydrography of hydrothermal plumes. *Eos, Trans. Am. Geophys. Union* 69: 1497 (Abstr.)

Merlivat, L., Pineau, F., Javoy, M. 1987. Hydrothermal vent waters at 13°N on the East Pacific Rise: isotopic composition and gas concentration. *Earth Planet. Sci. Lett.* 84: 100–8

Metz, S., Trefry, J. H. 1988. Scavenging of V by iron oxides in hydrothermal plumes. *Eos, Trans. Am. Geophys. Union* 69: 1489 (Abstr.)

Michard, G., Albarede, F., Michard, A., Minster, J.-F., Charlou, J. L., Tan, N. 1984. Chemistry of solutions from the 13°N East Pacific Rise hydrothermal site. *Earth Planet. Sci. Lett.* 67: 297–307

Mottl, M. J. 1983. Metabasalts, axial hot springs, and the structure of hydrothermal systems at mid-ocean ridges. *Geol. Soc. Am. Bull.* 94: 161–80

Olivares, A. M., Owen, R. M. 1988. Rare earth element scavenging by hydrothermal precipitates. *Eos, Trans. Am. Geophys. Union* 69: 1489 (Abstr.)

Palmer, M. R., Edmond, J. M. 1989. The strontium isotope budget of the modern ocean. *Earth Planet. Sci. Lett.* 92: 11–26

Peter, J. M., Scott, S. D. 1988. Mineralogy, composition, and fluid-inclusion microthermometry of seafloor hydrothermal deposits in the Southern Trough of Guaymas Basin, Gulf of California. *Can. Mineral.* 26: 567–87

Potter, R. W. II, Brown, D. L. 1977. The volumetric properties of aqueous sodium chloride solutions from 0° to 500°C at pressures up to 2000 bars based on a regression of available data in the literature. *US Geol. Surv. Bull. 1421C.* 36 pp.

Rodgers, P. S. Z., Pitzer, K. S. 1982. Volumetric properties of aqueous sodium chloride solutions. *J. Phys. Chem. Ref. Data* 11: 15–81

Rucklidge, J. C., Patterson, G. C. 1977. The role of chlorine in serpentinization. *Contrib. Mineral. Petrol.* 65: 39–44

Seyfried, W. E. Jr., Bischoff, J. L. 1979. Low temperature basalt alteration by seawater: an experimental study at 70°C and 150°C. *Geochim. Cosmochim. Acta* 43: 1937–47

Seyfried, W. E. Jr., Berndt, M. E., Janecky, D. R. 1986. Chloride depletions and enrichments in seafloor hydrothermal fluids: constraints from experimental basalt alteration studies. *Geochim. Cosmochim. Acta* 50: 469–75

Shanks, W. C. III, Seyfried, W. E. Jr. 1987. Stable isotope studies of vent fluids and chimney minerals, southern Juan de Fuca Ridge: sodium metasomatism and seawater sulfate reduction. *J. Geophys. Res.* 92: 11,387–99

Sourirajan, S., Kennedy, G. C. 1962. The system NaCl-H_2O at elevated temperatures and pressures. *Am. J. Sci.* 260: 115–242

Spivack, A. J., Edmond, J. M. 1987. Boron isotope exchange between seawater and the oceanic crust. *Geochim. Cosmochim. Acta* 51: 1033–43

Spivack, A. J., Palmer, M. R., Edmond, J. M. 1987. The sedimentary cycle of the boron isotopes. *Geochim. Cosmochim. Acta* 51: 1939–49

Stakes, D., Vanko, D. A. 1986. Multistage hydrothermal alteration of gabbroic rocks from the failed Mathematician Ridge. *Earth Planet. Sci. Lett.* 79: 75–92

Thompson, G. 1983. Hydrothermal fluxes in the ocean. In *Chemical Oceanography*, ed. J. P. Riley, R. Chester, 8: 271–337. New York: Academic

Thompson, G., Humphris, S. E., Schroeder, B., Sulanowska, M., Rona, P. A. 1988. Active vents and massive sulfides at 26°N (TAG) and 23°N (Snakepit) on the Mid-Atlantic Ridge. *Can. Mineral.* 26: 697–711

Tivey, M. K., Delaney, J. R. 1986. Growth of large sulfide structures on the Endeavour segment of the Juan de Fuca Ridge. *Earth Planet. Sci. Lett.* 77: 303–17

Vanko, D. A. 1986. High-chlorine amphiboles from oceanic rocks: product of highly-saline hydrothermal fluids? *Am. Mineral.* 71: 51–59

Vanko, D. A. 1988. Temperature, pressure, and composition of hydrothermal fluids, and their bearing on the magnitude of tectonic uplift at mid-ocean ridges, inferred from fluid inclusions in oceanic layer 3 rocks. *J. Geophys. Res.* 93: 4595–4611

Von Damm, K. L. 1988. Systematics of and postulated controls on submarine hydrothermal solution chemistry. *J. Geophys. Res.* 93: 4551–61

Von Damm, K. L., Bischoff, J. L. 1987. Chemistry of hydrothermal solutions from the southern Juan de Fuca Ridge. *J. Geophys. Res.* 92: 11,334–46

Von Damm, K. L., Edmond, J. M., Grant, B., Measures, C. I., Walden, B., Weiss, R. F. 1985a. Chemistry of submarine hydrothermal solutions at 21°N, East Pacific Rise. *Geochim. Cosmochim. Acta* 49: 2197–2220

Von Damm, K. L., Edmond, J. M., Measures, C. I., Grant, B. 1985b. Chemistry of submarine hydrothermal solutions at Guaymas Basin, Gulf of California. *Geochim. Cosmochim. Acta* 49: 2221–37

Welhan, J. A. 1981. *Carbon and hydrogen gases in hydrothermal systems: the search for a mantle source.* PhD thesis. Univ. Calif., San Diego. 194 pp.

Welhan, J. A., Craig, H. 1979. Methane and hydrogen in East Pacific Rise hydrothermal fluids. *Geophys. Res. Lett.* 6: 829–31

Wolery, T. J. 1983. EQ3NR a computer program for geochemical aqueous speciation-solubility calculations: user's guide and documentation. *Rep. UCRL-53414*, Lawrence Livermore Natl. Lab., Livermore, Calif.

Woodruff, L. G., Shanks, W. C. III. 1988. Sulfur isotope study of chimney minerals and vent fluids from 21°N, East Pacific Rise: hydrothermal sulfur sources and disequilibrium sulfate reduction. *J. Geophys. Res.* 93: 4562–72

Annu. Rev. Earth Planet. Sci. 1990. 18:205–56
Copyright © 1990 by Annual Reviews Inc. All rights reserved

FORMATION OF THE EARTH

George W. Wetherill

Department of Terrestrial Magnetism, Carnegie Institution of Washington, Washington, DC 20015

INTRODUCTION

Probably the most fundamental problem of geology is that of understanding the physical and chemical processes and events that controlled the formation of the Earth and determined its initial state. In many ways our planet's subsequent evolution and present state were determined by this initial state, although striking differences between Earth and its sister planet Venus suggest that additional, and perhaps unpredictably stochastic, events only slightly later in their early history may have had major long-term consequences (e.g. Kaula 1990).

The traditional and well-established approach to understanding the history of the Earth is examination of the geological record. Application of this approach to the formation and earliest history of the Earth is confounded, however, by the geological activity of our planet. Hutton (1788) was nearly correct when he wrote that in the rocks of the Earth, "We find no sign of a beginning." Fortunately, the rocks do reveal some important evidence relevant to the beginning of the Earth, such as its 4500 ± 50 m.y. age (Patterson 1956). Progress during the last several decades in the development of a variety of isotopic methods for measuring the age of igneous and metamorphic events has lifted the veil that previously obscured the correlation and interpretation of Precambrian events, encompassing 87% of Earth history. As a consequence, some important evolution in tectonic and petrologic styles is now apparent, even though the moderately well-preserved terrestrial record extends back only as far as about 3.8 b.y. ago (reviewed by Kröner 1985). This ancient and limited terrestrial evidence has now been extended by well-dated records of earlier geological events in the solar system obtained from less active planetary bodies: the Moon and the asteroidal sources of meteorites (Wetherill 1975, Tera et al 1981, Van Schmus 1981, Carlson & Lugmair 1988).

0084–6597/90/0515–0205$02.00

A complementary source of information regarding the formation and initial state of the Earth is provided by astronomical theory and observations. The formation of the Earth and other planets was a by-product of the formation of the Sun and is thereby linked to observational evidence and theoretical understanding of the processes by which stars like the Sun form. A considerable, but by no means definitive, body of theoretical work also exists regarding the manner in which residual material, remaining after the formation of the Sun, may be accumulated to form planets.

The origin and initial state of the Earth can therefore be approached from two directions: looking backward into the geological record, and working forward from the protosolar environment to planet formation. Achievement of a satisfactory understanding requires that the conclusions reached by these two approaches be at least compatible and, better yet, reinforce one another. To some extent this has occurred, but in other ways, this dual approach appears at present to lead to paradoxes that represent challenges requiring further investigation.

In this review, principal emphasis is placed on the formation of the Earth. It can be considered an update of an earlier review on the formation of the terrestrial planets (Wetherill 1980) and a more specialized version of a recent review of planet formation (Wetherill 1989). A general review of solar system formation is that of Cameron (1988). Many aspects of the subject are treated in more detail in individual chapters of the book *Protostars and Planets II* (Black & Matthews 1985). A collection of articles providing extensive theoretical treatments of many dynamical problems highly relevant to the formation of the Earth may be found in a volume assembled by Hayashi and his colleagues, mostly in Tokyo and Kyoto (Hayashi et al 1988). This publication was not available until well after the deadline for submission of this review, but an effort has been made to mention some of the important contributions contained therein. The forthcoming proceedings of the Conference on the Origin of the Earth, held in Oakland, California, in December 1988, may be expected to contain important reviews and research results, particularly with regard to chemical aspects of the formation of the Earth (Jones & Newsom 1990). An effort is made to organize the present review on a conceptual basis, rather than emphasizing adversarial positions that have been taken by various workers at various times.

Standard Model of Planet Formation

A discussion of the origin of the Earth must be placed in the context of the formation of the Sun and the planets. This is done in an abbreviated way here, with references to other recent reviews. This context is provided by the emergence during the past 20 years of what can be considered to be a standard model for the formation of the sun and planets that encompasses the work of the Moscow, Kyoto and Tokyo workers, the Tucson Planetary Science Institute, the Carnegie Institution of Washington, and a number of others. The sequential course of events that took place during the formation of the solar system, according to the standard model, is outlined briefly. The model is quite flexible, and alternate pathways occur at several junctures. Many of these alternatives are skipped over in this introduction, but, when highly relevant to Earth formation, they are considered in later sections. Although for expository purposes the model is described in a declarative manner, it must be remembered that for most of these events definitive, or even quantitative, demonstration is lacking, and this catalog of events should be thought of more as a research agenda than as a list of conclusions.

According to this model, the Sun formed, together with a number of stars, as a result of gravitational instability in a dense interstellar molecular cloud. Probably the most common mode of gravitational collapse of a rotating system of this kind leads to formation of double- and multiple-star systems. However in those cases where the rotation is slow, i.e. angular momentum of the collapsing system is relatively low, it is possible for single stars like the Sun to form. Even in these cases, a mechanism that transfers excess angular momentum away from the central star is required. Primary candidates for this mechanism are angular momentum transport associated with a viscous gaseous accretion disk (Lynden-Bell & Pringle 1974) surrounding the central star, and angular momentum transport by gravitational torques associated with asymmetries in the collapse (Larson 1984, Boss 1984). In either case, a small (2–10%) portion of the gas and dust will form a flattened disk surrounding the star that contains most of the angular momentum of the system. This disk of dust and gas is termed the "solar nebula."

According to the standard model, planets grow by agglomeration of planetesimals that form in the solar nebula. In the vicinity of the Earth's orbit, temperatures are expected to be low enough to permit crystallization of (Fe, Mg)-silicates and metallic iron but not ices of the volatile elements H, C, and N. Particles grow into these rocky planetesimals, either by nongravitational cohesion or by gravitational instabilities, up to diameters of a few kilometers. Larger bodies ("planetary embryos") in the mass range of 10^{25}–10^{26} g ($\sim$Moon- or Mercury-size) form as a result of collision and merger of these small planetesimals. The rate of growth of the planetesimals into embryos is determined by their relative velocity, which in turn is determined by mutual gravitational perturbations and, therefore, by the mass distribution of the growing planetesimal swarm. The significance of this self-regulated coupling of the mass and velocity distribution was quantitatively identified by Safronov (e.g. Safronov 1969) and later developed further by others (Nakagawa et al 1983, Wetherill & Stewart 1989). During the growth of planetesimals into embryos, the growing planetesimals will be sufficiently closely packed to permit collisions between them, a necessary condition for their growth and merger. Calculations of this stage of growth indicate that bodies as large as 10^{26} g can form in $\sim 10^5$ yr.

The present total mass of the terrestrial planet region is 1.18×10^{28} g, of which 51% is in the Earth-Moon system. Therefore, the growth of the Earth and terrestrial planets would require the merger of ~ 100 10^{26}-g planetary embryos. These larger preplanetary bodies will usually be more distant from one another, the rate of growth will slow down, and this final stage of terrestrial planet growth will require 10^7–10^8 yr for its completion, probably long after the loss of the gaseous solar nebula in $\lesssim 3 \times 10^6$ yr. This stage of growth is likely to be characterized by the emergence, after $\sim 5 \times 10^6$ yr, of the dominant embryos of the two large terrestrial planets, Earth and Venus, in orbits of relatively low eccentricity and inclination. As a result of gravitational perturbations associated with close encounters of the smaller 10^{26}–10^{27}-g embryos with Earth and Venus, the remainder of the material will probably be found in more eccentric orbits ($e \sim 0.1$–0.3) that span the terrestrial planet region, eliminating local "feeding zones" for each planet. Final accumulation of Earth and Venus then includes "giant impacts," collisions with residual smaller planet-size embryos. Such impacts have been associated with special events in terrestrial planet history, such as the formation of the Moon (reviewed by Stevenson 1987), the loss of Mercury's silicate mantle (Wetherill 1988, Cameron et al 1988, Vityazev et al 1988), and the loss of the Earth's original atmosphere (Cameron 1983).

The same model must account, in a natural way, for the mechanism and time scale for the formation of other bodies in the solar system [i.e. Jupiter and the other giant planets (Lissauer 1987, Stevenson & Lunine 1988)], the depletion of mass in the asteroid belt (Weidenschilling 1988, Wetherill 1989, 1990a), and the formation of the source regions of long- and short-period comets (Fernandez & Ip 1981). Our understanding of all these events is, at best, rudimentary. A complete theory of Earth and terrestrial planet formation will require that the formation of these other solar system bodies be understood more adequately.

Among these other problems, the way in which Jupiter was formed and its formation time scale are most critical. Because almost all of Jupiter's mass consists of hydrogen and helium, the gaseous solar nebula must have still been present when Jupiter (and Saturn) were formed. Observations of pre-main-sequence stars (Walter 1986, Strom et al 1989) show that protostellar gas envelopes are removed on time scales $<10^7$ yr, and sometimes $\ll 3 \times 10^6$ yr. For this reason, the $\sim 15\ M_{\oplus}$ (Earth mass) cores of Jupiter and Saturn must have formed rapidly in order to initiate the accretion of nebular gas while it was still available. These events and their timing are highly relevant to the formation of the Earth. They proscribe the nebular gas concentration during the final stages of Earth formation, which in turn determines the size and nature of Earth's primordial atmosphere (Hayashi et al 1979, Nakazawa et al 1985) and the role played by gas-embryo gravitational torques in the radial migration of planetary embryos in the inner solar system (Ward 1986, 1988). In addition, the regions occupied by Jupiter and the other giant planets, as well as the asteroid belt, may have been source regions for volatile and oxidized constituents of the Earth (Wänke 1981, Dreibus & Wänke 1989). Determination of the time scale and quantity of this transferred material requires an understanding of the formation of these bodies that lie beyond the orbit of the Earth.

The most significant variant of the standard model is the "Kyoto model" (Hayashi et al 1985) in which the loss of the nebular gas occurred *after* the formation of the Earth, in contrast to the assumption of most other workers that the loss of nebular gas at 1 AU (astronomical unit) occurred well before the final stages of the Earth's growth. Some principal differences between the outcome of the gas-free and gas-rich models are chemical in nature, inasmuch as accumulation of Earth in a nebula that had time to cool to ~ 300 K would provide a major admixture of volatile compounds, perhaps similar to those found in some carbonaceous meteorites. In addition, a large ($>10^{26}$ g) H_2-rich primordial terrestrial atmosphere, $\sim 10^5$ times as massive as the present atmosphere, is predicted by the gas-rich model (Hayashi et al 1979). As a result of the work of Takeda et al (1985) and Takeda (1988), which predicts a gravitationally enhanced and therefore very large gas drag coefficient for a planet-size body, it may be expected that important dynamical differences (Ohtsuki et al 1988) may also exist that have not yet been fully studied.

Some workers have termed the standard model a "paradigm." I regard this appellation to be premature. At present, when compared with what needs to be known, the model tells us rather little that can be considered at all certain. Some specific problems have been explored relatively thoroughly and quantitatively, but most are in the qualitative "scenario" stage. At present the model can be regarded as a very useful working hypothesis, one that can provide a focus for workers of varying backgrounds and experience, facilitate communication between them, and permit a ra-tional division of labor, rather than requiring everyone to make up a personal grand model. Only time will tell if it will succeed as a unifying theory.

There are of course alternatives to the standard model. The most prominent of these has been the concept that planetary formation took place in parallel with the formation of the Sun as a consequence of massive gas-dust instabilities in the solar nebula, termed "giant gaseous protoplanets." Theories of this general type have a venerable history (e.g. von Weizsäcker 1944, Kuiper 1951, Cameron 1978), but their more recent development has been rather quiescent. References to these and other alternative models can be found in the conference report *The Origin of the Solar System* (Dermott 1978).

FORMATION OF THE SUN AND SOLAR NEBULA

Radio and infrared observations show that $\sim 1\text{-}M_{\odot}$ (solar mass) star formation occurs in cold (10 K), dense (10^4–10^5 H_2 molecules cm^{-3}) cores found within larger dark molecular cloud complexes with masses of 10^3–10^4 $M_{\odot}$. Low-mass stars are also formed, along with more massive stars, in giant molecular clouds with masses up to 10^6 $M_{\odot}$ (Shu et al 1987). It is reasonable to use observational and theoretical work on present-day star-forming regions of this kind as models for the formation of the Sun 4.5 b.y. ago. It may also be recognized, however, that star formation in galaxies often occurs in episodic "starbursts," perhaps associated with galactic collisions and mergers (Schweizer 1986). Scalo (1988) has presented evidence that such a starburst occurred in our Galaxy about 5 b.y. ago, approximately the age of the Sun.

Calculations of the collapse of a rotating molecular cloud core with specific angular momentum $\gtrsim 5 \times 10^{18}$ cm^2 s^{-1} show that a disk forms that is centrifugally supported, i.e. that does not rapidly collapse onto the central protostellar object. When the specific angular momentum is high enough ($\gtrsim 2 \times 10^{20}$ cm^2 s^{-1}) and a strong central condensation is absent, fragmentation into a binary or multiple-star system will occur. Even in the absence of turbulent viscosity, a core with intermediate specific angular momentum and/or initial strong central condensation can probably evolve into a central single star, surrounded by a circumstellar disk, as required for the solar nebula in the standard model (Boss 1989, 1990).

A calculated surface density distribution for one such case is shown in Figure 1. The mass of the initial $1.04\text{-}M_{\odot}$ system is separated into a stellar core of mass 0.99 $M_{\odot}$ and a circumstellar disk shown with a mass of 0.05 $M_{\odot}$. The decrease in surface density with heliocentric distance does not follow a simple power law. The calculated surface densities are considerably greater than those required to form the terrestrial planets and

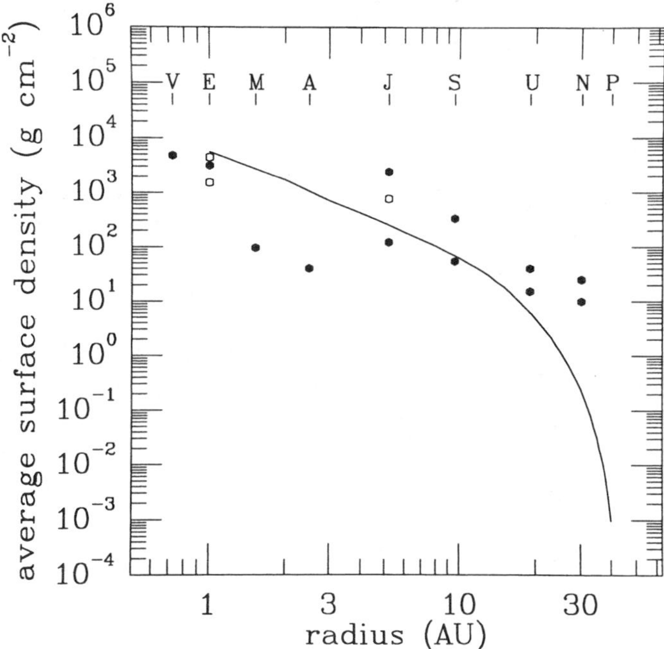

Figure 1 Calculated total (gas + solid) density distribution in a model residual nebula of 0.05 $M_\odot$ surrounding a 0.99-$M_\odot$ star (Boss 1989). The *open symbols* at the position of Earth (E) represent two estimates (7.5 and 22.4 g cm^{-2}) of the surface density solid matter required to form the Earth. The *open symbol* at the position of Jupiter (J) represents the value of the surface density at Jupiter's distance needed to form Jupiter in about 0.5 m.y. (Lissauer 1987). The *solid symbols* bracket the "reconstruction" of the gaseous solar nebula using present solid-matter densities by Weidenschilling (1977a), based on present masses and positions.

asteroids but less than those required to form Jupiter rapidly (Lissauer 1987) and Uranus and Neptune at all. This system has not yet reached centrifugal equilibrium, but gravitational torques are already transferring angular momentum rapidly outward from the inner part of the disk. It is plausible that a star-disk system with these, or not very different, parameters will evolve into a solar nebula resembling those usually assumed in theories of planet formation in about one free-fall time ($\sim 10^5$ yr). No tendency to form massive gas-dust instabilities leading to giant gaseous protoplanets was found. The midplane temperature variations in this same nebula are shown in Figure 2. These temperatures (~ 1500 K out to 3 AU) are considerably higher than those found in some other solar nebula models (Cameron & Pine 1973). This difference arises from including in the three-dimensional (3D) models of Boss (1990) the heating associated

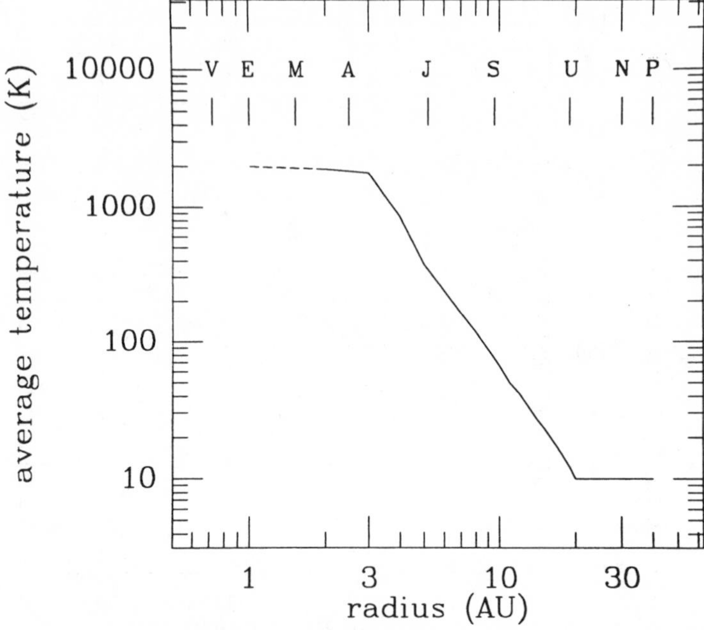

Figure 2 Variations of temperature with distance for the model of Figure 1 (Boss 1989). Temperatures in the asteroid belt reach 1500 K, high enough to vaporize solid grains of silicate and iron. The *dashed line* is an extrapolation into the terrestrial planet region, where the temperature must be at least as high as in the asteroid belt.

with compression of the collapsing gas. The time scale for cooling of this nebula is estimated to be $\sim 10^5$ yr. These temperatures are sufficiently high to volatilize iron grains (1420 K; Pollack et al 1985) and all but the most refractory silicate and oxide minerals. It is possible that the temperature of the solar nebula was self-regulated by the opacity of iron grains.

There is considerable observational evidence for embedded infrared sources in dark cloud cores (Beichman et al 1986) that are likely to be protostars, surrounded by disks. Observations of young pre-main-sequence stars constitute observational evidence that this obscuration can be removed on a short ($\sim 10^6$–10^7 yr) time scale (Lada 1985, Strom et al 1989). Observation of residual, strongly flattened dust extending to ~ 500 AU surrounding the main-sequence star β Pictoris (Smith & Terrile 1984) provides additional generally supporting but qualitative evidence that the picture of solar system formation outlined by the standard model has some basis in reality.

COAGULATION OF GRAINS AND FORMATION OF SMALL (1–10 km DIAMETER) PLANETESIMALS

Settling of Grains to the Midplane of the Nebula and Nongravitational Coagulation

At least on average, the interstellar grains associated with the infalling gas that formed the solar nebula were originally quite small (<0.1 μm). It is not clear to what extent the nonvolatile constituents of the Earth then experienced a stage of volatilization and recondensation or coagulated to form larger dust grains without vaporization. The ~ 1500 K theoretically based temperatures discussed in the previous section support extensive vaporization. On the other hand, there is abundant evidence for isotopic disequilibrium in both refractory and volatile constituents of meteorites (reviewed by Anders 1988, Zinner 1988). This is not what one would expect from a well-mixed gaseous solar nebula, but continued infall of dust and gas at later times could "contaminate" a once hot solar nebula with isotopically heterogeneous interstellar matter.

In either case, formation of planetesimals, planetary embryos, and final planets required the aggregation of newly condensed grains and/or refractory residues into larger bodies. The present status of this problem of early coagulation has been reviewed by Weidenschilling (1988), Sekiya & Nakagawa (1988), and Weidenschilling et al (1989). The principal problem appears to be the need to grow fairly large ($\gtrsim 10$ m) planetesimals before the time that nebular gas is removed from the terrestrial planet region ($\sim 10^6$ yr); otherwise, the material needed to form the Earth will be removed along with the gas.

Because of gas drag, in the absence of turbulence, small grains will spiral down toward the central plane of the solar nebula. The characteristic settling time τ_z of a small ($\lesssim 1$ cm) grain is (Weidenschilling 1988)

$$\tau_z = \frac{\rho c}{s \rho_s \Omega^2}, \tag{1}$$

where s is the particle radius, ρ_s the particle's material density, Ω the Keplerian orbital angular velocity, ρ the gas density, and c the mean molecular speed. At 1 AU, for $T = 1000$ K, $c = 3.1 \times 10^5$ cm s^{-1}, $\rho = 1.18 \times 10^{-9}$ g cm^{-3}, $\rho_s = 1$ g cm^{-3}, the characteristic settling time of a 1-μm grain will be 2.9×10^6 yr. This is very long, comparable to the time scale for removal of the gas in the solar nebula. Turbulence will lengthen the settling time even more by entraining the small particles in the turbulent flow.

In order to retain solid matter in the solar nebula during the loss of nebular gas, it is necessary that settling proceed more rapidly, presumably because the grains grew larger. For example, if chondrules of $\sim$1-mm radius, as found in most stony meteorites, formed in the solar nebula (Grossman 1988), the corresponding settling time of these objects in a nonturbulent nebula will be shorter ($\sim 3 \times 10^3$ yr). Even in the absence of chondrule formation, from the theoretical point of view it may be expected that significant coagulation of grains will result from particle-to-particle collisions during descent to the central plane, provided that any turbulence in the solar nebula has decreased sufficiently to preclude mutual collisional fragmentation of particles. For the low values of turbulent velocity calculated by Cabot et al (1987) of $\sim$0.01 times the sound speed, particles could grow without disruption. For these values of turbulent velocity and for 100% "sticking efficiency" at collision, coagulation during descent could possibly concentrate solid bodies as large as 10 m in the central plane of the nebula on a 10^3-yr time scale without major loss of material into the Sun. Further growth can occur, primarily by sweep up of material during radial migration within the central dust layer. Weidenschilling (1988) estimates that kilometer-sized planetesimals could grow in this manner in $\sim 10^4$ yr, but this estimate is lengthened by one or two orders of magnitude if the nebular grains are fluffy aggregates, rather than compact objects like chondrules (Donn & Meakin 1988, Weidenschilling et al 1989).

Introduction of the concepts of turbulence and porous fluffy aggregates into the discussion of formation of planetesimals gives rise to paradoxes. Both a moderate degree of turbulence and soft porous target bodies facilitate the nongravitational growth of planetesimals. On the other hand, they both delay the settling of bodies to the nebular midplane, perhaps excessively (Mizuno 1989). It has been proposed (Völk 1983) that the "best of both worlds" may have occurred: oscillation of the nebula between turbulent and nonturbulent states. Growth would occur during turbulent times, followed by falling of objects to the central plane when turbulence was absent.

In addition to turbulence-induced velocities, during the stage of growth from 1 to 10 cm up to 1–10 km, relative velocity differences between bodies of different sizes will also be caused by size-dependent gas drag. Large bodies ($\gtrsim$1 m) will move in Keplerian orbits at Keplerian velocity. The nebular gas will be supported by its buoyancy, partially reducing solar gravity, and will move at a velocity about 0.2% slower than the Keplerian velocity (Whipple 1973, Adachi et al 1976, Weidenschilling 1977b). Because small bodies ($\lesssim$10 cm) move with the gas, a relative velocity between large and small bodies of magnitude $\sim$60 m s^{-1} will result. Although the physical processes are not well understood, it has been proposed that even at these velocities, and at high temperatures, compressible or porous bodies could grow. The relatively high-velocity small particles may be embedded within the large bodies during impact (Leliwa-Kopystynski et al 1984), and collisions between more similar-sized bodies will be at low relative velocity, also permitting growth.

The nongravitational "coagulative" stage of growth will end when the relative velocity of the larger bodies becomes dominated by their mutual gravitational perturbations, rather than by the small residual turbulence of the nebula. The size at which this transition occurs is uncertain, primarily because of uncertainties in the estimated turbulent velocities. Planetesimal radii of 0.1–10 km correspond to plausible values of the relevant parameters.

Growth of Planetesimals by Gravitational Instability in the Central Dust Layer of the Nebula

An alternative mode of growth of small bodies into kilometer-size bodies is dust-layer gravitational instability in the central plane of the nebula (Edgeworth 1949, Safronov 1969, Goldreich & Ward 1973, Sekiya 1983). This mechanism has been considered appealing in that it appears to involve no assumption regarding the sticking efficiency of colliding bodies. In fact, it now appears that "stickiness" is necessary in any case. In the presence of even modest turbulence, bodies will have to grow to $\gtrsim 1$ m in size if they are to be concentrated sufficiently in the central plane to cause gravitational instability. On the other hand, if because of unexpectedly low turbulence, <1-m-size bodies do concentrate in the central plane, shear between the central dust layer and the gas above and below it will produce a turbulent Ekman boundary layer that will stir the dust sufficiently to preclude the gravitational instability (Weidenschilling 1980). For this reason, acceptance or rejection of dust-layer instability must be decided on its own merits and not on the dubious basis of simplicity.

Although various authors have formulated the problem in different ways, there is general agreement that the mass of the instabilities (m_c) will have a value of about

$$m_c = \frac{\xi \pi^5 \sigma^3 G^2}{\Omega^4}, \tag{2}$$

where ξ is a factor ≈ 1, σ is the surface dust density, and G is the gravitational constant. For $\sigma = 10$ g cm^{-2} and $\xi = 1$ at 1 AU, m_c has a value of 8.6×10^{17} g, corresponding to a diameter of 11.8 km for a planetesimal material density of 1 g cm^{-1}. Because the angular velocity Ω decreases with heliocentric distance, probably more rapidly than the surface density σ, Equation (2) shows that the mass of the planetesimals can be strongly dependent on heliocentric distance. On the other hand, because the mechanism is purely gravitational, no major dependence on the physical properties of the aggregating material would be expected once the conditions for achieving the instability were reached.

A problem with this mechanism is the very low dust particle and turbulent velocities required for the instability to develop. Even in the absence of turbulence, dust velocities can be no more than ~ 10 cm s^{-1} at 1 AU. This is only $\sim 3 \times 10^{-5}$ times the sound speed. For bodies smaller than ~ 1 m, even lower velocities are necessary, because otherwise turbulent motion will limit achieving the dust concentration required for the gravitational instability to develop. Even after allowing for our present ignorance of the extent of turbulence in the solar nebula, it would be surprising if the turbulent velocity could become so low. On the other hand, if the objects are large enough (~ 1 m) they could fall toward the central plane to form a layer thin enough to possibly produce a gravitational instability (Weidenschilling 1988). In any case, nongravitational coagulation is required to produce the 1-m bodies. Purely gravitational formation of planetesimals seems very unlikely.

Because of our near-ignorance of the actual physical state of the solar nebula and of the solid grains that comprise its dust component, despite serious work by several workers, it is not really possible at present to say whether formation of ~ 0.1–10 km planetesimals in the central plane of the nebula would be expected to proceed by gravitational instability or by continuing nongravitational agglomeration. If one simply assumes that somehow or other planetesimals must have formed or we would not be here, the problem can be set aside, and one can go on to a discussion of the way planetesimals grew into planets. There is some merit to this view, but it is excessively facile. It may be that differences in the mode of formation of planetesimals at different distances and times in the solar nebula may be the key to a number of obscure problems, such as the time scales for giant-planet formation and the depletion of the region between 1 and 5 AU, as well as that interior to Venus.

GRAVITATIONAL ACCUMULATION OF PLANETESIMALS INTO PLANETARY EMBRYOS

Transition From the Nongravitational to the Gravitational Growth Regime

Accumulation of micron-size dust grains into $\sim$ 1-km bodies was discussed in the previous section. In that grain-size range, the relative velocities of the accumulating bodies were primarily determined by nongravitational forces, those forces resulting from coupling of small bodies to the turbulent velocity field of the nebula, and the differential velocity resulting from the size-dependent buoyancy caused by the radial gas pressure gradient of the nebula. The mutual cohesion of particles required for bodies to grow was also of nongravitational origin, because even weak Van der Waals binding energies of 10^3 ergs g^{-1} are comparable to the gravitational binding energy of a 1-km-diameter body.

When the bodies become larger than 10 km diameter, this situation changes. For bodies of this size, mutual gravitational perturbations associated with mutual close encounters, balanced with collisions, are sufficient to maintain a steady-state velocity comparable to their escape velocity of 13 m s^{-1}. In comparison, the turbulence-induced velocity of a body this size will be lower, $\sim$2 m s^{-1} (Völk et al 1980) at a Mach number of 0.05. The gravitational binding energy will be $\gtrsim 5 \times 10^5$ ergs g^{-1}, approaching the mechanical strength of coherent but weak bodies. For this reason, a theoretical discussion of the accumulation of bodies larger than this must include gravitational effects.

This transition is not sudden, however, and the regions dominated by gravitational effects overlap those in which nongravitational effects are of primary importance. For example, it cannot be ruled out that collective dust-layer gravitational instability could allow agglomeration of $\sim$1-m-diameter objects into larger bodies to occur in the central dust layer. On the other hand, as long as a significant fraction of the material encountered is in the $\lesssim$1-m size range, possibly as a result of continuing infall, impacts with 10-km bodies will occur at the relative velocity of the gas ($\sim$60 m s^{-1}), far above the escape velocity of these larger bodies. Depending on the mechanical properties of the colliding bodies, this could lead either to nongravitational growth by embedding of the small particles in their targets or to "sandblasting" of the planetesimal's surface, inhibiting growth. If the 10-km body is growing by accumulation from high-velocity small particles entrained in such a "head wind," the associated angular momentum transfer will tend to make the orbits of the larger bodies circular and coplanar.

Runaway vs. Nonrunaway Gravitational Growth: General Considerations

Considerable work has been done on the gravitational accumulation of initially 1–10 km diameter bodies into much more massive $\sim$4000-km objects. For the terrestrial planets to be assembled from $\sim$10-km-diameter ($\sim 10^{18}$ g) planetesimals, $\sim 10^{10}$ initial bodies are required. Following the orbital evolution of so large a swarm of objects by the methods of celestial mechanics is not feasible. The alternative of treating the evolution of the swarm by the methods of gas dynamics has had significant success. In this approach, the individual planetesimals are the analogues of gas molecules in the kinetic theory of gases. This technique has been used quite successfully in stellar dynamics (Chandrasekhar 1942). From this point of view, the planetesimals constitute a swarm of bodies moving in nearly circular and coplanar concentric heliocentric orbits. In a reference frame rotating with the objects at a given heliocentric distance, the relative velocities of the bodies will be very low in comparison with their circular Keplerian velocity of 30 km s^{-1} at 1 AU. Their mutual gravitational perturbations will cause the orbits to have small but nonzero eccentricities and inclinations. This introduces small random and disordered components of velocity relative to that associated with the velocity characteristic of the near-circular Keplerian motion. In the rotating reference frame, this disordered motion is analogous to the random thermal motion of gas molecules confined in a container with walls at a fixed temperature. For this reason, this approach to planetesimal growth is often termed "particle in a box," to distinguish it from a treatment in which the celestial mechanical orbits are explicitly introduced.

In the simplified case where a planetesimal grows by sweeping up bodies considerably smaller than itself, the rate of growth of its mass (M) will be

$$\frac{dM}{dt} = \pi R^2 \bar{V}_{rel} \rho F_g \tag{3}$$

where R is the radius of the growing body, $\bar{V}_{rel}$ the average relative velocity between the large and small bodies, ρ the mass density of small bodies per unit volume of the nebula, and F_g a factor representing the extent to which the gravitational field of the growing body enhances its effective cross section above the geometric value of πR^2. When the growth results from nongravitational cohesion, as considered in the previous section, $F_g = 1$. In the approximation that the encounter between the bodies can be treated as a two-body interaction between the colliding bodies, rather than as a three-body interaction in which the solar gravity appears as a perturbation during the encounter, the gravitational enhancement factor assumes the familiar form

$$F_g = (1 + V_e^2/\bar{V}_{rel}^2) = (1 + 2\vartheta) \tag{4}$$

where V_e is the surface escape velocity of the larger body, and ϑ is the dimensionless Safronov number, $V_e^2/2V_{rel}^2$. This factor arises from the ratio of the distance of close approach to the asymptotic unperturbed impact parameter in a two-body hyperbolic encounter (illustrated by Wood 1961). In this approximation, any object that in the absence of gravity would have a distance of closest approach $RF_g^{1/2}$ will be gravitationally perturbed to an approach distance $< R$, i.e. it will collide.

The factor F_g is of considerable conceptual importance because its value determines whether the growth of a planetesimal swarm is either to be "orderly," in which case the mass distribution evolves smoothly and continuously as the bodies grow, or "runaway," in which case a single largest body in a local region of the swarm grows so rapidly relative to the others that the mass distribution becomes grossly discontinuous, with eventually most of the mass being concentrated in the largest body. In the more detailed discussion later in this section, this distinction between orderly and runaway growth appears as a bifurcation in the numerical solutions of the system of coupled nonlinear equations that determines the mass and velocity evolution of a swarm consisting of bodies of different mass and with mass-dependent relative velocities. The physical basis for this bifurcation can be illustrated by consideration of a simpler case, however.

When the surface escape velocity V_e is expressed in terms of its radius, i.e.

$$V_e^2 = \frac{2GM}{R} = \frac{8}{3}\pi\rho_p R^2, \tag{5}$$

where ρ_p is the material density of the planetesimal, Equation (3) then becomes

$$\frac{dM}{dt} = \pi R^2 \bar{V}_{rel}\rho\left(1 + \frac{8}{3}\frac{\pi\rho_p R^2}{V_{rel}^2}\right). \tag{6}$$

When V_{rel} is large compared with V_e, the growth rate is approximately proportional to R^2; when V_{rel} is small compared with V_e, dM/dt is nearly proportional to R^4. For any given intermediate value of V_{rel}/V_e, the dependence of dM/dt on R will vary continuously between R^2 and R^4. The curve describing this dependence will have a varying slope, but at any point its slope can be represented locally by a power-law dependence on R:

$$\frac{dM}{dt} = kR^q, \tag{7}$$

where the exponent q varies between 2 and 4, depending on the value of R. At what value of q does the bifurcation take place?

Consider two large bodies with masses M_1, M_2 and radii R_1, R_2 ($M_1 > M_2$, $R_1 > R_2$). The rate of change of the ratio of their masses will be

$$\frac{d}{dt}\left(\frac{M_1}{M_2}\right) = \frac{1}{M_2}\left(\frac{dM_1}{dt} - \frac{M_1}{M_2}\frac{dM_2}{dt}\right). \tag{8}$$

If we express dM_1/dt and dM_2/dt in terms of Equation (7), make use of the R_1^3-dependence of M_1, and factor out positive-definite terms, Equation (8) becomes

$$\frac{d}{dt}\left(\frac{M_1}{M_2}\right) = \frac{kM_1}{M_2^2}R_2^q\left[\left(\frac{R_1}{R_2}\right)^{q-3} - 1\right]. \tag{9}$$

When $q > 3$, $d/dt(M_1/M_2)$ is positive and M_1 will become even larger relative to M_2. When $q < 3$, $d/dt(M_1/M_2)$ is negative and M_2 will tend to "catch up" with the larger body. When $q = 3$ (i.e. the growth rate is proportional to the mass of the growing body) $d/dt(M_1/M_2)$ is zero and the ratio of the two masses neither decreases nor increases. This is the boundary between orderly and runaway growth for this simple case.

Safronov's Solution of the Coagulation Equation

Safronov (1962, 1969) obtained an analytical solution for the integro-differential coagulation equation that represents the extension of Equation (3) to a continuous distribution in which the masses of all the bodies are growing by collisions with those smaller than themselves, subject to the condition that the growth rate is proportional to the mass [i.e. obeys Equation (8) with $q = 3$, the limiting value for nonrunaway growth]. In that work, the initial mass distribution was assumed to be of the form

$$n(m) = \frac{N_0 e^{-m/m_0}}{m_0}, \tag{10}$$

where $n(m)\,dm$ is the number density of bodies between m and $m+dm$, N_0 is the total number density at the initial time, and m_0 is the mean mass of the initial mass distribution. After an initial transient, a solution in dimensionless form is found to be

$$n(m,\tau) = \frac{(1-\tau)}{2\pi^{1/2}\tau^{3/4}} m^{-3/2} e^{-(1-\tau^{1/2})^2}, \tag{11}$$

where τ is a dimensionless timelike variable relating the total number density of bodies N to the initial number density N_0, i.e.

$$\tau = 1 - N/N_0, \tag{12}$$

and where n is in units of N_0 and m is in units of m_0.

Under the circumstances considered by Safronov, the ratio (N/N_0) decays exponentially:

$$\left(\frac{N}{N_0}\right) = (1-\tau) = e^{-\eta}. \tag{13}$$

The dimensionless scaled time η is given by

$$\eta = A_1 N_0 m_0 t \tag{14}$$

where t is actual time, and A_1 is the constant coagulation coefficient that represents the assumed proportionality of the growth rates to the combined masses, in accordance with the value $q = 3$ in Equation (9). At the initial time $t = 0$, it follows that $N = N_0$ and thus that $\tau = 0$. At later times, the number of bodies decreases as a result of coagulation and $\tau \to 1$ for very long times.

Safronov's result is plotted in cumulative form in Figure 3, including the initial transient. These curves describe the archetypal form of nonrunaway growth. Although some differences are found when the assumptions are varied, all calculations of orderly growth have the general morphology of

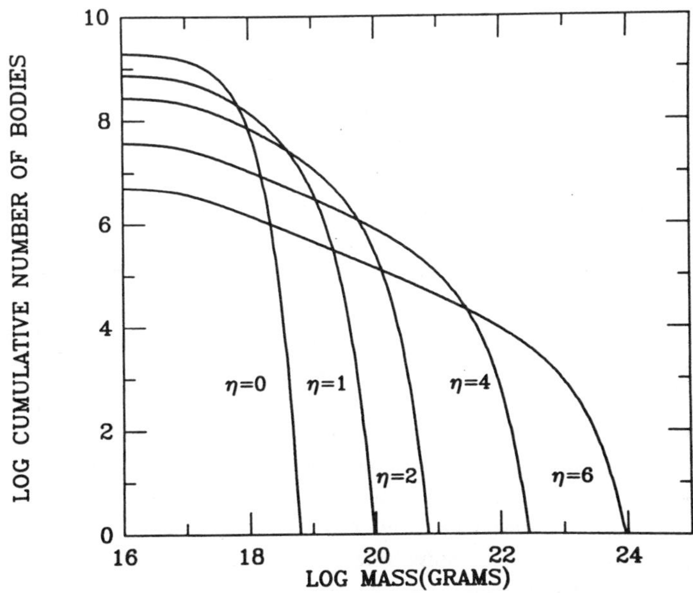

Figure 3 Solution of the coagulation equation with linear kernel (growth rate proportional to sum of masses) found by Safronov (1962, 1969). The initial conditions correspond to a reasonable planetesimal swarm in which the total mass of 6×10^{26} g is distributed over a 0.02-AU-wide zone centered at 1 AU, and the mean mass of the initial planetesimals is 3×10^{17} g. The equivalence to actual time of the dimensionless time variable η is model dependent, but $\eta = 1$ approximately corresponds to 3×10^5 yr. With the passage of time, the initial steep distribution evolves into a power law, as described in the text.

the results plotted in this figure. These results are presented for several values of η and for a reasonable choice of m_0 of 3×10^{17} g and a total mass of 6×10^{26} g in a local zone 0.02 AU in width centered at 1 AU. Although the decay in N is not actually exponential in real physical time as a consequence of the thickness of the swarm expanding with time, roughly $\eta = 1$ corresponds to $t = 3 \times 10^5$ yr.

The initial very steep mass distribution of Equation (10) evolves into a rather flat power-law mass dependence ($dn/dm \propto m^{-3/2}$) truncated at the high-mass end by an exponential remnant of the original steep distribution. With the passage of time, the upper end of the distribution "marches" forward in mass, but with diminishing amplitude. At the same time the power-law portion of the distribution includes an increasing range of masses. In this way an initial steep exponential distribution evolves into a power-law distribution in which most of the mass is concentrated in the largest bodies. Power-law distributions of this kind, with varying exponents, form the basis for much of Safronov's treatment of planetary growth.

Although supported by other theoretical calculations, orderly growth was "built in" to Safronov's result by the assumption that $q = 3$. This constraint that the growth rate is proportional to the mass was dropped in the subsequent work on this same problem by Nakagawa et al (1983). Consistent with the earlier studies of the Kyoto "school" [see Hayashi et al (1985) for references], this work differed from that of Safronov in that the velocity damping that balances the gravitationally induced increase in velocity was assumed to result from gas drag rather than from mutual collisions. Both of these phenomena can be expected to play a role in decreasing the velocity of the swarm. Safronov did not include gas drag, whereas Hayashi and his associates did not include collisional damping. The final effect of this considerable difference in assumptions is not as great as one might guess. The reason is that the steady-state velocity is approximately proportional to only the fourth root of the coefficients of the expressions for the energy changes associated with collisional and gas drag damping of velocity. Omission of one or the other of these terms will require only a somewhat higher steady-state velocity to establish the steady-state balance. It is sometimes stated that Hayashi and colleagues erred by considering gas drag to be important, whereas "it is well-known" that gas drag is of minor importance for large bodies. This criticism is unjustified, however. It is true that for the same relative velocity with respect to the gas, the effect of the gas drag force, proportional to R^2, falls off as $1/R$ when the acceleration is obtained by dividing the force by the mass. In the absence of gravitational enhancement of the collisional cross-section factor F_g in Equation (3), this is also true for the acceleration caused by collisional damping. In both cases the system will regulate itself by increasing the velocity v of the large bodies, thereby offsetting this dependence of dv/dt on the radius. When gravitational focusing causes dM/dt to be proportional to R^3, this $1/R$-dependence is no longer present for collisional damping, but the fourth-root dependence of the steady-state velocity requires a velocity difference of only a factor of two relative to the unfocused case when the radius of the body increases from 100 km to lunar size. For this reason, the basic results of Nakagawa et al (1983) are not very different from those of Safronov.

Equipartition of Energy in a Swarm Consisting of Bodies With Unequal Masses

The effect of the swarm containing bodies of different mass is more important than differences associated with the relative importance of collisional and gas drag damping of velocity. In both the Tokyo-Kyoto and Moscow work, as well as the tutorial illustrative calculations of Wetherill (1980), the expressions for the gravitational "pumping up" of velocities were always positive. This is actually true only when the bodies in the swarm are equal in mass. Even if the initial masses were all equal, stochastic dispersion of the swarm's mass distribution would quickly occur, if only because of combinatorial statistics. Stewart & Kaula (1980) showed that when the bodies were unequal in mass, the effect of their mutual gravitational perturbations could decrease their disordered components of velocity as well as increase them. This effect arises from the tendency of gravitational perturbations to equipartition energy between bodies of different size, analogous to the familiar nongravitational equipartition of energy in the kinetic theory of gases. This phenomenon is related to the "dynamical friction" introduced by Chandrasekhar (1943) in stellar dynamics. Its effect is to introduce additional terms into the theoretical expressions describing the gravitationally induced change of velocity of a body of mass m_1 and velocity v_1 by perturbations due to a second body of mass m_2 and v_2.

These terms are of the form

$$\frac{dv_1}{dt} \propto (m_2 v_2^2 - m_1 v_1^2). \tag{15}$$

If v_1 is approximately equal to v_2, and $m_1 > m_2$, these terms tend to cause the velocity v_1 to decrease. The analogous terms for dv_2/dt will have the opposite sign. Thus, a dispersion in the mass distribution will lead to velocity changes that decrease the average velocity of the larger bodies of the swarm, increase the average velocities of the smaller bodies, and leave relatively unchanged the velocity of bodies intermediate in mass.

The work of Stewart & Kaula (1980), Stewart (1980), and Hornung et al (1985) has been extended by Stewart & Wetherill (1988) using Boltzmann and Fokker-Planck equations to describe the velocity distribution of an assemblage of bodies in near-Keplerian orbits. Expressions for $\mathrm{d}v/\mathrm{d}t$ as a function of the masses and velocities of the interacting bodies were obtained that include mutual gravitational perturbations, collisions, and gas drag. These expressions were then used in numerical calculations in which the evolution of the coupled velocities and masses of a swarm of bodies was followed in successive time steps, during each of which bodies grew in accordance with an expression equivalent to Equation (3).

Because both the changes in mass and in velocity during a single time step are nonlinear functions of all the other masses and velocities, a general solution of these coupled equations cannot be obtained. On the other hand, it is found that for physically realistic initial conditions, the numerical solutions bifurcate into two qualitatively distinct types: runaway vs. orderly growth. These calculations are more complex analogues of the oversimplified system described by Equation (9), in which two noninteracting large masses both grew by sweeping up many much smaller bodies at a constant value of the parameter q and the Safronov number ϑ [Equation (4)]. In the general case the mass distribution is nearly continuous, and ϑ is neither a free parameter nor constant with either time or mass but rather is determined by the evolving mass and velocity distribution.

Numerical Calculations Corresponding to Runaway and Nonrunaway Growth

In the earlier work of the Moscow and Kyoto groups, their solutions followed the branch of orderly growth (Safronov 1969, Nakagawa et al 1983). These results were reproduced by the calculations of Wetherill & Stewart (1989) when the equipartition-of-energy terms [Equation (15)] were omitted, in accordance with the formulation of the problem by these workers. For orderly growth, $\sim 10^{25}$-g sublunar-size bodies formed at 1 AU in about 10^6 yr (Figure 4). In contrast, when these terms were included, the same initial conditions used by these authors led to a rapid runaway in which objects larger than the Moon formed in $\sim 10^5$ yr, and a marked discontinuity of two orders of magnitude appeared between the masses of the large runaway body and those of the remainder of the swarm (Figure 5). The difference between these two calculations was caused by the much lower velocities of the larger bodies attributable to the equipartition-of-energy terms. As shown in Figure 6, after only 3×10^4 yr, the velocities of all the bodies greater than about 10^{21} g in mass have dropped below their initial values, whereas those of the smaller bodies have increased. As a consequence, the relative velocities of the largest body and bodies of

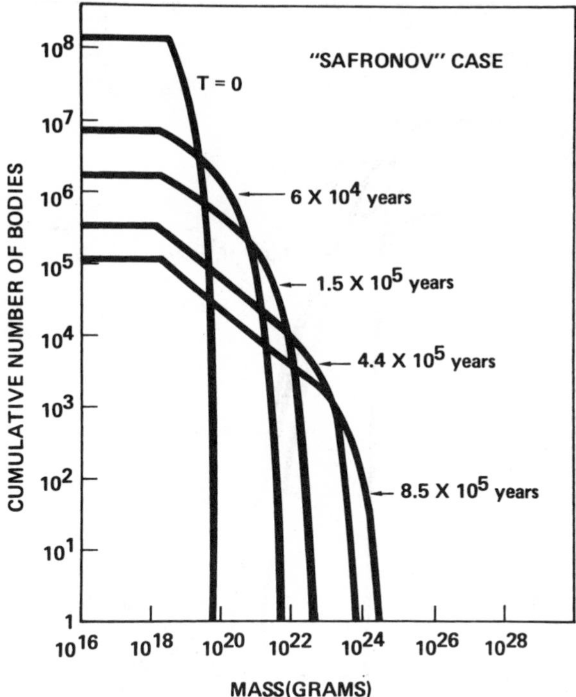

Figure 4 Numerical calculation of the evolution of the mass distribution of a swarm of planetesimals, for which the velocity distribution is determined entirely by the balance between positive-definite gravitational "pumping up" of velocity and collisional damping (Wetherill & Stewart 1989). The initial mass of the swarm was 6×10^{26} g, distributed between 0.99 and 1.01 AU with mean initial mass of 3×10^{18} g. The growth is found to be "orderly" and generally similar to the analytical result of Safronov, shown in Figure 3. No runaway is found; the evolution leads to a final mass distribution in which most of the mass is concentrated in 10^{24}–10^{25} g bodies at the upper end of the distribution.

intermediate mass will decrease, and the gravitational focusing factor F_g [Equation (4)] will become larger, increasing the rate at which the largest body sweeps up bodies of intermediate mass. This faster increase in the mass and, consequently, in the escape velocity of the largest body causes it to grow even more rapidly. This leads to the runaway instability seen in Figure 5. Because these calculations refer only to local zones 0.02 AU in width, it may be anticipated that many similar runaways lead to an assemblage of runaway bodies in concentric circular orbits around the Sun.

Similar runaways were reported by Greenberg et al (1978) under circumstances for which Wetherill & Stewart (1989) found no runaway. Those familiar with the computer code used by Greenberg et al believe

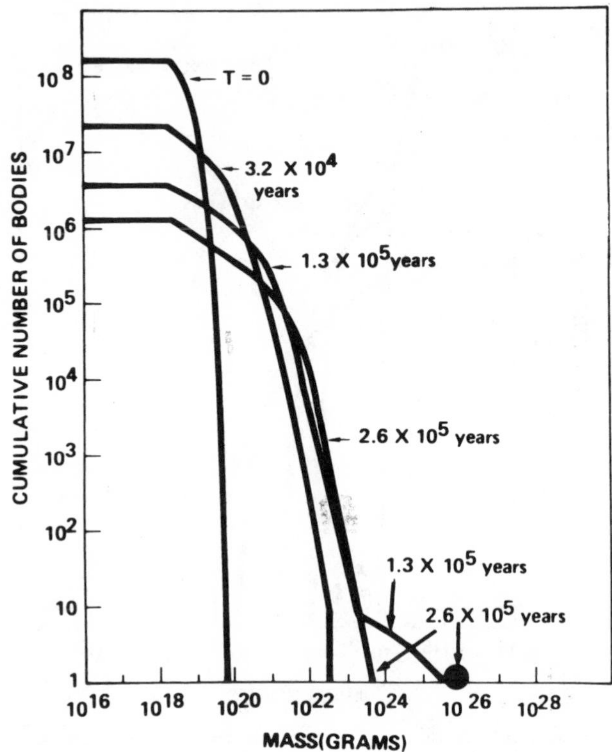

Figure 5 Effect of including the equipartition-of-energy terms [Equation (15)] in the calculation of the evolution of the same initial swarm as that in Figure 4 (Wetherill & Stewart 1989). The tendency toward equipartition of energy results in a velocity dispersion in which the velocity (with respect to a circular orbit) of the massive bodies falls below that of the swarm. After about 10^5 yr, a "multiple runaway" appears as a bulge in the mass distribution in the mass range 10^{24}–10^{25} g. After 2.6×10^5 yr, a single largest body of mass $\sim 10^{26}$ g (*large solid circle*) has swept up the bulge, and the continuous distribution of remaining bodies have masses $< 10^{24}$ g.

that numerical inaccuracies in the calculations of Greenberg et al are the most likely explanation of this difference (S. Weidenschilling, C. Patterson, D. Spaute, private communication, 1989). This difficulty is likely to be related to considerations of mass-binning in numerical solutions of coagulation equations discussed by Ohtsuki et al (1989).

Other Factors That are Likely to Influence Runaway Accumulation

Both the model and the theoretical expressions used in all of these calculations are obviously oversimplified. In an actual circumstellar nebula,

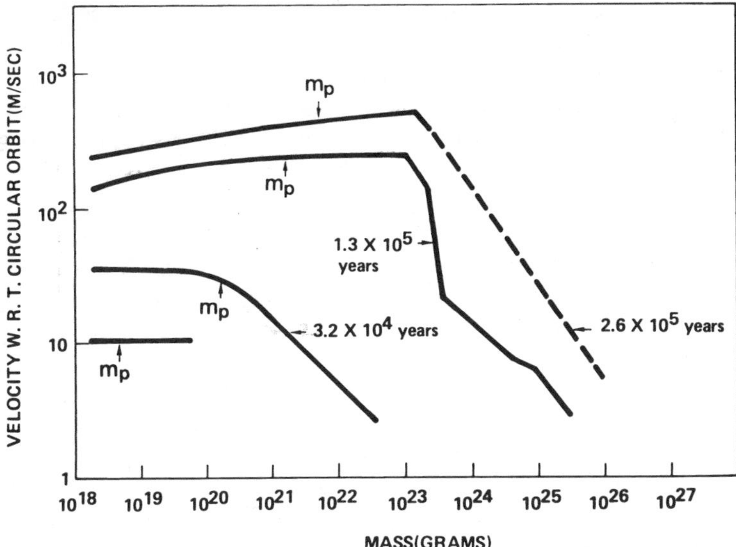

Figure 6 Velocity distribution resulting from inclusion of equipartition-of-energy terms (Wetherill & Stewart 1989). After 3×10^4 yr, the relative velocities of the largest bodies drop well below the midpoint mass m_p, the value for which half the mass of the swarm is in smaller bodies. This leads to a rapid growth of the largest bodies and ultimately to a runaway.

significant initial discontinuities in the mass distributions may have existed. Deviations from the two-body expression [Equation (4)] for gravitational focusing are found at low values of $\bar{V}_{rel}/V_e$ by numerical integration (Nishida 1983, Wetherill & Cox 1985, Ida & Nakazawa 1989, Greenzweig & Lissauer 1990). The smaller bodies of the swarm have relative velocities so large that they will fragment when they collide; this feature was introduced into such calculations for the first time by Greenberg et al (1978). The rate of growth of the runaway body may also increase because the body's gravitational field concentrates nebular gas in its vicinity, increasing gas drag and lowering velocity (Takeda et al 1985, Ohtsuki et al 1988).

The complications introduced by inclusion of these additional effects are formidable, and they have not been considered in anything like an exact way. Nevertheless, it is reasonably clear that all the phenomena listed above will increase both the likelihood and the rate of runaway rather than decrease them. Approximate inclusion of several of these effects by Wetherill & Stewart (1989) indicates that at 1 AU, bodies as large as 1.2×10^{26} g (about two lunar masses) may grow in less than 10^5 yr. Such rapid growth encroaches on the settling and growth time scales of grains in the solar nebula, and these could turn out to be the rate-limiting factors during this first stage of planetesimal growth.

Are there also factors that may limit the runaway, or inhibit it and lead to orderly growth? Three such phenomena are now discussed, the last two of which seem unlikely to be of much significance for the formation of the Earth at 1AU, even though they may be important for the formation of asteroids.

1. As the planetesimals grow into embryos, the supply of available smaller bodies necessary for their continued growth is exhausted, either because these bodies are beyond the "gravitational reach" of the embryos or because neighboring embryos accumulated them first.

During the orderly planetesimal growth depicted in Figures 3 and 4 and the early stage of that shown in Figure 5, the relative velocities are found to be high enough for the gravitational interactions of the bodies to be dominated by two-body close encounters of bodies in orbits that cross one another. Collisions leading to growth represent a subset of these close encounters that are unusually close. Under these circumstances, it was valid to use two-body dynamics and simply consider the evolution of the planetesimal swarm by the methods of the kinetic theory of gases, with each planetesimal being equivalent to a "gas molecule" moving in a gas comprised of other planetesimals. The encounter and growth rates are determined by the local density of the planetesimal "gas." At later times, much of the mass of the swarm will be concentrated in larger ($\sim 10^{25}$-g) bodies in neighboring heliocentric zones, and continued growth requires that these larger bodies still collide and merge with one another. When the growth is orderly, the relative velocities of even the largest bodies remain high enough to permit the orbits of neighboring bodies to cross. Under these circumstances, there is little tendency for the embryos to isolate themselves from their neighbors.

The situation is different when runaway growth occurs. Because of equipartition of energy, the orbits of the larger bodies that form the "bulge" in Figure 5 are found to become nearly circular; as a consequence, the bodies tend to become isolated, in the sense that as long as their orbits remain unchanged, close encounters and collisions cannot take place. Nevertheless, as shown by Dole (1962) and Giuli (1968), objects in apparently isolated concentric circular orbits can under some circumstances collide with one another. This is possible because at the low relative velocities associated with concentric circular orbits, gravitational interaction continues for a long time as bodies at slightly different heliocentric distances slowly pass one another as a consequence of their circular Keplerian motion. In these cases, the integrated effect of the gravitational perturbations caused by the prolonged, relatively distant approach turns out to be sufficient to change the isolated orbits into crossing orbits, permitting growth to continue. On the other hand, if noncrossing orbits are too distant from one another, the mutual perturbations will be insufficient to end their isolation.

Birn (1973) found that isolation will continue if the semimajor axes of the concentric circular orbits are separated by more than about

$$\delta = 2\sqrt{3}r_H. \tag{16}$$

The quantity r_H (the "Hill sphere radius") is given by

$$r_H = a(m/3M_\odot)^{1/3}, \tag{17}$$

where a is the heliocentric radius, and m is the sum of the masses of the two planetesimals or embryos. At 1 AU, r_H will have values of about 10^{-3} AU and 10^{-2} AU for 10^{25} g and 1-$M_\oplus$ bodies, respectively. A similar result was found by Graziani & Black (1981) and by Artymowicz (1987). Although the coefficient of r_H in Equation (16) depends somewhat on the ratio of the masses of the interacting bodies, as well as on their eccentricities and inclinations, similar results are found for small deviations from circularity (Wetherill & Cox 1984, 1985, Artymowicz 1987, Ida & Nakazawa 1989, Greenzweig & Lissauer 1990). The smaller bodies of the residual swarm will have eccentricities of ~ 0.007, and therefore their orbits will span the positions of several larger bodies in the bulge, at least as long as these larger bodies avoid colliding with the largest body. A possible quasi-steady-state outcome of this stage of growth would be a series of embryos in concentric circular orbits, separated by the minimum distance δ [Equation (17)], each embryo having a mass

$$M_e = m/2 = 2\pi a\sigma\delta, \tag{18}$$

where σ is the mass surface density of the annulus. Equations (16) and (17) can be substituted into (18), which can then be solved for the mass of the embryo:

$$\begin{aligned} M_e &= 8\sqrt{2} \times 3^{1/4}\pi^{3/2}a^3\sigma^{3/2}M_\odot^{-1/2} \\ &= 6.21 \times 10^{24}a^3\sigma^{3/2}, \end{aligned} \tag{19}$$

where in the second expression a is expressed in astronomical units and σ in grams per square centimeter. If this configuration develops, at 1 AU, runaway embryos will grow to masses of 1.1×10^{26} g to 5.6×10^{26} g for the range of surface densities (7–20 g cm^{-2}) required to supply the mass of the present terrestrial planets, depending on the range of heliocentric distance from which their material was obtained. Although the upper end of the mass range is similar to the mass of Mars, it is only 10% of the mass of the Earth, and merger of a fairly large number of embryos will be required to complete the growth of the Earth. It is also possible that equipartition of energy will inhibit the merger of the bodies in the bulge until the smaller bodies are nearly all accumulated. This would effectively reduce the coefficient of r_H in Equation (16) and result in a larger number of smaller embryos. Merger of runaway embryos is the final stage in the formation of the Earth and is discussed in the next section.

2. Another factor that under some circumstances may inhibit or even prevent runaway growth is the size distribution of the initial planetesimals. The equipartition-of-energy terms that are responsible for the runaway [Equation (15)] require that the swarm contains bodies that differ in mass in order to establish the differences in velocity seen in Figure 5. Even if all the bodies in the swarm had identical masses at the outset, the mass distribution will disperse simply as a result of stochastic variations in the collision rates of the very large number of bodies in the initial swarm. For example, for a swarm of originally equal-mass bodies, as soon as two bodies collide and merge, the mass of the largest body will be twice that of those comprising the rest of the swarm. If the initial swarm contains 10^{10} bodies, about 8 of them will collide and merge 12 times by the time most of the bodies have collided once. The ratio of the mass of the single largest body to that of the mean mass of the swarm will increase as the number of bodies in the swarm increases, and decrease when the number of bodies decreases, simply as a consequence of combinatorial probabilities. If the initial swarm consists of a small enough number of large bodies, the dispersion will be so small that the runaway will develop so slowly that the largest bodies will still be of similar mass when they grow to about lunar size.

It is not expected that this phenomenon will be important for the Earth and terrestrial planets. The initial masses for the runaway case shown in Figure 5 were about those expected by the expression of Goldreich & Ward (1973) for the dust-layer gravitational instability. As discussed by Weidenschilling et al (1989), gravitational instability is not particularly likely, and the first-formed planetesimals in the central plane of the solar nebula would probably be smaller (0.1–1.0 km), leading to greater dispersion and a more rapid runaway. The Goldreich-Ward bodies expected at the distance of Jupiter would be ~ 80 km in diameter. It is conceivable, but not likely, that a nearly uniform-size population of such large planetesimals could form at 5 AU and greatly inhibit the runaway growth of Jupiter's core. But in the terrestrial planet region, this is even more implausible.

3. The calculations described in the foregoing discussion were all made

under the assumption that, apart from solar gravity, the only forces acting on the planetesimals were those associated with their mutual gravitational perturbations, collisions, and gas drag. It is also possible that long-range perturbations by massive, distant, external perturbers could increase the velocities of the growing planetesimals. If these perturbations could offset sufficiently the decrease in velocity of the incipient runaway embryos caused by equipartition of energy, the runaway might be aborted. It is also possible that external perturbations could increase the velocities of the smaller bodies of the swarm so much that fragmentation, rather than growth, would dominate.

The principal candidate external perturbers in the early solar system are Jupiter (and Saturn) and perturbations associated with loss of the gas of the residual solar nebula (Heppenheimer 1979, Ward 1980). Because the time scale for runaway growth of embryos at 1 AU is so short ($\sim 10^5$ yr), it does not seem likely that Jupiter would exist at that time, nor that the gas would be lost so soon, inasmuch as the gas is required to form Jupiter and Saturn. For this reason, external perturbations are not expected to be of major importance in inhibiting runaways in the terrestrial planet region, even though they could be of great importance in the outer solar system and the asteroid belt.

FINAL STAGE OF ACCUMULATION: EMBRYOS TO PLANETS

The Onset of the Final Stage of Accumulation

Runaway growth of planetesimals is expected to yield planetary embryos in the mass range of 10^{26}–10^{27} g, separated from one another by 0.01–0.02 AU on a time scale of about 10^5 yr. In accordance with Equations (16) and (19) the spacing and mass of the embryos increase with heliocentric distance. Collision and merger of about 30 such embryos, distributed between about 0.7 and 1.1 AU, can provide the mass, energy, and angular momentum of the present terrestrial planets. A less likely alternative is that for some unexpected reason, runaway growth of embryos did not take place. In this case, orderly growth from planetesimals to embryos would be expected to produce a larger number (e.g. 1000) of smaller bodies with sublunar masses of about 10^{25} g.

Actual formation of the Earth and other terrestrial planets requires that in either case these embryos can make close encounters with one another; otherwise, collisions and growth will not occur. For the case of orderly growth this is not a problem because the calculated eccentricities of the embryos are well above the values necessary to allow their orbits to cross one another. For the case of runaway growth the situation is more complicated and less well studied. Nevertheless, the onset of the final stage of growth can be discussed in a semiquantitative way.

In the early stages of runaway growth, when the incipient runaway body has a mass of $\sim 10^{24}$ g, almost all the mass of the swarm is still in small, $\lesssim 10^{21}$-g bodies (Figure 5). Under these circumstances, equipartition of energy with these smaller bodies is found to maintain the incipient runaways in very nearly circular and coplanar orbits. As the runaway embryos grow into the $\sim 10^{26}$–10^{27} g range, the total mass in the residual swarm will be depleted, reducing the constraint imposed by equipartition of energy. At the same time, the masses of adjacent embryos, separated by 0.01–0.02 AU, will have become larger, increasing the strength of the long-range gravitational perturbations between embryos. When the long-range perturbations begin to dominate, the changes in eccentricities caused by these perturbations will average between about 0.005 and 0.01, approximately that required to permit the orbits of adjacent embryos to cross. The magnitude of these perturbations often exceeds the average by more than a factor of two, ensuring orbital crossing (Greenzweig & Lissauer 1990). Once orbital crossing of embryos is achieved, at least after nebular gas has been removed from the terrestrial planet region, close encounters between embryos will increase their relative velocities to values comparable to their escape velocities of about 3 km s^{-1}. Such velocities correspond to eccentricities of about 0.1, which are sufficient to permit every embryo to cross the orbits of several other embryos. In addition, as embryos approach a mass of $\sim 10^{26}$ g, radial migration resulting from gravitational interaction between the embryos and any nebular gas still present (Ward 1986) can also be expected to lead to mutual orbital crossing, particularly if there are fairly large stochastic variations in the masses of adjacent embryos. For these reasons, it is not expected that the runaway embryos will become gravitationally or collisionally isolated from one another. If anything, it seems more likely they will begin to interact with one another somewhat before the runaway is completed.

Multiplanet vs. Single-Planet Accumulation

The task of understanding the formation of the Earth and terrestrial planets then becomes that of understanding the subsequent dynamical evolution of these embryos moving in crossing orbits. Because the masses of the embryos are comparable to the masses of Mercury and even Mars, this final stage of accumulation will involve collisions of very massive bodies with one another. As orbital evolution proceeds, the number of embryos will decrease and their semimajor axes will become more widely separated. At some point, determined by their eccentricities, the bodies will become permanently gravitationally isolated from one another, and the evolution will cease. The remaining embryos can then be designated "final planets."

Thus, during this final stage of planetary growth, the formation of the Earth necessarily becomes a problem in multiplanet, rather than single-planet, accumulation. In some investigations this fact has been set aside and the accumulation of the Earth was formally allowed to proceed as if at any early stage a single embryo was designated "Earth" and the remaining planetesimals were assumed not to grow larger, at least within a designated "feeding zone" surrounding the planet. Calculations of this kind have contributed much to our understanding of planetary formation. The very important pioneering investigations of Safronov (1969) and the Kyoto-Tokyo group (e.g. Nakagawa et al 1983) fall into this category.

To some extent, the successes of these contributions are attributable to the use of positive-definite expressions for the mutual gravitational perturbations of the embryos, i.e. they omitted the equipartition-of-energy terms introduced by Stewart & Wetherill (1988). The resulting orderly growth has the consequence that the distinction between the earlier stage during which planetesimals grow into embryos and the later one of embryos growing into planets is weakened. At all times, the planetesimals/embryos are crossing the orbits of many neighbors, and relative velocities remain near the escape velocity of the larger bodies throughout their growth. Appropriate long time scales (10^7–10^8 yr) for complete accumulation of the Earth are found. Within the scope of single-planet formation theories of this kind, it is even possible to infer the possible occurrence of giant impacts (Wetherill 1976), although it must be admitted that the theory provides no way of distinguishing between a giant impactor and an isolated second largest planet.

This situation is quite different in the case of runaway growth of embryos, now recognized to be physically plausible. Oversimplified continuation of such calculations leads to a single Earth-sized planet in $\sim 10^5$ yr. Of course, a single-planet formation theory can tell nothing about the final number, heliocentric distances, or relative masses of the final planets. With the exception of the work of Horedt (1985), all studies of multiplanet accumulation have been numerical. Treatment of the orbital evolution of embryos into final planets is in one way difficult and in another way relatively easy. One difficulty stems from inapplicability of the kinetic theory approximation to systems containing only from four to several hundred bodies, distributed over a wide range of heliocentric distance. A greater difficulty is the present uncertainty regarding the possible importance of nebular gas during the beginning of the final stage of accumulation. The relative ease is due to the possibility of following the orbital history of each individual body.

Effect of Nebular Gas During the Final Stage of Accumulation

The question of whether or not nebular gas played an important role during the final stage of accumulation is a vexing one. It involves two questions: whether or not a major fraction of the nebular gas was still present at 1 AU at the time the planetary embryos were formed, and if so, whether or not this may be expected to make any difference either to the dynamics of accumulation or to the chemical nature of the final planets. It is not possible to give a clear answer to either of these questions at present.

Before the likelihood of runaway formation of embryos was established, it seemed reasonably safe to ignore the complexities of embryo-nebula interaction and to simply assume that nebular gas would be lost within the first million years of the 10^7–10^8 yr accumulation time of the final planets from embryos. Gas drag forces on $> 10^{25}$-g embryos were expected to be small compared with other forces that can decrease relative velocities, i.e. collisions and equipartition of energy ("dynamical friction"). This view fits in well with the idea that initially well-collimated bipolar flows from pre-main-sequence stars may continually erode the nebular gas at the edges of the more or less conical boundaries of the flow. Removal of this gas, proposed to be responsible for the collimation, then permits the cone to open further and eventually evolve into a more isotropic stellar wind (Boss 1987). It might be expected that in the course of this evolution of the outflow, a "hole" nearly devoid of gas would develop in the innermost part of the nebula on a time scale that is short (e.g. $\sim 10^5$ yr at 1 AU) compared with the time scale for removal of all of the nebular gas. Thus, gas-free final accumulation of the terrestrial planets might proceed before the giant planets had completed their growth.

Now, however, it seems likely that runaway embryos are likely to form quite rapidly, in $\sim 10^5$ yr, and this places more strain on this hypothesized chain of events. In addition, it has also been proposed (Takeda et al 1985) that for planetary embryos the gas drag coefficient may be greatly enhanced for a gravitating body at low values of the Reynolds number (Re $\sim$ 20). Ohtsuki et al (1988) have pursued the consequences of this enhanced gas drag. They find, among other things, that gas drag could then be the dominant "damping" force during the coalescence of embryos; that gas-drag-induced radial migration of large embryos may be important, in addition to that caused by tidal density wave effects (Ward 1986); and that relative velocities during the final stage of accumulation may be significantly lower than in the more popular gas-free case. Provided that all these inadequately studied effects can actually produce a system of terrestrial planets similar to that observed, it is also quite likely that the time scale for the final growth of the Earth could be as short as 10^7 yr, which suggests that capture of a very massive ($>10^{26}$ g) and very hot (~ 3000 K) primitive atmosphere (Hayashi et al 1979) might then be expected if the nebular gas were still present at 1 AU for as long as 10^7 yr.

Ohtsuki et al (1988) conclude that the effects described above may be overestimated because of oversimplifications in the theoretical treatment of Takeda et al (1985), e.g. the assumption of Reynolds numbers lower than those expected in the solar nebula and the neglect of solar gravity. Nevertheless, subject to much more quantitative scrutiny, an evolution of this kind may be considered to be an alternative to the more conventional gas-free final accumulation of the terrestrial planets. This alternative could have observationally distinguishable consequences, but these have not yet been worked out completely enough to be discussed in a responsible way at present.

Numerical Simulations of the Final Stage of Accumulation

A number of calculations have been made of the gas-free evolution of 50 to 1000 embryos in the terrestrial planet region, both in two dimensions (Cox & Lewis 1980, Wetherill 1980, Lecar & Aarseth 1986, Ipatov 1987, Beaugé & Aarseth 1990) and in three dimensions (Wetherill 1980, 1985, 1986, 1988). All of these calculations should be considered "Monte Carlo calculations," both because of the chaotic nature of the orbital changes caused by close encounters of the embryos with one another and also because of the impossibility of completely numerically integrating, even in two dimensions, the orbits of ~ 100 bodies, all perturbing one another for $>10^6$ orbital periods. The two-dimensional (2D) calculations raise the additional problem of interpreting the results of 2D phenomena in terms of the real, three-dimensional (3D) world. There are essential differences between the 2D and 3D situations:

1. In two dimensions crossing orbits are always intersecting orbits, whereas in three dimensions crossing orbits will not in general intersect; they can simply be looped, like links in a chain.
2. In three dimensions the ratio of close encounters to actual collisions is much greater than in two dimensions, for the simple geometrical reason that in three dimensions the number of "misses" within a distance R of collision increases with R^2, whereas in two dimensions it increases only linearly with R.

On the other hand, there are potential advantages to 2D calculations, in that some phenomena, such as the effects of distant perturbations and possibly resonances, can be addressed in two dimensions but are computationally prohibitive in three dimensions. In this regard, Lecar & Aarseth (1986) and Beaugé & Aarseth (1990) have carried out a total of four simulations of the final stages of terrestrial planet growth in which numerical integration provides some insight regarding the importance of distant vs. close encounters.

The most extensive available investigations are those of the author, who has reported the results of 45 3D Monte Carlo simulations of the final stages of terrestrial planet growth and has also carried out more than 100 unpublished calculations, studying the effects of a wide range of initial conditions, assumptions, and physical parameters involving swarms of up to 1000 bodies. These variations include exploratory studies of the probable effects of collisional fragmentation, tidal disruption by encounters within the Roche limit, irregular mass distributions as a function of semimajor axis, and commensurability resonances with the orbital periods of Jupiter and the largest embryos. Although the techniques used to simulate these phenomena were rudimentary, it does seem that these studies show that none of these effects are likely to qualitatively modify the terrestrial planet evolution found when these phenomena are not included. Perhaps the most interesting conclusion of these studies is the general similarity of the final results, despite their other differences. In all cases the large number of initial embryos spontaneously decreased to a small number of final planet-size bodies, almost always between 3 and 5, distributed throughout the region occupied by the observed terrestrial planets. In all cases, the growth of Earth-size planets was punctuated by giant impacts, of the size proposed by Cameron & Ward (1976) and Hartmann & Davis (1975) as responsible for the formation of the Moon. Another general feature is the widespread radial migration of the smaller embryos as they are passed back and forth between the larger planetary bodies "Earth" and "Venus" and into the terrestrial planet space beyond these planets. A ubiquitous result is the extensive heating of the larger bodies. Bodies more massive than 0.5 $M_\oplus$ are invariably melted, as are most Mars-size final planets. All of these same phenomena are seen in the 2D calculations as well. The only obvious difference between the 3D and 2D calculations are the time scales for accumulations. In three dimensions the Earth requires about 25 m.y. to grow to half its present mass and about 100 m.y. to grow to 99% of its present mass. In two dimensions the time scales are shorter by about a factor of 100, an artifact of the geometrical difference in orbital intersections mentioned earlier.

All of these calculations, as well as all those that will be made in the foreseeable future, are inevitably subject to potentially serious limitations. Little will be gained if authors simply annihilate one another by emphasizing the deficiencies in the calculations of other people. A more fruitful and positive approach, whereby significance of limitations in one mode of calculation are tested by alternative approaches that involve limitations of a different kind, may be expected to sort out the serious problems from those that constitute only sophisticated nit-picking.

The Monte Carlo calculations carried out by the author represent an effort to break through the barrier presented by the overwhelming difficulty of integrating in three dimensions the orbits of hundreds of interacting bodies for more than 10^8 orbital periods. Even if such integration were possible, the orbital interactions in three dimensions would be determined in large part by precessional motion caused by external distant perturbers, such as the giant planets, which would also have to be included in the calculations in some way. Of course, a price must be paid for this simplification that permits the problem to be studied at all. Probably the most serious is the limitation to perturbations by bodies in crossing orbits. Comparison of the 2D calculations of Cox & Lewis (1980) and Wetherill (1980) with those of Lecar & Aarseth (1986) and Beaugé & Aarseth (1990) shows that in two dimensions neglect of distant encounters leads to premature isolation of embryos and an excessive number of final planets.

A possible deficiency of the 3D Monte Carlo calculations is the use of two-body expressions in calculating the mutual perturbations and collisions of planetary embryos. Most likely, this is not a serious problem. The calculations of Wetherill & Cox (1984, 1985) were motivated by this question. In view of the orbital complexity found by numerical integrations of close encounters at low velocity, it seemed possible that neglect of solar gravity would cause gross changes in both the collision rates and the averaged perturbations. This was found not to be the case. Individual restricted three-body trajectories deviated considerably from those found by two-body calculations when the velocity ratio

$$R_v = \frac{V}{V_e} = \frac{(e^2+i^2)^{1/2}}{V_e} V_K \lesssim 0.35, \tag{20}$$

where e and i are, respectively, the unperturbed eccentricity and inclination of the smaller body, V_K is its circular Keplerian velocity, and V_e is the surface escape velocity of the larger body. Nevertheless, as long as $V/V_e > 0.14$, the averaged perturbations were within a factor of two of those found by the approximate Öpik Monte Carlo algorithm and only a modest enhancement in the collision probability was found. These relatively small effects have been confirmed by more complete investigations of the same problem (Ida & Nakazawa 1989, Greenzweig & Lissauer 1990). Approximate corrections for these effects have been introduced into the Monte Carlo algorithms, but because the corrections are small and values of $V/V_e < 0.1$ are rare, the effect of these changes is not noticeable. It is also sometimes stated that the use of random numbers in these calculations represents a deficiency, particularly since different choices of random numbers can lead to very different final planetary configurations. This is not a deficiency, because it is an essential characteristic of the chaotic nature of close-encounter orbits, as also demonstrated by numerical integration of planet-crossing orbits (Milani et al 1989).

Of the 45 published 3D simulations of the final stages of terrestrial planet growth, only about 10% lead to a final configuration quite similar to that of the observed terrestrial planets. Detailed illustrations of the evolution of the swarm have been presented for three such cases (Wetherill 1985, 1986, 1988). More often, the final configuration differs from the observed solar system in some significant way. For example, equivalents to one or both of the small planets (Mercury and Mars) may be absent, three $>2 \times 10^{27}$-g planets may form, or a small planet may remain between the orbits of "Earth" and "Venus." It is not known to what extent these variations simply reflect the stochastic nature of planetary accumulation, or inadequacies in the model and technique of calculation. Probably both of these factors are important.

An Example of a 3D Simulation of the Final Stage of Accumulation

Results of a previously unpublished calculation are presented here. This case illustrates a qualified success, in which no "Mars" was produced, even though several potential such objects were formed but later lost. The initial conditions are a variation from those presented earlier and represent only one of a large number of such variations that have been calculated (for example, see Wetherill 1990b). The initial swarm of 222 bodies extended from 0.44 to 1.32 AU, with most of the mass between 0.7 and 1.1 AU, in order to obtain a final state that matched the observed specific energy and angular momentum of the observed planets within a few percent (Figure 7*a*). The size distribution was that of a runaway, determined by an equation of the form of (19), except that the minimum separation of the embryos was taken as 1.6 r_H instead of the value of Birn (1973) of $2\sqrt{3}\ r_H$. The time at which the local runaways reached their final masses as a function of heliocentric distance was taken to vary as $a^{3/2}$, to be inversely proportioned to the local surface mass density, and had a value of 3×10^4 yr at 0.4 AU. The initial eccentricities were much lower than those of the previously published calculations, being barely those necessary to permit the orbits of the initial embryos to cross one another. Fragmentation, as described in Wetherill (1988) for initial runaway embryos, was included.

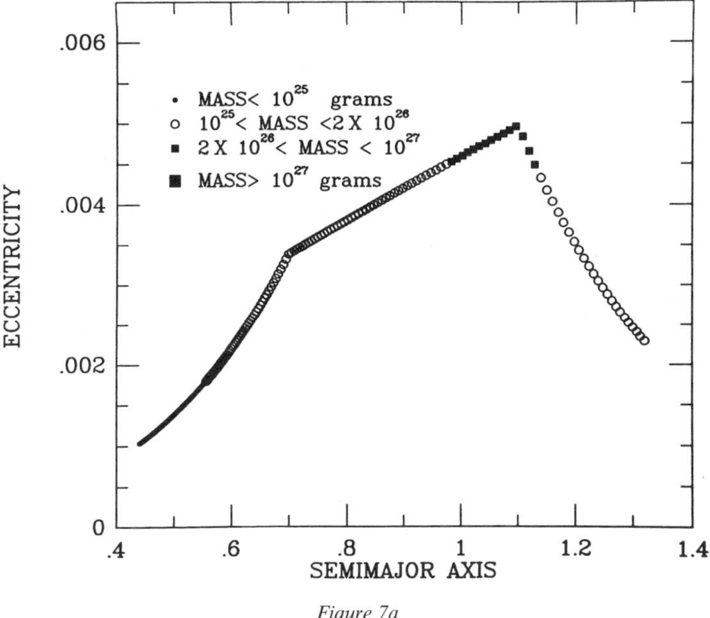

Figure 7a

Figure 7 Simulation of the final stage of accumulation of terrestrial planets from runaway embryos. (*a*) Initial configuration of embryos using a surface density distribution that will evolve into one matching the present mass, energy, and angular momentum of the terrestrial planets. The symbols representing the masses of the bodies apply to Figures 7*a*–*g*. Initial eccentricities correspond to marginal orbit crossing of adjacent embryos. (*b*) Evolution of swarm after 1.22 m.y. Symbols same as Figure 7*a*. As discussed in text, radial mixing of bodies has occurred, and eccentricities have increased by an order of magnitude. (*c*) Swarm at 3.8 m.y. Symbols same as Figure 7*a*. *Large solid squares* indicate very large bodies with masses $>10^{27}$ g. (*d*) Swarm at 13.3 m.y. The large object at 0.95 AU will become the simulated "Earth." The large body at 0.7 AU originally formed at 1.08 AU and migrated to 0.7 AU. This object will become the simulated "Venus." (*e*) Swarm at 46 m.y. The growth of "Venus" continues with the merger of the two large bodies at 0.6–0.7 AU in Figure 7*d*. A potential "Mercury" may be seen at 0.45 AU, having a mass of 2.0×10^{26} g. It does not survive because of a collision at 56 m.y. with a 2.5×10^{26}-g body that is at 1.35 AU in this figure. (*f*) Swarm at 77 m.y. "Earth" has grown to 99% of its final mass, and Venus to 87%. A swarm of moderate-size collision debris now occupies Mercury's position. After a complex chain of fragmentation and merger, some of these objects will form a final "Mercury," completed at 178 m.y., and three potential Mars' have $a \gtrsim 1.5$ AU. None of these "Mars" objects survived. Two were ejected from the solar system, and the other (6×10^{26} g) collided with Venus in an unusually late giant impact at 416 m.y. (*g*) Swarm at 879 m.y. Three final planets remain. Earth, Venus, and Mercury have masses of 5.8×10^{27} g, 4.8×10^{27} g, and 3.2×10^{26} g, respectively. The last impact on Earth was a 0.25-lunar-mass object at 220 m.y.

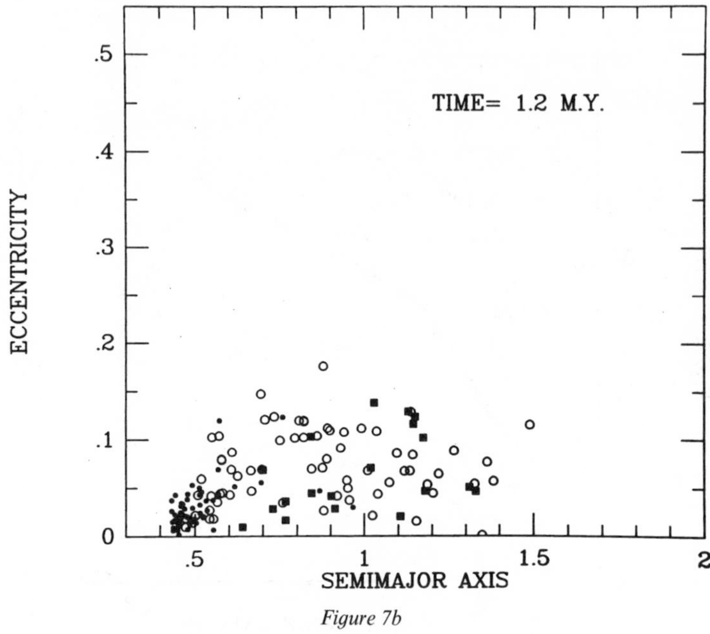

Figure 7b

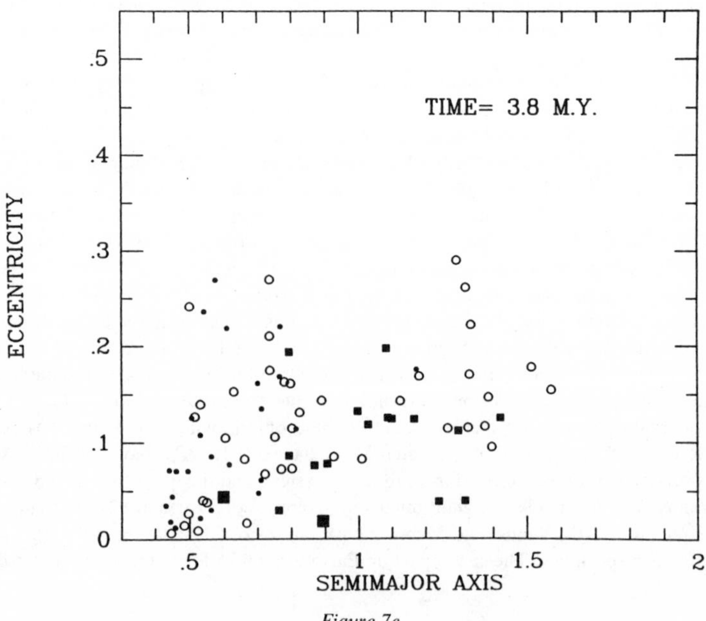

Figure 7c

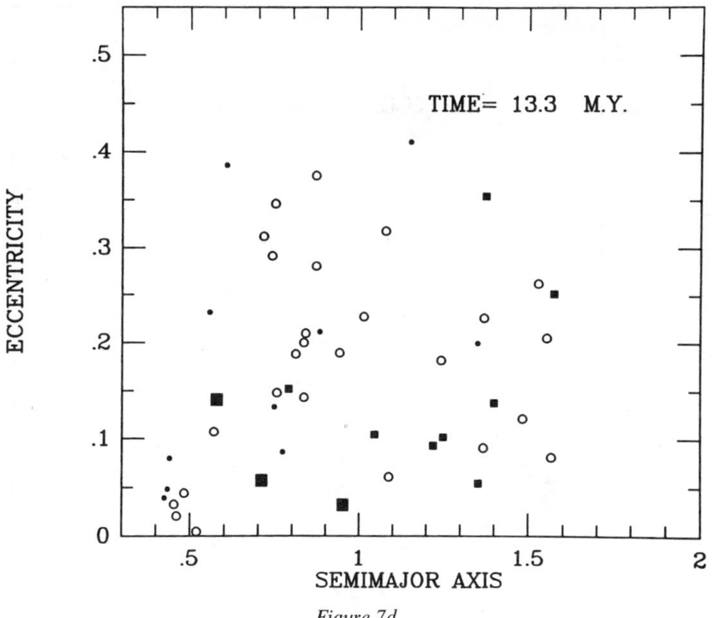

Figure 7d

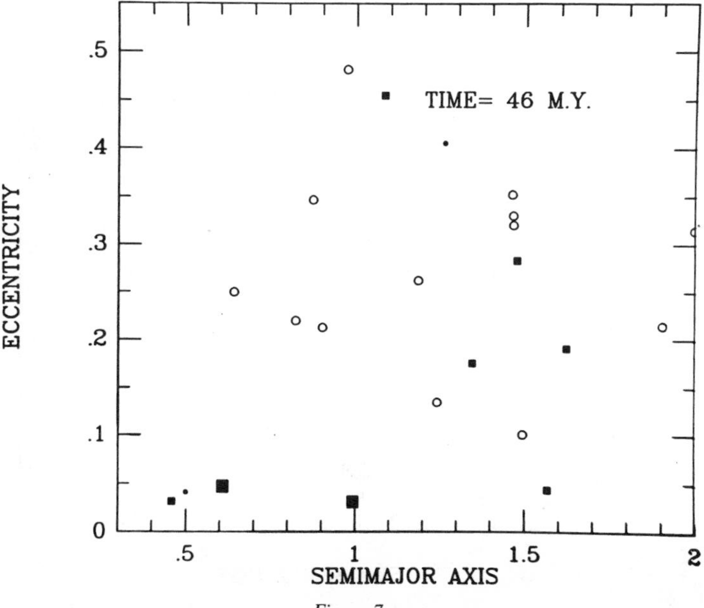

Figure 7e

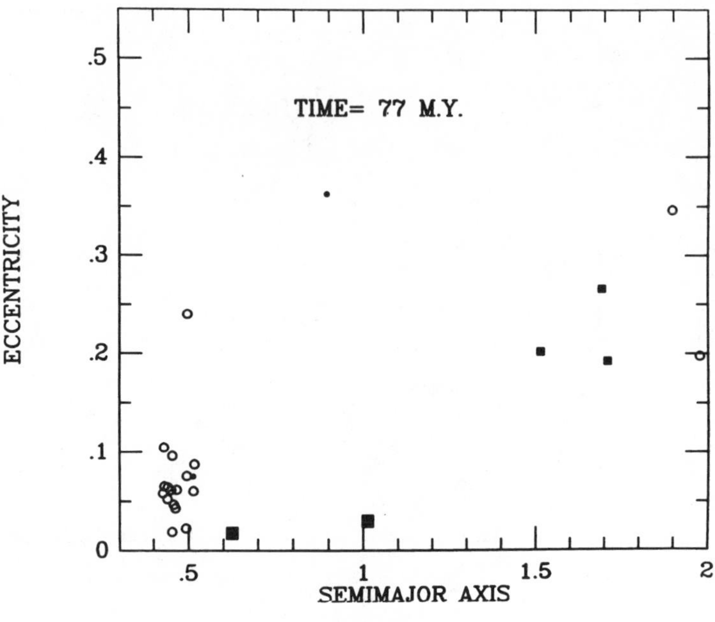

Figure 7f

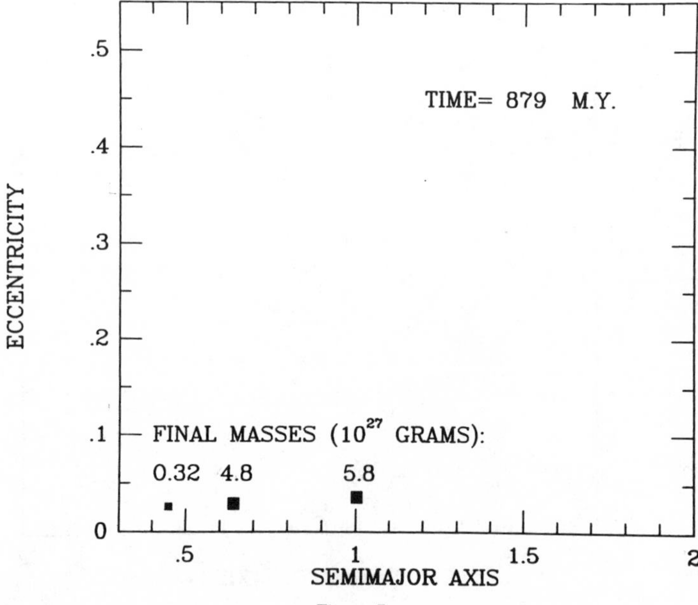

Figure 7g

These details are given in order to document what was done. In fact, it is found that the outcome of such calculations is not in any obvious way dependent on variations of this kind, and insofar as the results may be deemed unsatisfactory, they will not be significantly altered by changes of this kind.

The state of evolution of the swarm at selected times is shown in Figures 7*a–g*. The semimajor axis and eccentricity of each object is shown, with a symbol representing the mass range into which each body falls. The initial eccentricities are very low (Figure 7*a*) and were chosen to represent the situation when the small-body end of the swarm has been consumed by the runaway embryos sufficiently to relax the effect of equipartition of energy that earlier constrained the orbits of the embryos to be nearly circular. The largest initial embryos, indicated by solid squares, are between 1.13 and 0.99 AU and range in mass from 2×10^{26} g to 2.8×10^{26} g.

The mutual gravitational perturbation of the embryos caused their eccentricities to rise tenfold during the first 1.2 m.y. (Figure 7*b*). The larger initial bodies (solid squares) have been scattered between ~0.6 AU and 1.35 AU and are well mixed with bodies that range in mass between 10^{25} g and 2×10^{26} g. At this time the mass of the largest body is 7.9×10^{26} g, larger than the present mass of Mars, with a heliocentric distance of 0.64 AU. The initial mass of this body was only 8.2×10^{25} g, about lunar size. Its prominence is, in part, caused by the more rapid growth of runaways at smaller heliocentric distances, and in part stochastic. Six other bodies, all at greater heliocentric distances, have at least half the mass of this body.

At 3.8 m.y. (Figure 7*c*), the eccentricities have increased to values comparable to the escape velocities of the largest bodies. There are now two bodies with masses of $\sim 1.2 \times 10^{27}$ g and with eccentricities well below the average (large solid squares). Both of these bodies have grown to this mass by collision and merger with smaller bodies. The provenance of these larger bodies spans a wide range of heliocentric distances, rather than being derived from individual feeding zones.

At 13.3 m.y. the spread in eccentricities and semimajor axis has increased further (Figure 7*d*). The massive body at 0.6 AU has increased its eccentricity and grown slightly to 1.4×10^{27} g. Two additional objects with masses $>10^{27}$ g are now found. The object near 0.95 AU was produced by the rapid growth of a body having a mass 0.6×10^{27} g at 3.8 m.y. into a 3.75×10^{27}-g body. In the course of its growth, at 11 m.y., it merged with the 1.2×10^{27}-g body shown at 0.9 AU on Figure 7*c*. This body is destined to become the simulated equivalent of "Earth." The other new body is at 0.7 AU with a mass of 2.0×10^{27} g. It is destined to become the final "Venus." This body was initially at 1.08 AU, with a mass of

2.7×10^{26} g. As a result of gravitational scattering by other bodies, it migrated to 0.7 AU, colliding with a number of smaller bodies during its migration.

By 46 m.y. (Figure 7*e*), the swarm is greatly depleted in bodies more massive than 10^{24} g (Ceres size). The two $>10^{27}$-g bodies seen earlier around 0.6–0.7 AU have collided in a giant impact to form a single body with a mass of 3.8×10^{27} g. A body of mass 2.0×10^{26} has appeared near the position of Mercury. This body was formed by a complex series of breakup, melting, and reassemblage of objects (mostly bodies with original masses $<10^{25}$ g) that populated the inner portion of the initial swarm (Figure 7*a*). A similar history is illustrated in Figure 11 of Wetherill (1988) for a body that eventually grew to become a simulated "Mercury." In the present calculation, this body does not survive. At 56 m.y. it is disrupted by collision with the 2.5×10^{26} g body plotted in Figure 7*e* at $a = 1.35$ AU and $e = 0.18$. The perihelion of this body evolved inward as a consequence of a series of chance encounters with other bodies, the most important of which were the two $>10^{27}$-g bodies "Earth" and "Venus." Thus, a potential "Mars" and a potential "Mercury" were both lost by a single collision.

At 77 m.y. (Figure 7*f*), "Earth" at 1.0 AU has reached 99% of its final mass of 5.83×10^{27} g and "Venus" has grown to 87% of its final mass of 3.84×10^{27} g. An assemblage of melted collision fragments, mostly in the 10^{25}-g size range, populate the region where $a < 0.6$ AU. The largest of these has a fairly large mass, 1.7×10^{26} g. This body grew by reassemblage of smaller collision products, mostly resulting from the large collision described in the previous paragraph. It grew to a mass of 3.1×10^{26} before it was again disrupted at 161 m.y., after which it reassembled itself to provide the simulated "Mercury." At 77 m.y. three potential candidates for Mars still remain between 1.5 and 1.7 AU, with masses of 2.0×10^{26} g, 5.8×10^{26} g, and 7.6×10^{26} g. None of these will survive. The largest and smallest of these bodies are to be ejected into solar system escape orbits, and the other object will strike "Venus" in an unusually late giant impact at 416 m.y.

At 879 m.y., three final planets remain (Figure 7*g*). Earth, Venus, and Mercury have final masses of 5.8×10^{27} g, 4.8×10^{27} g, and 3.2×10^{26} g, respectively. The last impact on "Earth" took place at 220 m.y. (1.9×10^{25} g). The last impact on Venus was the giant impact at 416 m.y., and the reassemblage of "Mercury" was completed at 178 m.y. No "Mars" remains. Under different stochastically determined conditions, one of the planet-size bodies still present at $\sim$1.6 AU at 77 m.y. could have survived.

All three of the final planets have "melted," in the sense that under the assumption that 75% of the center-of-mass impact energies were dissipated internally as heat, the accumulated heat energy would be $>3 \times 10^{10}$ ergs g^{-1}. This result is not sensitive to the assumption of 75% dissipation. Large planets like Earth and Venus are melted even for less than 25% dissipation, and the collision fragments that assembled to produce Mercury are also melted.

The growth of "Earth" was dominated by giant impacts: Six impacts took place with bodies intermediate in mass between Mercury and Mars, and two occurred with masses greater than Mars. The largest of these (1.2×10^{27} g) occurred at 10.8 m.y. Finding so many giant impacts is unusual, but this result lies well within the range found for the more than 150 simulations that have been calculated. For the case considered here, the giant impact history of "Venus" is more typical: two impacts of bodies with masses intermediate between Mercury and Mars, and one more massive than Mars (1.4×10^{27} g).

Comments on the Initial Distributions Used in These Simulations

Although considerable progress has been made in understanding the processes that are likely to have accompanied the formation of the Earth, much remains to be understood. For example, attention should be drawn to the initial distribution assumed in the above simulation in which the semimajor axes of the swarm were concentrated in a fairly narrow range, narrower than that represented by the present range in heliocentric distance of the terrestrial planets. If the physical processes considered here are all that were of major significance, a restricted distribution of this kind is essential to the formation of these planets. The reason is that mass, specific energy, and specific angular momentum are nearly conserved during the growth. For bodies no larger than the Earth and so deep in the Sun's gravitational well, only $\sim 5\%$ of the mass is lost from the system by ejection into hyperbolic solar system escape orbits. Except for angular momentum carried by escaping bodies, angular momentum is strictly conserved. Only several percent of the initial gravitational energy is dissipated as heat during the collisions responsible for the growth. For this reason, only a swarm with a mass, specific angular momentum, and specific energy nearly the same as that of the present solar system can evolve into the present planets (Wetherill 1978). This result is not only obvious but has been easily confirmed by use of inappropriate initial distributions.

Clearly, this confinement of the initial embryos to a narrow band of semimajor axes requires an explanation, but a satisfactory one is not available at present. The problem is that of explaining the absence of bodies $\gtrsim 1000$ km in diameter ($\sim 10^{24}$ g) between the orbits of Mars and Jupiter, and the small masses of Mars and Mercury. Inside the orbit of

Venus, it is possible that condensation of solid grains was precluded by high temperatures up to the time when solar outflow removed the nebular gas and small grains from the solar system.

In the middle solar system, between the orbits of Mars and Jupiter, a taxonomy of hypothetical explanations can be given for the deficiency of material:

1. There was a gap in the density distribution in the solar nebula, in contrast to the smooth decrease shown in Figure 1. Present theoretical models of the solar nebula give no support to this hypothesis.

2. Planetesimals $\gtrsim 10$ m in diameter failed to form in this region, and the smaller bodies were removed from the region by gas drag (Weidenschilling 1988).

3. There was no marked initial depletion of material from this region, and planetesimals at least began to form there in the same way as at 1AU, but on a longer time scale. This time scale was proportional to the Keplerian period, varying as $a^{3/2}$, and to the reciprocal of the surface density. For a surface density that varies as $a^{-3/2}$, the time scale for growth at 3 AU would be 27 times that at 1 AU, and a runaway could require >1 m.y. to develop. If Jupiter formed before the asteroidal runaways were completed, there are several ways that the presence of Jupiter (and Saturn) and the loss of nebular gas may have aborted planet formation in this region. These are discussed by Wetherill (1990a). Although some theory of this kind may turn out to explain the evolution of the asteroid belt, at least at present it does not seem very promising for the region between 1 and 2 AU.

4. Some other physical process, not included in the 3D simulations of the kind illustrated in Figure 7, may have "compressed" the embryo swarm into the narrow band required. One such possibility has been explored numerically (Wetherill 1990b). This is transfer of angular momentum to the residual solar nebula via spiral density waves (Ward 1986, 1988) generated by resonant interactions between embryos and the gas. By using these published coefficients for radial migration, together with a model for the removal of nebular gas interior to 2 AU, it seems that the necessary compression of the embryo swarm might be achieved. It now appears, however, that as a result of further development of the theory, the coefficients that I used are a factor of 10 or more too high because the effect is nearly canceled by additional resonances not previously included (W. R. Ward, private communication, 1989). If so, this process seems unlikely to provide the answer to the problem.

Because no satisfactory quantitative theory yet exists for the processes that depleted the region between Earth and Jupiter, it is not yet possible to properly address the important question of a possible contribution of material removed from this region to a "late veneer" of more volatile-rich and oxidized material to the Earth and other terrestrial planets, particularly Mars (Dreibus & Wänke 1989).

INITIAL STATE AND EARLY HISTORY OF THE EARTH

Initial High Temperatures

It seems most likely that the stage of terrestrial planet growth in which planetesimals coalesced to form embryos led to the formation of many more lunar- to Mars-size planetary embryos than the four terrestrial planets seen today. The difference between the outcomes of orderly and runaway growth turns out to be only one of degree rather than kind in this regard. Orderly growth leads to a larger number of embryos, which of course must therefore be on average smaller.

The final stage of accumulation consists of the collision of these embryos. Because of their large size, particularly after their further growth, these collisions represent giant impacts. These effects have been illustrated for both the case of orderly (Wetherill 1985, 1986) and runaway growth (Wetherill 1988; see also Figures 7, 8 herein). In the case of runaway formation of embryos, even without further growth the larger embryos are in the size range considered in theoretical and numerical models of lunar formation involving giant impacts (Benz et al 1986, 1987, Melosh & Sonett 1986, Kipp & Melosh 1986). Furthermore, all of the embryos are vulnerable to merger with one another, and most of the giant impacts will be between embryos that have already grown far beyond their initial mass.

A consequence of these giant impacts is that the Earth should have been extensively heated and melted during all but the earliest part of its growth. Even if the Earth and the impacting embryo were in orbits with negligible components of relative velocity, when the Earth had reached half its present mass, its gravitational attraction alone would result in impact velocities of about 9 km s^{-1}. This corresponds to a kinetic energy of about 4×10^{11} ergs g^{-1} of impactor, far above the $<3 \times 10^{10}$ ergs g^{-1} required for melting. Impacts between planet-size bodies must be far from elastic, and much (or even most) of this energy would be deposited and distributed within the growing Earth, as found in the numerical simulations of Benz et al (1986, 1987).

In fact, the relative velocities of the colliding bodies will probably not be small as a consequence of the embryos undergoing many more near-misses with one another than actual collisions. Both the analytical and numerical calculations discussed in the previous section show that the

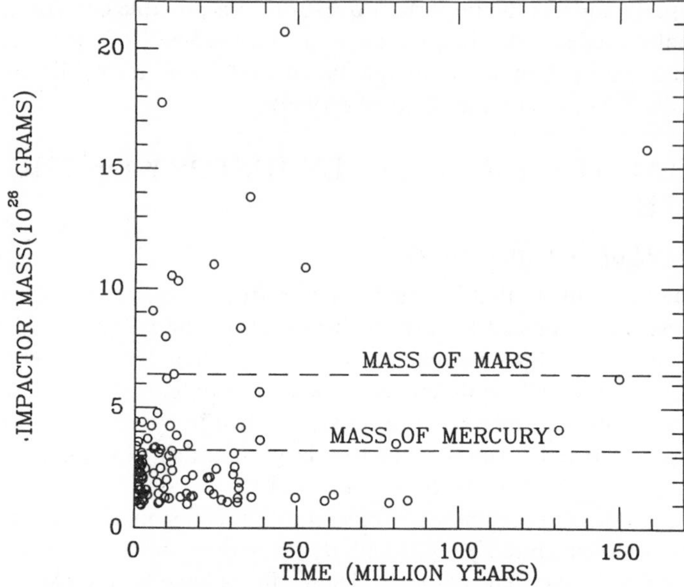

Figure 8 Distribution in size and time of giant impacts for 10 simulations of the final accumulations of runaway embryos, under the same assumptions used for the simulation shown in Figure 7. On average, each simulation is associated with one impact on "Earth" with an object more massive than the present mass of Mars and about two impacts with objects intermediate in mass between those of present-day Mercury and Mars.

random-walk accumulation of these close encounters results in relative velocities comparable to the escape velocities of the larger embryos, increasing impact heating by a factor of 2 to 4.

Radial Mixing of Planetary Material in the Terrestrial Planet Region

Relative orbital velocities of ~ 10 km s^{-1}, resulting from close encounters, are about one third of the circular Keplerian velocity at 1 AU. Therefore it should be expected that prior to their elimination by collision with a nearly formed planet, the smaller embryos will have eccentricities of ~ 0.3, sufficient to span the heliocentric distance between Earth and Venus. This causes these smaller embryos to be passed from the neighborhood of one planet to another, vitiating the concept of a chemically unique feeding zone for each final terrestrial planet. Numerical simulations of the final stage of accumulation show that this mixing does not entirely erase the memory of the initial provenance of the embryos that form the final planet, but it argues strongly against any planet gathering to itself a unique chemical or isotopic component (Wetherill 1988). A possible, but by no means expected, exception to this conclusion regarding widespread mixing could be the gas-rich final stage of accumulation discussed by Ohtsuki et al (1988).

The chemical effects expected to accompany the growth of terrestrial planets from planetesimals have not yet been considered in any detail. Because the Sun and average solar nebula may be plausibly expected to be "chondritic," it is reasonable to suppose that the more refractory elements would be found in the Earth in nearly chondritic (or solar) proportions. On the other hand, there is no good reason to think the Earth was principally formed by the accumulation of any particular type of meteorite found in meteorite collections. It is extremely likely (e.g. Wetherill & Chapman 1988) that almost all undifferentiated meteorites are relatively recently fractured fragments of asteroidal bodies that have never been larger than about 1000 km in diameter. During formation of these asteroids, it seems likely that they experienced physical and chemical processing similar to that of the planetesimals that are supposed to have grown into the terrestrial planet embryos. Insofar as there may have been chemical and/or temperature gradients in the solar nebula, the outcome of this first stage of accumulation in the asteroid belt may be expected to have been different from that in the terrestrial planet region. In addition, the terrestrial planetary material was then subjected to the vicissitudes of accumulation into full-scale planets, accompanied by its own peculiar set of chemical and physical processes.

In view of the foregoing, is it reasonable to expect there has been some significant degree of mixing between the asteroidal meteoritical material and that of the Earth? The answer is probably yes, but so little quantitative work has been done in connection with this question that any answer may be considered to be speculative, albeit interesting. All calculations of the orbital evolution of the material from which the Earth formed show that a significant quantity of terrestrial material is transferred to the inner portion of the asteroid belt (Wetherill 1977, Ip 1989). Thus, it is plausible that to some extent the asteroid belt, greatly depleted in its own indigenous mass, was vulnerable to "pollution" with terrestrial matter. On the other hand, it is also by no means out of the question that the Earth was contaminated with large amounts of material from greater heliocentric distances—from the region of Mars, the asteroid belt, and the outer planets—as suggested on geochemical grounds (Dreibus & Wänke 1989). Chemical and isotopic differences between meteorites and the Earth demonstrate that chemical mixing of these regions was not complete. Progress in understanding these important matters from a dynamical point of view awaits replacement of uncertain preliminary calculations with serious quantitative work.

Theoretical vs. Observational Evidence Regarding the Initial State of the Earth

The picture of the early 4.4–3.8 b.y. ago Earth that emerges from consideration of all even moderately quantitative theories of its formation is that of an initially extremely hot and melted planet, surrounded by a fragile atmosphere and subject to violent impacts by bodies of the size of Ceres and even the Moon, during the hundreds of millions of years that separate Earth's formation from the oldest, reasonably well-preserved geological record (about 3.8 b.y. in age). This picture is supported by the observational evidence for extensive early melting of a much smaller body, the Moon (Taylor 1982), regardless of how it was formed.

Can this picture be reconciled with the planetological, geological, and geochemical record? The answer seems to be: not easily. A proper discussion of this subject would require at least an additional review, for which I am certainly not the most prominent candidate author, and, more importantly, requires much further theoretical, observational, and experimental attention. Nevertheless, a little should be said.

There are things to be said on the positive side. The existence of the Moon, the gross difference in the inert gas composition of the atmospheres of Earth and Venus, the high I-Xe age of the Earth (Wetherill 1975, Pepin & Phinney 1976), and the large iron core of Mercury can be explained in a reasonably natural way in terms of the version of the standard model emphasized in this review. The Pb isotope age of the Earth's liquid core (Patterson 1956) and the enrichment in radiogenic ^{143}Nd found in the most ancient rocks (Shirey & Hanson 1986), suggesting that the melted rocks were extracted from the Earth's mantle even earlier in its history, are at least supportive of the concept of a high-temperature primitive Earth.

Apparently serious problems arise, however, when one considers the chemical fractionation expected during the solidification of a once melted silicate mantle. Crystallization of minerals from large bodies of liquid silicate is invariably associated with some degree of major-, minor-, and trace-element chemical fractionation. The major minerals have different chemical compositions and densities from one another. The various minor and trace elements substitute with varying compatibility into the crystal structure of some of these major minerals. Differential gravitationally driven settling of major mineral grains causes the crystallization of large magma chambers to be compositionally layered. A melted Earth would constitute a magma chamber of global dimensions, and a layered terrestrial mantle would be expected to develop during the cooling and crystallization of the Earth's mantle during the first several hundred million years of Earth history.

Major-element data from samples of the Earth's mantle, combined with laboratory experimental data, have been interpreted as supporting this picture of a layered early Earth (Agee & Walker 1988). An important result of these laboratory studies is the inference that silicate melts of olivine-rich composition ($\sim Mg_2SiO_4$) at upper-mantle pressure are likely to be more dense than the solid silicates crystallizing from these melts. Agee & Walker conclude that during the initial crystallization of a liquid Earth, this phenomenon may have produced a "buried magma ocean" of olivine composition, and that its subsequent crystallization, followed by convective mixing, may explain the elevated (above chondritic) Mg/Si ratio of the present upper mantle.

Trace-element data are difficult to interpret in terms of this model [but see Agee & Walker (1989) for a contrary view]. It is inferred from geochemical studies of mantle-derived rocks that a number of refractory trace elements have chondritic ratios in the mantle, e.g. Sm/Hf, Sc/Sm, Ir/Au, and Co/Ni. The principal minerals expected to form in a global crystallizing magma chamber as envisaged by Agee & Walker are olivine (Mg_2SiO_4); $MgSiO_3$ with the dense, high-pressure perovskite structure; and the high-pressure garnet majorite. Laboratory data exist regarding the partitioning of these trace elements between these high-pressure minerals and the residual liquid that one would expect to finally crystallize in the upper mantle (Kato et al 1988a,b, 1989, McFarlane & Drake 1990). The data are interpreted by these workers as predicting observable fractionation of these elements in the upper mantle. Even though these experimental trace-element partition coefficients are not as precise as desirable, it seems unlikely that they are totally incorrect, or that observable trace-element fractionation would fail to accompany separation of at least some of these high-pressure minerals from a silicate liquid, given the differences in their atomic structures.

There are only a few general ways to resolve this paradox. One is that although the Earth was melted, the global-scale fractionation did not occur, perhaps because of mixing of residual liquid with the settling crystals (Tonks & Melosh 1990). It is also conceivable that although the fractionation did occur, it has subsequently been erased. The physical processes that accompanied the early Earth—giant and not-so-giant Moon-size impacts causing both mixing and more local partial melting and fractional crystallization, and possibly complex convective and solid-liquid separation processes—have been barely discussed. It is possible, but not necessarily expected, that the geochemical and dynamical pictures of Earth formation may converge when these processes are understood.

Finally, it is possible that the dynamical history of Earth formation described by the standard model is simply incorrect. This admission should not be seized upon lightly. At the present time there are not even any semiquantitative theories extant that lead to an unmelted Earth. One must have faith that reliance on basic physical principles in the development of models of planet formation is the only way that the correct answer can ultimately be found. The presently available body of serious quantitative work, although quite incomplete, should not be cast aside lightly in favor of some qualitative "scenario" that may seem more pleasing. But it also should be remembered that doubt is the other side of the coin of faith, and one must clasp them both at the same time. This is natural, because in the absence of doubt there is no need for faith. It can then be replaced with certainty. We haven't reached that point yet.

ACKNOWLEDGMENTS

I wish to thank Alan Boss, Richard Carlson, and Michael Drake for helpful discussions, and Janice Dunlap and Mary Coder for converting this to readable form. This work was supported by NASA grant NSG 7437 and forms a part of a larger departmental program that was supported by NASA grant NAGW 398.

Literature Cited

Adachi, I., Hayashi, C., Nakazawa, K. 1976. The gas drag effect on the elliptic motion of a solid body in the primordial solar nebula. *Prog. Theor. Phys.* 56: 1756–71

Agee, C. B., Walker, D. 1988. Mass balance and phase density constraints on early differentiation of chondritic mantle. *Earth Planet. Sci. Lett.* 90: 144–56

Agee, C. B., Walker, D. 1989. Comments on "Constraints on element partition coefficients between $MgSiO_3$ perovskite and liquid determined by direct measurements" by T. Kato, A. E. Ringwood, and T. Irifune. *Earth Planet. Sci. Lett.* 94: 160–61

Anders, E. 1988. Circumstellar material in meteorites: Noble gases, carbon and nitrogen. In *Meteorites and the Early Solar System*, ed. J. F. Kerridge, M. S. Matthews, pp. 927–55. Tucson: Univ. Ariz. Press

Artymowicz, P. 1987. Self-regulating protoplanet growth. *Icarus* 70: 303–18

Beaugé, C., Aarseth, S. J. 1990. *N*-body simulations of planetary accumulation. *Mon. Not. R. Astron. Soc.* In press

Beichman, C. A., Myers, P. C., Emerson, J. P., Harris, S., Mathieu, R. D., Benson, P. J., Jennings, R. E. 1986. Candidate solar-type protostars in nearby molecular cloud cores. *Astrophys. J.* 307: 339–49

Benz, W., Slattery, W. L., Cameron, A. G. W. 1986. The origin of the moon and the single impact hypothesis I. *Icarus* 66: 515–35

Benz, W., Slattery, W. L., Cameron, A. G. W. 1987. The origin of the moon and the single impact hypothesis II. *Icarus* 71: 30–45

Birn, J. 1973. On the stability of the solar system. *Astron. Astrophys.* 24: 283–93

Black, D. C., Matthews, M. S., eds. 1985. *Protostars and Planets II*. Tucson: Univ. Ariz. Press. 1293 pp.

Boss, A. P. 1984. Angular momentum transfer by gravitational torque and the evolution of binary protostars. *Mon. Not. R. Astron. Soc.* 209: 543–67

Boss, A. P. 1987. Bipolar flows, molecular gas disks, and the collapse and accretion of rotating interstellar clouds. *Astrophys. J.* 316: 721–32

Boss, A. P. 1989. Evolution of the solar nebula. I. Nonaxisymmetric structure during nebula formation. *Astrophys. J.* 345: 544–71

Boss, A. P. 1990. 3D solar nebula models: implications for Earth origin. See Jones & Newsom 1990. In press

Cabot, W., Canuto, V. M., Hubickyj, O., Pollack, J. B. 1987. The role of turbulent convection in the primitive solar nebula. I. Theory, II. Results. *Icarus* 69: 387–457

Cameron, A. G. W. 1978. Physics of the primitive solar accretion disk. *The Moon and the Planets* 18: 5–40

Cameron, A. G. W. 1983. Origin of the atmospheres of the terrestrial planets. *Icarus* 56: 195–201

Cameron, A. G. W. 1988. Origin of the solar system. *Annu. Rev. Astron. Astrophys.* 26: 441–72

Cameron, A. G. W., Fegley, B., Benz, W., Slattery, W. L. 1988. The strange density of Mercury: theoretical considerations. In *Mercury*, ed. C. Chapman, F. Vilas, pp. 692–708. Tucson: Univ. Ariz. Press

Cameron, A. G. W., Pine, M. R. 1973. Numerical models of the primitive solar nebula. *Icarus* 18: 377–406

Cameron, A. G. W., Ward, W. R. 1976. The origin of the Moon. *Lunar Sci. VII*, pp. 120–22. Houston: Lunar Planet. Inst. (Abstr.)

Carlson, R. W., Lugmair, G. W. 1988. The age of ferroan anorthosite 60025: oldest crust on a young Moon? *Earth Planet. Sci. Lett.* 90: 119–30

Chandrasekhar, S. 1942. *Principles of Stellar Dynamics*. Chicago: Univ. Chicago Press

Chandrasekhar, S. 1943. Dynamical friction. I. General considertions: the coefficient of dynamical friction. *Astrophys. J.* 97: 255–62

Cox, L. P., Lewis, J. S. 1980. Numerical simulation of the final stages of terrestrial planet formation. *Icarus* 44: 706–21

Dermott, S. F., ed. 1978. *The Origin of the Solar System*. New York: Wiley. 668 pp.

Dole, S. H. 1962. The gravitational concentration of particles in space near the Earth. *Planet. Space Sci.* 9: 541–53

Donn, B., Meakin, P. 1988. Collisions of macroscopic fluffy aggregates in the primordial solar nebula. *Lunar Planet Sci. XIX*, pp. 281–82. Houston: Lunar Planet. Inst.

Dreibus, G., Wänke, H. 1989. Supply and loss of volatile constituents during the accretion of terrestrial planets. In *Origin and Evolution of Planetary and Satellite Atmospheres*, ed. S. K. Atreya, J. B. Pollack, M. S. Matthews, pp. 268–88. Tucson: Univ. Ariz. Press

Edgeworth, K. E. 1949. The origin and evolution of the solar system. *Mon. Not. R. Astron. Soc.* 109: 600–9

Fernandez, J. A., Ip, W.-H. 1981. Dynamical evolution of a cometary swarm in the outer planetary region. *Icarus* 47: 470–79

Giuli, R. T. 1968. On the rotation of the Earth produced by gravitational accretion of particles. *Icarus* 8: 301–23

Goldreich, P., Ward, W. R. 1973. The formation of planetesimals. *Astrophys. J.* 183: 1051–61

Graziani, F., Black, D. C. 1981. Orbital stability constraints on the nature of planetary systems. *Astrophys. J.* 251: 337–41

Greenberg, R., Wacker, J. F., Hartmann, W. K., Chapman, C. R. 1978. Planetesimals to planets: numerical simulation of collisional evolution. *Icarus* 35: 1–26

Greenzweig, Y., Lissauer, J. 1990. Accretion cross-sections of protoplanets. *Icarus*. Submitted for publication

Grossman, J. N. 1988. Formation of chondrules. In *Meteorites and the Early Solar System*, ed. J. F. Kerridge, M. S. Matthews, pp. 680–96. Tucson: Univ. Ariz. Press

Hartmann, W. K., Davis, D. R. 1975. Satellite-sized planetesimals and lunar origin. *Icarus* 24: 504–15

Hayashi, C., Nakazawa, K., Miyama, S. M., eds. 1988. *Origin of the Solar System. Prog. Theor. Phys. Suppl.*, Vol. 96

Hayashi, C., Nakazawa, K., Mizuno, H. 1979. Earth's melting due to the blanketing effect of the primordial dense atmosphere. *Earth Planet. Sci. Lett.* 43: 22–28

Hayashi, C., Nakazawa, K., Nakagawa, Y. 1985. Formation of the solar system. See Black & Matthews 1985, pp. 1100–53

Heppenheimer, T. A. 1979. Secular resonances and the origin of the eccentricities of Mars and the asteroids. *Lunar Planet. Sci. X*, pp. 531–33. Houston: Lunar Planet. Inst.

Horedt, G. P. 1985. Late stages of planetary accretion. *Icarus* 64: 448–70

Hornung, P., Pellat, R., Barge, P. 1985. Thermal velocity equilibrium in the protoplanetary cloud. *Icarus* 64: 295–307

Hutton, J. 1788. Theory of the Earth. *Trans. R. Soc. Edinburgh* 1: 216–304

Ida, S., Nakazawa, K. 1989. Collisional probability of planetesimals revolving in the solar gravitational field. III. *Astron. Astrophys.* 221: 342–47

Ip, W.-H. 1989. Dynamical injection of Mars-sized planetoids into the asteroidal belt from the terrestrial planetary accretion zone. *Icarus* 78: 270–79

Ipatov, S. I. 1987. Solid-body accumulation of terrestrial planets. *Astron. Vestn.* 21: 207–15

Jones, J., Newsom, H., eds. 1990. *Origin of the Earth*. Oxford: Oxford Univ. Press. In press

Kato, T., Ringwood, A. E., Irifune, T. 1988a. Experimental determination of element partitioning between silicate perovskites, garnets, and liquids: constraints on early differentiation of the mantle. *Earth Planet. Sci. Lett.* 89: 123–45

Kato, T., Ringwood, A. E., Irifune, T. 1988b. Constraints on element partition coefficients between $MgSiO_3$ perovskite and liquid determined by direct measurements. *Earth Planet. Sci. Lett.* 90: 65–68

Kato, T., Ringwood, A. E., Irifune, T. 1989. Constraints on element partition coefficients between $MgSiO_3$ perovskite and liquid determined by direct measurements—reply to C. B. Agee and D. Walker. *Earth Planet. Sci. Lett.* 94: 162–64

Kaula, W. M. 1990. Differences between the Earth and Venus arising from origin by large planetesimal infall. See Jones & Newsom 1990. In press

Kipp, M. E., Melosh, H. J. 1986. Short note: a preliminary numerical study of colliding planets. In *Origin of the Moon*, ed. W. Hartmann, R. J. Phillips, G. J. Taylor, pp. 643–46. Houston: Lunar Planet. Inst.

Kröner, A. 1985. Evolution of the Archean continental crust. *Annu. Rev. Earth Planet. Sci.* 13: 49–74

Kuiper, G. P. 1951. On the origin of the solar system. In *Astrophysics*, ed. J. A. Hynek, Ch. 8. New York: McGraw-Hill

Lada, C. J. 1985. Cold outflows, energetic winds, and enigmatic jets around young stellar objects. *Annu. Rev. Astron. Astrophys.* 23: 267–317

Larson, R. B. 1984. Gravitational torques and star formation. *Mon. Not. R. Astron. Soc.* 145: 271–95

Lecar, M., Aarseth, S. J. 1986. A numerical simulation of the formation of the terrestrial planets. *Astrophys. J.* 305: 564–79

Leliwa-Kopystynski, J., Taniguchi, T., Kondo, K., Sawaska, A. 1984. Sticking in moderate velocity impact: application to planetology. *Icarus* 57: 280–93

Lissauer, J. J. 1987. Time scales for planetary accretion and the structure of the protoplanetary disk. *Icarus* 69: 249–65

Lynden-Bell, D., Pringle, J. E. 1974. The evolution of viscous discs and the origin of the nebular variables. *Mon. Not. R. Astron. Soc.* 168: 603–37

McFarlane, E. A., Drake, M. J. 1990. Element partitioning and the early thermal history of the Earth. See Jones & Newsom 1990. In press

Melosh, H. J., Sonett, C. P. 1986. When worlds collide: jetted vapor plumes and the Moon's origin. In *Origin of the Moon*, ed. W. Hartmann, R. J. Phillips, G. J. Taylor, pp. 621–42. Houston: Lunar Planet. Inst.

Milani, A., Hahn, G., Carpino, M., Nobili, A. M. 1989. Dynamics of planet-crossing asteroids: classes of orbital behaviour: Project SPACEGUARD. *Icarus* 78: 212–69

Mizuno, H. 1989. Grain growth in the turbulent accretion disk solar nebula. *Icarus* 80: 189–201

Nakagawa, Y., Hayashi, C., Nakazawa, K. 1983. Accumulation of planetesimals in the solar nebula. *Icarus* 54: 361–76

Nakazawa, K., Mizuno, H., Sekiya, M., Hayashi, C. 1985. Structure of the primordial atmosphere surrounding the early Earth. *J. Geomagn. Geoelectr.* 37: 781–99

Nishida, S. 1983. Collisional processes of planetesimals with a protoplanet under the gravity of proto-Sun. *Prog. Theor. Phys.* 70: 93–105

Ohtsuki, K., Nakagawa, Y., Nakazawa, K. 1988. Growth of the Earth in nebular gas. *Icarus* 75: 552–65

Ohtsuki, K., Nakagawa, Y., Nakazawa, K. 1989. Artificial acceleration in accumulation due to coarse mass-coordinate divisions in numerical simulation. *Icarus*. Submitted for publication

Patterson, C. C. 1956. Ages of meteorites and the Earth. *Geochim. Cosmochim. Acta* 10: 230–37

Pepin, R. O., Phinney, D. 1976. The formation interval of the Earth. *Lunar Sci. VII*, pp. 612–19. Houston: Lunar Planet. Inst.

Pollack, J. B., McKay, C. P., Christofferson, B. M. 1985. A calculation of the Rosseland mean opacity of dust grains in primordial solar system nebulae. *Icarus* 64: 471–92

Safronov, V. S. 1962. A particular solution of the coagulation equation. *Dokl. Akad. Nauk SSSR* 147(1): 64–66

Safronov, V. S. 1969. *Evolution of the Protoplanetary Cloud and Formation of the Earth and Planets*. Moscow: Nauka. Transl., 1972, for NASA and NSF by Isr. Program Sci. Transl. as *NASA TT F-677*

Scalo, J. M. 1988. The initial mass function, starbursts, and the Milky Way. In *Starbursts and Galaxy Evolution (22nd Recontre de Moriond)*, ed. T. Montmerle, pp. 445–66. Paris: Ed. Front.

Schweizer, F. 1986. Colliding and merging galaxies. *Science* 231: 227–34

Sekiya, M. 1983. Gravitational instabilities in a dust-gas layer and formation of planetesimals in a solar nebula. *Prog. Theor. Phys.* 69: 1116–30

Sekiya, M., Nakagawa, Y. 1988. Settling of dust particles and formation of planetesimals. See Hayashi et al 1988, pp. 141–50

Shirey, S. B., Hanson, G. N. 1986. Mantle heterogeneity and crustal recycling of Archean granite-greenstone belts: evidence from Nd isotopes and trace elements in the Rainy Lake area, Superior Province,

Ontario, Canada. *Geochim. Cosmochim. Acta* 50: 2631–51

Shu, F. H., Adams, F. C., Lizano, S. 1987. Star formation in molecular clouds: observation and theory. *Annu. Rev. Astron. Astrophys.* 25: 23–81

Smith, B. A., Terrile, R. J. 1984. A circumstellar disk around β Pictoris. *Science* 226: 1421–24

Stevenson, D. J. 1987. Origin of the moon—the collision hypothesis. *Annu. Rev. Earth Planet. Sci.* 15: 271–315

Stevenson, D. J., Lunine, D. I. 1988. Rapid formation of Jupiter by diffusive redistribution of water vapor in the solar nebula. *Icarus* 75: 146–55

Stewart, G. R. 1980. Planetesimal acceleration by transfer of energy from circular motion to random motion. *Lunar Planet. Sci. XI*, pp. 1094–96. Houston: Lunar Planet. Inst.

Stewart, G. R., Kaula, W. M. 1980. Gravitational kinetic theory for planetesimals. *Icarus* 44: 154–71

Stewart, G. R., Wetherill, G. W. 1988. Evolution of planetesimal velocities. *Icarus* 74: 542–53

Strom, K., Strom, S. E., Edwards, S., Cabrit, S., Strutskie, M. F. 1989. Circumstellar material associated with stellar-type pre-main-sequence stars: a possible constraint on the timescale for planet building. *Astron. J.* 97: 1451–70

Takeda, H. 1988. Drag on a gravitating body. See Hayashi et al 1988, pp. 196–210

Takeda, H., Matsuda, T., Sawada, K., Hayashi, C. 1985. Drag on a gravitating sphere moving through a gas. *Prog. Theor. Phys.* 74: 272–87

Taylor, S. R. 1982. *Planetary Science: A Lunar Perspective*. Houston: Lunar Planet. Inst. 481 pp.

Tera, F., Eberhardt, P., Wetherill, G. W. 1981. Chronology of lunar volcanism. In *Basaltic Volcanism on the Terrestrial Planets*, pp. 948–74. New York: Pergamon

Tonks, W. B., Melosh, H. J. 1990. The physics of crystal settling and suspension in turbulent magma ocean. See Jones & Newsom 1990. In press

Van Schmus, W. R. 1981. Chronology and isotopic studies on basaltic meteorites. In *Basaltic Volcanism on the Terrestrial Planets*, pp. 935–47. New York: Pergamon

Vityazev, A. V., Pechernikova, G. V., Safronov, V. S. 1988. Formation of Mercury and removal of its silicate shell. In *Mercury*, ed. C. Chapman, F. Vilas, pp. 667–69. Tucson: Univ. Ariz. Press

Völk, H. 1983. Formation of planetesimals in turbulent protoplanetary accretion disks. *Meteoritics* 18: 412–13

Völk, H., Jones, F., Morfill, G., Röser, S. 1980. Collisions between grains in a turbulent gas. *Astron. Astrophys.* 85: 316–25

von Weizsäcker, C. F. 1944. Über die Entstehung des Planetensystems. *Z. Astrophys.* 22: 319–55

Walter, F. M. 1986. X-ray sources in regions of star formation. I. The naked T Tauri stars. *Astrophys. J.* 306: 573–86

Wänke, H. 1981. Constitution of the terrestrial planets. *Philos. Trans. R. Soc. London Ser. A* 303: 287–302

Ward, W. R. 1980. Scanning secular resonances: a cosmogonical broom? *Lunar Planet. Sci. XI*, pp. 1199–1201. Houston: Lunar Planet. Inst.

Ward, W. R. 1986. Density waves in the solar nebula: differential Lindblad torque. *Icarus* 67: 164–80

Ward, W. R. 1988. On disk-planet interactions and orbital eccentricities. *Icarus* 73: 330–48

Weidenschilling, S. J. 1977a. The distribution of mass in the planetary system and solar nebula. *Astrophys. Space Sci.* 51: 153–58

Weidenschilling, S. J. 1977b. Aerodynamics of solid bodies in the solar nebula. *Mon. Not. R. Astron. Soc.* 180: 57–70

Weidenschilling, S. J. 1980. Dust to planetesimals: settling and coagulation in the solar nebula. *Icarus* 44: 172–89

Weidenschilling, S. J. 1988. Formation processes and time scales for meteorite parent bodies. In *Meteorites and the Early Solar System*, ed. J. F. Kerridge, M. S. Matthews, pp. 348–71. Tucson: Univ. Ariz. Press

Weidenschilling, S. J., Donn, B., Meakin, P. 1989. Physics of planetesimal formation. In *The Formation and Evolution of Planetary Systems*, ed. H. A. Weaver, L. Danly, pp. 131–50. Cambridge: Univ. Press

Wetherill, G. W. 1975. Radiometric chronology of the early solar system. *Annu. Rev. Nucl. Part. Sci.* 25: 283–328

Wetherill, G. W. 1976. The role of large bodies in the formation of the Earth and Moon. *Proc. Lunar Sci. Conf., 7th*, pp. 3245–57. New York: Pergamon

Wetherill, G. W. 1977. Evolution of the Earth's planetesimal swarm subsequent to the formation of the Earth and Moon. *Proc. Lunar Sci. Conf., 8th*, pp. 1–16

Wetherill, G. W. 1978. Accumulation of the terrestrial planets. In *Protostars and Planets*, ed. T. Gehrels, pp. 565–98. Tucson: Univ. Ariz. Press

Wetherill, G. W. 1980. Formation of the terrestrial planets. *Annu. Rev. Astron. Astrophys.* 18: 77–113

Wetherill, G. W. 1985. Giant impacts during

the growth of the terrestrial planets. *Science* 228: 877–79

Wetherill, G. W. 1986. Accumulation of the terrestrial planets and implications concerning lunar origin. In *Origin of the Moon*, ed. W. K. Hartmann, R. J. Phillips, G. J. Taylor, pp. 519–50. Houston: Lunar Planet. Inst.

Wetherill, G. W. 1988. Accumulation of Mercury from planetesimals. In *Mercury*, ed. C. Chapman, F. Vilas, pp. 670–91. Tucson: Univ. Ariz. Press

Wetherill, G. W. 1989. The formation of the solar system: consensus, alternatives, and missing factors. In *The Formation and Evolution of Planetary Systems*, ed. H. A. Weaver, L. Danly, pp. 1–30. Cambridge: Univ. Press

Wetherill, G. W. 1990a. Origin of the asteroid belt. In *Asteroids II*, ed. R. Binzel, T. Gehrels, M. S. Matthews, pp. 661–80. Tucson: Univ. Ariz. Press

Wetherill, G. W. 1990b. Formation of the terrestrial planets from planetesimals. *Proc. US-USSR Workshop Planet. Sci.* Washington, DC: Natl. Acad. Sci. Press. In press

Wetherill, G. W., Chapman, C. R. 1988. Asteroids and meteorites. In *Meteorites and the Early Solar System*, ed. J. F. Kerridge, M. S. Matthews, pp. 35–67. Tucson: Univ. Ariz. Press

Wetherill, G. W., Cox, L. P. 1984. The range of validity of the two-body approximation in models of terrestrial planet accumulation. I. Gravitational perturbations. *Icarus* 60: 40–55

Wetherill, G. W., Cox, L. P. 1985. The range of validity of the two-body approximation in models of terrestrial planet accumulation. II. Gravitational cross-sections and runaway accretion. *Icarus* 63: 290–303

Wetherill, G. W., Stewart, G. R. 1989. Accumulation of a swarm of small planetesimals. *Icarus* 77: 330–57

Whipple, F. L. 1973. Radial pressure in the solar nebula as affecting the motions of planetesimals. In *Evolutionary and Physical Properties of Meteoroids*, ed. C. L. Hemenway, P. M. Millman, A. F. Cook, pp. 355–61. *NASA SP-319*

Wood, J. A. 1961. Stony meteorite orbits. *Mon. Not. R. Astron. Soc.* 122: 79–88

Zinner, E. 1988. Interstellar cloud material in meteorites. In *Meteorites and the Early Solar System*, ed. J. F. Kerridge, M. S. Matthews, pp. 956–83. Tucson: Univ. Ariz. Press

Annu. Rev. Earth Planet. Sci. 1990. 18:257–86
Copyright © 1990 by Annual Reviews Inc. All rights reserved

SEISMIC DISCRIMINATION OF NUCLEAR EXPLOSIONS

Paul G. Richards

Lamont-Doherty Geological Observatory and Department of Geological Sciences, Columbia University, Palisades, New York 10964

John Zavales

School of International and Public Affairs, Columbia University, New York, New York 10027[1]

INTRODUCTION

The monitoring of underground nuclear explosions has been discussed for over 30 years in connection with verification of test ban treaties, and monitoring is also one of the basic ways to learn about nuclear weapons development in different countries. Explosion monitoring is thus an important technical exercise, occasionally attracting attention from high-level policymakers and generating front-page news. We can expect this level of interest to continue as long as there is debate on the merits of modernizing nuclear weapons.

The underlying issues here are related to the uses of nuclear weapons, their cost, and the strategic capabilities and vulnerabilities of the superpowers and other countries. Because these are some of the largest and most contentious questions of our time, arguments sometimes spill over from the policy arena into the technical subject of monitoring capability itself.

As a side effect of the debate, geophysics (especially seismology) continues to be shaped by significant funding for research and development in verification/monitoring technologies. New funds at the level of about

[1] Now at US Department of the Army, Security Affairs Command, Alexandria, Virginia 22332.

0084–6597/90/0515–0257$02.00

\$30 million came into seismic monitoring programs in the United States in 1989.

In the sections that follow we review the history of nuclear testing and test ban negotiations, introduce some basic terminology, give examples of detection capability, describe characteristics of different seismic sources and methods of explosion identification, and briefly review different evasion scenarios. We conclude with a summary statement of identification capability and its implication for future possibilities in verifying new nuclear testing limitations.

BRIEF HISTORY OF NUCLEAR TESTING AND NEGOTIATIONS

The rate at which nuclear explosions have been conducted is characterized in Figure 1 by a pair of histograms giving the numbers carried out each year by the superpowers, beginning in 1945 with TRINITY. The World War II explosions in Japan at Hiroshima and Nagasaki are included. Atmospheric explosions are shown plotted above the zero line, and underground explosions are plotted below.

Note the extensive series of atmospheric tests in the 1950s and again in the early 1960s. The first underground nuclear explosion, code-named RAINIER, was carried out in 1957 in the United States at the Nevada Test Site. The gap for the years 1959 and 1960 represents the moratorium on testing that began in 1958 and ended in 1961, during a time of severe antagonism between the US and the USSR over nuclear testing issues, when the Soviets quite suddenly began a major atmospheric test program. In 1963 this type of testing was banned by the Limited Test Ban Treaty (LTBT), which prohibited nuclear explosions in the atmosphere, oceans, and space, but permitted them underground.

Since 1963, the US and the USSR have each carried out about 10 to 30 underground nuclear explosions per year. More recently, a unilateral moratorium was announced by the Soviet Union, which began in 1985 and ended in 1987 (Figure 1).

The main Soviet sites for nuclear weapons testing are at Novaya Zemlya (an island north of the Eurasian land mass) and in eastern Kazakhstan, near the city of Semipalatinsk. The eastern Kazakhstan site includes both Degelen and Shagan River as separate test sites. All explosions of concern with respect to the 150-kt Threshold Test Ban Treaty (TTBT) have occurred since 1976 at Shagan River. [A kiloton (kt) is a unit of energy used to quantify the yield of a nuclear explosion. Originally defined as the energy equivalent to exploding a thousand tons of TNT, it is now defined as 10^{12} cal.] The USSR has also set off about 100 peaceful nuclear

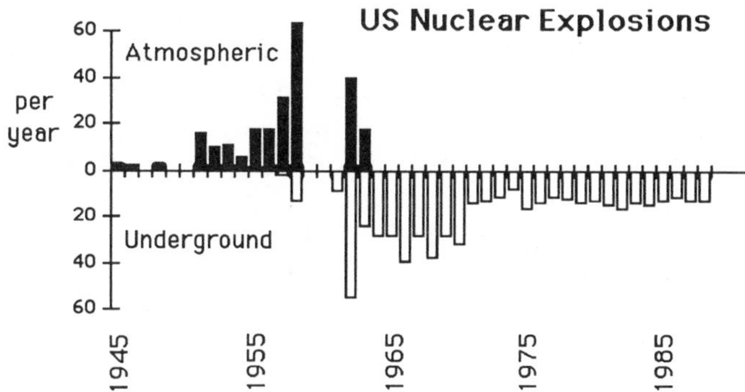

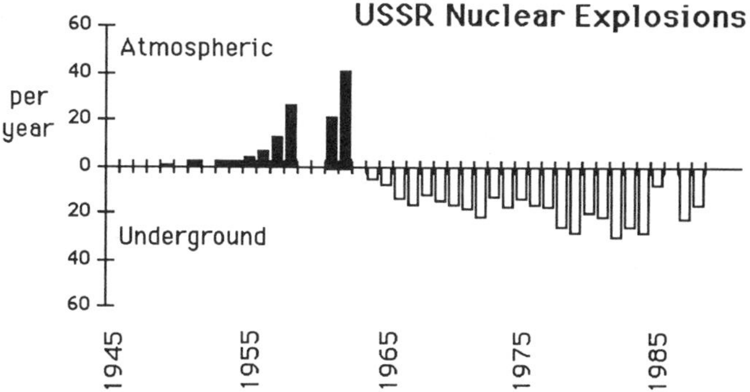

Figure 1 Numbers of nuclear explosions conducted each year by the US and USSR to 1988. These numbers include explosions announced as nuclear by the nation conducting the test, plus explosions routinely identified as "presumed nuclear" by organizations such as the US Geological Survey and the Swedish Ministry of Defence.

explosions, away from weapons test sites, for a variety of purposes that include creation of underground cavities for storage of chemicals, putting out oil-well fires, damming a river, and providing seismic signals for geophysical surveys of the structure of the continent (Scheimer & Borg 1984). The US now restricts nuclear testing to the Nevada Test Site, though in the past a few underground explosions were carried out elsewhere in the western US, in Mississippi, and in the Aleutian Islands of Alaska.

Underground nuclear explosions were conducted in the Sahara, in what is now part of Algeria, by France in the 1960s. In recent years the French have continued testing underground at sites in the South Pacific, notably on the Mururoa Atoll in the Tuamoto Archipelago. The United Kingdom has a test program, now conducted in the US at the Nevada Test Site. The Chinese have also tested nuclear weapons underground, with explosions at Lop Nor in the Xinjiang region of northwest China, and India carried out one underground nuclear explosion in 1974.

Nuclear testing was one of the leading international political issues of the 1950s and 1960s. During the presidencies of Dwight Eisenhower and John Kennedy, negotiations were conducted with the Soviet Union for a ban on underground tests above seismic magnitude 4.75 (associated at the time with tests above about 20 kt TNT equivalent), together with a moratorium on smaller underground tests. A joint seismic research program with the Soviets was also proposed in order to improve the monitoring of these smaller tests so that they too could eventually be prohibited in a Comprehensive Test Ban Treaty (Seaborg 1981). But, ostensibly because of disagreements regarding on-site inspections, how the operation of seismic monitoring would be conducted, and who would control the organization that carried out the monitoring, the treaty that eventually emerged in 1963 placed no limits whatsoever on testing underground, other than a ban on nuclear explosions that cause radioactive debris to cross national borders. As a part of the political process of securing the advice and consent of the US Senate, President Kennedy made a commitment, under the so-called safeguards associated with the LTBT, to promote a vigorous program of underground nuclear testing.

Widespread interest in nuclear testing resumed during the presidency of Richard Nixon, when quite suddenly in 1974 the Threshold Test Ban Treaty (TTBT) was negotiated and signed at a summit meeting in Moscow. This treaty introduced a new technical issue of importance in treaty monitoring, namely the question of yield estimation, because it banned underground nuclear weapons tests greater than 150 kt. In addition to the requirements of detecting these explosions and of identifying them as nuclear, it now became important to estimate their size. Some would say the TTBT is a fairly minor arms control treaty, but insofar as it was intended to terminate the practice of high-yield testing, it can be called successful. Both superpowers had been carrying out weapons tests prior to 1974 at yield levels in excess of 1 Mt. Though still not ratified as of 1989, this treaty is in effect and has been the focus of political debate on nuclear testing in both the Reagan and Bush administrations. The companion to the TTBT is the Peaceful Nuclear Explosions Treaty (PNET) of 1976, which was concluded after 18 months of intensive discussion and which has a very extensive protocol for verification. This treaty bans nonmilitary underground nuclear explosions that individually exceed 150 kt, though in certain circumstances, and subject to special in-country monitoring, it permits a salvo of nuclear explosions with total yield exceeding the TTBT limit.

The Carter administration made a commitment to achieve a comprehensive test ban. In trilateral negotiations (1977–80) that included the United Kingdom, agreement was eventually reached on the monitoring network of seismometers to be deployed in the US and USSR, on allowing an unlimited number of "challenge-type" on-site inspections, and on a moratorium for peaceful nuclear explosions. Acceptance of a request for on-site inspection was not to be mandatory, but it was agreed that the rejection of a request would be a serious matter (see Heckrotte 1988). The monitoring of those treaties now in effect (LTBT, TTBT, PNET) does not require access to "in-country" seismic data because seismic signals for explosions approaching the 150-kt threshold are large enough to be picked up by seismometers all over the globe. However, for a Comprehensive Test Ban Treaty (CTBT) or a Low Yield Threshold Test Ban Treaty (LYTTBT), in-country data would be essential. The Carter administration's test ban negotiations were eventually held up by debate over larger political issues (such as SALT II), became mired in detail over the number of seismic stations required on United Kingdom territory, and never reached their intended goal (York 1987, Chap. 14).

On July 20, 1982, early in the Reagan administration, a press briefing was given at the White House announcing that trilateral test ban negotiations would not resume, and that there were problems with verification of the TTBT and PNET. Specifically, it was stated that "on several occasions, seismic signals from the Soviet Union have been of sufficient magnitude to call into question Soviet compliance with the threshold of 150 kt, and it's because of the uncertainties of our yield estimation process that the United States cannot prove beyond any reasonable doubt that the Soviets have violated the Threshold Test Ban Treaty. . . . The Soviets have always asserted, when challenged on this point, that they have not violated the agreement." Subsequently, the Reagan administration stated that a comprehensive treaty remained a long-term goal of the United States, but provided a series of restrictions governing the circumstances under which a CTBT would be favored. The main issue concerning nuclear testing in the Reagan years became the assessment of Soviet compliance with the 150-kt yield threshold, with the Administration reporting several times to the US Congress that "while, in view of ambiguities in the pattern of Soviet testing and in view of verification uncertainties, the available evidence is ambiguous and we have been unable to reach a definitive conclusion, this evidence indicates that Soviet nuclear-testing activities for a number of tests constitute a likely violation of legal obligations under the TTBT."

Such claims became contentious because no technical evidence was publicly offered in support, and those scientists who had presumably participated in governmental review of the evidence typically reached a different conclusion. For example, the Director of the Lawrence Livermore National Laboratory, with responsibilities for designing and certifying US nuclear warheads, stated in 1987 (Batzel 1987) that "the Soviets appear to be observing some yield limit. Livermore's best estimate of this yield limit . . . is that it is consistent with TTBT compliance. However . . . , we cannot rule out the possibility that a few Soviet tests may have exceeded the limit."

Between 1986 and 1988, the US Congress threatened several times to cut off funds (amounting to about $800 million per year) for the US nuclear test program, and in September 1987 a US-Soviet joint statement announced the resumption of negotiations on limitations on nuclear testing with the following words: "In these negotiations the sides as the first step will agree upon effective verification measures which will make it possible to ratify the US-USSR Threshold Test Ban Treaty of 1974 . . . and [will] proceed to negotiating further intermediate limitations on nuclear testing."

OVERVIEW OF TECHNICAL ISSUES

There are three main steps in seismic monitoring of underground nuclear explosions. The first is that of *detection*, usually taken to mean detection of a signal by enough stations to enable location of the seismic source. The second step is that of *identification*, the main technical subject under discussion in this review paper. And the third step is that of *yield estimation* (Richards 1988, Sykes & Ruggi 1989). Because of the technical problems introduced by the 1974 TTBT, most research in explosion monitoring in the years 1975–85 was concerned with achieving accurate and credible yield estimates for large explosions. (It is remarkable that the US and the USSR each sent substantial teams for the first time in 1988 to the other side's main weapons test site to improve techniques of yield estimation. Negotiations on improving verification of the TTBT and PNET continue under the Bush administration.) However, it appears that yield estimation for large explosions is no longer the most challenging frontier in monitoring research. Promoted in part by the superpowers' joint statement of 1987, the research focus has shifted back to the question of what identification capability for small explosions can be achieved by novel seismic instrumentation and data analysis.

Associated with each of the three main technical steps in explosion monitoring (detection, identification, yield estimation) is a set of questions about what can be done to evade, or "spoof," the side doing the monitoring. Thus, one asks how well a nuclear explosion could be hidden in various environments. This debate has to be developed in terms of thresholds: What size nuclear explosion might go undetected; what size unidentified; and at what size can yield estimation be spoofed? Nuclear explosions can be conducted even down to such small sizes as pounds or even a few grams of TNT equivalent (Thorn & Westervelt 1987). It is becoming more widely recognized that there is an interaction between policy options and seismic monitoring capability that is driven by the following two technical points: (*a*) It is not possible to monitor a truly comprehensive test ban; and (*b*) a low-yield threshold test ban would require of monitoring programs the capability to estimate yield, which is increasingly difficult to do as yield level is reduced.

There is therefore a fundamental trade-off between the stringency of desired limitations on testing and the stringency of achievable verification standards. The monitoring objective in general is to drive downward the yield level at which even well-organized and well-funded attempts at clandestine testing might be successful. Looking at this another way, monitoring might be associated with some type of requirement to demonstrate that clandestine testing would be detected and identified if it occurs above a yield level that is significant from the perspective of weapons design.

Somewhat different conclusions can be reached on the three main steps in seismic monitoring regarding the quality of current capabilities. For example, it appears that detection capability in general is very good. Some examples to illustrate this capability are given in the next section. Current identification capability is also very good for events above a certain size, including those that raise questions of compliance with the 150-kt limit of the TTBT; is good even for events 10 times smaller; and is adequate in most but not all situations down to around 1 kt. No agreement exists, however, on how well nuclear explosions can now be identified at very low magnitude, and on whether the US could, with access to seismic data from the Soviet Union, identify all underground nuclear explosions down to 1 kt. As part of the debate on how well one would be able to estimate very low yields, there is a need for discussion of what restraints on allowed testing environment might be negotiated to give the monitoring side confidence that a new low-yield threshold was not being exceeded.

BASIC TERMINOLOGY AND DETECTION CAPABILITY

Three different types of seismic waves are described in this review—namely, body waves (specifically the *P*-wave), surface waves (specifically

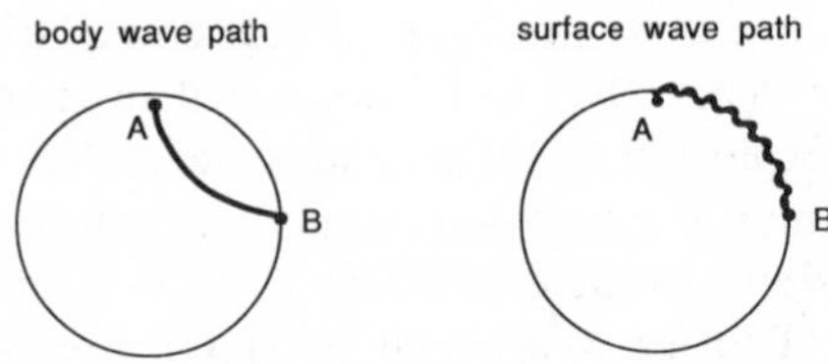

Figure 2 (*Left*) The path of a *P*-wave through the Earth, from a source at point A to a receiver at point B. This is a type of body wave, meaning that it travels through the Earth's deep interior. It is also a type of sound wave, traveling as a sequence of alternating compressions and rarefactions. An explosion at A may be recorded on a seismometer at B. (*Right*) This is a schematic for surface waves. A Rayleigh-wave path is shown, going from a source at point A to a receiver at point B. This wave travels around the Earth's surface and is conceptually similar to the set of ripples spreading away from a disturbance over the surface of a pond.

the Rayleigh wave), and regional waves (specifically the *Lg*-wave). Each of these may be made the basis of a seismic method for yield estimation. *P*-waves and *Lg*-waves are important for assessing detection capability, since they are observable even for relatively small explosions.

A body wave is a seismic wave that propagates through the body of the spherical Earth [see Figure 2 (*left*)]. A surface wave travels a completely different path [Figure 2 (*right*)]. *P*-waves and Rayleigh waves (discovered by Lord Rayleigh in 1887) have long been used to estimate the size of seismic sources. Body waves and surface waves are said to be "teleseismic," meaning that they propagate to distances beyond 2000 km. In contrast, seismic waves such as *Lg* (see Figure 3), are called "regional" not only because they do not propagate all over the globe, but also because they are different from region to region. The *Lg*-wave can be the most important of the regional waves for purposes of explosion monitoring because it is typically the largest wave observed in a seismogram at a regional distance. The *Lg*-wave has most of its energy trapped in the Earth's crust by a series of reflections at the crust-mantle interface (the "Moho"), very similar to the reflections of light occurring within an optical waveguide.

The basic data for seismic monitoring is a seismogram, a record of

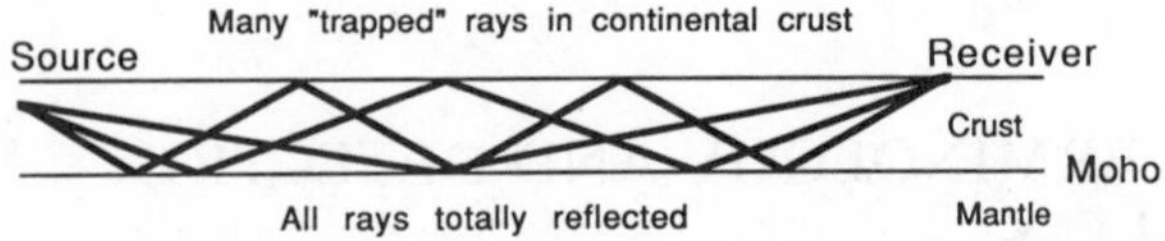

Figure 3 Schematic illustration of multiple total reflections, constituting the *Lg*-wave propagating in the low-velocity crustal waveguide.

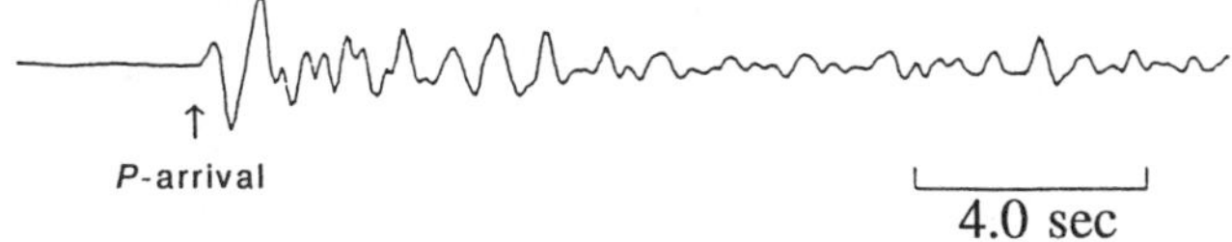

Figure 4 A teleseismic *P*-wave.

ground motion at a particular location. In Figure 4 is a typical seismogram showing a *P*-wave (which is always the first wave to arrive, traveling faster than, for example, surface waves). This seismogram records the vertical motion of the ground, measured in Scotland (July 4, 1976), due to a Shagan River explosion. Note the time scale, indicating that the period of the *P*-wave is roughly 1 s. A seismologist would call this a short-period wave.

In order to obtain the seismic magnitude, one measures both the amplitude A of the maximum ground motion in the first few seconds and the period T of this motion. This leads into the concept of body-wave magnitude, symbolized as m_b, which is assigned by evaluating the formula

$$m_b = \log(A/T) + \text{distance correction},$$

where A is the maximum amplitude of the body wave, read off a seismogram at a particular distance and converted to ground motion by allowing for the instrumental gain; and T is the period (often around 1 s) of the wave at maximum trace amplitude.

The logarithm of ground motion is employed because of the wide range of sizes of observed seismic signals. The distance correction makes allowance for the distance between the receiver and the source of the seismic waves. In the seismogram above, for example, the correction is that appropriate for the distance between Scotland and Shagan River. The magnitude m_b has been used in the study of earthquakes since 1935. The scale is directly related to signal levels, and thus is a phenomenological quantity.

The teleseismic *P*-wave signal from a 1-kt underground nuclear explosion corresponds roughly to $m_b = 4$. Since signal strength is approximately proportional to yield and the magnitude scale is logarithmic, it follows that a 100-kt explosion has a signal strength of about $m_b = 6$, and that factors of 2 in yield (up or down) correspond to about 0.3 magnitude units (plus or minus).

In practice the relationship between magnitude and yield is complicated by naturally occurring variability in both the efficiency with which nuclear yield couples into a seismic signal at the shot point and the efficiency with which the seismic signal propagates to teleseismic distances. The degree of coupling depends on rock type at the shot point, on the amount of water saturation there, and on whether the explosive device is "tamped" (i.e.

well packed in the emplacement hole) or placed within some type of cavity. The efficiency of propagation to teleseismic distances is thought to depend basically on temperatures of the solid material in the upper layers of the Earth within which the seismic wave propagates for part of its path. The resulting effects on wave attenuation may be estimated for different propagation paths, using a wide variety of seismic data that may be quite separate from the data on the explosion under study.

For surface waves, the surface-wave magnitude M_s is assigned by

$$M_s = \log(A/T) + \text{distance correction}$$

where A is the amplitude of surface-wave ground motion, and T is the period at which this motion takes place. The main differences between this scale and the m_b scale stem from the two ways in which surface waves (see Figure 5) appear very different from body waves in a seismogram.

First, surface waves are dispersed, meaning that different frequency components travel at different speeds. The effect is to make a surface wave spread out in time over several minutes. The second way in which surface waves differ from body waves is that they are a long-period signal. The period T is typically in the wave range 18–20 s for the part of the well-dispersed wave train where amplitudes are at their maximum (where A is measured). Care must be taken to avoid measuring A at the so-called Airy phase, a poorly dispersed signal composed of a range of frequencies all arriving at the same time, creating a more impulsive motion.

It should be noted that in monitoring for small explosions under a new test ban regime, reliance would be placed not so much on teleseismic signals, propagating as illustrated in Figures 2, 4, and 5, but on regional waves as recorded at distances less than about 1000 km between source and receiver. There are several such seismic waves guided by layering in the crust and upper mantle—we have described Lg in Figure 3. Although for a given yield they are typically stronger than teleseismic waves, they are significantly harder to interpret because crustal and upper mantle

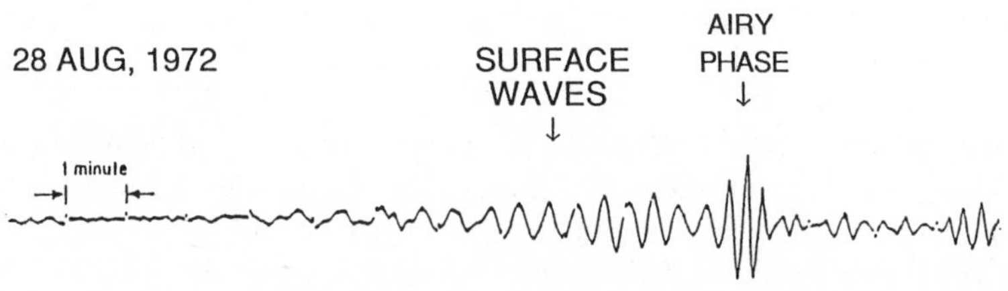

Figure 5 Example of a surface wave, teleseismically recorded at Eilat, Israel, from a large underground nuclear explosion ($m_b = 6.3$) at Novaya Zemlya.

properties vary from region to region. Measures of regional-wave amplitude often show much more scatter than is the case for teleseismic observations. However, such scatter may be greatly reduced in specialized studies of a single region. This is important both for assessing detection and identification capabilities using regional waves and for purposes of yield estimation.

Some Examples of Detection Capability

The Earth's surface continuously undergoes small motions, stimulated by ocean tides, winds, surf, train and truck traffic, etc. Seismic signals from earthquakes and explosions can be detected to the extent that they are stronger than this background noise, which varies greatly at different sites and in different frequency bands. Deployment of seismometers is more effective if sites can be found that are very quiet.

The signals of interest in seismic monitoring have a relatively impulsive nature. It follows that explosion signals can be detected by comparing signal levels averaged over a short time (about 2 s or less) with levels averaged over a longer period (about 30 s) and continually updating the comparison. Such analysis has traditionally been done by eye, but it is now routinely automated.

With many different detectors, the next technical problem is recognizing an association between signals collected at different geographic locations. That is, the signals must be related to some presumed common source, such as an earthquake or an explosion (whose signals have been picked up by a network deployed over some region of the Earth or perhaps over the whole globe); and that source must be located. The basic method of location is the same as that by which we use our ears to tell from which direction a sound is coming. One analyzes the slight difference in times of arrival of signal at different places. For a good enough network and a strong enough source, the method has considerable redundancy. *P*-wave signals detected at four or more stations are usually sufficient to give a useful source location. A large underground explosion (150 kt) would be picked up at hundreds of seismometers around the world. A small explosion (1 kt) would be picked up on relatively few.

With many years of experience it has proven possible to assess the capabilities of a hypothetical network of detectors, given the location and sensitivity of each seismometer as well as a characterization of noise levels at each site in the network. A useful way to describe these capabilities, both actual and predicted, is by contouring the globe in magnitude units that give the threshold value of source strength above which a seismic source will be detected with high probability and at enough stations to be located by the network. Figure 6 shows the actual and the predicted value

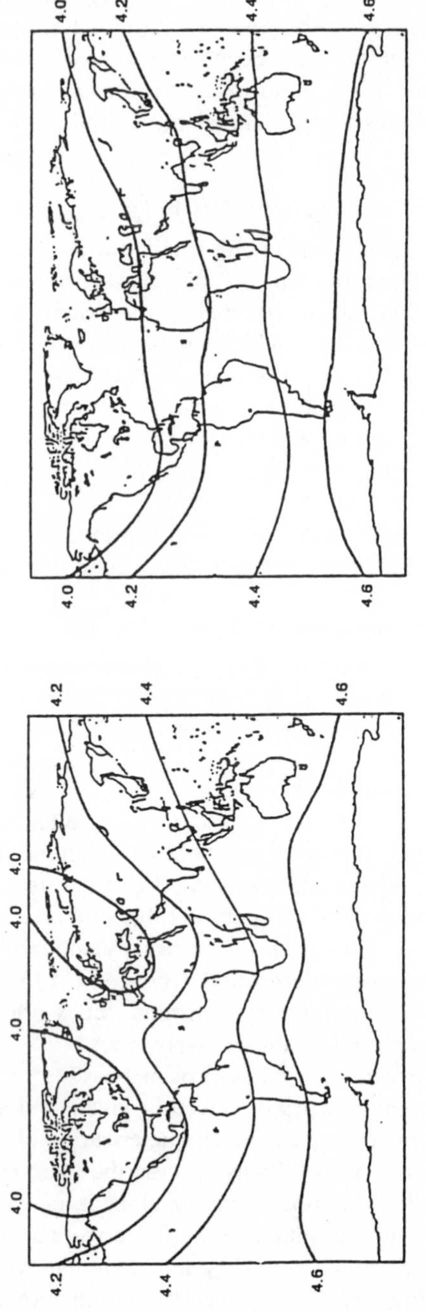

Figure 6 Contours showing detection capability of a particular seismic network of existing stations. For planning purposes, it is important that the estimated capability is in good agreement with the actual capability. Here, the differences are only about 0.1-m_b unit (Ringdal 1985).

of these contours, for a particular network of 115 globally distributed stations, requiring the quite stringent standard of 90% probability of detection (for events near the threshold level) at a minimum of four stations, at least one of which must be at a teleseismic distance.

Though Figure 6 shows a "globally distributed" network, it suffers from relatively poor coverage in the Southern Hemisphere. Thus, for this network, although the detection threshold has m_b in the range 3.9–4.3 in the Northern Hemisphere, the range is 4.2–4.8 in the south. To improve detection capability for sources in some particular part of the world, one would need to add stations located at distances favorable for picking up signals from that region.

Another practical element is introduced when one considers the capabilities of a group of seismometers in the same general region, all operated together as an "array," with central processing facilities that allow the multichannel signals to be filtered, delayed, and summed. Such an array is the seismologist's version of a phased array radar, permitting significant improvement in seismic signal-to-noise ratio as compared with this ratio for a single sensor. By adding the *P*-waves recorded at individual sensors (after an appropriate time shift to give all these first arrivals a common start time), the signal is enhanced if the *P*-wave first motion is coherent across the array. The relatively incoherent noise (which is apparent prior to the *P*-waves on each sensor) is suppressed. The outcome of such processing is often referred to as the "beam" formed by the array, which may be "steered" to "look" at different source regions by using the set of delays at each sensor appropriate for each source region, prior to signal summation. An array operated since about 1972 in southern Norway, known as NORSAR, with several tens of seismometers in a region about 200 km across, shows remarkable teleseismic detection capability for seismic sources in the USSR. The detection threshold for Semipalatinsk explosions, about 4000 km away, is about $m_b = 2.5–3.0$, and it is about 2.0–2.5 for the Caspian Sea area (Ringdal 1984). In June 1985, another seismic array became operational in southern Norway. Known as NORESS, it consists of 25 seismometer sites over an area about 3 km across, linked by optical fiber to a central processing facility. NORESS was designed for signals from sources at "regional" distances, and the detection threshold for this array is down to about $m_b = 2.0–2.5$ at 1500-km epicentral distance. At regional distances, the seismic waves *Pg*, *Pn*, *Sn*, and *Lg* are commonly seen, with *Lg* commonly being the strongest signal. [*Pg* propagates wholly within the crust; *Pn* and *Sn* travel in the top of the mantle, just below the Moho; and *Lg* is guided largely by the crustal layer (see Figure 3).] NORESS *P*-wave spectra extend up to about 18 Hz for quarry blasts about 1000 km away, and it has also been shown for a single sensor that NORESS signals have important information in even higher frequency bands. Remarkably, for a magnitude 3.0 event 500 km away, the signal is about 20 dB above noise, all the way out to 50 Hz (Ringdal et al 1986).

From the above examples of detection capability, it is apparent that the research frontier for seismic monitoring is represented by events roughly in the range $2 \leq m_b \leq 4$, corresponding to fractions of a kiloton for typical explosions. We turn next to a characterization of different types of seismic sources.

SEISMIC SOURCES

Most seismic signals from located events are caused by earthquakes, although chemical explosions are also recorded in large numbers. Nuclear explosions are relatively rare, and significant rockbursts even rarer. A monitoring network, however, will receive seismic signals created by all of these sources, and methods must be demonstrated not only to detect but also to identify with high confidence a clandestine nuclear test against the background of other seismic sources. In this section we review the basic properties of different types of seismic sources, contrasting them (where appropriate) with properties of nuclear explosions, in order to provide a physical basis for source identification.

Nuclear Explosions

For reference, we first comment on the production of seismic signals from the energy in a nuclear explosion—energy that is released in less than a microsecond. This energy produces a bubble of gas in which pressure and temperature rise steeply, reaching several million atmospheres and about 10^6 K within a few microseconds. A spherical shock wave expands outward from the bubble into the surrounding rock, crushing the rock as it travels and gradually weakening as it travels farther outward and away from the shot point (the point of emplacement of the nuclear device). As the hot gases expand and rock is vaporized near the shot point, a cavity is created. The size of this cavity is somewhat dependent on rock properties. While the cavity is forming, the shock wave continues to propagate away into the surrounding medium, eventually weakening to the point where rock no longer stays crushed but instead is merely compressed and returns (after a few cycles) essentially to its original state, thus becoming the propagating *P*-wave observed seismically. Because the first motion to occur in an explosion-generated *P*-wave is compressive, the first motion of the ground at the instant of the *P*-wave arrival is away from the source. It follows that the first motion of the ground is always upward for a well-recorded explosion.

Earthquake Sources

Earthquake activity is a global phenomenon, though most of the larger earthquakes are concentrated in active tectonic regions along the edges of the Earth's continental and oceanic plates. In the USSR, most earthquakes are located in a few active regions along the border of the country. For example, in the Kurile Islands region, shallow earthquakes generally occur on the ocean side of the islands, whereas the deeper earthquakes occur beneath the islands and toward the USSR landmass. Large areas of the USSR have no earthquake activity and hence are referred to as *aseismic*. However, some activity may occur in these regions at magnitudes below $m_b = 3.0$.

Worldwide, about 7000 to 7500 earthquakes occur annually with $m_b \geq 4.0$, and about 7% of these occur in the USSR. About two thirds of Soviet earthquakes, however, occur under the ocean around the coast. The smaller the magnitude, the more earthquakes there are. Table 1 provides a summary of roughly how many earthquakes can be expected each year above a given magnitude.

Earthquakes and their associated seismograms have been studied on a quantitative basis for about 100 years. Since 1959, seismologists have engaged in a substantial research effort to discriminate between earthquakes and explosions by analyzing seismic signals. They have relied on a growing body of seismic observations to develop empirical source identification criteria, and on theoretical models to promote a better understanding of the underlying physical phenomena.

Table 1 Two estimates of the numbers of earthquakes occurring per year at or above various magnitude levels (m_b) of relevance for CTBT and low-yield TTBT monitoring

Seismic magnitude	Numbers of earthquakes, worldwide, estimated by:		Earthquakes in USSR land areas[a]
	Ringdal (1985)	Lilwall & Douglas (1985)	
4.5	2600	2300	67
4.0	7400	7100	183
3.5	21,000	19,000	500
3.0	59,000	68,000	1400
2.5	166,000	209,000	4000

[a] About 2–3% of these events occur on Soviet continental territory (Bache 1985). This column (from US Congress 1988) gives numbers of earthquakes per year estimated for Soviet land areas.

From such studies, three fundamental differences between nuclear explosions and earthquakes have been recognized:

1. differences in characteristic source location,
2. differences in source geometry,
3. differences in efficiency of seismic-wave generation at different wavelengths.

With regard to the first difference, many earthquakes occur deeper than 10 km, whereas the deepest underground nuclear explosion to date appears to have been around 2.5 km, and weapons tests under the 150-kt threshold all appear to be conducted at depths less than 1 km. Testing at depths much greater than current practice adds significantly to both expense and difficulty. Almost no holes, for any purpose, have been drilled to more than 10-km depth. The exceptions are few, well known, and are too substantial to hide. Because nuclear explosions are restricted to shallow depths, discrimination between many earthquakes and explosions on the basis of depth is possible in principle, although in practice the uncertainty in depth determination makes the division less clear. To compensate for the possibility of very deep emplacement of an explosion, the depth must be determined to be greater than 15 km with high confidence before the event can be identified unequivocally as an earthquake.

With regard to source geometry, the fundamental differences are due to how the seismic signals are generated. An underground nuclear explosion is a highly concentrated source of compressional seismic waves (*P*-waves), sent out with approximately the same strength in all directions. This type of signal occurs because the explosion products apply a relatively uniform pressure to the walls of the cavity created by the explosion. In constrast, an earthquake is generated as a result of massive rock failure that typically accomplishes a net shearing motion across a fault between two blocks. Because of this shearing motion, an earthquake can be thought of as radiating predominantly transverse motions, called *S*-waves, from all parts of the fault that rupture. Although *P*-waves from an earthquake are generated at a substantial fraction of the *S*-wave level, they are associated with a four-lobed radiation pattern of alternating compressions and rarefactions in the radiated first motions, rather than the relatively uniform pattern of *P*-wave compressions radiated in all directions from an explosion source. The idealized radiation patterns for *P*-waves are shown schematically in Figure 7.

With regard to the efficiency of seismic-wave generation, we note that earthquakes tend to have larger source dimensions than explosions because they involve a larger volume of rock. The bigger the source volume, the longer the wavelength of seismic waves set up by the source. As a result,

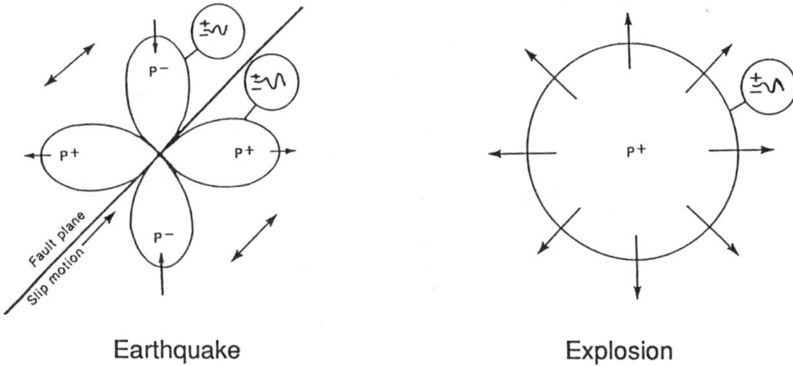

Figure 7 Radiation patterns of *P*-waves.

in comparing seismic signals of about the same strength from an earthquake and from an explosion, the earthquake signal is stronger at longer wavelengths and the explosion signal is stronger at short wavelengths. Instead of stating the outcome in terms of wavelength, an equivalent description can be given in terms of seismic-wave frequency. Thus, explosions, which entail a smaller but more intense source region, typically send out stronger signals at high frequency (see Figure 8). However, exceptions have been observed, and the distinction may diminish or even reverse for small events. The difference in frequency content of the respective signals, particularly at high frequencies, is a topic of active research, and as new types of seismic data become available (for example, data from within the USSR) it may be expected that methods of identification will be found that work at smaller magnitude levels. In fact, the success of identification methods for distinguishing between nuclear explosions and earthquakes is such that the remaining fundamental difficulties (which arise only at the lower yields) are in several respects less than the difficulties of distinguishing seismically between nuclear explosions and chemical explosions.

Chemical Explosions

Chemical explosions are used routinely in the mining and construction industries and also occur in military programs and on nuclear test sites. In general, from the seismic monitoring perspective, a chemical explosion can be considered to be a small spherical source of energy very similar to a nuclear explosion. The magnitudes represented by chemical explosions are below $m_b = 4.0$, with few exceptions (which are well known). The fact that large numbers of chemical explosions in the yield range 0.001–0.01 kt

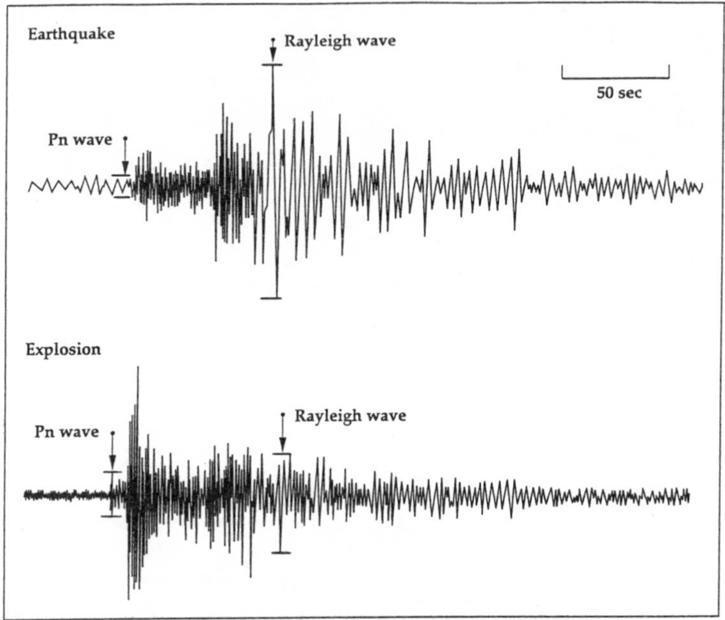

Figure 8 Seismograms for an earthquake and an explosion located near each other at the Nevada Test Site. Note the larger size of the (high-frequency) *Pn*-wave for the explosion, and the larger size of the (low-frequency) Rayleigh wave for the earthquake (figure from Jarpe 1989).

are detected and located seismically is a testament to the capability of seismic networks to work with signals below $m_b = 3.0$.

There is little summary information available on the number and location of chemical explosions in the USSR, although it appears that useful summaries could be prepared from currently available seismic data. In the US, chemical explosions of yield around 0.2 kt are common at about 20 mines. At each of these special locations, tens of such explosions may occur each year. Presuming that similar operations are mounted in the USSR, this activity is clearly a challenge when monitoring for nuclear explosions with yields below around 1 kt. It is also a challenge when considering the possibility that a nuclear-testing nation might try to muffle a larger nuclear explosion (say, around 5 kt), so that its seismic signals resembled those of a much smaller (say, around 0.1 kt) chemical explosion.

Progress in identifying chemical explosions can be based on three approaches:

1. Almost all chemical explosions for industrial purposes listed with yields at 0.1 kt or greater are in fact a series of more than 100 small explosions spaced a few meters apart and fired with small time delays between individual detonations. An effect of such "ripple firing" is to generate seismic signals that are like those of a small earthquake, in that both these types of sources (ripple-fired chemical explosion, and small earthquake) are spatially extended and thus lose some of the characteristics of a highly concentrated source, such as a nuclear explosion. Recent research has shown the importance of acquiring seismic data at high frequency (30–50 Hz); indeed, such data suggest the existence of a distinctive signature for ripple-fired explosions (Baumgardt & Ziegler 1988, Hedlin et al 1989).
2. The rare chemical explosions above about 0.5 kt that are not ripple fired can be expected to result in substantial ground deformation (e.g. cratering). Such surface effects, together with the absence of a radiochemical signal, would indicate a chemical rather than a nuclear explosion. The basis for this discriminant is that chemical explosions are typically very shallow.
3. To the extent that remote methods of monitoring chemical explosions are deemed inadequate (for example, in a low-yield threshold or comprehensive test ban regime), solutions could be sought in requiring prior announcement of certain types of chemical explosion. Note that chemical explosions above 0.1 kt occur routinely only at a limited number of sites. Some type of minimal on-site inspection and/or a program of in-country radiochemical monitoring could also make a contribution.

For an explosion that is well coupled into the ground (a "tamped" explosion), a magnitude of $m_b = 2.0$ can be achieved with a yield of about 0.01 kt. At this level, a monitoring program would confront perhaps 1000 to 10,000 chemical explosions per year. Given the difficulty of discriminating between chemical and nuclear explosions, it is clear that seismic monitoring fails at some levels of low but still detectable yield. Thus, a truly comprehensive ban on all nuclear testing could not be strictly verified seismically. However, recognizing that relatively few chemical explosions occur at the upper end of this m_b range (2.0–4.0), where discriminants between chemical and nuclear explosions do exist, data recorded within the USSR can clearly contribute substantially to an improved monitoring capability.

Rockbursts

In underground mining involving tunneling activities, the rock face in the deeper tunnels may occasionally burst suddenly into the tunnel. This is referred to as a rockburst and results from the difference between the low pressure existing within the tunnel and the much greater pressure within the surrounding rock.

In terms of magnitude, rockbursts are all small. They occur over very restricted regions in the Earth and generally have a seismic magnitude of less than $m_b = 4.0$.

The source mechanisms of rockbursts are very similar to those of small earthquakes. In particular, the direction of the first seismic motion from a rockburst will have a pattern similar to that observed for earthquakes. Therefore, for the seismic identification problem, rockbursts can be considered small earthquakes that occur at very shallow depths.

IDENTIFICATION METHODS

Over the years, a number of identification methods have been shown to be fairly robust. Some of these methods perform the identification process by identifying certain earthquakes as being earthquakes (but not identifying explosions as being explosions). Other methods identify certain earthquakes as being earthquakes and certain explosions as being explosions. The identification process is therefore a type of winnowing operation.

Location

The principal identification method is based on the location of a detected seismic source.

If the epicenter is determined to be in an oceanic area but no explosion hydroacoustic signals are recorded, then the event is identified as an earthquake. Large numbers of seismic events can routinely be identified in this way, since so much of the Earth's seismic activity occurs beneath or adjacent to ocean basins. If the location is determined to be in a land region, then in certain cases the event can still be identified as an earthquake on the basis of location alone [for example, if the site is clearly not suitable for nuclear explosions (such as near population centers) or if there is no evidence of human activity in the area].

Depth

With the exception of epicenter interpretation, interpretation of the seismic source depth is probably the most useful identification technique. A seismic event can be identified as an earthquake if its depth is determined with high confidence to be below 15 km.

The procedure for determining source depth is a part of the process of finding the event location using the arrival time of four or more *P*-wave signals. A depth estimate can be obtained by measuring the time difference between the first arriving P-wave energy and the arrival time of signals reflected from the Earth's surface above the source.

An advantage of depth as an identification method is that it is not dependent on magnitude—it will work for small events as well as large ones, provided that the basic data are of adequate signal quality. It will not alone, however, distinguish between chemical and nuclear explosions unless the nuclear explosion is relatively large and deep; nor will it distinguish between underground nuclear explosions and earthquakes unless the earthquakes are sufficiently deep.

$M_s : m_b$

Underground nuclear explosions generate signals that tend to have surface-wave magnitudes M_s and body-wave magnitudes m_b that differ systematically from earthquake signals. This is basically an expression of the efficiency of explosions in exciting body waves (short period) and of earthquakes in exciting surface waves (long period). The phenomenon is often apparent in the original seismograms (see Figure 8), and an example is shown in terms of an $M_s : m_b$ diagram in Figure 9. These diagrams can be thought of as separating the population of explosions from that of earthquakes. For any event that is clearly in one population or another, the event is identified. For any event that lies between the two populations (as occasionally happens, particularly with lower quality seismic data for explosions at the Nevada Test Site), this method does not provide reliable identification.

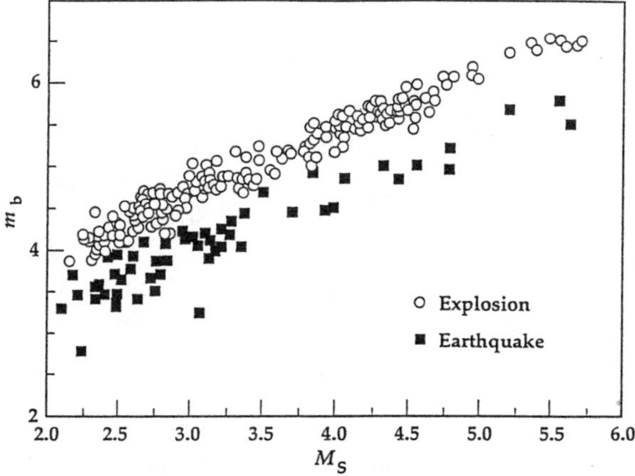

Figure 9 $M_s : m_b$ diagram for seismic sources at or near the Nevada Test Site. Note the separation into two populations (figure from Jarpe 1989).

To use this identification method, both m_b and M_s values are required for the event. This is no problem for the larger events, but for smaller events (below $m_b = 4.5$) it can be very difficult to detect long-period surface waves using external stations alone. This difficulty arises particularly for explosions, with the result that the $M_s : m_b$ method intrinsically works better for identifying small earthquakes than it does for identifying small explosions. Internal stations would provide important additional capability for obtaining M_s values of events down to $m_b = 3.0$ and perhaps below.

Spectral Methods

The basis of the success of $M_s : m_b$ diagrams is that the ratio of low-frequency waves to high-frequency waves is typically different for earthquakes and explosions. Thus, M_s is a measure of signal strength at around 20 s (frequency 0.05 Hz), and m_b is a measure of signal strength at around 1 s (frequency 1 Hz). As a way of exploiting such differences, $M_s : m_b$ diagrams are quite crude compared with methods that use a more complete characterization of the frequency content of seismic signals.

What then can be learned from analysis of earthquake and explosion signals across the whole spectrum of frequencies that they contain? This question drives much of the current research and development efforts in seismic monitoring.

Because $M_s : m_b$ diagrams work so well when surface waves are large enough to be measured, the main contribution required of more sophisticated analysis is that it make better use of the information contained in short-period *P*-wave signals. This offers the prospect of improved identification capabilities just where they are most needed, namely for smaller events. Thus, instead of boiling down the *P*-wave signals at all stations simply to a network m_b value, one can measure the amplitude at a variety of frequencies and seek a discriminant that works on systematic differences in the way that earthquake and explosion signals vary with frequency. One such procedure is the variable frequency magnitude (VFM) method, which measures the short-period body-wave signal strength from seismograms filtered to pass energy at two different frequencies, f_1 and f_2. Discrimination is based on a comparison of $m_b(f_1)$ and $m_b(f_2)$, which in many cases shows a clear separation of earthquake and explosion populations (Stevens & Day 1985).

High-Frequency Signals

The use of high-frequency signals in seismic monitoring is strongly linked to the question of what can be learned from seismic stations in the USSR whose data are made available to the US. Such in-country or internal stations were considered in CTBT negotiations in the late 1950s and early 1960s, and the technical community concerned with monitoring in those years was well aware that seismic signals up to several tens of hertz could be observed to propagate from explosions out to distances of several hundred kilometers—what we have referred to previously as "regional distances."

However, this early interest in regional waves lessened once the Limited Test Ban Treaty of 1963 was signed, because it was then recognized that subsequent programs of underground nuclear explosions conducted by different countries could be monitored teleseismically. At teleseismic distances, seismic body-wave signals are usually simpler, are usually not seen with frequency content above 10 Hz, and have indeed proven adequate for most purposes of current seismic monitoring, even under the treaty restrictions on underground nuclear testing that have developed since 1963. It is in the context of considering further restraints on testing, such as a CTBT or LYTTBT, that the seismic monitoring community is required to evaluate high-frequency/regional seismic signals. In such a hypothetical testing regime, the greatest challenge to monitoring will come from very small nuclear explosions, and for them the most favorable signal-to-noise ratios will often be at high frequency (Evernden et al 1986). Internal stations, producing data of high quality and at quiet sites, will be essential in addressing the monitoring issues associated with such events.

The claim that high-frequency seismic data from internal stations in the USSR can lead to great improvements in detection capability has been supported by the data recorded within that country since 1987. Nongovernmental agreements have allowed US seismologists to deploy sophisticated monitoring equipment at sites within Soviet Central Asia (Given et al 1990, Hansen et al 1989).

The daily recording of very small seismic events, made possible by improvements in detection capability, focuses attention on the discrimination problem for events in the magnitude range $2.0 \leq m_b \leq 4.0$. In practice, the main use of in-country high-frequency signals (i.e. above 10 Hz) in improving identification capability may lie in interpretation of the improved locations, including depth estimates (Ruud et al 1988, Thurber et al 1989).

TEST BAN EVASION

Several methods have been suggested by which a low-threshold test ban might be evaded. These include testing behind the Sun or in deep space; detonating a series of explosions to simulate an earthquake; testing during or soon after an earthquake; testing in large underground cavities, in nonspherical cavities, or in low-coupling material such as deposits of dry alluvium; and masking a test with a large, legitimate industrial explosion.

Although the majority of proposed evasion scenarios have been shown to be readily defeated, a few concepts have evoked serious concern and analysis. We describe them below, in what we believe roughly to be the order of increasing effectiveness.

Testing Behind the Sun or in Deep Space

The concern is that one or more vehicles sent into space could detonate a nuclear device and record the testing information to relay back to Earth. Some feel that such a testing scenario is both technically and economically feasible. Others believe that the technical and economic resources required, when viewed in terms of the uncertainty and risk of discovery, would greatly exceed the utility of any information that could be obtained. [Note the logistical effort implied by the fact that testing operations currently engage 8000 employees at the Nevada Test Site, plus many more of the 10,000 employees at each of the three US national weapons laboratories (Los Alamos, Livermore, and Sandia); and that many thousands of people were needed to execute the US atmospheric test program in the Pacific up to 1963.] A nuclear test explosion in space would be an unambiguous violation of several treaties, making discovery very costly to the cheater. If testing can be demonstrated on technical grounds to be a concern, the risk could be addressed, although at high cost, by deploying satellites to orbit the Sun. Alternatively, the risk could be addressed politically by negotiating an agreement to conduct simple inspections of the rare vehicles that go into deep space.

Simulating an Earthquake

It has been suggested that a series of nuclear explosions could be sequentially detonated over a period of a few seconds to mimic the seismic signal created by a naturally occurring earthquake. However, it would be a poor mimic, working only if the *P*-wave amplitude, over just one cycle of motion, were measured at stations within a certain distance range and if reliance were placed mainly in the $M_s : m_b$ discriminant. Consequently, a network that records over a variety of distances and/or records *P*-wave amplitudes measured over several cycles could be used to monitor against this scenario.

Testing During an Earthquake

If a small nuclear weapons test were to be detonated shortly after a nearby naturally occurring earthquake, the signals of the explosion would to some extent be obscured by earthquake signals. The logistical difficulties of holding a test in readiness for months or years would be substantial, and strong nerves and prompt good judgment would be required to identify the suitability of a candidate earthquake within less than a minute of its occurrence. (Note that seismologists currently have no reliable techniques for the short-term prediction of the time, location, and size of earthquakes, and this limitation is unlikely to be overcome in the near future.)

The feasibility of this scenario is further reduced by using networks that record high-frequency waves. The obscuring effects of an earthquake are much reduced in a high-frequency record. For this reason, and for the logistical difficulties, the hide-in-earthquake scenario is no longer considered a credible evasion threat.

Testing in Low-Coupling Materials

The fraction of energy from a nuclear explosion that is converted into seismic energy (and thus into possibly detectable signals) depends on the type of rock in which the explosion occurs. Low-coupling materials such as dry porous alluvium have air-filled pore spaces that absorb much of the explosive energy. This has led to concern that a monitoring network could be evaded by detonating an explosion in such material. However, the opportunities for such evasion are thought to be limited in the Soviet Union, where no great thicknesses of dry alluvium are known to exist. This contrasts with the geology of the Nevada Test Site, which, with its unconsolidated sediments and low water table, has some of the most unusual geology in the world. Estimates of the maximum thickness of alluvium in the USSR indicate that it would be sufficient to muffle explosions only up to 1 or 2 kt. In addition, because alluvium is easily disturbed, an explosion could create a subsidence crater or other telltale evidence.

Masking a Test With a Chemical Explosion

Unlike an earthquake, large chemical explosions, such as those used routinely in mining and construction, could be timed to coincide with a clandestine nuclear test. If done in combination with cavity decoupling, to hide a small nuclear test as discussed below, this scenario presents perhaps the most serious challenge to a monitoring network.

Cavity Decoupling

If a nuclear explosion is set off in a sufficiently large underground cavity, it will emit seismic waves that are much smaller than those from the same-size explosion detonated in a tamped underground test. The physical principles underlying this scheme, called cavity decoupling, have been verified at small yields. It appears that opportunities exist within the USSR and other countries to construct underground cavities suitable for so muffling low-yield explosions. The easiest method would probably be via solution mining in a salt dome. Consequently the capability to conduct clandestine decoupled nuclear tests determines the yield threshold below which treaty verification, by seismic means alone, is no longer possible with high confidence.

An underground nuclear explosion creates seismic waves with a broad range of frequencies. For purposes of seismic detection and discrimination of small explosions, frequencies from about 1 Hz to as high as 30 or 50 Hz may be important.

For the lower end of this frequency range, the amplitude of the waves created by the explosion is approximately proportional to the total amount of new cavity volume created by the explosion. A standard test is conducted in a hole whose initial volume is negligible compared with its posttest volume. Because the initial hole is small, the rock surrounding the explosion is strained at high levels and flows plastically. This flow results in large displacements just outside the cavity, leading to a large net increase in the volume of rock in the vicinity of the shot point, and hence to efficient generation of seismic waves.

If, on the other hand, the explosion occurs in a hole of much greater initial volume, the explosive stresses at the cavity wall will be much smaller, resulting in less flow of rock and hence less volumetric expansion and reduced excitation of seismic waves. If the initial hole is so large that the stresses in the surrounding rock never exceed the limit at which rock flow begins to occur, the seismic coupling is minimized. Further increase of the emplacement hole size will not further reduce coupling at low frequencies, and the explosion is said to be "fully decoupled." For strong salt or granite, the cavity radius required for full decoupling has been estimated as somewhat greater than 20 m times the cube root of the yield (in kilotons). For weak salt it is about 50% greater. However, the signal strength is not nearly so strongly reduced by decoupling at frequencies above that associated with resonances of the internal surface surrounding the shot point at which rock flow ceases. In practice, the frequency above which decoupling is likely to be substantially less effective is around 10 to 20 Hz, divided by the cube root of the yield.

From the above discussion, we can distinguish between the efficiency of full decoupling at low frequencies and the efficiency at high frequencies (Larson 1985). At low frequencies, the amount by which seismic-wave amplitudes are reduced by full decoupling is about a factor of 70, described equivalently as a reduction by 1.85 m_b units. At high frequencies, the reduction is about a factor of 7 (i.e. 0.85 m_b units).

Of course, a thorough discussion of decoupling as an evasion scenario would have to include a number of nonseismological considerations. These include the military significance of being able to carry out nuclear tests up to various different yield levels (e.g. 0.1, 1, or 10 kt), and the political consequences if a clandestine test program were uncovered. Technical considerations include methods of (clandestine) cavity construction, and the capabilities of nonseismological surveillance techniques.

An impression of the scale of effort required to fully decouple a 5-kt underground nuclear explosion is conveyed in Figure 10. Such an explosion would still generate detectable seismic signals on a well-designed in-country network and would thus be likely to attract some attention.

SUMMARY OF IDENTIFICATION CAPABILITY

It is difficult to reach precise but general conclusions on what future low-yield levels could be monitored because much would depend on the degree

Figure 10 To fully decouple a 5-kt explosion in salt (the easiest medium in which to construct a large cavity), a spherical cavity with a diameter of at least 85 m would be required. The Statue of Liberty (height 74 m, including pedestal) is shown for scale (US Congress 1988).

of effort put into new seismic networks and new data analysis. It would depend, too, on judgments about the level of effort that might be put into clandestine nuclear testing.

The identification capability of any network will always be poorer than the detection capability for that network. Experience with teleseismic signals indicates that the difference amounts to about 0.5 m_b units—that is, a detection threshold of about 3.5 indicates an identification threshold of about 4.0 (Hannon 1985). However, for regional signals from events with magnitude less than $m_b = 3.5$, the difference may be larger than 0.5 m_b. A summary statement conveyed in a May 1988 Office of Technology Assessment Report (US Congress 1988) was as follows:

> . . . there appears to be agreement that, with internal stations that detect down to m_b 2.0–2.5, identification can be accomplished in the U.S.S.R. down to *at least* as low as m_b 3.5. This cautious identification threshold is currently set by the uncertainty associated with identifying routine chemical explosions that occur below this level. Many experts claim that this identification threshold is too cautious and that with an internal network, identification could be done with high confidence down to m_b 3.0.

This cautious identification threshold corresponds to a subkiloton yield for a tamped explosion. But if we are doubly cautious and go on to interpret it in terms of a substantially decoupled shot, the yield would be about 5 kt. By this argument, which can hardly be refuted on seismological grounds alone until one can build confidence in the routine achievability of an identification threshold lower than $m_b = 3.5$, the bottom line is a capability to monitor above about 5 kt.

FUTURE DIRECTIONS

Once the yield estimation issues related to the 150-kt Threshold Test Ban Treaty are politically resolved, it is required under the declared agenda of current US-USSR negotiations that attention turn to "further limitations on nuclear testing." Given the difficulty, if not the impossibility, of monitoring a truly comprehensive test ban, it appears from the perspective of verification that there is merit in reducing the allowed yield threshold in stages. Quoting again from the 1988 Office of Technology Assessment Report, which has guided much of the presentation of material in this review, such a staged approach "would begin with a limit that can be monitored with high confidence using current methods, but would establish the verification network for the desired lowest level. The threshold would then be lowered as information, experience, and confidence increase" (US Congress 1988).

ACKNOWLEDGMENTS

We thank Karen Fischer, Bill Haxby, and Won-Young Kim for reviews of the manuscript, as well as our many colleagues whose work we have drawn upon extensively, especially Steve Day, Frode Ringdal, Lynn Sykes, and Gregory van der Vink. We acknowledge a grant from the MacArthur Foundation to the School of International and Public Affairs at Columbia. This article is Lamont-Doherty Geological Observatory Contribution No. 4578.

Literature Cited

Bache, T. C. 1985. Seismic identification of small events. See Larson 1985, pp. V-101–V-142

Baumgardt, D. R., Ziegler, K. A. 1988. Spectral evidence for source multiplicity in explosions: applications to regional discrimination of earthquakes and explosions. *Bull. Seismol. Soc. Am.* 78: 1773–95

Batzel, R. 1987. In *US Congress, Senate Foreign Relations Committee, Hearing 100–115*, pp. 220–21. Washington, DC: US Gov. Print. Off.

Evernden, J. F., Archambeau, C. B., Cranswick, E. 1986. An evaluation of seismic decoupling and underground nuclear test monitoring using high-frequency seismic data. *Rev. Geophys.* 24: 143–215

Given, H. K., Tarasov, N. T., Zhuravlev, V., Vernon, F. L., Berger, J., Nersesov, I. L. 1990. High-frequency seismic observations in eastern Kazakhstan, USSR, with emphasis on chemical explosion experiments. *J. Geophys. Res.* 95: 295–307

Hannon, W. J. 1985. Seismic verification of a comprehensive test ban. *Science* 227: 251–57

Hansen, R. A., Ringdal, F., Richards, P. G. 1989. Analysis of IRIS data for Soviet nuclear explosions. In *NORSAR Semiannual Technical Summary, 1 Oct 1988–31 Mar 1989*, pp. 121–40. Kjeller, Nor: NTNF/NORSAR

Heckrotte, W. 1988. On-site inspection to check compliance. In *Nuclear Weapons Tests: Prohibition or Limitation*, ed. J. Goldblat, D. Cox, pp. 248–49. Oxford: Oxford Univ. Press. 423 pp.

Hedlin, M., Minster, J. B., Orcutt, J. A. 1989. The time-frequency characteristics of quarry blasts and calibration explosions recorded in Kazakhstan U.S.S.R. *Geophys. J. Int.* 99: 109–21

Jarpe, S. P. 1989. Lawrence Livermore National Laboratory Nevada Test Site Regional Seismic Network. *Brochure LLL-TB-100*, Lawrence Livermore Natl. Lab., Livermore, Calif.

Larson, D. B., ed. 1985. *Proceedings of DOE-Sponsored Cavity Decoupling Workshop, Pajaro Dunes, Calif.* Washington, DC: US Dep. Energy

Lilwall, R. C., Douglas, A. 1985. Global seismicity in terms of short-period magnitude m_b based on individual station magnitude-frequency distributions. *Rep. O 23/84*, UK At. Weapons Res. Establ. London: Her Majesty's Stationery Off.

Richards, P. G. 1988. Seismic methods for verifying test ban treaties. In *Nuclear Arms Technologies in the 1990's. AIP Conf. Proc. No. 178*, ed. D. Schroeer, D. Hafemeister, pp. 54–108. New York: Am. Inst. Phys.

Ringdal, F. 1984. Teleseismic detection at high frequencies using NORSAR data. In *NORSAR Semiannual Technical Summary, 1 Apr–30 Sept 1984*, pp. 54–62. Kjeller, Nor: NTNF/NORSAR. 69 pp.

Ringdal, F. 1985. Study of magnitudes, seismicity, and earthquake detectability using a global network. In *The VELA Program: A Twenty-Five Year Review of Basic Research*, ed. A. U. Kerr, pp. 611–24. Washington, DC: Exec. Graph. Serv. 964 pp.

Ringdal, F., Hokland, B. Kr., Kværna, T. 1986. Initial results from the NORESS high frequency seismic element, (HFSE). In *NORSAR Semiannual Technical Summary, 1 Oct 1985–31 Mar 1986*, pp. 31–39. Kjeller, Nor: NTNF/NORSAR. 93 pp.

Ruud, B. O., Husebye, E. S., Ingate, S. F., Christoffersson, A. 1988. Event location at any distance using seismic data from a single, three-component station. *Bull. Seismol. Soc. Am.* 78: 308–25

Seaborg, G. T. 1981. *Kennedy, Khrushchev and the Test Ban.* Berkeley: Univ. Calif. Press. 320 pp.

Scheimer, J. F., Borg, I. Y. 1984. Deep seismic sounding with nuclear explosives in the Soviet Union. *Science* 226: 787–92

Stevens, J. L., Day, S. M. 1985. The physical basis of $m_b : M_s$ and variable frequency magnitude methods for earthquake/explosion discrimination. *J. Geophys. Res.* 90: 3009–20

Sykes, L. R., Ruggi, S. 1989. Soviet nuclear testing. In *Nuclear Weapons Databook*, Vol. 4, *Soviet Nuclear Weapons*, ed. T. B. Cochran, W. A. Arkin, R. S. Norris, J. I. Sands, pp. 332–82. Cambridge, Mass: Ballinger

Thorn, R. N., Westervelt, D. R. 1987. Hydronuclear experiments. *Rep. LA-10902-MS, UC-2*, Los Alamos Natl. Lab., Los Alamos, N. Mex.

Thurber, C., Given, H. K., Berger, J. 1989. Regional seismic event location with a sparse network: application to eastern Kazakhstan, USSR. *J. Geophys. Res.* 94: 17,767–80

US Congress 1988. *Seismic Verification of Nuclear Testing Treaties. OTA-ISC-361*, Off. Technol. Assess. Washington, DC: US Gov. Print. Off. 139 pp.

York, H. 1987. *Making Weapons, Talking Peace*. New York: Basic Books. 359 pp.

Annu. Rev. Earth Planet. Sci. 1990. 18:287–315
Copyright © 1990 by Annual Reviews Inc. All rights reserved

BRINE MIGRATIONS ACROSS NORTH AMERICA— THE PLATE TECTONICS OF GROUNDWATER

Craig M. Bethke and Stephen Marshak

Department of Geology, University of Illinois, Urbana, Illinois 61801

INTRODUCTION

In recent years geologists have come to appreciate that warm, saline groundwaters have migrated for many hundreds of kilometers across the North American craton. The migrating brines left their diagenetic signatures on the sediments through which they passed, giving rise to modern petrographic and geochronologic mysteries. More importantly, the migrations created many of the economic resources found in the continent's sedimentary cover. As migrating brines entered shallowly buried sediments, they precipitated metallic minerals that form world-class ore districts. Petroleum migrated with the brines to accumulate, far from oil source rocks, in reservoirs from which it is produced today.

Increasingly it is clear that the migrating brines originated in the forelands of North American tectonic belts, and that the migrations coincided in time with the intervals during which the belts were deformed. In this paper we describe the evidence that the brine migrations occurred as giant hydrothermal systems operating on regional scales, paying greatest attention to the Ouachita-Arkoma belt and the neighboring area of the midcontinent to the north (parts of Arkansas, Missouri, Oklahoma, Iowa, and Kansas). In addition, we investigate the link between tectonic deformation and deep groundwater flow. The resulting picture shows groundwater migration as an important, albeit generally neglected, aspect of plate tectonic theory.

0084–6597/90/0515–0287$02.00

EVIDENCE OF LONG-RANGE MIGRATION

Migrating brines chemically and thermally altered the sedimentary cover of the continental interior, leaving behind evidence of the extent and nature of the migration events. The evidence includes (*a*) ore bodies that were deposited from metal-bearing brines, (*b*) thermal histories that are anomalous for shallowly buried sediments, (*c*) regional potassium metasomatism, (*d*) epigenetic dolomite cements in ore bodies and deep aquifers, and (*e*) correlations of petroleum to distant source rocks.

Mississippi Valley–Type Ores

The earliest consensus that brines migrate over considerable distances arose from study of Mississippi Valley–type ore deposits. These deposits, which are commonly found in shallow sediments on the margins of sedimentary basins, account for a significant fraction of the world's reserves of Pb and Zn, as well as economic quantities of Ba, F, and other elements (Sverjensky 1986). Study of fluid inclusions within ore minerals by White (1958) and Hall & Friedman (1963) showed that the ores precipitated from saline groundwaters with chemical and isotopic compositions typical of brines found deep in sedimentary basins. The presence of petroleum within fluid inclusions in some districts further indicated that ore-forming fluids had migrated considerable distances from deep basin strata.

Regional migration of ore-forming brines is best demonstrated in the midcontinent area north of the Arkoma basin. Ore districts here include the Viburnum Trend, Old Lead Belt, and Tri-State, as well as the smaller Northern Arkansas, Central Missouri Barite, and Southeast Missouri Barite districts. Several lines of evidence indicate that mineralization formed from a single hydrothermal system that flowed northward from the Ouachita Mountains and Arkoma basin across the continental platform during the Ouachita orogeny of the late Paleozoic (Leach 1979, Leach & Rowan 1986). Most notably, it is becoming clear that the districts are not isolated packets of mineralization but rather are part of a continuum that ranges in grade upward from the minor amounts of sulfide and barite deposited pervasively through Paleozoic strata (Erickson et al 1981). In terms of the homogenization temperatures and salinities of fluid from fluid inclusions, the trace mineralization found in country rock is identical to ores in the major districts (Leach 1979, Coveney et al 1987). These results imply that the brines completely replaced indigenous groundwaters from the platform and reached thermal equilibrium with an enormous volume of rock. The K/Na ratio found in fluid inclusions from ore minerals increases northward following a linear trend (Viets & Leach 1990). Similar trends (discussed later) are evident for temperatures of ore formation and the minor-element and isotopic compositions of epigenetic dolomite, suggesting that the brine progressively cooled and reacted with the country rock as it migrated northward.

As first described by Oliver (1986), the midcontinent hydrothermal system forms part of a continent-wide pattern of inferred brine migrations from the tectonic belts of North America onto the continental interior (Figure 1). Leach & Rowan (1986) associate deposits all along the Appalachian fold belt (e.g. Daniels Harbour, Friedensville, Austinville, and East

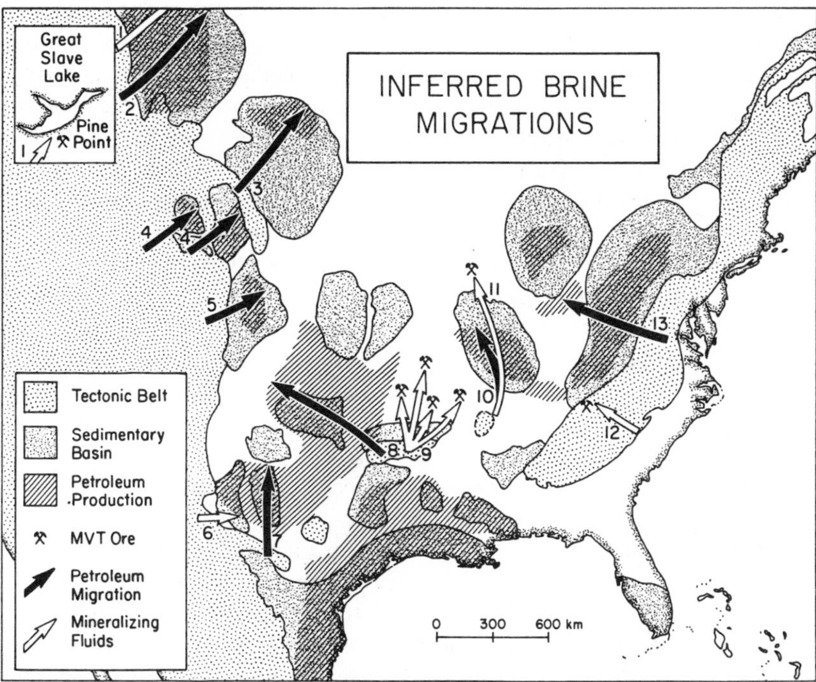

Figure 1 Map showing the inferred migration pathways of ore-forming fluids and petroleum across the continental interior of North America. Patterns show major sedimentary basins, tectonic belts, and the distribution of petroleum production; symbols show some major Mississippi Valley–type (MVT) ore districts. Key: (1) Western Canada basin and Pine Point district (Garven 1985); (2) Alberta Tar Sands (Moshier & Waples 1985, Garven 1989); (3) Williston basin (Dow 1974); (4) Big Horn basin (Stone 1967, Sheldon 1967); (5) Denver basin (Clayton & Swetland 1980); (6) mineralization in Permian basin (Mazzullo 1986); (7) Permian basin (Salisbury 1968); (8) Anadarko basin (Rice et al 1988); (9) Tri-State, Central Missouri Barite, Northern Arkansas, and Viburnum Trend districts (Sharp 1978, Leach 1979, Leach & Rowan 1986); (10) Illinois basin (Bethke et al 1990); (11) Upper Mississippi Valley district (Bethke 1986a); (12) East Tennessee districts (Taylor et al 1983); (13) Appalachian Foreland basin (Woodward 1958, Oliver 1986). Base map after Wilkerson (1982).

Tennessee) with flow from the Appalachian orogen, and Duane & de Wit (1988) include in this pattern deposits found in the Caledonides of Scandinavia, Greenland, Ireland, Scotland and Canada. Similarly, a chain of deposits in the Northwest Territories and British Columbia (Gayna River, Robb Lake, Monarch, Kicking Horse) might be related to development of the Canadian Cordillera.

Thermal Anomalies

Studies of fluid inclusions from Pb-Zn deposits demonstrate that the ores of the North American continental interior were deposited at temperatures between about 75° and 150°C, well above those expected in the shallowly buried sediments that host the ores (Sverjensky 1986). The high temperatures most likely reflect heat carried into the districts by mineralizing brines—alternative heat sources such as igneous intrusions seem to be absent near the ore districts.

Mineralization temperatures in the midcontinent reflect a thermal anomaly that spanned the entire region, not just sediments near the ore districts. Coveney et al (1987) showed that trace mineralization across the midcontinent and far from known ore bodies occurred at temperatures within about 10°C of mean mineralization temperatures in the main districts. Fluid inclusion studies by Rowan & Leach (1990) revealed no thermal trends near the Viburnum Trend in southeast Missouri. Formation temperatures for both ore (Leach & Rowan 1986) and trace (Coveney et al 1987) sphalerite, however, follow a cooling trend on a regional scale northward from the Arkoma basin (Figure 2). The trend probably reflects the cooling of brines migrating through deep aquifers as their excess heat was conducted to the land surface.

Organic sediments in the Arkoma basin and mineralogical evidence from the midcontinent also record an anomalous thermal history. Thermal maturity increases regionally from west to east (Figure 3). Sediments in the southern basin, which were buried most deeply by younger sediments and thrust sheets, are less mature than sediments found at moderate depth farther north. Hathon & Houseknecht (1987, 1990) attribute this unusual pattern to heat redistributed eastward and into shallow strata by migrating fluids. Well-crystallized kaolinite and dickite from veins and partings also seem to reflect past regional heating by hydrothermal fluids (Keller 1988).

Potassium Metasomatism

Lower Paleozoic sediments of the North American interior have been altered by the formation of authigenic feldspar and clay minerals, greatly enriching the potassium content of these strata. Cambrian and Ordovician carbonate strata from the Valley and Ridge Province of the central and

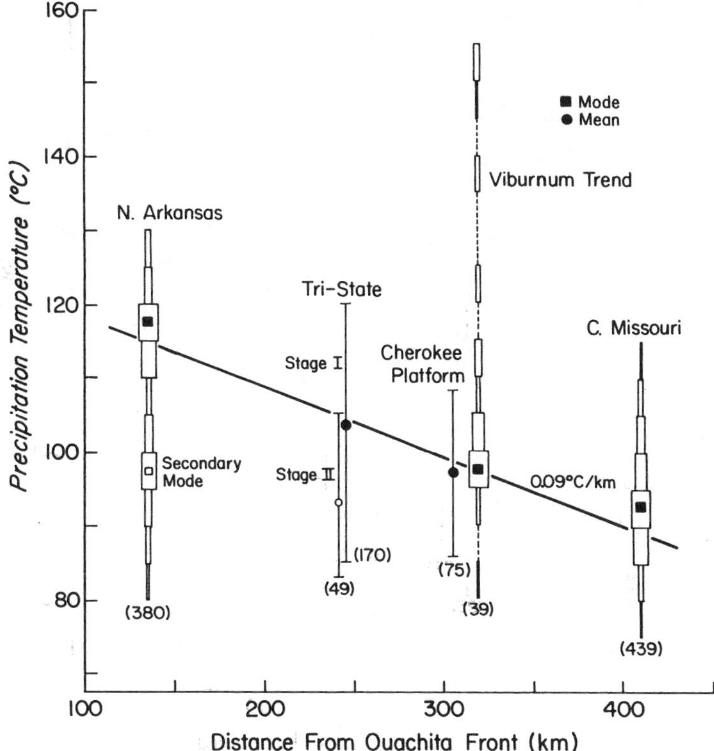

Figure 2 Homogenization temperatures of fluid inclusions in sphalerite ores from mid-continent Pb-Zn districts (Bethke et al 1988). Width of vertical bars is proportional to number of measurements in each temperature range; number of data points is shown in parentheses. Data compiled by E. L. Rowan; original sources are listed in Bethke et al (1988).

southern Appalachians contain as much as 25% authigenic potassium feldspar (Buyce & Friedman 1975). Elliott & Aronson (1987) found that shales throughout the Appalachian foreland have been enriched by potassium illite. In the Mississippi Valley region, Ordovician layers of volcanic ash have been converted to K-feldspar and K-bentonite (Hay et al 1988; Figure 4 herein), and feldspar overgrowths are common in Cambrian and Ordovician strata of Wisconsin and Illinois (Odom et al 1979). In the midcontinent, feldspar overgrowths are found in the Lamotte Sandstone and sandy zones of the Bonneterre Formation (Hearn et al 1986).

The origin of the potassium enrichment has long been enigmatic (e.g. Buyce & Friedman 1975, Mazzullo 1976, Ranganathan 1983), but recent studies suggest that the minerals precipitated from migrating brines long

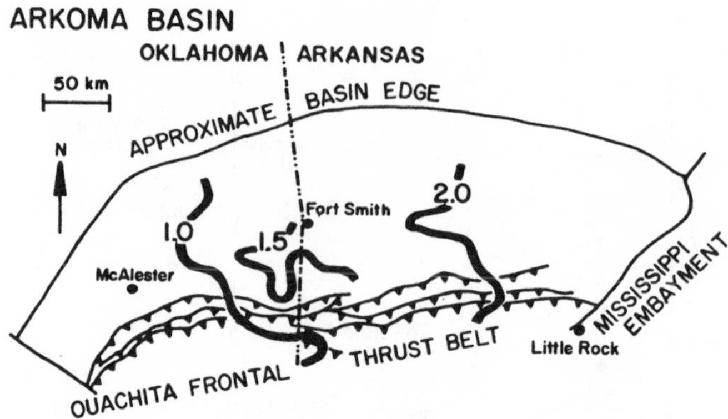

Figure 3 Vitrinite reflectance contours (percent reflectance in oil) for Hartshorne Coal in the Arkoma basin (Hathon & Houseknecht 1990). Thermal maturity increases from east to west. Along a south-north transect (for example, along the Oklahoma-Arkansas border), the highest levels of maturity are in the basin center at moderate burial depth, rather than in the most deeply buried sediments along the basin's southern margin. The unusual maturity pattern apparently reflects past heat redistribution by migrating brines.

after the sediments were buried. Hay et al (1988) found that the isotopic compositions of feldspars and clays from the Mississippi Valley are consistent with the minerals having formed from warm brines. Potassium to form feldspar overgrowths in the St. Peter Sandstone of Wisconsin and Illinois must have been supplied by migrating fluids (Odom et al 1979). In the Appalachian carbonates, mass balance also requires a source of potassium from outside the host beds (Hearn et al 1987). Hearn et al (1987), furthermore, found a close association between feldspar overgrowth and Pb-Zn ores in the Appalachian foreland: K-feldspar occurs in high concentrations near known ore districts, and the fluids that formed the K-feldspar are typical of mineralizing fluids from Mississippi Valley–type deposits.

Epigenetic Dolomite Cement

The ore districts and deep aquifer systems of the continental interior contain epigenetic dolomite cements that precipitated from migrating brines. The cements typically occur as sparry pore fillings. Dolomitization is characteristic of rocks that have been altered by sedimentary brines. For example, carbonate host rocks of Mississippi Valley–type deposits worldwide are ubiquitously converted to dolomite. In the vicinity of the Viburnum Trend, dolomite is widespread as a gangue mineral within

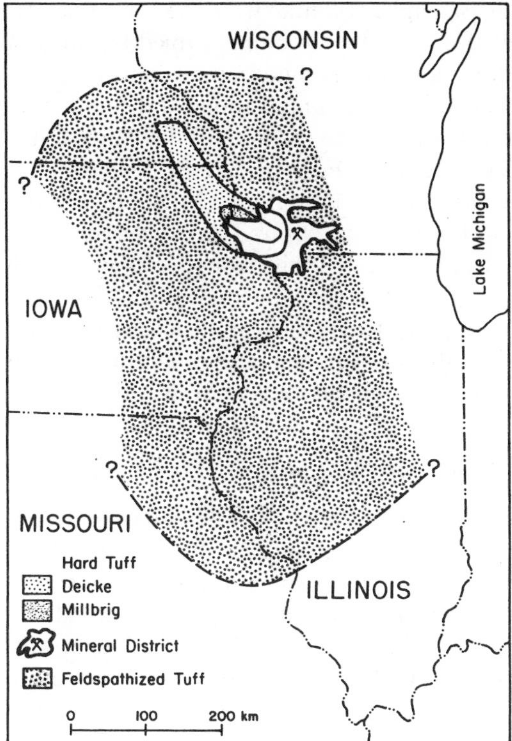

Figure 4 Map showing the distribution of authigenic potassium feldspar within Ordovician tuffs of the Mississippi Valley (Hay et al 1988). Also shown are areas where about half the beds in the Deicke and overlying Millbrig tuffs are composed of hard feldspar, and the Upper Mississippi Valley Zn-Pb district.

deposits and as a pore-filling cement in unmineralized rock outside the district.

The basal layer of the Bonneterre Formation immediately above its contact with the Lamotte Sandstone has been altered from limestone to dolomite, probably by reaction with fluids from the underlying Lamotte (Gregg 1985). The basal layer averages 6 m in thickness and extends over a broad area of southern Missouri. The dolomite cement must have been formed at the same time as the Viburnum Trend ores because individual bands of cement can be traced into the ore bodies, where they interfinger with layers of sulfide (Voss & Hagni 1985). The cement's texture and isotopic composition suggest that the dolomite precipitated from a warm sedimentary brine. Studies of fluid inclusions in pore-filling dolomite cements as well as gangue dolomite from the Viburnum Trend (Rowan & Leach 1990) confirm that the dolomite formed from warm, highly saline brines such as the mineralizing fluid in the ore district.

The chemical and isotopic compositions of the dolomite cements in the Bonneterre Formation seem to record past flow directions. Gregg & Shelton (1989) showed that the Fe, Mn, and Sr contents of dolomite at the Bonneterre-Lamotte contact in southeast Missouri vary in a regional trend northward from the Arkoma basin. The oxygen and strontium isotopic compositions of epigenetic dolomite from the Bonneterre also fall along a northward trend. Except in the immediate area of the Viburnum Trend, where other fluids seem to have left a diagenetic signature, these trends seem to record south-to-north migration of a brine that was progressively depleted in Fe and Mn and enriched in Sr as it reacted with limestone to form dolomite.

Banding in Ores and Cements

Sphalerite from Mississippi Valley–type deposits worldwide contains cyclic banding patterns that reflect changes in Fe content. The banding, which Roedder (1968) compared to varves in lake sediments, formed as the minerals precipitated under varying chemical or thermal conditions. Sphalerite banding has been studied carefully in the Upper Mississippi Valley district in Wisconsin and Illinois, where McLimans et al (1980) traced centimeter-scale banding among a series of deposits for 23 km across the district. This correlation indicates that the district formed from a widespread hydrothermal system rather than a series of local systems.

Similarly, cathodoluminescence microscopy reveals a banding in dolomite cements that arises from variations in the concentrations of minor elements, such as Fe and Mn. The bands likely reflect the processes that produced banding in sphalerite because, at least in the Viburnum Trend, the dolomite bands interfinger with layers of sulfide ore (Voss & Hagni 1985). The cements extend through aquifers far from economic mineralization, allowing correlations to be made among widely separated ore districts. Individual bands that extend among the ore bodies of the Viburnum Trend (Voss & Hagni 1985) correlate to cements in the Bonneterre Formation 80 km from the district (Gregg 1985). Rowan (1986) traced this banding for more than 350 km along the northern margin of the Arkoma basin from the Viburnum Trend southwest to the margins of the Tri-State and Northern Arkansas districts. Banded dolomite cements in Lower Mississippian strata of the Mississippi River arch in the vicinity of the Upper Mississippi Valley district can be traced through parts of Illinois, Missouri, and Iowa (Calder et al 1984); dolomite bands extend throughout the Mascot–Jefferson City district in Tennessee and for 50 km northeast to the Copper Ridge district (Ebers & Kopp 1979). Such long-range correlations across ore districts and aquifer systems provide strong evidence of the regional nature of brine migrations.

Banding in sphalerite has been attributed to pulses of warm brine flowing along migration pathways into ore districts (e.g. Cathles & Smith 1983). The Fe content of sphalerite from the Upper Mississippi Valley district, however, varies from <0.2 to 20 mol% FeS (McLimans et al 1980), which is too extreme a range to attribute to swings in temperature alone. Mason (1987) used time series analysis to show that banding in the district has several discrete periodicities ranging from thousands to more than 100,000 yr. These periods correlate to the Milankovitch cycles observed in paleoclimate and the Earth's orbit; cycles with similar periodicities are recorded in cyclotherms, lake varves, and the isotopic compositions of arctic ice cores and deep-sea sediments and fossils. Because the redox state of the fluid at the time of precipitation exerts a strong control on the Fe content of sphalerite, variations in sphalerite composition probably reflect the degree to which ore-forming fluids mixed with local oxygenated groundwater. Mason (1987) concluded that banding in the district resulted from climatic variations, such as cycles in the amount of meteoric precipitation at ore districts.

Theoretical Model of Brine Diagenesis

As a brine migrates from deep strata and cools, its composition and the mineralogy of deep strata are altered by reaction between rock and fluid (e.g. Sverjensky 1984). To predict the diagenetic reactions expected during migration, we take a brine at 200°C and cool it to 20°C while maintaining equilibrium with the minerals in a hypothetical deep aquifer. Reaction in the calculation is driven by variation in the thermodynamic stability of species and minerals with temperature.

The calculation accounts for all possible reactions among species in solution and a large number of minerals with tabulated thermodynamic properties (Delany & Lundeen 1989). On the basis of analyses from Illinois basin brines (Meents et al 1952), we assume an initial fluid composition of 2.3 molal Cl, 1.7 molal Na, and 0.15 molal Ca; other species are set by equilibrium with quartz, K-feldspar, muscovite (a proxy for illite), dolomite, calcite, and Ca-smectite (which reacts to form phengite during the calculation). Activity coefficients for aqueous species were calculated from extended Debye-Huckel theory (Helgeson 1969).

As the brine cools, quartz, calcite, muscovite, and smectite react to form dolomite and K-feldspar (Figure 5). The overall reaction

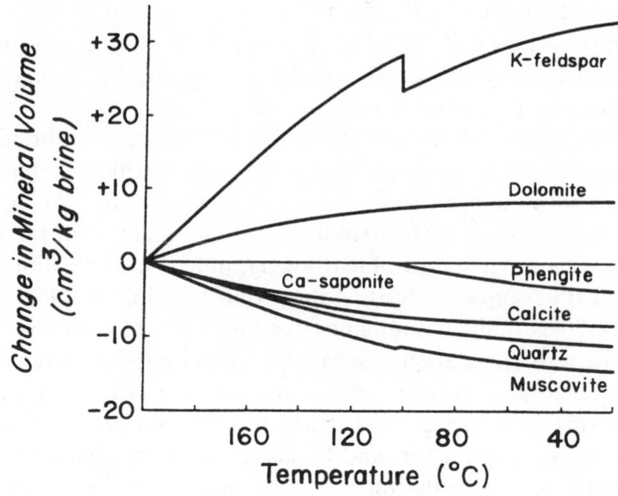

Figure 5 Theoretical prediction of diagenetic reactions accompanying brine migration, showing cumulative volumes of minerals dissolved and precipitated as a kilogram of brine cools in contact with minerals typical of a deep aquifer.

$$3.8\ SiO_2 + 2\ CaCO_3 + 0.4\ Ca_{0.165}Mg_3Al_{0.33}Si_{3.67}O_{10}(OH)_2$$
$$+ 0.8\ KAl_3Si_3O_{10}(OH)_2 + 1.8\ K^+ + 0.4\ HCO_3^- + 0.1\ H_4SiO_4$$
$$+ 0.3\ H^+ \rightarrow 2.6\ KAlSi_3O_8 + 1.2\ CaMg(CO_3)_2 + 0.9\ Ca^{2+} + 1.8\ H_2O \quad (1)$$

is driven in part by the retrograde solubility of calcite and in part by the thermodynamic drive to form potassium silicates from their Na and Ca counterparts at lower temperatures (e.g. Giggenbach 1984).

Cumulatively, the calculation predicts that about 33 cm^3 of feldspar and 8 cm^3 of dolomite will precipitate from each kilogram of brine. Hence, even if all of the feldspar could be deposited in one spot, highly altered sediments such as the Mississippi Valley tuffs must have reacted with more than 100 pore volumes of brine to account for their feldspar contents. Such high water-rock ratios are consistent with evidence already cited that large quantities of brine were involved in the migration events. It is also clear from the calculation that although considerable dolomite cement could be produced by a migrating brine, such epigenetic dolomite is unlikely to account for more than a small fraction of the massive amounts of dolomite found in interior basins.

Petroleum Migration

Exploration geologists working in many basins commonly assume that oil can be found within 10 km of its source rock. This rule of thumb, however, holds poorly in the North American interior. Here oil can be correlated geochemically to source rocks many tens or hundreds of kilometers from present-day reservoirs, providing strong evidence of long-range migration.

Migration directions and distances can be inferred in many cases by comparing the distribution of reservoirs with the extent of oil source beds. In more recent migration studies, oganic compounds in oils are correlated with those in residual bitumen from source rocks (e.g. Tissot 1984). For example, petroleum is produced in the Illinois basin today over an area that extends more than 100 km beyond likely source beds (Figure 6). Bethke et al (1990) showed that oils from outlying reservoirs in central Illinois are nearly identical in their gas chromatograms and carbon isotopic compositions to bitumen from the New Albany Shale in southern Illinois.

Such long-range migration is not unusual in interior basins. Demaison (1977) suggested that the oils that formed the immense Alberta Tar Sands deposits migrated eastward, away from the Canadian Cordillera, for more than 100 km. Dow (1974) and Clayton & Swetland (1980) estimated that 150 km of eastward migration occurred in the Williston and Denver basins. Salisbury (1968) attributed migration of natural gas in the Permian basin to overthrusting along the Marathon thrust belt, and Woodward (1958)

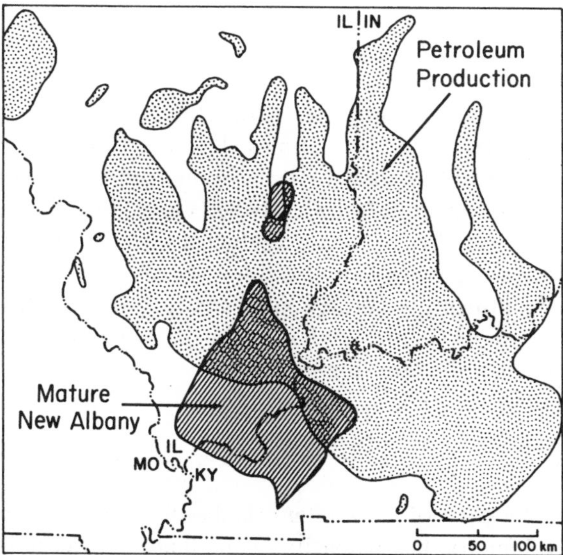

Figure 6 Distribution of petroleum production in the Illinois basin (Howard 1990) and location of the area where the Devonian-Mississippian New Albany Shale Group is inferred to be thermally mature and oil prone (Barrows & Cluff 1984).

argued that oil has migrated westward through the Appalachian basin. Oil from the Phosphoria Formation in western Wyoming seems to have migrated 500 km to present-day reservoirs in the Big Horn basin (Sheldon 1967). The Phosphoria oil probably accumulated in early reservoirs, only to remigrate eastward in response to the Laramide orogeny (Stone 1967). Interestingly, Haynes & Kesler (1989) identified the remnants of pre-Alleghenian petroleum reservoirs near the Eastern Tennessee lead-zinc district. Oil that accumulated along the Pre-Knox unconformity was later displaced, perhaps during the Alleghenian orogeny.

Together, the inferred pathways of migration form a continent-wide pattern of flow away from tectonic belts and across the continental interior (Oliver 1986; Figure 1 herein). The relationship between oil and brine migration, however, is not entirely straightforward. Oil migrates in response not only to the hydrodynamic forces that drive groundwater flow, but also to its buoyancy and the capillary effects of moving through microscopic pores (Hubbert 1953). Interior basins, unlike basins along continental margins, contain carrier beds and overlying capillary seals that are laterally extensive and nearly horizontal. Because the component of buoyant force acting along the lateral migration pathways is small, oil migration in interior basins should indeed reflect past groundwater migration (Bethke et al 1990).

RATES OF MIGRATION

The rates at which brines migrated across the craton can be constrained using the principles of energy and mass balance. Minimum rates can be estimated from evidence that the brines remained warm as they flowed through shallow sediments. If the brines had moved too slowly, they would have had time to cool as heat was conducted through overlying strata to the land surface.

In the midcontinent, the brine cooled at a rate of about 0.1°C km^{-1} as it flowed northward (Figure 2). Balancing the brine's enthalpy loss with the rate of heat conduction (Figure 7) gives

$$bq\rho C_P \frac{\partial T}{\partial x} = K_T \frac{\partial T'}{\partial z}. \tag{2}$$

Here b is the thickness of the aquifer, q is specific discharge [$cm^3\,(cm^2\,s)^{-1}$], ρC_P is the volumetric heat capacity of the fluid, and K_T is thermal conductivity. The derivatives $\partial T/\partial x$ and $\partial T'/\partial z$ give the cooling rate along the flow path and the geothermal gradient in excess of the conductive gradient from basement, respectively. Rearranging gives the rate of migration:

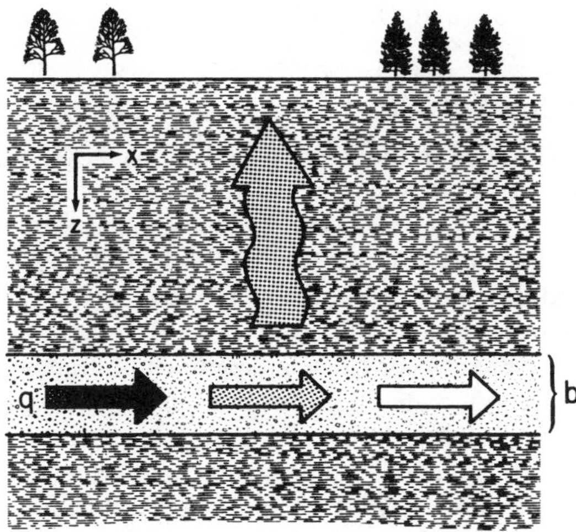

Figure 7 Simple model for estimating flow rate during brine migration. Enthalpy loss of the migrating fluid balances the heat conducted to the land surface in excess of background heat flow.

$$q = \frac{K_T}{b\rho C_P} \cdot \frac{\partial T'/\partial z}{\partial T/\partial x}. \tag{3}$$

If we take the Lamotte Sandstone to be 200 m thick, the thermal conductivity to be 2×10^{-3} cal (cm s °C)$^{-1}$, and a fluid at 100°C migrating at the maximum likely depth of 1.5 km, the discharge predicted by Equation (2) is about 10 m yr^{-1}. This result agrees with calculations that balance the amount of brine needed to carry metals to the Upper Mississippi Valley (Barnes 1983) and Pine Point (Garven 1985) ore districts, as well as with the results of numerical simulations of heat transfer in basins open to groundwater flow (Garven & Freeze 1984b, Bethke 1986a).

TIMING OF MIGRATION

As warm brines migrate into shallower strata, they react with the sediments and precipitate new minerals. A history of brine migration can be inferred when the absolute ages of these diagenetic minerals can be determined. Two tools have been used to determine the ages of diagenetic minerals in deep aquifers of the continental interior: paleomagnetic analysis and radiometric dating. The age determinations argue strongly that regional migration events occurred during the Pennsylvanian-Permian time period of the Alleghenian and Ouachita orogenies, and they hint at earlier migrations.

Paleomagnetic Studies

Paleomagnetic studies of the Paleozoic strata that cover the North American craton were initially conducted to determine the apparent polar-wander path for the continent. The studies were based on the assumption that a sample's magnetization was established during or shortly after sediment deposition, so that a sediment would record the Earth's magnetic pole at the time it was deposited. Almost from the onset it was recognized that the pole positions recorded in midcontinental strata were problematic because they did not match pole positions determined for like-age rocks from other regions of North America. Few if any magnetic reversals could be found, indicating that the magnetization of the sedimentary section was acquired over a relatively short interval of geologic time. A debate in the literature began in 1958 about whether the magnetization of the lower Paleozoic of the midcontinent and the Appalachian fold belt had been reset after the strata were deposited (McElhinny & Opdyke 1973, Kent & May 1987). By the late 1970s it was clear that the pole positions reflected a time significantly later than the depositional age of the strata (Al-Khadfaji & Vincenz 1971, Van der Voo & French 1977, Wu & Beales 1981, Scotese et al 1982, Wisniowiecki et al 1983, McCabe et al 1984, Kent & Opdyke 1985, Bachtadse et al 1987).

The source of the magnetic signal in the midcontinental strata remains a subject of debate. Magnetic minerals can be identified by analyzing how a rock acquires remanent magnetization and by using a scanning electron microscope (SEM) to examine insoluble residues extracted from strata. Magnetization studies (Kean 1981, McCabe et al 1983, Marshak et al 1989) for most reset limestones suggest that magnetite carries the magnetic signal; the signal in the remaining rocks is characteristic of hematite. SEM study (Lu et al 1990) reveals that rocks displaying a magnetite signal contain $\sim$25 ppm of Fe-rich spherules about 10 to 20 μm in diameter, some of which can be seen to be composed of 1- to 3-μm-wide octahedra of an iron mineral (Figure 8). Energy-dispersive analysis indicates that these grains contain neither S nor Ti and thus are likely magnetite; X-ray diffraction confirms the presence of magnetite in the insoluble residue. Hematite, when observed, occurs in a flaky habit.

Increasingly it is clear that the magnetic minerals formed diagenetically by reaction of the rocks with migrating brines, as proposed by Van der Voo & French (1977; see also Wisniowiecki et al 1983, McCabe et al 1983, Elmore et al 1985). The lack of Ti in the magnetite grains indicates that they formed at low temperature and are neither igneous nor metamorphic

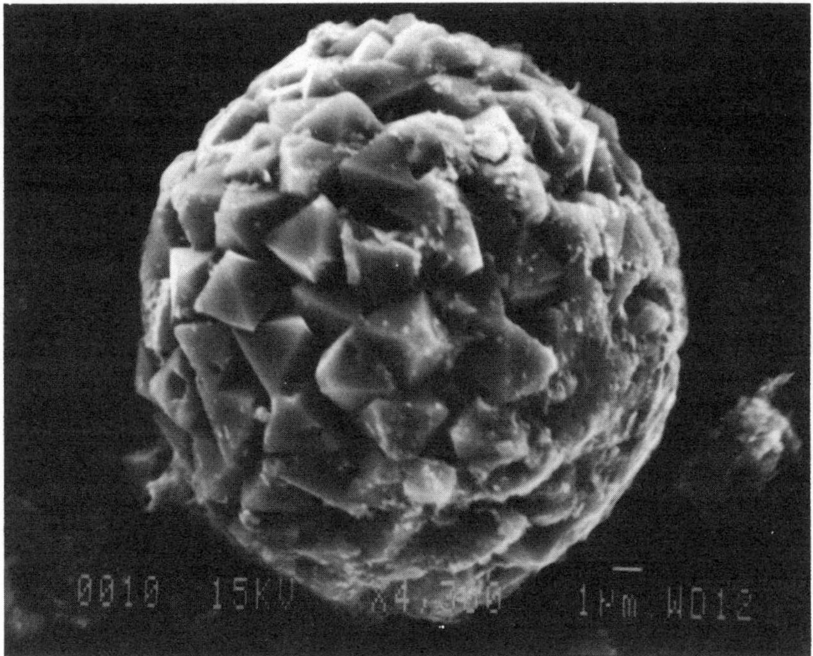

Figure 8 Scanning-electron photomicrograph of a magnetite framboid composed of octahedral grains (Lu et al 1990). The magnetite was extracted from Middle Ordovician limestone collected in the northwestern corner of Iowa. Photo taken by G. Lu, University of Illinois.

in origin (Jackson et al 1988, Lu et al 1990). Marshak et al (1989) report that magnetic intensity, which varies regionally, is strongest near the domes and arches of the midcontinent. These structures host Pb-Zn mineralization and extensive dolomitization and hence are likely to have experienced diagenetic reaction among Fe-bearing minerals. On the other hand, there is no reason to expect detrital magnetite to have been concentrated in sediments deposited in these areas. Finally, the magnetic minerals commonly occur with euhedral crystal faces, which would not have been preserved on detrital grains.

Although the reset paleopoles created confusion in reconstructing the tectonic history of North America, they provide a key to dating regional diagenesis and, by implication, brine migrations. A survey of pole positions measured for Paleozoic strata of all ages (Figure 9) indicates that the measured pole positions at localities throughout the US midcontinent coincide with the Kiamian reference pole for the Pennsylvanian-Permian of North America. In rocks where the Kiamian signal is not detected,

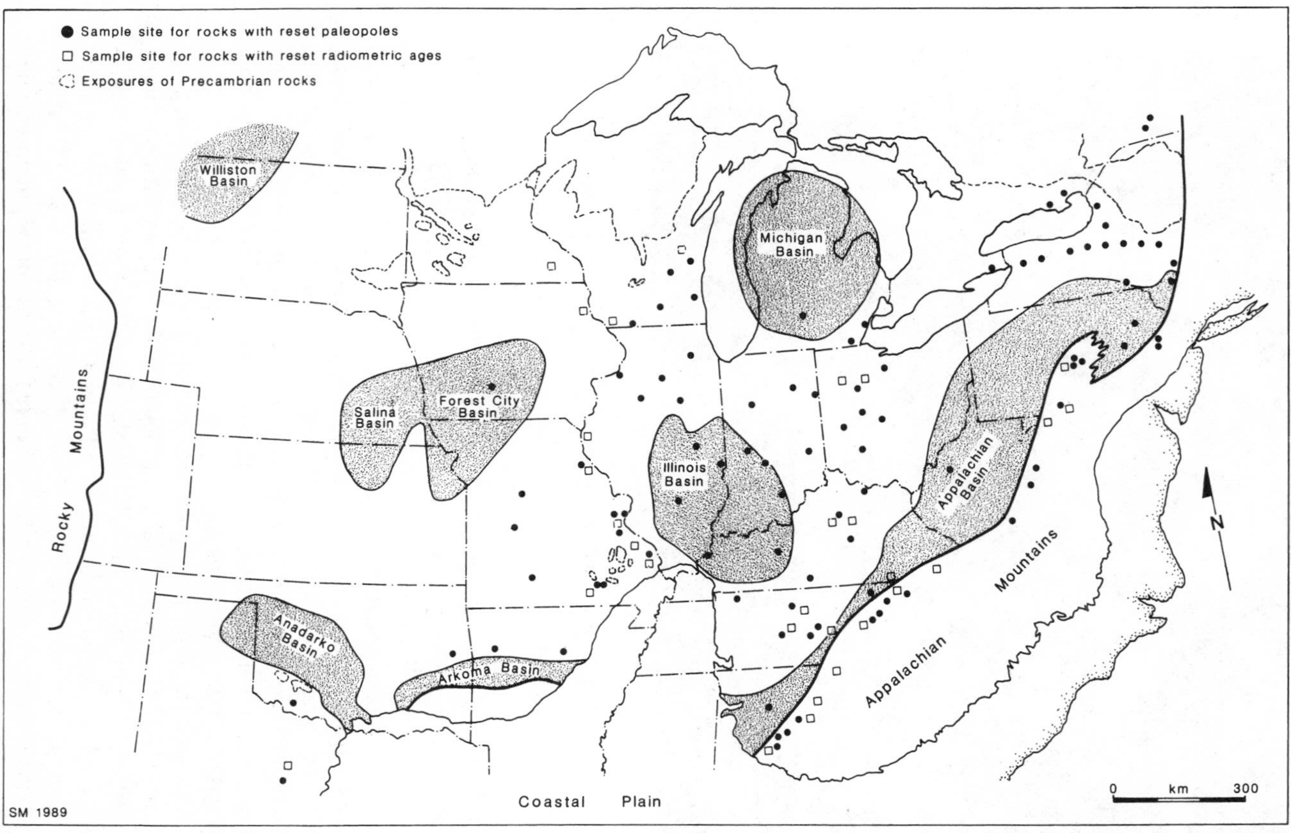

Figure 9 Generalized map showing the principal intracratonic basins of the midcontinent and the locations at which reset paleomagnetic poles and reset radiometric ages have been documented (base map: King 1969).

the signal is typically unstable or very weak. Thus, the paleomagnetic studies provide strong evidence that brine migration was coeval with the Alleghenian/Ouachita orogeny. This coincident timing is confirmed by study of magnetic poles measured from the opposite limbs of folds in the Appalachian foreland (Van der Voo 1979, Scotese et al 1982). Comparing pole positions shows that remagnetization occurred during the course of the orogeny when the folds were only partially developed. In addition, high-resolution analysis along the Appalachian fold belt shows that remagnetization began in the south and proceeded northward over a period of about 60 m.y. (Miller & Kent 1988), apparently reflecting the migration of the orogeny itself. Taken together, the evidence is overwhelming that brine migration during the Alleghenian/Ouachita orogeny remagnetized rocks across the North American midcontinent.

Radiometric Ages of Diagenetic Minerals

The ages of K-bearing minerals that form from migrating brines can sometimes be determined radiometrically, but separating the authigenic minerals from detrital grains can be difficult: Authigenic K-feldspar commonly forms as overgrowths on grains of the same mineral, and authigenic and detrital clays may be intermixed. For these reasons, there are fewer radiometric than paleomagnetic data, and the data are harder to interpret. The $^{40}Ar/^{39}Ar$ age spectra of K-feldspar overgrowths argue most convincingly for diagenetic alteration during Pennsylvanian-Permian time. Hearn & Sutter (1985) and Hearn et al (1986, 1987) used the age-spectrum technique to separately date overgrowths and cores from carbonate aquifers in the Appalachian foreland and midcontinent. Although some of the determinations have large ranges of uncertainty, their results suggest that the overgrowths formed between 325 and 275 Ma, during the Pennsylvanian and Early Permian. Elliott & Aronson (1987) found similar ages [303–272 Ma ($\pm$15 m.y.)] in K-Ar studies of illite from bentonite beds of the Appalachian basin, and Hay et al (1988) determined a Permian age (265 Ma) for illite from bentonites on the Mississippi River arch in Missouri. Desborough et al (1985) found Permian (278 ± 25 Ma) fission-track ages in detrital zircon from the Pennsylvanian of the Arkansas River valley. These analyses strongly support the relationship of Pennsylvanian-Permian tectonism to brine migration.

In contrast to the paleomagnetic ages, some radiometric studies also yield ages that are significantly older than Pennsylvanian. There is, in fact, a clustering of ages in the Devonian, including Rb-Sr determinations from glauconite from the midcontinent and the Appalachian foreland (Grant et al 1984, Stein & Kish 1985), and K-Ar ages of feldspar and illite from the Upper Mississippi Valley (Hay et al 1988). The Devonian ages either may represent a mixture of the ages of detrital and authigenic minerals or may record earlier brine migrations not yet revealed by paleomagnetic study.

TECTONIC ORIGIN OF BRINE MIGRATIONS

The mechanistic link between tectonic deformation and the brine migrations remains poorly understood, largely because of the difficulty in quantifying groundwater flow in the deformed rocks of orogenic belts. Discovery of the sedimentary origin of the fluids that mineralized Pb-Zn deposits focused attention on the processes by which brines could migrate for hundreds of kilometers from deep strata to the shallow sediments on basin margins. Early work related migration to the pore waters expelled as basin sediments compact during burial. More recent evidence that the fluids migrated from active tectonic belts both has expanded concepts of ore genesis to include the basin margins opposite those that host ore districts and has linked basin hydrology to plate tectonic theory.

In this section we consider the paleohydrology of the Arkoma basin and Ouachita fold belt, and try to constrain the roles of tectonic compression and thrusting, sediment compaction, and topographic uplift in driving Pennsylvanian migration across the midcontinent.

Thrusting and Tectonic Compression

The apparent association of brine migrations with periods of tectonic deformation led Oliver (1986) to suggest that the compression and deformation of tectonic orogenies drive migration directly. He envisioned that the migrating brines were fluids expelled from orogens, perhaps by thrust sheets overriding the sediments of foreland basins and acting as "squeegees."

The role that tectonic compression can play in driving brine migrations is limited by the slow rates at which tectonic belts deform. This point can be illustrated by considering limiting cases of the flow rates that would result as crustal plates converge during an orogeny (Figure 10). As the plates converge into a tectonic belt, their volume can be accommodated either by thickening the crust within the orogen or where thrust faults override the craton, or to a lesser extent by compressing fluid from the pore space in rocks and sediments (Figure 10*a,b*). In a hypothetical limiting case (Figure 10*c*), all of the plate convergence is accommodated by compressing pore volume, and all of the expelled fluid migrates laterally away from the orogen. In such a case, which would maximize the flow rates predicted, fluids would migrate away from the orogen at a specific discharge q [$\mathrm{cm}^3\ (\mathrm{cm}^2\ \mathrm{yr})^{-1}$ or $\mathrm{cm\ yr}^{-1}$] that is numerically equal to the velocity v_{plate} of plate convergence or v_{thrust} of thrusting.

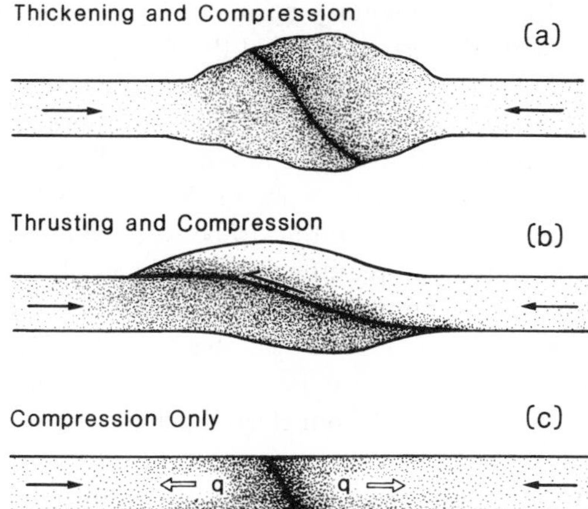

Figure 10 Simple models of (*a*) continental collision and (*b*) thrust faulting, and (*c*) limiting case of most rapid groundwater flow arising from lateral tectonic compression. In (*c*) the sediment volume added to the orogen is entirely accommodated by compression of pore volume, and all fluids migrate laterally.

Plates move at rates of 1–10 cm yr^{-1} or less, and thrust sheets are emplaced at comparable rates (Cello & Nur 1988). Thus, fluids driven through the sedimentary cover would migrate at average discharges of 10 cm yr^{-1} or less, since the assumptions of the limiting case are unlikely to be realized. Flow rates might be greater than the average value if flow was concentrated within aquifers, but the thick aquifer systems of the midcontinent comprise a large fraction of the sedimentary cover. Thus, compression and thrusting acting alone would be unlikely to drive migrations at the discharge of about 10 m yr^{-1} inferred from geologic evidence and transport theory.

Detailed numerical simulations confirm these simple calculations. Ge & Garven (1989) considered the effects of compression and vertical loading by thrust sheets as thick as 10 km in driving flow through a basal aquifer system of a foreland basin. Their results predict flow rates of about 1 cm yr^{-1} or less, in line with the estimate derived above. Ge & Garven conclude that compression and thrusting can drive groundwater flow on regional scales, but that such flow is subordinate to that resulting from the topographic relief of the orogen.

Sediment Compaction

To account for the origin of Mississippi Valley–type deposits, early workers (Noble 1963, Jackson & Beales 1967) called on the compaction of sediments deposited in sedimentary basins to drive brines toward basin margins. Recent calculations indicate that the slow process of compaction is unlikely to drive migration more than a few centimeters per year (Cathles & Smith 1983, Bethke 1985). Sharp (1978) and Cathles & Smith (1983) suggested that overpressured zones in compacting sediments, such as those that form in the Gulf of Mexico basin, might discharge brines in more rapid bursts. To evaluate the relationship between sediment compaction and brine migration, we use quantitative techniques (Bethke et al 1988, Bethke 1989) to reconstruct flow within the Arkoma basin as it compacted.

The calculations were made on the basis of a simplified hydrostratigraphy (Table 1). The basal Paleozoic formations, including the Lamotte Sandstone aquifer and Bonneterre Formation, are represented in the lower two units. The units are assumed to extend beneath the entire basin, although the deep stratigraphy is poorly known. The basal units are overlain by the Pennsylvanian clastic sequence, which is dominantly composed of shale. Porosity and permeability correlations (Table 2) of Bethke et al (1990) were used for all sediments except for the Pennsylvanian shales, which were assigned smaller permeabilities representative of overpressured shales in the Gulf of Mexico basin (Harrison et al 1990). Faults in the deep basin are not represented because their hydrologic properties are unknown. The time interval during which Pennsylvanian sediments accumulated is also poorly constrained because sediments of the lower Atokan contain few fossils. Estimates of the duration of sedimentation range from 5 to about 20 m.y.; calculations here assume a 5-m.y. interval

Table 1 Arkoma basin stratigraphy assumed in paleohydrologic models

Hydrostratigraphic unit	Interval of deposition (m.y.)	Greatest thickness (km)	Lithology assumed
Upper Desmoinesian (Pennsylvanian)	1	1.0	Shale, with discontinuous sandstone and carbonate lenses
Lower Desmoinesian (Pennsylvanian)	1.5	2.0	Same as above
Hartshorne Formation (Pennsylvanian)	0.5	0.2	Sandstone, siltstone, carbonate at base of Desmoinesian Group
Atokan (Pennsylvanian)	2	2.6	Shales of Atoka Group, with hydraulically discontinuous sandstone lenses
Ordovician–Mississippian	280	0.8	Carbonates, with sandstone and shale interbeds
Cambrian	—[a]	0.2	Lamotte Sandstone

[a] Present at onset of calculation.

Table 2 Correlations used to calculate porosity and permeability

	Porosity[a]			Permeability[b]		
	ϕ_0	b (km^{-1})	ϕ_1	A	B	k_x/k_z
Sandstone	0.40	0.50	0.05	15	−3	2.5
Carbonate	0.38	0.57	0.05	6	−4	2.5
Siltstone	0.47	0.65	0.05	5	−4	2.5
Shale	0.55	0.85	0.05	8	−7	10
Shale (during compaction)	0.55	0.85	0.05	8	−8	10

[a] $\phi = \phi_0 \exp(-bz_e) + \phi_1$, where z_e is effective depth.
[b] $\log k_x$ (μm^2) $= A\phi + B$; k_x is everywhere $\leq 1\ \mu m^2$.

to provide for the most rapid sedimentation and hence to maximize flow rates and overpressures.

Overpressures are likely to develop when about 1 km or more of shaly sediments accumulate per million years (Bethke 1986b). Such rapid burial rates were probably attained in the southern basin during the Pennsylvanian. The Arkoma basin differs in this way from the other interior basins, such as the Illinois basin, which were infilled too slowly and contain sediments too permeable to develop overpressures by sediment compaction. Figure 11 shows the numerical reconstruction of the flow resulting from sediment compaction at the end of Pennsylvanian deposition. The reconstructions indicate that fluid migrated slowly in response to compaction, and that moderate overpressures developed along the southern margin of the basin during its infilling.

Pressures predicted in the Pennsylvanian sediments are about 10–20 MPa (100–200 atm) greater than hydrostatic and occur along a relatively narrow band north of the Choctaw fault. The mass of fluid in the overpressured sediments (about 10^{18}–10^{19} g) is small when compared with the approximately 10^{20}–10^{21} g of sediments north of the basin that were invaded by migrating brines. Thus, it seems unlikely that fluids discharged from overpressured zones could carry sufficient heat to basin margins to heat the sedimentary cover even if no heat was lost by conduction. In addition, as noted by Leach & Rowan (1986), discharges from overpressured sediments could not explain evidence that the midcontinental brine migration occurred as a single flow system unless discharge occurred simultaneously across the basin.

Topography of the Orogen

Although topographic relief has long been known to drive groundwater flow through interior basins over regional scales (e.g. Darton 1909),

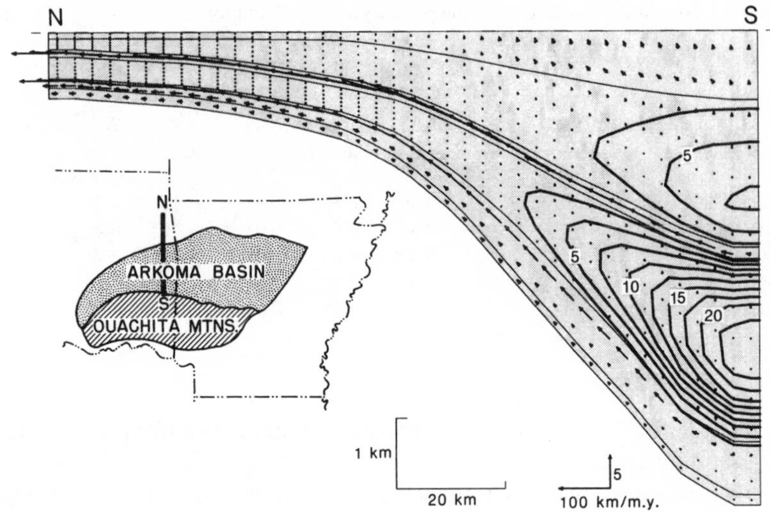

Figure 11 Paleohydrologic model of compaction-driven groundwater flow during the subsidence and infilling of the Arkoma basin, displayed along a north-south cross section extending from the Choctaw fault northward along the Arkansas-Oklahoma border. Arrows show fluid migration velocities. Bold lines are equipotentials (MPa) that show magnitude of excess fluid pressures developed by sediment compaction. Fine lines separate hydrostratigraphic units (Table 1).

recently there has been a resurgence of interest in the role of topography in driving brine migrations. For example, Garven & Freeze (1984a,b) and Garven (1985, 1989) related uplift of the Canadian Cordillera to brine migration in western Canada, and Bethke (1986a) and Bethke et al (1990) attributed brine and oil migration in the Illinois basin to past uplift along the southern basin margin.

Groundwater that recharges at high elevation migrates through the subsurface in response to its high potential energy toward areas where the water table is lower. Figure 12 shows a flow regime and temperature distribution calculated at steady state for the Arkoma basin after the Ouachita Mountains were uplifted in the Pennsylvanian. The calculation was made by assuming that basin strata extended into the fold belt (e.g. Houseknecht 1986), and that the mountains rose 4 km above the continental interior. The strata within the orogen were assumed to have been deformed, so that neither aquifers nor aquitards remained hydraulically continuous.

Reconstructing the hydrology of the ancestral Ouachitas and Arkoma basin is only possible in the most general sense because critical variables

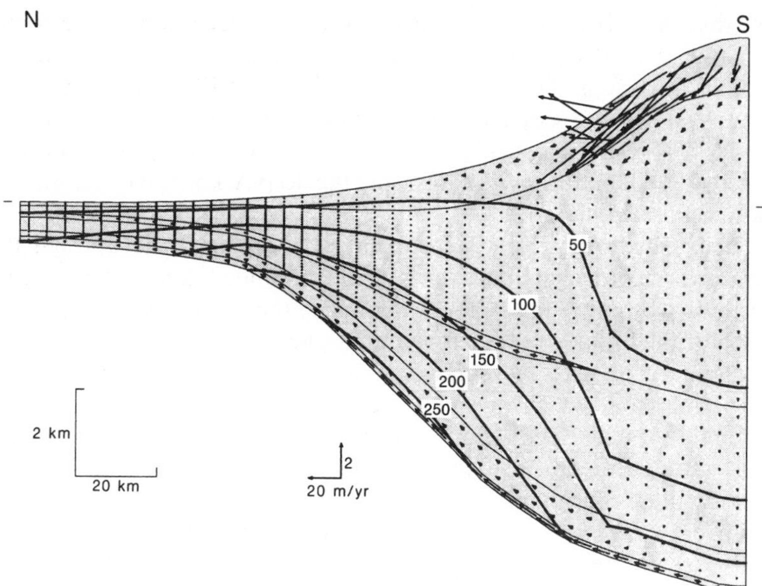

Figure 12 Model of groundwater flow and heat transport resulting from uplift of the Ouachita Mountains in the Pennsylvanian. Cross section is that of Figure 11, extended 50 km southward into the uplift. Bold lines are isotherms (°C).

can only be estimated. Unknowns include the amount and rate of uplift, the permeability of the orogen and deep sediments, the extent and hydraulic connectivity of the lower Paleozoic aquifers, the past heat flow, the hydraulic properties of faults in the uplift and basin, and the timing of fault development. Hence, the calculation is useful in depicting likely flow patterns and thermal structures but cannot accurately determine past flow rates and absolute temperatures.

The calculation shows the development of both shallow and deep groundwater flow that recharges on the Ouachita uplift. The thermal structure in the calculation reflects how heat conducted from the lower crust is distributed by conduction and groundwater advection. A basal heat flow of 100 mW m^{-2} is assumed, and varying this value affects the absolute temperatures predicted. The deep flow occurs rapidly enough to effectively transport heat, forming a regional hydrothermal system that discharges north of the Arkoma basin.

Heat is entrained by moving groundwaters in the calculation and is redistributed toward the discharge area to the north. The Ouachita core is cooled by descending groundwater, as is common in mountain belts, but this effect is probably exaggerated in the calculations by restricting recharge to a narrow band of the uplift and by ignoring the heat added by igneous intrusions. The thermal pattern along the Hartshorne Sandstone is unusual because sediments are warmest in the central rather than deep basin. This result may explain the unusual pattern in the Hartshorne in which thermal maturity increases from the deep Arkoma basin toward the basin center (Figure 3).

Such a hydrothermal system might also explain the past high temperatures recorded in shallow strata of the midcontinent. Heat carried northward by migrating groundwater warms the sediments on the platform north of the basin, so that temperatures greater than 100°C occur within a kilometer of the land surface. This result is in agreement with the temperatures at which ore and trace mineralization and hydrothermal dolomite formed. Such a flow system could also explain the high water-rock ratios inferred from diagenetic evidence. In addition, the calculations agree with evidence that the midcontinental brine migration occurred as a single hydrothermal system, and they offer an explanation of the high flow rates inferred for the migration.

CONCLUDING REMARKS

Plate tectonic theory successfully explains the deformation of the Earth's crust along plate margins. Recently, tectonic theory has recognized that deformation along continental margins can affect the interiors of continents in various ways, including causing basins to subside and arches to rise. In this paper we have presented evidence from a variety of disciplines to demonstrate that tectonism along the continental margins of North America also caused deep, saline groundwaters to migrate over long distances through the sedimentary cover of the continent's interior.

Geochronologic data best record migration accompanying the Alleghenian-Ouachita orogeny, the final stage of closing the proto-Atlantic Ocean in Pennsylvanian-Permian time. This orogeny was the most recent of three major tectonic events that shaped the eastern continental margin. Radiometric data hint at earlier migrations, but there is little definitive evidence of the hydrologic response to the Taconic or Acadian orogenies. A number of mechanisms may have contributed to flow away from the orogen, but quantitative modeling indicates that the topographic relief of the tectonic belt provided the dominant driving force for migration. This result raises the possibility that uplift of arches and plateaus in continental interiors or even continental rifting may also drive brine migration.

Like plate tectonic theory, the concept of brine migration over long distances provides a framework for understanding a number of significant but seemingly unrelated phenomena: the origin of metallic ores; high temperatures recorded in shallow sediments; diagenesis of interior sediments, including enrichment in potassic feldspar and clays, and sparry dolomite cements that correlate over hundreds of kilometers; paleomagnetic signatures and radiometric ages unrelated to the ages of host strata; and oil reservoirs found far from source beds. The scale of migration—over the lengths of orogenic belts and across plate interiors for many hundreds of kilometers—is that of plate tectonics. We argue that brine migration is the response of deep fluids to tectonic deformation, and that it should become an integral part of plate tectonic theory.

ACKNOWLEDGMENTS

This paper is in large part the result of discussions with many colleagues. We especially thank Mary Barrows, Dennis Kent, Richard Hay, David Houseknecht, Dennis Kolata, David Leach, Jackie Reed, Lanier Rowan, and Howard Schwalb. Matt Mankowski collected stratigraphic data, and Thomas Corbet helped make the hydrologic calculations. Gang Lu provided unpublished SEM data and the photomicrograph, and Dennis Kent shared unpublished paleomagnetic data. The comments of Lanier Rowan, Philip Bethke, Thomas Corbet, Ming-Kuo Lee, Charles Norris, and Patrick O'Boyle improved the manuscript; Joan Apperson and Jesse Knox drafted the figures.

Our research was supported by the Petroleum Research Fund of the American Chemical Society (grant 18680-G2) and the National Science Foundation (grants EAR 85-52649 and EAR 86-01178). We thank ARCO Oil and Gas Company, Amoco Production Company, British Petroleum, Exxon Production Research Company, Mobil Oil Company, Texaco USA, Shell Development Company, and Unocal for sponsoring aspects of this project.

Literature Cited

Al-Khadjafi, S. A., Vincenz, S. A. 1971. Magnetization of the Cambrian Lamotte Formation in Missouri. *Geophys. J. R. Astron. Soc.* 24: 175–202

Bachtadse, V., Van der Voo, R., Haynes, F. M., Kesler, S. E. 1987. Late Paleozoic magnetization of mineralized and unmineralized Ordovician carbonates from east Tennessee: evidence for a post-ore chemical event. *J. Geophys. Res.* 92: 14,165–76

Barnes, H. L. 1983. Ore-depositing reactions in Mississippi Valley–type deposits. *Proc. Int. Conf. Mississippi Valley–Type Lead-Zinc Depos.*, ed. G. Kisvarsanyi, S. K. Grant, W. P. Pratt, J. W. Koenig, pp. 75–85. Rolla: Univ. Mo. 603 pp.

Barrows, M. H., Cluff, R. M. 1984. New Albany Shale Group (Devonian-Mississippian) source rocks and hydrocarbon generation in the Illinois basin. In *Petroleum Geochemistry and Basin Evaluation*, ed. G. Demaison, R. J. Murris, pp. 111–38. Tulsa: Am. Assoc. Pet. Geol. 426 pp.

Bethke, C. M. 1985. A numerical model of compaction-driven groundwater flow and heat transfer and its application to the paleohydrology of intracratonic sedimentary basins. *J. Geophys. Res.* 80: 6817–28

Bethke, C. M. 1986a. Hydrologic constraints on the genesis of the Upper Mississippi Valley mineral district from Illinois basin brines. *Econ. Geol.* 81: 233–49

Bethke, C. M. 1986b. Inverse hydrologic analysis of the distribution and origin of Gulf Coast–type geopressured zones. *J. Geophys. Res.* 91: 6535–45

Bethke, C. M. 1989. Modeling subsurface flow in sedimentary basins. *Geol. Rundsch.* 78: 129–54

Bethke, C. M., Harrison, W. J., Upson, C., Altaner, S. P. 1988. Supercomputer analysis of sedimentary basins. *Science* 239: 261–67

Bethke, C. M., Reed, J. D., Barrows, M. H., Oltz, D. F. 1990. Long-range petroleum migration in the Illinois basin. *Am. Assoc. Pet. Geol. Bull.* In press

Buyce, M. R., Friedman, G. M. 1975. Significance of authigenic K-feldspar in Cambrian-Ordovician carbonate rocks of the proto-Atlantic shelf in North America. *J. Sediment. Petrol.* 45: 808–21

Calder, H. S., Kaufman, J., Hanson, G. N., Meyers, W. J. 1984. Two-stage cementation of the Burlington-Keokuk Limestone in Illinois, Missouri, and Iowa. *Geol. Soc. Am. Abstr. with Programs* 16: 631 (Abstr.)

Cathles, L. M., Smith, A. T. 1983. Thermal constraints on the formation of Mississippi Valley–type lead-zinc deposits and their implications for episodic basin dewatering and deposit genesis. *Econ. Geol.* 78: 983–1002

Cello, G., Nur, A. 1988. Emplacement of foreland thrust systems. *Tectonics* 7: 261–71

Clayton, J. L., Swetland, P. J. 1980. Petroleum generation and migration in Denver basin. *Am. Assoc. Pet. Geol. Bull.* 64: 1613–33

Coveney, R. M. Jr., Goebel, E. D., Ragan, V. M. 1987. Pressures and temperatures from aqueous fluid inclusions in sphalerite from midcontinent country rocks. *Econ. Geol.* 82: 740–51

Darton, N. H. 1909. Geology and underground waters of South Dakota. *US Geol. Surv. Water-Supply Pap. 227.* 156 pp.

Delany, J. M., Lundeen, S. R. 1989. The LLNL thermochemical database. *Univ. Calif. Rep. 21658*, Lawrence Livermore Natl. Lab., Livermore, Calif. 150 pp.

Demaison, G. J. 1977. Tar sands and supergiant oil fields. *Am. Assoc. Pet. Geol. Bull.* 61: 1950–61

Desborough, G. A., Zimmermann, R. A., Elrick, M., Stone, C. 1985. Early Permian thermal alteration of Carboniferous strata in the Ouachita region and Arkansas River valley, Arkansas. *Geol. Soc. Am. Abstr. with Programs* 17: 155 (Abstr.)

Dow, W. G. 1974. Application of oil-correlation and source-rock data to exploration in Williston basin. *Am. Assoc. Pet. Geol. Bull.* 58: 1253–62

Duane, M. J., de Wit, M. J. 1988. Pb-Zn ore deposits of the northern Caledonides: products of continental-scale fluid mixing and tectonic expulsion during continental collision. *Geology* 16: 999–1002

Ebers, M. L., Kopp, O. C. 1979. Cathodoluminescent microstratigraphy in gangue dolomite, the Mascot–Jefferson City district, Tennessee. *Econ. Geol.* 74: 908–18

Elliott, W. C., Aronson, J. L. 1987. Alleghenian episode of K-bentonite illitization in the southern Appalachian basin. *Geology* 15: 735–39

Ellmore, R. D., Dunn, W., Peck, C. 1985. Absolute dating of dedolomitization by means of paleomagnetic techniques. *Geology* 13: 558–61

Erickson, R. L., Mosier, E. L., Odland, S. K., Erickson, M. S. 1981. A favorable belt for possible mineral discovery in subsurface Cambrian rocks in southern Missouri. *Econ. Geol.* 76: 921–33

Garven, G. 1985. The role of regional fluid flow in the genesis of the Pine Point deposit, Western Canada sedimentary basin. *Econ. Geol.* 80: 307–24

Garven, G. 1989. A hydrogeologic model for the formation of the giant oil sands deposits of the Western Canada sedimentary basin. *Am. J. Sci.* 289: 105–66

Garven, G., Freeze, R. A. 1984a. Theoretical analysis of the role of groundwater flow in the genesis of stratabound ore deposits: 1. Mathematical and numerical model. *Am. J. Sci.* 284: 1085–1124

Garven, G., Freeze, R. A. 1984b. Theoretical analysis of the role of groundwater flow in the genesis of stratabound ore deposits: 2. Quantitative results. *Am. J. Sci.* 284: 1125–74

Ge, S., Garven, G. 1989. Tectonically induced transient groundwater flow in foreland basins. In *The Origin and Evolution of Sedimentary Basins and Their Energy and Mineral Resources. Geophys. Monogr.*, ed. R. A. Price, 48: 145–57. Washington, DC: Am. Geophys. Union

Giggenbach, W. F. 1984. Mass transfer in hydrothermal alteration systems—a conceptual approach. *Geochim. Cosmochim. Acta* 48: 2693–2711

Grant, N. K., Laskowski, T. E., Foland, K. A. 1984. Rb-Sr and K-Ar ages of Paleozoic glauconites from Ohio-Indiana and Missouri, U.S.A. *Isot. Geosci.* 2: 217–39

Gregg, J. M. 1985. Regional epigenetic dolomitization in the Bonneterre Dolomite (Cambrian), southeastern Missouri. *Geology* 13: 503–6

Gregg, J. M., Shelton, K. L. 1989. Minor- and trace-element distributions in the Bonneterre Dolomite (Cambrian), southeast Missouri: evidence for possible multiple basin fluid sources and pathways during lead-zinc mineralization. *Geol. Soc. Am. Bull.* 101: 221–30

Hall, W. E., Friedman, I. 1963. Composition of fluid inclusions, Cave-In-Rock fluorite district, Illinois, and Upper Mississippi Valley zinc-lead district. *Econ. Geol.* 58: 886–911

Harrison, W. J., Summa, L. 1990. The paleohydrology of the Gulf of Mexico basin. Submitted for publication

Hathon, L. A., Houseknecht, D. W. 1987. Hydrocarbons in an overmature basin: I. Thermal maturity of Atoka and Hartshorne Formations, Arkoma Basin, Oklahoma and Arkansas. *Am. Assoc. Pet. Geol. Bull.* 71: 993 (Abstr.)

Hathon, L. A., Houseknecht, D. W. 1990. Thermal maturity of Atokan strata, Arkoma basin. Submitted for publication

Hay, R. L., Lee, M., Kolata, D. R., Matthews, J. C., Morton, J. P. 1988. Episodic potassic diagenesis of Ordovician tuffs in the Mississippi Valley area. *Geology* 16: 743–47

Haynes, F. M., Kesler, S. E. 1989. Pre-Alleghenian (Pennsylvanian-Permian) hydrocarbon emplacement along Ordovician Knox unconformity, eastern Tennessee. *Am. Assoc. Pet. Geol. Bull.* 73: 289–97

Hearn, P. P. Jr., Sutter, J. F. 1985. Authigenic potassium feldspar in Cambrian carbonates: evidence of Alleghenian brine migration. *Science* 228: 1529–31

Hearn, P. P. Jr., Sutter, J. F., Evans, J. R. 1986. Authigenic K-feldspar in the Bonneterre Dolomite: evidence for a Carboniferous fluid migration event. *Proc. Symp. Bonneterre Formation (Cambrian), Southeast. Mo.*, p. 11 (Abstr.)

Hearn, P. P. Jr., Sutter, J. F., Belkin, H. E. 1987. Evidence for late-Paleozoic brine migration in Cambrian carbonate rocks of the central and southern Appalachians: implications for Mississippi Valley–type sulfide mineralization. *Geochim. Cosmochim. Acta* 51: 1323–34

Helgeson, H. C. 1969. Thermodynamics of hydrothermal systems at elevated temperatures and pressures. *Am. J. Sci.* 267: 729–804

Houseknecht, D. W. 1986. Evolution from passive margin to foreland basin: the Atoka Formation of the Arkoma basin, south-central U.S.A. *Spec. Publ. Int. Assoc. Sedimentol.* 8: 327–45

Howard, R. H. 1990. Hydrocarbon reservoir distribution in the Illinois basin. In *Interior Cratonic Sag Basins. Pet. Basins Ser.*, ed. M. W. Leighton. Tulsa: Am. Assoc. Pet. Geol. In press

Hubbert, M. K. 1953. Entrapment of petroleum under hydrodynamic conditions. *Am. Assoc. Pet. Geol. Bull.* 37: 1954–2026

Jackson, M., McCabe, C., Ballard, M. M., Van der Voo, R. 1988. Magnetite authigenesis and diagenetic paleotemperatures across the northern Appalachian basin. *Geology* 16: 592–95

Jackson, S. A., Beales, F. W. 1967. An aspect of sedimentary basin evolution, the concentration of Mississippi Valley–type ores during the late stages of diagenesis. *Bull. Can. Pet. Geol.* 15: 393–433

Kean, W. F. 1981. Paleomagnetism of the Late Ordovician Neda Iron Ore from Wisconsin, Iowa, and Illinois. *Geophys. Res. Lett.* 8: 880–82

Keller, W. D. 1988. Authigenic kaolinite and dickite associated with metal sulfides—probable indicators of a regional thermal event. *Clays Clay Miner.* 36: 153–58

Kent, D. V., May, S. R. 1987. Polar wander and paleomagnetic reference pole controversies. *Rev. Geophys.* 25: 961–70

Kent, D. V., Opdyke, N. D. 1985. Multicomponent magnetizations from the Mississippian Mauch Chunk Formation of the central Appalachians and their tectonic implications. *J. Geophys. Res.* 90: 5371–85

King, P. B. 1969. *Tectonic Map of North America.* Washington, DC: US Geol. Surv.

Leach, D. L. 1979. Temperature and salinity of fluids responsible for minor occurrences of sphalerite in the Ozark region of Missouri. *Econ. Geol.* 74: 931–37

Leach, D. L., Rowan, E. L. 1986. Genetic link between Ouachita foldbelt tectonism and the Mississippi Valley–type lead-zinc deposits of the Ozarks. *Geology* 14: 931–35

Lu, G., Marshak, S., Kent, D. V. 1990. Magnetic carriers responsible for late Paleozoic remagnetization in carbonate strata of the midcontinent, U.S.A. Submitted for publication

Marshak, S., Lu, G., Kent, D. V. 1989. Reconnaissance investigation of remagnetizations in Paleozoic limestones from the mid-continent of North America. *Eos, Trans. Am. Geophys. Union* 70: 310 (Abstr.)

Mason, S. E. 1987. *Periodicities in color banding in sphalerite of the Upper Mis-*

sissippi Valley district. MS thesis. Pa. State Univ., University Park. 110 pp.

Mazzullo, S. J. 1976. Significance of authigenic K-feldspar in Cambrian-Ordovician carbonate rocks of the proto-Atlantic shelf in North America: a discussion. *J. Sediment. Petrol.* 46: 1035–40

Mazzullo, S. J. 1986. Mississippi Valley–type sulfides in Lower Permian dolomites, Delaware basin, Texas: implications for basin evolution. *Am. Assoc. Pet. Geol. Bull.* 70: 943–52

McCabe, C. R., Van der Voo, R., Peacor, D. R., Scotese, C. R., Freeman, R. 1983. Diagenetic magnetite carries ancient yet secondary remanence in some Paleozoic sedimentary carbonates. *Geology* 11: 221–23

McCabe, C., Van der Voo, R., Ballard, M. M. 1984. Late Paleozoic remagnetization of the Trenton Limestone. *Geophys. Res. Lett.* 11: 979–82

McElhinny, M. W., Opdyke, N. D. 1973. Remagnetization hypothesis discounted: a paleomagnetic study of the Trenton Limestone, New York State. *Geol. Soc. Am. Bull.* 84: 3697–3708

McLimans, R. K., Barnes, H. L., Ohmoto, H. 1980. Sphalerite stratigraphy of the Upper Mississippi Valley zinc-lead district, southwest Wisconsin. *Econ. Geol.* 75: 351–61

Meents, W. F., Bell, A. H., Rees, O. W., Tilbury, W. G. 1952. Illinois oil-field brines. *Ill. State Geol. Surv. Ill. Pet. 66.* 38 pp.

Miller, J. D., Kent, D. V. 1988. Regional trends in the timing of Alleghenian remagnetization in the Appalachians. *Geology* 16: 588–91

Moshier, S. O., Waples, D. W. 1985. Quantitative evaluation of Lower Cretaceous Manville Group as source rock for Alberta's oil sands. *Am. Assoc. Pet. Geol. Bull.* 69: 161–72

Noble, E. A. 1963. Formation of ore deposits by water of compaction. *Econ. Geol.* 58: 1145–56

Odom, I. E., Willand, T. N., Lassin, R. J. 1979. Paragenesis of diagenetic minerals in the St. Peter Sandstone (Ordovician), Wisconsin and Illinois. In *Aspects of Diagenesis. Soc. Econ. Paleontol. Mineral. Spec. Publ.*, ed. P. A. Scholle, P. R. Schluger, 26: 425–43. Tulsa: SEPM

Oliver, J. 1986. Fluids expelled tectonically from orogenic belts: their role in hydrocarbon migration and other geologic phenomena. *Geology* 14: 99–102

Ranganathan, V. 1983. The significance of abundant K-feldspar in potassium-rich Cambrian shales of the Appalachian basin. *Southeast. Geol.* 24: 139–46

Rice, D. D., Threlkeld, C. N., Vuletich, A. K. 1988. Character, origin and occurrence of natural gases in the Anadarko basin, southwestern Kansas, western Oklahoma and Texas panhandle, U.S.A. *Chem. Geol.* 71: 149–57

Roedder, E. 1968. The noncolloidal origin of "colliform" textures in sphalerite ores. *Econ. Geol.* 63: 451–71

Rowan, E. L. 1986. Cathodoluminescent zonation in hydrothermal dolomite cements: relationship to Mississippi Valley–type Pb-Zn mineralization in southern Missouri and northern Arkansas. In *Process Mineralogy VI*, ed. R. D. Hagni, pp. 69–87. Warrenville, Pa: Metall. Soc. 503 pp.

Rowan, E. L., Leach, D. L. 1990. Constraints from fluid inclusions on sulfide precipitation mechanisms and ore fluid migration in the Viburnum Trend lead district, Missouri. *Econ. Geol.* In press

Salisbury, G. P. 1968. Natural gas in Devonian and Silurian rocks of Permian basin, West Texas and southeast New Mexico. *Am. Assoc. Pet. Geol. Mem.* 9: 1433–45

Scotese, C. R., Van der Voo, R., McCabe, C. 1982. Paleomagnetism of the Upper Silurian and Lower Devonian carbonates of New York State: evidence for secondary magnetizations residing in magnetite. *Phys. Earth Planet. Inter.* 30: 385–95

Sharp, J. M. Jr. 1978. Energy and momentum transport model of the Ouachita basin and its possible impact on formation of economic mineral deposits. *Econ. Geol.* 73: 1057–68

Sheldon, R. P. 1967. Long-distance migration of oil in Wyoming. *Mt. Geol.* 4: 53–65

Stein, H. J., Kish, S. A. 1985. The timing of ore formation in southeast Missouri: Rb-Sr glauconite dating at the Magmont Mine, Viburnum Trend. *Econ. Geol.* 80: 739–53

Stone, D. S. 1967. Theory of Paleozoic oil and gas accumulation in Big Horn basin, Wyoming. *Am. Assoc. Pet. Geol. Bull.* 51: 2056–2114

Sverjensky, D. A. 1984. Oil-field brines as ore-forming solutions. *Econ. Geol.* 79: 23–37

Sverjensky, D. A. 1986. Genesis of Mississippi Valley–type lead-zinc deposits. *Annu. Rev. Earth Planet Sci.* 14: 177–99

Taylor, M., Kelly, W. C., Kesler, S. E., McCormick, J. E., Rasnick, F. D., Mellon, W. V. 1983. Relationship of zinc mineralization in East Tennessee to Appalachian orogenic events. *Proc. Int. Conf. Mississippi Valley–Type Lead-Zinc Depos.*, ed. G. Kisvarsanyi, S. K. Grant, W. P.

Pratt, J. W. Koenig, pp. 271–78. Rolla: Univ. Mo. 603 pp.

Tissot, B. P. 1984. Recent advances in petroleum geochemistry applied to hydrocarbon exploration. *Am. Assoc. Pet. Geol. Bull.* 68: 545–63

Van der Voo, R. 1979. Age of the Alleghenian folding in the central Appalachians. *Geology* 7: 297–98

Van der Voo, R., French, R. B. 1977. Paleomagnetism of the Late Ordovician Juniata Formation and the remagnetization hypothesis. *J. Geophys. Res.* 82: 5796–5802

Viets, J. G., Leach, D. L. 1990. Genetic implications of the geochemical evolution of brines associated with Mississippi Valley–type lead-zinc deposits of the Ozark region, USA. *Geol. Soc. Am. Bull.* In press

Voss, R. L., Hagni, R. D. 1985. The application of cathodoluminescence microscopy to the study of sparry dolomite from the Viburnum Trend, southeastern Missouri. In *Mineralogy—Applications to the Minerals Industry*, ed. D. M. Hausen, O. C. Kopp, pp. 51–68. New York: Soc. Min. Eng./Am. Inst. Min. Eng. 1194 pp.

White, D. E. 1958. Liquid of inclusions in sulfides from Tri-State (Missouri-Kansas-Arkansas) is probably connate in origin. *Geol. Soc. Am. Bull.* 69: 1660 (Abstr.)

Wilkerson, R. M., ed. 1982. *Oil and Gas Fields of the U.S.* Tulsa: Penn Well Publ. 1 p.

Wisniowiecki, M., Van der Voo, R., McCabe, C., Kelly, W. C. 1983. A Pennsylvanian paleomagnetic pole from the mineralized Late Cambrian Bonneterre Formation, southeast Missouri. *J. Geophys. Res.* 88: 6540–48

Woodward, H. P. 1958. Emplacement of oil and gas in the Appalachian basin. In *Habitat of Oil*, ed. L. G. Weeks, pp. 494–510. Tulsa: Am. Assoc. Pet. Geol. 1384 pp.

Wu, Y., Beales, F. W. 1981. A reconnaissance study by paleomagnetic methods of the age of mineralization along the Viburnum Trend, southeast Missouri. *Econ. Geol.* 76: 1879–94

Annu. Rev. Earth Planet. Sci. 1990. 18:317–56
Copyright © 1990 by Annual Reviews Inc. All rights reserved

THE ORIGIN AND EARLY EVOLUTION OF LIFE ON EARTH

J. Oró

Department of Biochemical and Biophysical Sciences, University of Houston, Houston, Texas 77004

Stanley L. Miller

Department of Chemistry, University of California at San Diego, La Jolla, California 92093

Antonio Lazcano

Departamento de Biologia, Facultad de Ciencias—Universidad Nacional Autonoma de Mexico, Apartado Postal 70-407, 04511 Mexico, DF, Mexico

PERSPECTIVES AND SUMMARY

Although during the first decades of the twentieth century Darwin's explanation of the causes of evolution had fallen out of favor among biologists, his ideas provided the essential framework for the development of a large body of thought devoted during that period to the question of the origins of life. Of all the hypotheses that were suggested (Farley 1977, Kamminga 1988), few were as fruitful as those of Oparin (1924, 1936), who proposed a long period of chemical abiotic synthesis of organic compounds as a necessary precondition for the appearance of the first life forms. The first forms of life would then have been anaerobic heterotrophic microorganisms. Similar ideas were suggested independently by Haldane (1929).

0084–6597/90/0515–0317$02.00

In Table 1 a list of some of the major discoveries and scientific developments directly or indirectly related to the origin of life is given.

Following the model of Oparin's reducing primordial terrestrial atmosphere developed by Urey (1952), Miller (1953) showed that a number of protein amino acids and a diverse assortment of other small organic molecules of biochemical significance could be made in the laboratory under environmental conditions thought to be representative of the Hadean or early Archean Earth. Since then, a wide variety of organic compounds of biochemical significance have been experimentally formed from simple molecules such as water, methane, ammonia, and HCN (Miller 1987). The results of these experiments have been highlighted by the discovery of a large variety of organic molecules in the interstellar clouds of the Milky Way and other galaxies, in comets, and in carbonaceous chondrites (Oró & Mills 1989). These findings have been complemented with information from comparative planetology, Archean geology, and Precambrian micropaleontology. Nevertheless, it is unlikely that the early Archean geological record will provide evidence of the transition from the prebiotic organic molecules to the first cells. Therefore, to recon-

Table 1 Some scientific discoveries and developments related to the origin and early evolution of life on Earth

Discovery or development	Reference(s)
1. Concept of chemical and prebiological evolution	Oparin 1924, Haldane 1929
2. Quantitative models of the Earth's reducing atmosphere	Urey 1952
3. Prebiotic synthesis of amino acids	Miller 1953
4. Founding of Precambrian micropaleontology (Gunflint microfossils, 2000 Ma)	Tyler & Barghoorn 1954
5. Prebiotic synthesis of adenine	Oró 1960
6. Cometary contribution to the primitive Earth	Oró 1961
7. Swaziland microfossils (3400 Ma)	Barghoorn & Schopf 1965
8. Prebiotic synthesis of pyrimidines	Sanchez et al 1966
9. Early biological evolution and the origin of eukaryotes	Sagan (Margulis) 1967, Margulis 1970, pp. 45–68
10. Racemic mixtures of meteoritic amino acids	Kvenvolden et al 1970
11. Template-directed polymerization reactions of nucleotides	Orgel & Lohrmann 1974
12. Phylogenies based on ribosomal RNA	Woese & Fox 1977
13. Catalytic activity of RNA	Cech et al 1981
14. Warrawoona microfossils (3500 Ma)	Awramik 1981, Awramik et al 1983
15. Stereoselective aminoacylation of RNA	Usher & Needels 1984, Lacey et al 1988
16. Template-directed synthesis of novel nucleic acid–like structures	Schwartz & Orgel 1985

struct the possible characteristics of the first living systems, we must rely on studies of prebiotic chemistry as well as on information in contemporary biological systems. These include studies of macromolecular sequence data from proteins and nucleic acids of different organisms, and from comparisons of metabolic pathways and their phylogenetic distribution (Margulis & Guerrero 1986, Lazcano 1986, Woese 1987, Lazcano et al 1988a, Berry & Jensen 1988). A brief discussion concerning the time of the origin of life has been undertaken somewhat recently by Oró & Lazcano (1984). Oberbeck & Fogleman (1989) have calculated the maximum time required for life to originate and evolve to the level of complexity represented by the oldest microfossils; in short, most available calculations and evidence seem to indicate that the emergence of life occurred approximately 3.85 Gyr ago (e.g. Carlin 1980).

We do not have a detailed knowledge of the processes that led to the appearance of life on Earth. In this review we bring together some of the most important results that have provided insights into the cosmic and primitive Earth environments, particularly those environments in which life is thought to have originated. To do so, we first discuss the evidence bearing on the antiquity of life on our planet and the prebiotic significance of organic compounds found in interstellar clouds and in primitive solar system bodies such as comets, dark asteroids, and carbonaceous chondrites. This is followed by a discussion on the environmental models of the Hadean and early Archean Earth, as well as on the prebiotic formation of organic monomers and polymers essential to life. We then consider the processes that may have led to the appearance in the Archean of the first cells, and how these processes may have affected the early steps of biological evolution. Finally, the significance of these results to the study of the distribution of life in the Universe is discussed.

There are many books on the origins of life. We cannot mention all of them here, but among those that discuss aspects of interest to Earth scientists are the books by Calvin (1969), Kenyon & Steinman (1969), Rutten (1971), Miller & Orgel (1974), as well as those edited by Kvenvolden (1974), Ponnamperuma (1977), Halvorson & Van Holde (1980), Holland & Schidlowski (1982), Nagy et al (1983), and Schopf (1983).

THE ANTIQUITY OF TERRESTRIAL LIFE

"To the question of why we do not find rich fossiliferous deposits belonging to these assumed earliest periods prior to the Cambrian system," wrote Darwin in *On the Origin of Species by Natural Selection*, "I can give no satisfactory answer." The answer was provided almost a century later, when Tyler & Barghoorn (1954) and Barghoorn & Tyler (1965) published their description of the 2000-Myr-old microfossils from the Gunflint Formation. Thus the field of Precambrian micropaleontology was born, demonstrating that life is indeed a very ancient phenomenon on our planet. Some of the major problems in Precambrian paleobiology were discussed in this series some time ago by Schopf (1975) and, more recently, in the *Annual Review of Microbiology* by Knoll (1985), who emphasized Proterozoic microbial evolution.

The Archean Era (4.55–2.5 Ga) is generally divided into three main intervals: the Hadean, which comprises the initial period after the formation of the Earth (4.55–3.9 Ga); the early Archean (3.9–2.9 Ga); and the late Archean (2.9–2.5 Ga) (Ernst 1983). We have no definitive evidence of Hadean life, and evidence of early Archean life is sometimes contradictory and difficult to accept (Schopf & Walter 1983). For instance, the biological nature of the structures previously described as fossil prokaryotes by Barghoorn & Schopf (1965) in South African 3000-Myr-old sediments has been disputed (Schopf 1976). Previous reports of yeastlike microfossils (Pflug 1978, Pflug & Jaeschke-Boyer 1979) in the 3800-Myr-old carbonate-rich metasedimentary rocks of Isua, West Greenland (Moorbath et al 1973), are now considered to correspond to structures having nonbiological, inorganic origins [probably fluid inclusions (Bridgwater et al 1981)].

Ratios of stable carbon, nitrogen, and oxygen isotopes in the Isua metasediments have been interpreted to imply the existence of an Archean carbon cycle driven by photosynthetic microbes (Schidlowski et al 1979, 1983). However, since the $^{13}C/^{12}C$ values would have been deeply affected by high-grade metamorphic processes that have modified the Isua Supracrustal Belt samples (Awramik 1982, Knoll 1986), at the present time it is not possible to assert the existence of microbial communities based on photoautotrophs 3800 Myr ago, although such organisms may have existed then.

Nevertheless, there is ample morphological evidence documenting the existence of a complex microbiota in nonmetamorphosed sediments from the early Archean. It consists of many different morphotypes of structurally well-preserved fossil bacteria from the 3500-Myr-old Australian Warrawoona Group (Awramik 1981, 1982, Awramik et al 1983, Schopf 1983, Schopf & Packer 1987) and from 3400-Myr-old South African sediments from the Zimbabwe craton (Knoll & Barghoorn 1977). This evidence is further supported by several Archean occurrences of stromatolites, which are organo-sedimentary structures in whose formation microbial communities play an active role (Margulis et al 1980). The oldest known stromatolites from Archean sequences date back to 3500-Myr-old sediments in the Pilbara Group in northwestern Australia (Walter 1983).

Possible stromatolites have been reported by Byerly et al (1986) in 3300-Myr-old rocks from the Fig Tree Group in South Africa, and Orpen & Wilson (1981) have found structures in sediments of the Mushandike Sanctuary (Zimbabwe) that could also correspond to 3500-Myr-old stromatolites.

EXTRATERRESTRIAL ORGANIC COMPOUNDS

Interstellar Molecules

Although the hypothesis of a period of abiotic chemical syntheses of organic compounds prior to the emergence of life that Oparin (1924, 1936) and Haldane (1929) suggested was originally received with considerable skepticism, contemporary evidence clearly shows that organic molecules not only formed before the appearance of life, but even before the origin of the Earth itself. The existence of simple carbon combinations (C_2, CN, CH, CO) in the atmospheres of relatively cool stars had been known for a long time, and neutral molecules could be presumed to exist in interstellar space (Oró 1963a), but it was not until 1969 that microwave techniques led to the detection of interstellar ammonia, water, and formaldehyde, the first organic molecule to be discovered in space (Snyder et al 1969). Ever since, more than 80 different molecular species have been detected in interstellar or circumstellar clouds. A significant percentage of these compounds are organic, and they include most of the functional groups of biochemical compounds, with the exception of those involving phosphorus (Lazcano-Araujo & Oró 1981). Recent discoveries include the detection of phosphorus nitride (PN) achieved simultaneously by Ziurys (1987) and by Turner & Bally (1987), as well as cyclic C_3H (cf. Irvine 1989). The detection of the linear $HC_{11}N$ (Bell et al 1982) and C_3 molecules (Hinkle et al 1988) in the circumstellar spectrum of the carbon star IRC+10°216 shows that in our Galaxy, cool carbon stars are a source of complex interstellar molecules that are ejected by means of various processes into space.

Among the interstellar compounds, there are many that have been used in prebiotic organic synthesis to obtain most of the biochemical monomers found in contemporary living systems (Irvine 1989). However, as of May 1989, no interstellar amino acid, nucleotide derivative, or other metabolite has been reported. Searches for interstellar cyclic furan and imidazole (De Zafra et al 1971), pyrimidines and pyridines (Simon & Simon 1973), and glycine (Brown et al 1979, Hollis et al 1980) have been unsuccessful. A large number of interstellar emission lines in the 3–6 mm wavelength remain to be identified, and it is possible that some of them may correspond to biochemical compounds. This possibility is made plausible by the results of several laboratory simulations of the interstellar environment. In these studies, icy mixtures of water, methane, carbon dioxide, ammonia, and nitrogen have been irradiated by ionizing radiations (Oró 1963b, Lazcano-Araujo & Oró 1981) and ultraviolet light (Greenberg 1981, d'Hendecourt et al 1986), yielding, upon hydrolysis, significant amounts of amino acids, hydroxy acids, nitrogen bases, and other nonvolatile organic molecules, including polymers.

Although it is very unlikely that interstellar molecules played a direct role in the origin of life, the presence in interstellar space of a large array of many molecules of prebiotic importance, combined with the fact that their synthesis must differ from the chemical pathways that were possible in the primitive terrestrial environment, implies that some molecules are particularly easily synthesized when radicals and ions recombine.

Comets and Carbonaceous Chondrites

Since most of the interstellar organic compounds have been observed in dense, cool interstellar clouds where star formation is taking place, it is reasonable to assume that the primordial solar nebula had a somewhat similar composition. This hypothesis suggests that a certain amount of interstellar organic compounds, most likely incorporated on cosmic dust particles, became part of comets and planetesimals (Delsemme 1984, Kerridge & Anders 1988). This conclusion is supported by the observation of a number of organic molecules and radicals (C_2, C, CH, CO, CS, HCN, CH_3CN) and other simple chemical species (NH, NH_3, OH, H_2O) in cometary spectra. As shown by the recent observations of Comet Halley, cometary nuclei are relatively small bodies with diameters of a few kilometers in which frozen water comprises approximately 80% of the total volatile ices. The remaining 20% of the volatiles includes formic acid, formaldehyde, carbon dioxide, carbon monoxide, HCN, acetylene, and cyclopropenylidene in various proportions (Delsemme 1988, Oró & Mills 1989).

A recent detailed discussion of the chemistry of comets, including an analysis of elemental abundances in Comet Halley, has been given by Delsemme (1988), who argues that most of the chemical elements appear to exist in their cosmic abundances, with the exception of hydrogen (and, perhaps, of helium and neon), which is depleted by a factor close to 10^3. Cometary nuclei thus appear to be the most pristine bodies in the solar system. Recent analysis by Kissel & Krueger (1987) of the dust mass spectra of Comet Halley suggests that cometary dust has an organic component that includes adenine and other purines, together with other heterocyclic compounds and organic polymers. As argued by Matthews & Ludicky (1987), it is possible that the low albedo and other features of

Comet Halley could be explained by the presence of HCN polymers. A discussion of the possible routes of formation in cometary nuclei of these and other organic compounds of prebiotic significance may be found in Oró & Mills (1989). The biochemical monomers and properties that can be derived from interstellar and cometary molecules are summarized in Table 2.

The abiotic synthesis of biochemical compounds during the time of formation and early evolution of the solar system, or even during presolar events, is further indicated by the large array of proteinic and nonproteinic amino acids, carboxylic acids, purines, pyrimidines, hydrocarbons, and other molecules that have been found in carbonaceous chondritic meteorites. The earlier evidence has been reviewed by Hayes (1967) and Nagy (1975). Results on extraterrestrial D and L amino acids were published by Kvenvolden et al (1970, 1971) and Oró et al (1971a,b), and data on nucleic acid bases (Anders et al 1974, Stoks & Schwartz 1981) and amphiphilic compounds (Deamer 1985) have also been published. Of the 74 amino

Table 2 Biochemical monomers and properties that can be derived from interstellar and cometary molecules[a]

Interstellar and cometary molecules	Formulae	Biochemical monomers and properties
1. Hydrogen	H_2	Reducing agent
2. Water	H_2O	Universal solvent
3. Ammonia	NH_3	Catalysis and amination
4. Carbon monoxide	$CO(+H_2)$	Fatty acids
5. Formaldehyde	CH_2O	Ribose and glycerol
6. Acetaldehyde	$CH_3CHO(+CH_2O)$	Deoxyribose
7. Aldehydes	$RCHO(+HCN$ and $NH_3)$	Amino acids
8a. Hydrogen sulfide	H_2S(+other precursors)	Cysteine and methionine
8b. Thioformaldehyde (interstellar)	CH_2S	
9. Hydrogen cyanide	HCN	Purines and amino acids
10. Cyanacetylene (interstellar)	HC_3N(+cyanate)	Pyrimidines
11. Cyanamide (interstellar)	H_2NCN	Condensing agent for biopolymer synthesis
12a. Phosphorus nitride (interstellar)	PN	
12b. Phosphine (Jupiter and Saturn)	PH_3	Phosphates and nucleotides
12c. Phosphate[b]	PO_4^{3-}	

[a] From Oró et al (1978, 1990). With the exceptions of phosphine (PH_3) and phosphate (PO_4^{3-}), all other molecules have been detected in interstellar clouds, and most have also been detected in comets. Molecules that have only been detected in interstellar clouds are indicated by "(interstellar)." With the exception of the monosaccharides and some of the lipids, the prebiotic formation of the biochemical compounds (amino acids, purines, pyrimidines, etc) requires the participation of HCN or its derivatives.

[b] Found in meteorites and interplanetary dust particles.

acids found in samples of the Murchison meteorite, 8 are present in proteins, 11 have other biological roles, and the remaining 55 have only been found in extraterrestrial samples (Cronin 1989). All the meteoritic alpha amino acids are found as racemic mixtures, including the α-alkyl, α-amino acids such as isovaline, which cannot be racemized. This demonstrates that these compounds are indigenous to the meteorite and not terrestrial contaminants.

Recent studies on the chemical and isotopic compositions of the amino acids in the Murchison meteorite by Epstein et al (1987) and Cronin (1989) have provided major insights into their origins and probable mechanisms of synthesis. Although a comparison between the relative abundances of the α-amino acids found in the Murchison meteorite with those formed by electrical discharges in a Miller-Urey type of synthesis suggests that the latter process can account for their origin (Wolman et al 1972), an analysis of the deuterium/hydrogen ratio of the meteoritic amino acids shows that these compounds are highly enriched in deuterium compared with the average solar system D/H value (Epstein et al 1987, Cronin 1989).

From these measurements it seems likely that the amino acids could not have been formed from gasses of the solar nebula, but rather are more likely to have formed from highly deuterium-enriched molecules from presolar interstellar clouds. Accordingly, Oró & Mills (1989) have suggested that the aldehyde or ketone precursors to the amino acid R groups were part of a presolar interstellar cloud with a high D/H ratio, which became part of the material that condensed in the meteorite parent body (presumably a primitive dark asteroid). Further thermal evolution of these parental bodies may have led to transient hydrothermal phases during which the amino acids were synthesized following the Strecker-cyanohydrin mechanism involved in the synthesis by electrical discharges. This pathway supports the hypothesis of a direct relationship between organic-rich interstellar grains, comets, dark asteroids, and carbonaceous chondrites (Cruikshank 1989).

THE ENVIRONMENT OF THE HADEAN EARTH

The Origin of Terrestrial Volatiles

One of the major problems in attempting to reconstruct the conditions of the Hadean environment is that of the origin and detailed chemical composition on the early terrestrial atmosphere. Since there is general agreement that the solar system formed 4600 Myr ago as a result of the condensation of the solar nebula, whose elementary abundances and molecular composition were comparable to those of the dense gas and dust regions found in the Milky Way, it would be tempting to assume that the early Earth inherited not only simple organic molecules from the solar nebula, but also the complex carbon compounds that appear to exist in interstellar grains and comets, as well as the polymers found in carbonaceous chondrites like the Murchison meteorite.

It is unlikely, however, that the organic molecules from which it is assumed that the first living systems were formed were derived directly from the solar nebula. The amount of interstellar material trapped by the prebiotic Earth was very small, since the well-known terrestrial depletion of noble gases relative to solar abundances (Brown 1952, Suess & Urey 1956, Anders & Owen 1977, Cameron 1980) shows that the primordial atmosphere was almost completely lost, and that the Earth acquired a significant part of its secondary atmosphere from the release of internal volatiles (Walker 1977, 1986, Arrhenius 1987) and the late accretion of volatile-rich minor bodies such as cometary nuclei (Oró 1961, Lazcano-Araujo & Oró 1981).

According to the single-impact theory on the origin of the Earth-Moon system developed by Cameron and his collaborators (Benz et al 1986, 1987, Cameron & Benz 1989), a near-Mars-size body collided with the proto-Earth. Upon collision, all of the iron from the impactor was injected into the core of the proto-Earth. The impact of the collision was such that both the impactor and most of the Earth's mantle were melted, and a lump the size of the Moon was ejected into the Earth's orbit. This process explains the formation of the Earth-Moon system with all its physical and chemical characteristics, but it also implies that as a consequence of this very early catastrophic event the Earth must have lost into space practically all of the water and most of the biogenic elements (H, C, N, O, S, and P) that the protoplanet may have previously retained. Comets and other primitive solar system bodies captured by the primitive Earth must have been then the major source of the terrestrial volatiles (Oró 1961, Anders & Owen 1977, Lazcano-Araujo & Oró 1981, Delsemme 1984, Chyba 1987). Table 3 displays the various estimates of the amounts of cometary matter captured by the Earth.

In spite of all the uncertainties that exist in our current reconstruction of the first 800 Myr of the Earth (Chang et al 1983, Arrhenius 1987), most of the theoretical models developed to explain the formation of the terrestrial planets predict a molten early Earth (Wetherill 1980). This would produce an extensive pyrolysis of primordial organic compounds whose products would then form a gaseous mixture of CO_2, CO, CH_4, H_2, N_2, NH_3, and H_2O (Miller 1982). Moreover, the capture of large planetesimals—comparable both in size and in chemical composition to comets and dark asteroids—during the early and late phases of accretion would lead to their melting and complete vaporization at impact. The

Table 3 Estimates of cometary matter captured by the Earth

Cometary matter captured	Time period	Reference
2×10^{14} to 10^{18}	Initial 2×10^9 yr	Oró 1961
1×10^{25} to 10^{26}	Late accretion period	Whipple 1976
4×10^{21}	Late accretion period	Sill & Wilkening 1978
1×10^{23}	Initial 2×10^9 yr	Oró et al 1980
1×10^{24} to 10^{25}	Initial 1×10^9 yr	Delsemme 1981
2×10^{24}[a]	Initial 2×10^9 yr	Frank et al 1986
2.3×10^{24} to 10^{26}	Initial 7×10^8 yr	Chyba 1987
6×10^{24} to 10^{25}	Initial 1×10^9 yr	Ip & Fernandez 1988

[a] Calculated by extrapolating from current estimates of 10^{15} g yr^{-1} of incoming mass.

water vapor thus liberated would eventually condense, which implies that an atmosphere-ocean system was formed simultaneously with the increase of mass of the Earth (Arrhenius 1987). It is probable that a substantial fraction of the organic molecules present in the colliding volatile-rich minor bodies were destroyed due to the high temperatures and shock-wave energy generated during the impact with the Earth (Arrhenius 1987). However, under the anoxic conditions generally thought to prevail in the prebiotic atmosphere, the postcollisional formation of large numbers of excited molecules and radicals that are stable at high temperatures, such as CO, CN, and C_2 (Thomas et al 1989), could have led upon cooling to the formation of more complex organic compounds of biochemical significance (Lazcano-Araujo & Oró 1981, Mukhin et al 1989).

The Chemical Composition of the Archean Atmosphere

There is no general agreement on the detailed chemical composition of the Archean paleoatmosphere (Chang 1982, Holland 1984, Miller 1987). As shown below, there is considerable experimental evidence showing that the more reducing atmospheres (CH_4, NH_3, H_2O, H_2) yield a higher variety and amount of organic compounds than mildly reducing and nonreducing atmospheres, such as those in which CO and CO_2 are used instead of CH_4 (Schlesinger & Miller 1983a,b, Stribling & Miller 1987).

It is possible that if intense degassing took place during the accretional phase of our planet, then the Archean Earth could have had a highly reducing atmosphere (Pollack & Yung 1980). However, the presence of the 3800-Myr-old carbonate-rich metasedimentary rocks in Isua (Moorbath et al 1973) shows that even if methane was the predominant form of carbon in the prebiotic paleoatmosphere, a substantial amount of gaseous CO_2 was present in the Earth's atmosphere only 800 Myr after its formation

(Schidlowski 1978, Lazcano et al 1983). This conclusion is supported by models of planetary formation in which a rapid core formation leads to a short-lived, highly reducing terrestrial secondary atmosphere (Holland 1984). Because of the restrictions that high CO_2 and low H_2 pressures put on abiotic synthesis, we have suggested elsewhere (Lazcano et al 1983) that the nonenzymatic syntheses of organic compounds occurred very rapidly in very low concentrations of atmospheric CO_2. This conclusion is in agreement with the analysis that S. L. Miller and his collaborators performed using several equilibrium and rate constants for the formation of amino-, hydroxy-, and dicarboxylic acids from the Murchison meteorite. Their results suggest that these organic compounds were synthesized in the meteorite parent body and in the primitive Earth in less than 10^4 yr (Peltzer et al 1984).

Additional localized reducing environments that could have included NH_3 and CH_4 would be maintained by local gas production from crustal reservoirs (Arrhenius 1987) and from gases originating in the trailing edges of Archean plates in midocean spreading centers (Lazcano et al 1983). The latter possibility is supported by the geological evidence that points to the existence of continental masses and oceanic basins, midocean ridges, subduction, and other indicators of an active terrestrial crust in which plate tectonics was probably operating since the early Archean (Bickle 1978, 1986, Compston & Pidgeon 1986, Nisbet 1987). As discussed elsewhere (Miller & Bada 1988), this hypothesis does not suggest at all that life originated in high-temperature hydrothermal systems, but only that the latter could have been a source of reduced gases and of chemical recycling in the Archean environment.

Free-Energy Sources

The direct sources of energy available for prebiotic organic synthesis are listed in Table 4 (Miller 1987). With the exception of cosmic rays, it is clear that in the prebiotic Earth the energy values were higher than those shown in Table 4, which represent current estimates for our planet. This is particularly clear in the cases of heat from volcanic sources, shock waves, radioactivity, and solar ultraviolet light. For instance, the evidence for intense geological activity in the terrestrial crust during the Archean (Nisbet 1987), the collisional events occurring during the late accretion phase of the Earth (Lazcano-Araujo & Oró 1981), and the higher concentrations of unstable isotopes (which had not yet decayed) all imply that the amounts of free energy available in the prebiotic environment from these processes were much higher than the contemporary values. Moreover, although Welch et al (1985) have argued that the Sun was perhaps never massive enough to have shown the enhancement of ultraviolet light during its early

Table 4 Present sources of energy averaged over the Earth

Source	Energy	
	(cal cm^{-2} yr^{-1})	(J cm^{-2} yr^{-1})
Total radiation from Sun	260,000	1,090,000
Ultraviolet light:		
<3000 Å	3400	14,000
<2500 Å	563	2360
<2000 Å	41	170
<1500 Å	1.7	7
Electric discharges	4[a]	17
Cosmic rays	0.0015	0.006
Radioactivity (to 1.0-km depth)	0.8	3.0
Volcanoes	0.13	0.5
Shock waves	1.1[b]	4.6

[a] 3 cal cm^{-2} yr^{-1} of corona discharge + 1 cal cm^{-2} yr^{-1} of lightning.
[b] 1 cal cm^{-2} yr^{-1} of this is in the shock wave of lightning bolts and is also included under electric discharges.

stages of evolution that has been predicted by others (Gaustad & Vogel 1982, Canuto et al 1982), the absence of substantial amounts of free atmospheric oxygen and the subsequent lack of an ozone shield would account for the greater ultraviolet flux of the Hadean and early Archean (Lazcano et al 1983, Schopf 1983).

However, although it should be emphasized that neither a single energy source nor a single process can account for all of the organic molecules that were formed in the prebiotic Earth (Miller et al 1976), the importance of a given energy source is determined by the product of the energy available and its efficiency for synthesis of organic compounds or their intermediates, such as HCN (Oró & Lazcano-Araujo 1981, Miller 1987). Accordingly, the energy from the decay of radioactive elements was probably not an important energy source for abiotic organic synthesis, since most of the ionization events would have occurred in silicate rocks rather than in the reducing atmosphere (Miller 1987). Molten lava and other sources of volcanic heat have been suggested as playing a role in polymerization reactions (Fox & Dose 1977), but we believe that heat may have been important only in the pyrolytic synthesis of some amino acids (Friedmann et al 1971) and perhaps other organic compounds (Harada 1967). Ultraviolet light was certainly the largest source of energy on the primitive Earth but not necessarily the most effective one, since most of the photochemical synthetic reactions would occur in the upper atmosphere and the products formed would be decomposed by longer ultraviolet wavelengths before reaching the oceans (Miller & Orgel 1974, Miller 1987).

The most widely used sources of energy for laboratory simulations of prebiotic synthesis are electric discharges, including spark, semicorona, arc, and silent discharges (Figure 1). Electric discharges are very efficient in synthesizing hydrogen cyanide, whereas ultraviolet light is not. Furthermore, since electrical storms occur close to the surface of the Earth, the organic molecules and their intermediates thus formed could reach the oceans before being photodissociated (Miller & Orgel 1974). This quenching and the protective steps are critical, since the organic compounds are destroyed if subjected continuously to the energy source (Miller 1987).

CHEMICAL SYNTHESES OF MONOMERS

The wide variety of energy sources and chemical precursors that have been used in experiments simulating prebiological organic synthesis have been reviewed in considerable detail elsewhere (Abelson 1966, Miller & Orgel 1974, Lazcano et al 1983, Chang 1982, Chang et al 1983, Ferris & Usher 1983, Miller 1987, Ferris 1987). Several of these experiments have been performed under extreme concentrations of precursors, under high inputs of energy (Miller & Van Trump 1981, Chang 1981, 1982), or (in the case of the Fischer-Tropsch synthesis of fatty acids) under conditions of pressure and temperature that, although they may have been present in the solar nebula, are more difficult to envisage on the prebiotic Earth

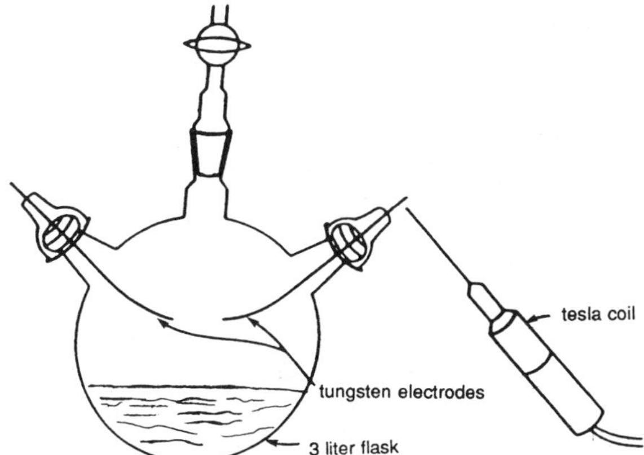

Figure 1 Electric discharge apparatus used to synthesize amino acids at room temperature. The 3-liter flask is shown with two tungsten electrodes and a spark generator (after Miller 1953, Oró 1963b).

(Miller et al 1976). In principle, however, the results of the laboratory nonenzymatic synthesis of organic compounds indicate that comparable processes took place on the early Earth, leading eventually to the accumulation of a complex heterogeneous mixture of monomeric and polymeric organic compounds.

Prebiotic Synthesis of Amino Acids

The first successful synthesis of organic compounds under plausible prebiotic conditions was accomplished by the action of electric discharges acting for a week over a mixture of CH_4, H_2, H_2O, and NH_3 (Miller 1953). Several amino acids, hydroxy acids, and other molecules were formed. A more detailed study of the electric discharge experiments demonstrated that essentially all the protein and nonprotein amino acids found in a sample of the Murchison meteorite were present among the products of the electric discharge (Ring et al 1972, Wolman et al 1972). Perhaps one of the most remarkable results of this comparison is that the ratios of amino acids relative to glycine in the Murchison sample are quantitatively similar to those found in the spark discharge experiments.

Miller (1957) has shown that the mechanism of synthesis of amino acids is the Strecker condensation, where HCN and aldehydes formed by the electrical discharges condense with each other in the presence of ammonia, forming the amino nitriles that upon hydrolysis yield the amino acids (Figure 2*a*). A similar mechanism can account for the synthesis of the reported hydroxy acids (Figure 2*b*). The study of the rate and equilibrium constants of these reactions (Miller & Van Trump 1981) has shown that both amino acids and hydroxy acids can be synthesized even at high dilutions of HCN and aldehydes in a primitive hydrosphere. The half-lives for the hydrolysis of the amino and hydroxy nitriles are less than 10^3 yr at 0°C (Miller 1987). Following previous studies on the formation of imidazole from glyoxal, ammonia, and formaldehyde (Oró et al 1984), and of imidazole-4-glycol and imidazole-4-acetaldehyde under prebiotic conditions (Shen et al 1987), Shen et al (1990a) have recently performed

$$\text{(a)}\quad RCHO + HCN + NH_3 \rightleftarrows RCH(NH_2)CN \xrightarrow{H_2O} RCH(NH_2)\overset{O}{\overset{\|}{C}}\text{-}NH_2 \xrightarrow{H_2O} RCH(NH_2)COOH$$

$$\text{(b)}\quad RCHO + HCN \rightleftarrows RCH(OH)CN \xrightarrow{H_2O} RCH(OH)\overset{O}{\overset{\|}{C}}\text{-}NH_2 \xrightarrow{H_2O} RCH(OH)COOH$$

Figure 2 Strecker synthesis of (*a*) amino acids and (*b*) hydroxy acids (after Miller 1957).

the nonenzymatic synthesis of histidine from erythrose and formamidine followed by a Strecker synthesis.

Additional routes for the abiotic synthesis of amino acids have also been suggested (Sanchez et al 1966, Ponnamperuma & Woeller 1967); a very important one appears to be the condensation of HCN from NH_4CN into polymers, which upon hydrolysis yield glycine, alanine, aspartic acid, and several other amino acids (Oró & Kamat 1961). The base-catalyzed condensation of HCN yields a very heterogeneous brown polymer. Hydrolysis of this polymer produces many of the α-amino acids commonly found in proteins (Lowe et al 1963, Harada 1967, Friedmann & Miller 1969). Moreover, Ferris et al (1978) have shown that at pH 9.2, dilute solutions (0.1 M) of HCN condense into oligomers, which upon hydrolysis yield several protein and nonprotein amino acids as well as purines and pyrimidines.

Prebiotic Synthesis of Bases, Sugars, and Their Derivatives

Several possible prebiotic routes for the formation of all the components of nucleic acids have been suggested. For instance, sugars are readily formed when formaldehyde polymerizes under alkaline conditions (Butlerow 1861). However, it is difficult to explain the presence of only ribose and deoxyribose in the prebiotic environment by this mechanism (Reid & Orgel 1967). Under the slightly basic conditions of the Butlerow synthesis, a complex mixture of more than 50 different pentoses, hexoses, and many other sugars is obtained (Decker et al 1982). Furthermore, under the reaction conditions in which it is formed, ribose tends to decompose into acidic compounds (Socha et al 1981). Accordingly, at the present time it is difficult to understand how ribose could have accumulated and separated from other sugars of abiotic origin in the prebiotic environment (Shapiro 1984, 1988, Ferris 1987).

Adenine, a purine that plays a central role in genetic processes and energy utilization, was first shown by Oró (1960) to form under prebiotic Earth conditions from HCN. When aqueous ammonia solutions containing large concentrations of HCN (1–10 M) are refluxed, adenine is formed together with other purines and biochemical compounds, including amino acids and pteridines (Oró 1960, Oró & Kimball 1961, 1962). The synthesis of adenine is thought to take place by a series of condensation and cyclization reactions involving imidazole derivatives. The overall reaction can be represented simply as five molecules of hydrogen cyanide producing one molecule of adenine (Figure 3).

There are other alternative pathways that have been suggested for the prebiotic synthesis of adenine. These include direct consecutive condensation of hydrogen cyanide (Oró & Kimball 1962), photochemical

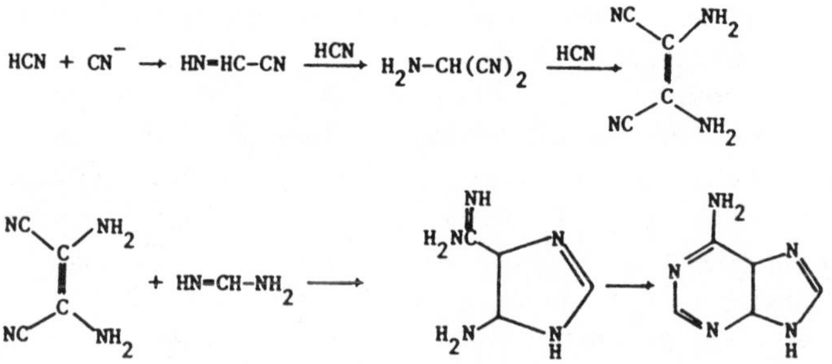

Figure 3 Prebiotic synthesis of adenine (after Oró 1960).

isomerization of *cis*-diaminomaleonitrile (which is one of the HCN tetramers) and condensation with formamidine (Sanchez et al 1966), as well as a more complex sequence of reactions in which no isomerization of the *cis*-diaminomaleonitrile is required (Schwartz & Voet 1984). It is likely that more than one of these different pathways may have occurred under the variable conditions of the primordial environment (Basile et al 1985).

Guanine and xanthine are formed by the condensation of 4-amino-imidazole-5-carboxamide with urea (Oró & Kimball 1962). It should be noted that these imidazole derivatives can all condense with monocarbon compounds to generate all biological purines (Oró 1965). The prebiotic synthesis of adenine was confirmed by Lowe et al (1963) and by Ponnamperuma et al (1963), who found adenine after irradiating a Miller-Urey type of reducing atmosphere (CH_4, NH_3, H_2O, and H_2) with an electron beam. The kinetics and mechanisms of HCN oligomerization, as well as the potential role of the HCN trimer and tetramer for purine synthesis, have been extensively studied by Ferris & Orgel and their co-workers (1965 et seq), who confirmed that the above sequence of reactions follows the same direction: first imidazole formation, and then pyrimidine cyclization, as in the biological synthesis of purines.

Pyrimidines can be formed under plausible prebiotic conditions by the condensation of cyanoacetylene with either urea or cyanate (Sanchez et al 1966), a method that yields considerable amounts of cytosine. Cytosine is readily converted into uracil by deamination (Miller & Orgel 1974). Thymine was produced in small yields from the reaction between uracil and formaldehyde in the presence of hydrazine, which acts as a reducing agent (Stephen-Sherwood et al 1971). As noted above, Ferris et al (1978) have found that orotic acid (the common biochemical precursor of pyrimidinic bases in contemporary organisms) is released by the hydrolysis of HCN oligomers, together with purines and amino acids. A detailed summary of recent work on the abiotic synthesis of both purines and pyrimidines may be found elsewhere (Ferris & Hagan 1984, Basile et al 1985).

Prebiotic Synthesis and Activation of Nucleotides

Although additional studies are required to demonstrate the prebiotic formation of nucleotides (Ferris 1987), the phosphorylated forms of the nucleosides of purine and pyrimidine bases can be produced non-enzymatically in the presence of linear and cyclic phosphates. However, a detailed discussion of the formation and subsequent activation of mononucleotides from individual sugars, phosphates, and bases is beyond the scope of this review and has been discussed elsewhere (Joyce 1989). Suffice it to say, as Joyce does, that "it is a long and difficult road from HCN and H_2CO to the activated mononucleotides" capable of providing a substrate for template-directed oligomerization (discussed below). In fact, evidence of a number of cross-inhibition phenomena occurring between mononucleotide isomers in template-directed syntheses attempted in various laboratories have led Joyce (1989) to the interpretation that life did not start with nucleotide polymers. Instead, there may have been a simpler genetic system that preceded nucleotides. As discussed in a later section of this review, this has led some researchers to a new area of research involving the study of polymerization reactions of nucleic acid–like molecules that might have preceded RNA. We return to a more detailed discussion of template-directed synthesis of nucleotides and their analogues following a brief but necessary discussion of the prebiotic synthesis of coenzymes and the role of prebiotic condensing agents.

Prebiotic Synthesis of Coenzymes

One of the most ubiquitous coenzymes in metabolic pathways is nicotinamide-adenine dinucleotide (NAD). This coenzyme functions as an electron and proton carrier and is required for activity by many mainstream metabolic enzymes. The functional part of this coenzyme is a nicotinamide ring, and the prebiotic synthesis of this cyclic nitrogen–containing vitamin has been achieved previously and is discussed by Miller & Orgel (1974). Additional significant work on the prebiotic synthesis of a number of coenzymes includes the synthesis of nitrogenous base derivatives such as the coenzymes adenosine diphosphate glucose, guanosine diphosphate glucose, uridine diphosphate glucose, and cytidine diphosphoenthanolamine (Mar et al 1987, Mar & Oró 1990). These syntheses were carried out by reaction of individual constituents under aqueous conditions, at moderate temperatures for short periods of time, using urea and cyanamide as condensing agents. The ease by which the synthesis of these compounds can be achieved supports the hypothesis of their presence in the prebiotic environment.

Prebiotic Synthesis of Other Compounds

There are a number of additional compounds that have been synthesized under possible primitive Earth conditions, but space limitations do not permit a lengthy discussion here. These include the mono-, di-, and tricarboxylic acids, branched- and straight-chain fatty acids (C_2–C_{10}, etc), and a number of membrane-forming lipids, to name a few. Of particular interest to the origin and early evolution of life is the synthesis of amphiphilic membrane-forming phospholipids through the condensation of fatty acids, glycerol, and phosphate under primordial conditions (Hargreaves & Deamer 1978a, Eichberg et al 1977, Epps et al 1978). However, a thorough discussion of the prebiotic formation of some of the specific phospholipids found in contemporary membranes, as well as a discussion of the role of membranes in the processes that led to the origin of life, is deferred until later in this review.

PREBIOTIC CONDENSATION REACTIONS

Prebiotic Condensing Agents

Several mechanisms for the condensation of amino acids and nucleotides have been studied under abiotic conditions (Oró & Stephen-Sherwood 1974). These include coupling reactions using activated derivatives, condensation reactions using polyphosphates, thermal polymerization, and condensation reactions using organic condensing agents derived from HCN, such as cyanamide (Oró 1963a, Hulshof & Ponnamperuma 1976, Oró & Lazcano-Araujo 1981). Cyanamide, which has been detected in the interstellar medium and was presumably synthesized on the primitive Earth from HCN, is one of the best condensing agents. The evaporating pond model provides a geologically plausible and realistic model by means of which cyclic changes in humidity and temperature in the presence of cyanamide may have yielded significant amounts of oligopeptides (Hawker & Oró 1981) [including the catalytic dipeptide histidyl-histidine (Shen et al 1990b,c)], oligonucleotides (Odom et al 1982), and phospholipids (Rao et al 1982). The prebiotic condensing reactions leading to the synthesis of other membrane components have been reviewed extensively by Oró et al (1978) and by Hargreaves & Deamer (1978a).

Some of the major prebiotic pathways that may have been responsible for the nonenzymatic synthesis of oligomers and polymers on the prebiotic Earth have been reviewed elsewhere (Miller & Orgel 1974, Hulshof & Ponnamperuma 1976, Oró & Lazcano-Araujo 1981). The use of cyanamide and other condensing agents in the polymerization of ribonucleotides yields a significant amount of "unnatural" 2′–5′ bonds in the phosphodiester backbone of the resulting oligomer (Hulshof & Ponnamperuma 1976, Odom et al 1982). However, the hydration/dehydration cyclical pathways for the nonenzymatic synthesis of polynucleotides suggested by Usher & McHale (1976) and Usher (1977) would have led to the preferential hydrolysis of the 2′–5′ bonds and the subsequent accumulation of molecules linked by 3′–5′ bonds.

Template-Directed Polymerization Reactions

Polynucleotide formation has been extensively studied by L. E. Orgel and his collaborators, who have shown that the nonenzymatic template-directed synthesis of activated nucleotides (Lohrman et al 1980) can occur in the presence of metallic ions. Of all the cations used, Zn^{++} leads to an extremely efficient incorporation of complementary bases (Bridson & Orgel 1980, Bridson et al 1981). This result is quite interesting from a biological point of view, since many DNA- and RNA-polymerases are known to be Zn-metalloenzymes (Mildvan & Loeb 1979). Since Orgel and his group have found no evidence for the direct activation of the 3′-OH group, they have suggested that the Zn ion functions by coordination with the N-7 position of guanine, thereby changing the detailed stereochemistry of the double helical complex (Inoue & Orgel 1981). A favorable orientation of the double-helix complex was also achieved using a nucleotide activated with 2-methyl-imidazole (Inoue & Orgel 1982). In the absence of metals, guanosine-5′-phospho-2-methyl imidazole derivatives that polymerize on a poly(C) template can give rise to over 90% of 3′–5′ linked oligo(G) nucleotides, with chain lengths from 2 to 40 bases (Joyce et al 1984). A simplified model of this system is shown in Figure 4. The kinetics and molecular evolutionary implications of these reactions have been studied and discussed recently by Kanavarioti and coworkers (Kanavarioti & White 1987, Kanavarioti et al 1989). A possible prebiotic synthesis of imidazole and the 2-methyl imidazole derivative has also been achieved (Oró et al 1984). However, the complete nucleotide containing the imidazole derivative together with sugar, phosphate, and nitrogenous base, such as in the guanosine-5′-phospho-2-methyl imidazolide mentioned above, has yet to be synthesized under reasonable prebiotic conditions. Condensation reactions involving deoxyribonucleotide substrates are much less efficient (Lohrmann & Orgel 1977). The polymerization of 2-methyl-imidazole derivatives using random poly(C,U) (Joyce et al 1984) and random poly(C,G) copolymers and templates has also been achieved (Joyce & Orgel 1986). From the point of view of simple chemical models, not necessarily prebiotic, an elegant demonstration of an autocatalytic

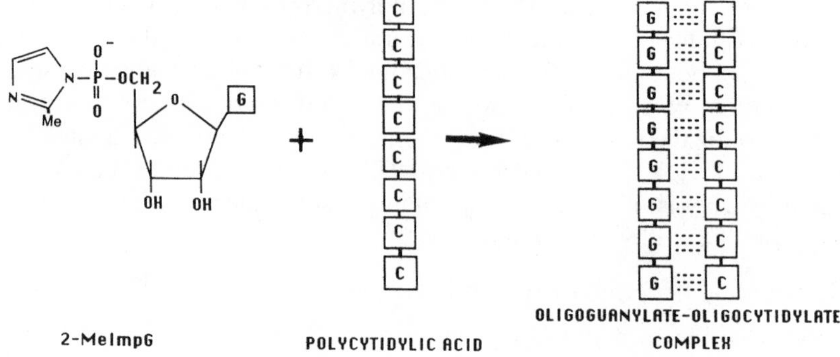

Figure 4 Template-directed oligonucleotide polymerization model.

self-replicating system (hexadeoxynucleotide) has been presented by von Kiedrowski (1986). A similar reaction with a tetranucleotide analogue has been reported by Zielinsky & Orgel (1987). Recent results on the nonenzymatic template-directed synthesis of polynucleotides have been reviewed by Joyce (1987).

As discussed above, it is extremely difficult at present to assert the presence of ribose in the prebiotic environment. This has led Orgel and his associates to a whole new area of research involving the study of polymerization reactions of nucleic acid–like molecules that could have preceded both RNA and DNA. These experiments have been recently discussed by Schwartz et al (1987a) and include polymerization reactions of nucleic acid analogues with unusual backbones, such as the Pb^{++}-catalyzed formation of 2′–5′ phosphodiester-linked oligoguanylic acid (Lohrmann et al 1980), the 3′–5′-linked phosphoramidates that result from the polymerization reaction of 3′-amino-3′-deoxynucleoside-5′-phosphoimidazolides (Zielinski & Orgel 1985), and the pyrophosphate-linked products of the abiotic condensation of deoxynucleoside-3′,5′-diphosphates (Schwartz & Orgel 1985).

The acyclic analog 9-(1,3-dihydroxy-2-propoxy)-methyl-guanine can polymerize in the absence of a template (Schwartz et al 1987b), but the reaction proceeds more efficiently in the presence of poly(C). This analogue is one of particular interest, since it could easily form under prebiotic conditions from guanine, formaldehyde, and glycerol (Schwartz & Orgel 1985). These results support the idea that the first ancestral genetic systems may have consisted of nucleic acid–like informational replicating macromolecules derived from prochiral monomeric subunits (Joyce et al 1987, Joyce 1989).

THE ORIGIN OF THE FIRST CELLS

Needless to say, there is no detailed scheme of the mechanisms that led gradually—but not necessarily slowly—from the prebiotically formed monomers and polymers we have described above to the earliest cells. Attempts to solve this problem have sometimes relied on the assumption that life arose as a result of the spontaneous formation of a single substance that embodied within itself all of the fundamental properties associated with life (Troland 1914 et seq, Muller 1955). But there is no such thing as a living molecule. Biology has been unable to produce a wholly satisfactory definition of life, but at least on phenomenological grounds it is clear that one cannot reduce (the catalytic properties of RNA molecules notwithstanding) all characteristics of living systems to a particular substance that arises suddenly by a lucky combination of atoms or simple monomers (Lazcano 1986).

The Origins of Replication and Translation: Two Major Unsolved Problems

How the first biological systems capable of replication and translation emerged is one of the major problems in the study of the origins of life. It is even possible that the genetic code was established prior to the origin of RNA itself (Orgel 1987). It has been hypothesized that a replicative system could have resulted from the interaction between abiotically synthesized polynucleotides and simple catalytic peptides such as His-His (Shen et al 1990c), inside liposomes (Oró & Lazcano-Araujo 1981, 1984). However, the mere coexistence of small peptides and polynucleotides (or nucleic acid–like molecules) does not guarantee by itself that protein synthesis will develop.

In fact, even if we assume that an extremely rich supply of abiotic polypeptides with the catalytic and structural characteristics necessary for the maintenance and reproduction of systems based on polyribonucleotide replication were available in the primitive environment, in the absence of some mechanism of translation sooner or later these polypeptides would be exhausted and eventually hydrolyzed. Regardless of how many catalytic polypeptides were formed abiotically in the primitive Earth, protein synthesis would not have evolved without a replicating mechanism ensuring the maintenance, stability, and diversification of its basic components. In other words, the synthesis of peptide bonds took place via the intermediate formation of aminoacylated oligoribonucleotides, which were presumably ancestral to contemporary aminoacylated tRNAs. The primitive tRNAs, or proto-tRNAs, could translate the information coded in the base sequence of nucleic acid molecules (ancestral to messenger RNAs), which until then had been only templates for nonenzymatic polynucleotide replication (Lazcano 1986).

The ease with which amino acids and their polymers are formed under plausible prebiotic conditions has led to the idea that the first replicating systems consisted of polypeptides (Fox & Dose 1977, Dyson 1982, Kauffman 1986), but this possibility appears to us to be extremely unlikely (Miller & Orgel 1974, Orgel 1987). A counterview is represented by the hypothesis that early biological systems based both their reproduction and their metabolism in RNA molecules, an idea that was first suggested some 20 years ago (Woese 1967, Crick 1968, Orgel 1968, Sulston & Orgel 1971).

The idea of the primordial primacy of nucleic acid molecules has received considerable attention due to the discovery of catalytic and autocatalytic properties of RNA molecules (Pace & Marsh 1985, Zaug & Cech 1986, Cech 1987). These findings led Alberts (1986), Gilbert (1986), and Lazcano (1986) to suggest that there was no prebiotic translational synthesis of proteins—i.e. that although there are many experiments in which the abiotic formation of peptide bonds has been demonstrated, translational processes mediated by tRNAs and ribosomes evolved in RNA cells in which reproduction and metabolism had been dependent until then on catalytic RNA molecules.

The RNA World

The idea that RNA preceded DNA as genetic material has been expressed independently by many authors (cf. Lazcano 1986). This hypothesis is supported by several lines of evidence, including (*a*) the central role that different RNA molecules play in protein biosynthesis (Crick 1968), and (*b*) the well-known fact that this process can take place in the absence of DNA but not of RNA (Spirin 1986). Additional support is provided by (*c*) the existence of replicating biological systems such as viroids (Diener 1982) and RNA viruses (Reanney 1982) that use either single- or double-stranded RNA molecules to store genetic information; and by (*d*) the fact that the biosynthesis of deoxyribonucleotides always proceeds via the enzymatic reduction of ribonucleotides, i.e. all dexoyribonucleotides are formed directly or indirectly (as in the case of dTTP) from a cellular pool of ribonucleotides (Sprengel & Follmann 1981, Lammers & Follmann 1983).

The existence of a primordial replicating and catalytic apparatus devoid of both DNA and proteins and based solely on RNA molecules was suggested some time ago by Woese (1967), Crick (1968), and Orgel (1968). Such a possibility has received considerable support from the recently discovered catalytic properties of RNA (Cech 1987) and led Gilbert (1986) to propose the existence of what he has termed the RNA world. Evidence supporting the hypothesis that catalytic RNA molecules preceded proteins includes (*a*) the involvement of the 2′-OH group of ribose in several prebiotically plausible phosphorylation (Halmann et al 1969), hydrolytic (Usher & McHale 1976), and condensation reactions (White & Erickson 1981); and (*b*) the ubiquity of pyridine nucleotide coenzymes and similar ribonucleotide cofactors in metabolic pathways, which has led to several independent suggestions that coenzymes represent pregenetic code catalysts that were employed by early Archean cells before the appearance of proteins (Sulston & Orgel 1971, White 1976, 1982).

Although it is not known how protein synthesis appeared, Alberts (1986), Gilbert (1986), and Lazcano (1986) suggested that the basic selective pressure for the origin and stabilization of the primitive translation apparatus (genetic code, tRNAs, ribosomes) was the enhancement of the catalytic activities of RNA-based cells in order to increase their dynamic stability and survival (see also Joyce 1989, Orgel 1989, and references therein). However, these models should be updated. They should be revised following an in-depth study of the catalytic and replicative nucleic acid–like molecules (Cech 1987) and should consider the difficulties in the synthesis of prebiotic RNA discussed by Joyce et al (1987).

The Appearance of Membranes and Liposomes

If the origin of life did not depend on the chance appearance of a single molecule but on the gradual evolution of systems formed by sets of different compounds of biochemical significance, then it is obvious that a mechanism for keeping them together was required from the very beginning, i.e. a decisive step toward the origin of life must have been the early appearance of membrane-bound polymolecular systems (Oparin 1936 et seq, Haldane 1954).

Although some authors have argued that the formation of membranes occurred after the development of a replicating system (Eigen & Schuster 1978), the lack of compartments would not only severely limit the possible cooperative interaction between the different molecules forming the replicating apparatus but would also lead eventually to their dispersal. Membranes are essential to life, not only because they allow cells to maintain an internal microenvironment different from the exterior, but also because any self-replicating systems lacking them would be unable to undergo preferential accumulation and, eventually, differential replication (Oparin 1972, 1975, Oró 1980, Ferris & Usher 1983, Oró & Lazcano-Araujo 1981, 1984).

There has been no lack of theoretical and experimental models developed to study the properties of putative precellular systems. To cite just a few, this list includes coacervates (Oparin 1936), protenoid microspheres (Fox & Dose 1977), and sulphobes (Herrera 1942). However, all of these systems are formed under conditions that tend to produce the condensation of small molecules into polymers (Deamer & Oró 1980), and their choice as laboratory models of prebiotic systems has rested mainly upon the idiosyncratic assumptions about the nature of early life made by each researcher. A critique of these and other models has been carried out by Day (1984), whose detailed analysis led him to give considerable support to liposomes as the best contemporary laboratory models of precellular systems. The experimental evidence leading to this view has been presented by Oró et al (1978), Hargreaves & Deamer (1978a), Deamer & Oró (1980), Stillwell (1980), Ferris & Usher (1983), and Oró & Lazcano (1984).

Liposomes are spheres with diameters of 5–50 μm whose amphiphilic components self-assemble into vesicles with a double-layered membrane in the absence of any polymerization process (Deamer & Oró 1980). Although the prebiotic formation of lipids has not been studied as extensively as that of amino acids or nitrogen bases, their presence in the prebiotic environment is indicated by the nonenzymatic synthesis of neutral lipids, including fatty acids up to C18 synthesized by a modification of a Fischer-Tropsch process that produces several hydrocarbons (Nooner & Oró 1979). The prebiotic synthesis of lipids, including the amphiphilic phosphatidic acids (Epps et al 1978), phosphotidylcholine (Rao et al 1982), and phosphotidyl ethanolamine (Rao et al 1987), has been summarized and discussed by Oró et al (1978) and Hargreaves & Deamer (1978a). This liposome hypothesis will receive additional support if isotopic analysis confirms the extraterrestrial origin of the nonpolar membrane-forming molecules that have been extracted from samples of the Murchison carbonaceous chondrite (Deamer 1985, Deamer & Pashley 1989).

Since a C10–C12 linear fatty acid is the minimum size required to form a bilayer membrane (Deamer 1986), it is possible that the first liposomes that formed in the prebiotic oceans were relatively simple structures comparable to those described by Hargreaves & Deamer (1978b). Encapsulation of DNA within liposomes of more complex chemical nature has been achieved by dehydration-hydration cycles that may have occurred in intertidal settings in the prebiotic Earth, as they do today (Deamer & Barchfeld 1982). The evidence supporting the hypothesis that RNA molecules preceded contemporary DNA cellular genomes (Alberts 1986, Gilbert 1986, Lazcano 1986, Lazcano et al 1988b) led Baeza et al (1986) to study the encapsulation of polyribonucleotides within liposomes. The efficiency of encapsulation of poly(C) and poly(U) was not affected by the presence of a number of prebiotic condensing and catalytic agents, such as urea, cyanamide, and Zn^{++} (Baeza et al 1987). As indicated by liposome models of primordial cells, basic polypeptides of abiotic origin may have played an important role in the encapsulation of nucleic acid–like molecules within the lipidic boundaries (Jay & Gilbert 1987).

Models of transport of amino acids, nucleotides, and sugars inside a liposome have been developed by Stillwell (1980) using simple lipid-soluble carriers that may have existed in the prebiotic hydrosphere. Unfortunately, very little is known about the origin and early evolution of transport mechanisms and membrane-bound bioenergetic systems (Holden 1968, Wilson & Maloney 1978, Morowitz et al 1988). In our view, this is an area that requires intense study in order that we may develop an understanding of the origin of ion pumps and bioenergetic processes and the development of metabolism (Maloney & Wilson 1985, Maloney 1986).

EARLY BIOLOGICAL EVOLUTION

The Origin of the Genetic Code

A direct correlation has been found between the hydrophobicity ranking of most amino acids and their anticodons (Weber & Lacey 1978, Lacey & Mullins 1983). Four codon assignments, corresponding to Trp, Tyr, Ser, and Ile, do not fit into this scheme, which has led Lacey et al (1985) to suggest that these amino acids are later additions to the code. This idea is consistent with Wong's (1981) hypothesis of a coevolution of the genetic code and amino acid biosynthesis, which proposes that the code evolved to accommodate new amino acids that were not present in the prebiotic terrestrial environment. The monophyletic origin of the triplet-based genetic code is supported by its universal distribution among contemporary forms of life. The minor exceptions to the code that have been detected among mycoplasms and protists and in animal, fungal, and plant mitochondria appear to be recent adaptations (Jukes 1983, 1984, Fox 1987). The synthesis of proteins is one of the most complex biochemical processes occurring in living systems. It requires the participation of a complicated translation apparatus (ribosome), three kinds of RNAs, and many structural proteins, enzymes, and cofactors. One can only conceive the emergence of the translation apparatus through a gradual sequence of steps that go from the relatively simple activation of amino acids to ever more complex phases of structural and functional molecular evolution. In one of its simplest expressions, the RNA-directed synthesis of proteins could be visualized by the interaction of a polynucleotide (acting as mRNA) with the loop regions of short hairpin oligonucleotides (acting as tRNAs), each having a carboxyl-activated amino acid at the 3′-OH group of the terminal adenosine (CCA). If the proximity of two of the neighboring activated amino acid derivatives (AA-proto-tRNA) were of the order of molecular interatomic distances, then the formation of a peptide bond could take place by a combined transesterification-transpeptidation, which is an isoergonic process.

Studies on different aspects of this process from a prebiotic point of view have been carried out by several investigators. We know from the work of Usher & Needels (1984) and more recently of Lacey et al (1988) that the chirality of the sugar (D-ribose) determines the preferred chirality of the incorporated amino acid (L-amino acid) in the activated 3′-OH position. We also know from modeling studies by Kuhn & Waser (1981) that the hairpin oligonucleotides, acting at the same time as adaptors and activators of the amino acids, can be quite small (5–10 units) and still be able to assemble into a three-dimensional structure that may facilitate the formation of the peptide bond. A recent review by Orgel (1989) on the origin of polynucleotide-directed protein synthesis discusses the evolution of the adaptor moeity of these hairpins to eventually produce a triplet code.

Another detailed scheme based on RNA molecules and simple basic peptides has been developed by Maizels & Weiner (1987) to explain the origin of protein synthesis. They suggest that tRNA-like molecules initially evolved as tags at the 3′-ends of primordial replicating RNA genomes, which implies that the first proto-tRNAs and tRNA synthetases emerged prior to the other components of the protein biosynthetic system. It is also possible that primitive tRNAs with catalytic activity acted as their own synthetases, in which case vestigial evidence of this activity could still be detected under modified laboratory conditions (Lazcano 1986).

A somewhat different model has been developed by Orgel (1989), who suggests that prior to protein synthesis the attachment of amino acids to the 2′ (3′) termini of RNA templates favored initiation of replication at the end of a template instead of at internal sequences. These aminoacylated RNA molecules would become the evolutionary ancestors of later tRNAs. A second step leading to triplet-coded protein synthesis would be the pairwise association of the aminoacylated RNA molecules in such a way as to favor noncoded synthesis of peptides.

The Early Evolution of Protein Biosynthesis

Contemporary organisms contain information on their origins and evolutionary development in the form of informational macromolecules (Zuckerkandl & Pauling 1965) and biochemical pathways (Berry & Jensen 1988). The development of nucleic acid sequencing techniques applied to the study of ribosomal RNA has provided evidence of an early divergence among microorganisms that led to archaebacteria, eubacteria, and eukaryotes (Woese & Fox 1977, Pace et al 1986). The universal distribution among all contemporary forms of life of what is essentially the same genetic code, together with all the other highly conserved features of DNA replication and protein biosynthesis that are shared by archaebacteria, eubacteria, and eukaryotes, imply that the basic features of both genome replication and gene expression were established before the three major cellular lines diverged. Therefore, features in common found in these three cellular lines must have been present in their common ancestor (Woese & Fox 1977, Margulis & Guerrero 1986, Pace et al 1986).

The above appears to be the case for the protein synthetic molecular machinery, which may have been originally dependent on ribosomes in which no proteins were present and RNA played all the catalytic and structural roles (Crick 1968, Woese 1980). Protein biosynthesis is today a complex process in which a large number of molecules participate in a highly orderly manner (Spirin 1986), but a Darwinian view of the molecular processes of living systems implies that primitive translation could not have been as complex and efficient as in contemporary cells (Spirin 1976). This conclusion is supported by a number of observations, including several experiments that have shown that under in vitro conditions a simplified process of protein biosynthesis can take place in the absence of elongation factors, initiation components, and several ribosomal proteins (Spirin 1986).

At least three major phases can be recognized in the evolution of cellular genetic material: first, genomes of nucleic acid–like molecules (Joyce et al 1987, Joyce 1989); second, genomes of RNA; and finally, double-stranded DNA genomes, such as those present in all contemporary cells and in many viruses (Lazcano et al 1988b). Thus, even if we do not know how protein synthesis originated, it is obvious that if RNA preceded DNA as an informational macromolecule, then RNA polymerase must be one of the oldest proteins (Lazcano 1986, Lazcano et al 1988a). Genes coding for RNA polymerases are present in many viruses and in all cells, and among the latter, several different types of DNA-dependent RNA polymerases may be recognized, depending on the number and size of their subunits, on their amino acid sequence, and on the type of RNA transcribed (Lazcano et al 1989). Can vestiges of the original replicase be identified among this large array of RNA polymerases?

The hypothesis that contemporary cellular RNA polymerases were originally RNA dependent, i.e. were replicases, is supported by the observation that under slightly modified laboratory conditions normally DNA-dependent eubacterial and eukaryotic RNA polymerases can use polyribonucleotides and RNA molecules as templates (Llaca et al 1990). This replicase activity in contemporary eubacterial RNA polymerases is interpreted as vestigial from before the appearance of double-stranded DNA molecules (Lazcano 1986, Lazcano et al 1988a, Llaca et al 1990). Based on the comparative immunological studies on contemporary eukaryotic, archaebacterial, and eubacterial DNA-dependent RNA polymerases performed by Zillig et al (1985), and on comparisons of the nucleotide sequence of the genes coding for the largest subunits of *Escherichia coli* and several archaebacterial and eukaryotic RNA polymerases, it has been argued that an important vestige of the original replicase is found in the contemporary eubacterial RNA polymerase β' subunit and its homologues (Lazcano et al 1988a, 1989). This hypothesis is strongly supported by the amino acid sequence analysis of a single-unit viral RNA-dependent RNA polymerase homologous to the eubacterial β' subunit (Roy et al 1988).

The Biological Origin of DNA

The "RNA prior to DNA" hypothesis can be further developed by understanding the selective pressures that led to the appearance of the biosynthetic pathways of deoxyribose, thymine, and DNA-dependent DNA polymerases with proofreading activity (Lazcano 1986). The arguments supporting this idea have been critically reviewed elsewhere (Lazcano et al 1988b) and include (*a*) the increased stability to basic hydrolysis of the 2′-deoxy-containing phosphodiester backbone of DNA as compared with its ribo-equivalent (Ferris & Usher 1983); (*b*) the absence of proofreading activity in RNA polymerases, which leads to a higher rate of mutation in RNA genomes relative to DNA (Reanney 1982); (*c*) the tendency of genetic information to degrade in RNA genomes owing to the hydrolytic reaction that deaminates cytosine into uracil and the lack of a correcting enzyme (Kornberg 1980); and (*d*) the fact that ultraviolet irradiation produces a larger number of photochemical changes in RNA molecules relative to double-stranded DNA. Thus, the absence of atmospheric ultraviolet attenuation during the early Archean must have imposed an intense selection pressure favoring duplex DNA over other genetic information storage systems.

The universal distribution among the three main cellular lineages (Woese & Fox 1977, Pace et al 1986) of the biosynthetic pathways leading to deoxyribose and thymine, together with all the highly conserved features of DNA replication (DNA polymerases, DNA primases, DNA topoisomerases that couple the unwinding of double-stranded DNA with ATP hydrolysis, etc), imply a monophyletic origin for DNA. Moreover, since changes in the template specificity of RNA polymerases can be easily achieved in the presence of Mn^{++} ions (Llaca et al 1990), the evolutionary transition from the ancestral RNA-dependent RNA polymerase into the present day DNA-dependent enzymes must have required only minor modifications of the active site. The functional similarities between all cellular nucleic acid polymerases (DNA- and RNA-polymerases, reverse-transcriptases, DNA primases) have been interpreted to imply that the genes coding for some of the proteins involved in DNA replication resulted from gene duplications and further differentiation from the ancestral replicase (Lazcano et al 1988b). This common origin is suggested by the high evolutionary conservation of regions of nucleotide sequences of the genes coding for these proteins (Lazcano et al 1989). Further evolutionary processes leading to contemporary differences in energy production, metabolism, morphology, and ecological setting among microbes are discussed elsewhere (Broda 1975, Margulis 1982, Margulis & Guerrero 1986, Woese 1987).

LIFE IN THE UNIVERSE

We have no evidence of extraterrestrial life. However, in spite of all the uncertainties involved in our description of the processes that led to the emergence of biological systems on our planet, it is clear that life is neither a miracle nor the result of a chance event but rather the outcome of a long evolutionary process. Whether or not this process can take place in any other place in the Universe is a matter for speculation, but the cosmic abundances (Cameron 1980) of carbon, nitrogen, oxygen, and other biogenic elements (Oró 1963a), the existence of extraterrestrial organic compounds (reviewed above), and the processes of stellar and planetary formation are certainly suggestive that at least some of the requirements for life are met elsewhere in our Galaxy (Oró et al 1982, Oró 1988).

More than a decade has passed since the Viking landers investigated the possibility of life on Mars. The data concerning this question have been summarized elsewhere (Flinn 1977, Ezell & Ezell 1984); the primary conclusion was that the biology experiments found no unequivocal evidence for the existence of a martian biota (Mazur et al 1978, Oró 1979, Bieman et al 1976, 1977, Horowitz et al 1976).

Organic and biological analyses of several samples of the martian regolith found no evidence of organic matter at the two Viking spacecraft landing sites. No organic matter was found at either site by the pyrolysis-gas chromatograph–mass spectrometer (GC–MS) at detection limits of the order of parts per billion and for a few substances closer to parts per million (Bieman et al 1976, 1977). On the other hand, two of the experiments designed to investigate the possibility of microorganisms in the martian regolith—the labeled release (LR; Levin & Straat 1976) and the gas exchange (GEx; Oyama & Berdahl 1977) experiments—gave what appeared to be positive results. The pyrolytic release experiment gave data that were interpreted not to be biological in nature (Horowitz et al 1976).

Two different possible explanations were suggested for the positive results of the LR and GEx experiments: Either very active microbial metabolic processes were taking place, or else chemical oxidation processes brought about by a chemically active martian regolith were responsible for the oxidation of the radioactively labeled nutrients. The latter view has been found to account for most of the observations (Oró 1976). However, since Mars appears to have been particularly rich in water in the past (Carr 1986, Pollack et al 1987), it is quite possible that the formation of biochemical compounds took place in the primitive Mars environment during the first 800–1000 Myr of its history. A more detailed discussion of this possibility has been reviewed recently by Oró & Mills (1989).

It is possible that conditions suitable for the origin and evolution of life have existed elsewhere in the solar system and may still persist in some of its minor bodies, such as Europa (Oró & Mills 1989, Oró et al 1990). The exobiological implications of the ocean of liquid water that some believe may exist under the cracked, icy surface of Europa, one of the Galilean satellites of Jupiter, is critically discussed elsewhere (Oró et al 1990). Finally, it is possible that chemical evolution and synthesis of biochemical compounds have occurred and are occurring now in Titan (Oró & Mills 1989). Titan has a reducing atmosphere whose origin and chemical composition have been discussed in considerable detail by Owen and his collaborators (Caldwell & Owen 1984, Owen & Gautier 1989). Even though the origin of the CH_4, CO, and N_2 present in Titan's atmosphere remains unresolved, Owen & Gautier (1989) have developed the hypothesis that Titan's major constituents, other than being equilibrated with the gases of the proto-saturnian nebula, are original or derived from primordial interstellar matter that accreted and formed the bulk of this satellite. Whereas the essential lack of liquid water renders Titan's atmosphere an unlikely model for prebiological synthesis of oxygen-bearing biochemical compounds, this satellite provides an excellent "anhydrous" model for the study of atmospheric chemical evolution (Oró & Mills 1989). In contrast to the essentially anhydrous condition of its atmosphere, it is possible that a thin ocean of a water-ammonia mixture (87% and 13%, respectively) is present in the subsurface of Titan if the scenario described by Lunine & Stevenson (1987) prevails in this satellite. If oceans and dissolved organic matter exist both in Titan and in Europa, a tremendous richness of subsurface organic synthetic reactions could be visualized for these satellites, which may explain, in part, some of the darker spots observed in Europa's outer surface and more recently in Triton, the remarkable satellite of Neptune.

This has been discussed recently by Carl Sagan and other members of the Voyager team (see Sagan 1989) following the successful exploration of

Neptune and its satellites by the *Voyager 2* robot spacecraft prior to its departure from the planetary frontiers of the solar system in search of other worlds.

This is indeed a fitting finale for the accomplishments of the National Aeronautics and Space Administration in this century. These accomplishments will include *Galileo* and other future missions to the terrestrial and outer planets, comets, and asteroids, and the long-awaited launch of the Hubble Space Telescope in a terrestrial orbit. Hopefully, the latter will enable us to see far-away stars and galaxies of our expanding universe.

Some of us hope to see the establishment of human colonies on the Moon and Mars by the next century in order for science to unravel pertinent questions concerning chemical evolution and the origin of life on our beautiful, blue terrestrial planet.

It is difficult to conceive from a scientific point of view that "intelligent" life on Earth is a singular phenomenon in the trillions of galaxies that comprise the observable universe.

Acknowledgments

This work was supported in part by NASA grants NGR 44-005-002 (to JO) and NAGW-20 (to SLM). We would also like to thank James Lacey for recent information on stereoselective aminoacylation, and Carl Sagan, Bishun Khare, W. Reid Thompson, Armand Delsemme, and Christopher Chyba for reprints and preprints pertinent to the chemical evolution of the outer solar system and the primitive Earth. We are also very grateful to Thomas Mills, Marianna O'Rourke, and the editors for their help in the final preparation of the manuscript.

Literature Cited

Abelson, P. H. 1966. Chemical events on the primitive Earth. *Proc. Natl. Acad. Sci. USA* 55: 1365–72

Alberts, B. M. 1986. The function of the hereditary materials: biological catalyses reflect the cell's evolutionary history. *Am. Zool.* 26: 781–96

Anders, E., Owen, T. 1977. Origins and abundance of volatiles. *Science* 198: 453–65

Anders, E., Hayatsu, R., Studier, M. H. 1974. Catalytic reactions in the solar nebula: implications for interstellar molecules and organic compounds in meteorites. In *Cosmochemical Evolution and the Origins of Life I*, ed. J. Oró, S. L. Miller, C. Ponnamperuma, R. S. Young, pp. 57–67. Boston: Reidel. 523 pp.

Arrhenius, G. 1987. The first 800 million years: environmental models for early Earth. *Earth, Moon and Planets* 37: 187–99

Awramik, S. M. 1981. The pre-Phanerozoic biosphere—three billion years of crises and opportunities. In *Biotic Crises in Ecological and Evolutionary Time*, ed. M. Nitecki, pp. 83–102. New York: Academic. 301 pp.

Awramik, S. M. 1982. The pre-Phanerozoic fossil record. See Holland & Schidlowski 1982, pp. 67–81

Awramik, S. M., Schopf, J. W., Walter, M. R. 1983. Filamentous fossil bacteria from the Archean of Western Australia. *Precambrian Res.* 20: 357–74

Baeza, I., Ibanez, M., Santiago, C., Wong,

C., Lazcano, A., Oró, J. 1986. Studies on precellular evolution: the encapsulation of polyribonucleotides by liposomes. *Adv. Space Res.* 6: 39–43

Baeza, I., Ibanez, M., Lazcano, A., Santiago, C., Arguello, C., Wong, C., Oró, J. 1987. Liposomes with polyribonucleotides as model of precellular systems. *Origins Life* 17: 321–31

Barghoorn, E. S., Schopf, J. W. 1965. Microorganisms three billion years old from the Precambrian of South Africa. *Science* 152: 758–63

Barghoorn, E. S., Tyler, S. A. 1965. Microorganisms from the Gunflint chert. *Science* 147: 563–77

Basile, B., Lazcano, A., Oró, J. 1985. Prebiotic syntheses of purines and pyrimidines. *Life Sci. Space Res.* 21: 125–31

Bell, M. B., Feldman, P. A., Kwok, S., Matthews, H. E. 1982. Detection of $HC_{11}N$ in IRC + 10° 216. *Nature* 295: 389–91

Benz, W., Slattery, W. L., Cameron, A. G. W. 1986. The origin of the Moon and the single impact hypothesis. I. *Icarus* 66: 515–35

Benz, W., Slattery, W. L., Cameron, A. G. W. 1987. The origin of the Moon and the single impact hypothesis. II. *Icarus* 71: 30–45

Berry, A., Jensen, R. A. 1988. Biochemical evidence for phylogenetic branching patterns. *BioScience* 38: 99–103

Bickle, M. J. 1978. Heat loss from the Earth: a constraint on Archean tectonics from the relation between geothermal gradients and the rate of plate production. *Earth Planet. Sci. Lett.* 40: 301–15

Bickle, M. J. 1986. Global thermal histories. *Nature* 319: 13–14

Bieman, K., Oró, J., Toulmin, P. III, Orgel, L. E., Nier, A. O., Anderson, D. M., Simmonds, D. R., Flory, D., Diaz, A. V., Rushneck, D. P., Biller, J. A. 1976. Search for organic and volatile inorganic compounds in two surface samples from the Chryse Planitia region of Mars. *Science* 194: 72–76

Bieman, K., Oró, J., Toulmin, P. III, Orgel, L. E., Nier, A. O., Anderson, D. M., Simmonds, D. P., Flory, D., Diaz, A. V., Rushneck, D. R., Biller, J. A., LaFleur, A. L. 1977. The search for organic substances and inorganic volatile compounds in the surface of Mars. *J. Geophys. Res.* 82: 4641–58

Bridgwater, D., Allaart, J. H., Schopf, J. W., Klein, C., Walter, M. R., Barghoorn, E. S., Strother, P., Knoll, A. H., Gorman, B. E. 1981. Microfossil-like objects from the Archean of Greenland: a cautionary note. *Nature* 289: 51–53

Bridson, P. K., Orgel, L. E. 1980. Catalysis of accurate poly(C)-directed synthesis of 3′-5′-linked oligoguanylates of Zn^{++}. *J. Mol. Biol.* 144: 567–77

Bridson, P. K., Fakhrai, H., Lohrmann, R., Orgel, L. E., Van Roode, M. 1981. Template-directed synthesis of oligoguanylic acids–metal ion catalysis. In *Origin of Life*, ed. Y. Wolman, pp. 233–39. Dordrecht: Reidel. 613 pp.

Broda, E. 1975. *The Evolution of the Bioenergetic Processes.* Oxford: Pergamon. 211 pp.

Brown, H. 1952. Rare gases and the formation of the Earth's atmosphere. In *The Atmospheres of the Earth and Planets*, ed. G. H. Kuiper, pp. 258–66. Chicago: Univ. Chicago Press. 428 pp.

Brown, R. D., Godfrey, P. D., Storey, J. W. V., Bassez, M. P., Robinson, B. J., Batchelor, R. A., McCulloch, M. G., Rydbeck, O. E. H., Hjlamarson, A. G. 1979. A search for interstellar glycine. *Mon. Not. R. Astron. Soc.* 185: 5P–8P

Butlerow, A. 1861. Formation sinthetique d'une substance sucree. *C. R. Acad. Sci.* 53: 145–47

Byerly, G. R., Lowe, D. R., Walsh, M. M. 1986. Stromatolites from the 3300–3500 Myr Swaziland Supergroup, Barberton Mountain Land, South Africa. *Nature* 319: 489–91

Caldwell, J., Owen, T. 1984. Chemical evolution on the giant planets and Titan. *Adv. Space Res.* 4: 51–58

Calvin, M. 1969. *Chemical Evolution.* Oxford: Oxford Univ. Press. 278 pp.

Cameron, A. G. W. 1980. A new table of abundances of the elements in the Solar System. In *Origin and Distribution of the Elements*, ed. L. A. Ahrens, pp. 125–43. New York: Pergamon. 280 pp.

Cameron, A. G. W., Benz, W. 1989. Possible scenarios resulting from the giant impact. *Lunar Planet. Sci. XX*, pp. 137–38. Houston: Luner Planet. Inst. (Abstr.)

Canuto, V. M., Levine, J. S., Augustssen, T. R., Imhoff, C. L. 1982. UV radiation from the young Sun and oxygen and ozone levels in the prebiological paleoatmosphere. *Nature* 296: 816–20

Carlin, R. K. 1980. Poly(A): a new evolutionary probe. *J. Theor. Biol.* 82: 353–62

Carr, M. H. 1986. Mars: a water-rich planet? *Icarus* 68: 187–216

Cech, T. R. 1987. The chemistry of self-splicing RNA and RNA enzymes. *Science* 236: 1532–36

Cech, T. R., Zaug, A. J., Grabowski, P. J. 1981. In vitro splicing of the ribosomal RNA precursor of *Tetrahymena*: involvement of a guanosine nucleotide in the excision of the intervening sequence. *Cell* 27: 487–96

Chang, S. 1981. Organic chemical evolution.

In *Life in the Universe*, ed. J. Billingham, pp. 21–46. Cambridge, Mass: MIT Press. 461 pp.

Chang, S. 1982. Prebiotic organic matter: possible pathways for synthesis in a geological context. *Phys. Earth Planet. Inter.* 29: 261–80

Chang, S., DesMarais, D., Mack, R., Miller, S. L., Strathearn, G. E. 1983. Prebiotic organic syntheses and the origin of life. See Schopf 1983, pp. 53–92

Chyba, C. F. 1987. The cometary contribution to the oceans of the primitive Earth. *Nature* 330: 632–35

Compston, W., Pidgeon, R. T. 1986. Jack Hills, evidence of more very old detrital zircons in Western Australia. *Nature* 321: 766–69

Crick, F. H. C. 1968. The origin of the genetic code. *J. Mol. Evol.* 38: 367–79

Cronin, J. R. 1989. Origin of organic compounds in carbonaceous chondrites. *Adv. Space Res.* 9: 54–64

Cruikshank, D. P. 1989. Dark surfaces of asteroids and comets: evidence for macromolecular carbon compounds. *Adv. Space Res.* 9: 65–71

Day, W. 1984. *Genesis on Planet Earth.* New Haven, Conn: Yale Univ. Press. 299 pp.

Deamer, D. W. 1985. Boundary structures are formed by organic components of the Murchison carbonaceous chondrite. *Nature* 317: 792–94

Deamer, D. W. 1986. Role of amphiphilic compounds on the evolution of membrane structure on the early Earth. *Origins Life* 17: 3–25

Deamer, D. W., Barchfeld, G. L. 1982. Encapsulation of macromolecules by lipid vesicles under simulated prebiotic conditions. *J. Mol. Biol.* 18: 203–6

Deamer, D. W., Oró, J. 1980. Role of lipids in prebiotic structures. *BioSystems* 12: 167–75

Deamer, D. W., Pashley, R. M. 1989. Amphiphilic components of the Murchison carbonaceous chondrite: surface properties and membrane formation. *Origins Life* 19: 21–38

Decker, P., Schweer, H., Pohlmann, R. 1982. Bioids. X. Identification of formose sugars, presumably prebiotic metabolites, using capillary gas chromatography/gas chromatography–mass spectrometry on *n*-butoxime trifluoroacetate on OV-225. *J. Chromatogr.* 244: 281–91

Delsemme, A. H. 1981. Are comets connected to the origin of life? In *Comets and the Origin of Life*, ed. C. Ponnamperuma, pp. 141–59. Dordrecht: Reidel. 282 pp.

Delsemme, A. H. 1984. The cometary connection with prebiotic chemistry. *Origins Life* 14: 51–60

Delsemme, A. H. 1988. The chemistry of comets. *Philos. Trans. R. Soc. London* 325: 509–23

De Zafra, R. L., Thaddeus, P., Kutner, M., Scoville, N., Solomon, P. M., Weaver, H., Williams, D. R. W. 1971. Search for interstellar furan and imidazole. *Astrophys. Lett.* 10: 1–3

d'Hendecourt, L. B., Allamandola, L. J., Greenberg, J. M. 1986. Time-dependent chemistry in dense molecular clouds. II. Ultraviolet photoprocessing and infrared spectroscopy of grain mantles. *Astron. Astrophys.* 158: 119–34

Diener, T. O. 1982. Viroids and their interactions with host cells. *Annu. Rev. Microbiol.* 36: 239–58

Dyson, F. J. 1982. A model for the origin of life. *J. Mol. Evol.* 18: 344–50

Eichberg, J., Sherwood, E., Epps, D. E., Oró, J. 1977. Cyanamide-mediated synthesis under plausible primitive Earth conditions. IV. The synthesis of acylglycerols. *J. Mol. Evol.* 10: 221–30

Eigen, M., Schuster, P. 1978. The hypercycle: a principle of natural self-organization. *Naturwissenschaften* 65: 341–69

Epps, D. E., Sherwood, E., Eichberg, J., Oró, J. 1978. Cyanamide-mediated synthesis under plausible primitive Earth conditions. V. The synthesis of phosphatidic acids. *J. Mol. Evol.* 11: 279–92

Epstein, S., Krishnamurthy, R. V., Cronin, J. R., Pizzarello, S., Yuen, G. U. 1987. Unusual stable isotope ratios in amino acid and carboxylic acid extracts from the Murchison meteorite. *Nature* 326: 477–79

Ernst, W. G. 1983. The early Earth and the Archean rock record. See Schopf 1983, pp. 41–52

Ezell, E. C., Ezell, L. N. 1984. On Mars: exploration of the Red Planet, 1958–1978. *NASA SP 4212.* 537 pp.

Farley, J. 1977. *The Spontaneous Generation Debate From Descartes to Oparin.* Baltimore: Johns Hopkins Univ. Press. 225 pp.

Ferris, J. P. 1987. Prebiotic synthesis: problems and challenges. *Cold Spring Harbor Symp. Quant. Biol.* 52: 29–35

Ferris, J. P., Hagan, W. J. Jr. 1984. HCN and chemical evolution: the possible role of cyano compounds in prebiotic synthesis. *Tetrahedron* 40: 1093–1120

Ferris, J. P., Orgel, L. E. 1965. Aminomalonitrile and 4-amino-5-cyanoimidazole in hydrogen cyanide polymerization and adenine synthesis. *J. Am. Chem. Soc.* 87: 4976–77

Ferris, J. P., Usher, D. A. 1983. Origins of life. In *Biochemistry*, ed. G. Zubay, pp. 1191–1241. Reading, Mass: Addison-Wesley. 1268 pp.

Ferris, J. P., Joshi, P. C., Edelson, E., Lawless, J. G. 1978. HCN: a plausible source of purines, pyrimidines and amino acids on the primitive Earth. *J. Mol. Evol.* 11: 293–311

Flinn, E. A., ed. 1977. *Scientific Results of the Viking Project. J. Geophys. Res. Spec. Iss.*, Vol. 82

Fox, S. W., Dose, K. 1977. *Molecular Evolution and the Origin of Life*. New York: Marcel Dekker. 370 pp.

Fox, T. D. 1987. Natural variation in the genetic code. *Annu. Rev. Genet.* 21: 67–91

Frank, L. A., Sigwarth, J. B., Craven, J. D. 1986. On the influx of small comets into the Earth's upper atmosphere. *Geophys. Res. Lett.* 13: 303–6

Friedmann, N., Miller, S. L. 1969. Synthesis of valine and isoleucine in primitive Earth conditions. *Nature* 221: 1152–53

Friedmann, N., Haverland, W. J., Miller, S. L. 1971. Prebiotic synthesis of the aromatic and other amino acids. In *Chemical Evolution and the Origin of Life*, ed. R. Buvet, C. Ponnamperuma, pp. 123–35. Amsterdam: North-Holland. 560 pp.

Gaustad, J. E., Vogel, S. N. 1982. High energy solar radiation and the origin of life. *Origins Life* 12: 3–8

Gilbert, W. 1986. The RNA world. *Nature* 319: 618

Greenberg, J. M. 1981. Chemical evolution of interstellar dust. In *Comets and the Origin of Life*, ed. C. Ponnamperuma, pp. 111–27. Dordrecht: Reidel. 282 pp.

Haldane, J. B. S. 1929. The origin of life. *Ration. Ann.* 148: 3–10

Haldane, J. B. S. 1954. The origins of life. *New Biol.* (*Penguin*) 16: 12–27

Halmann, M., Sanchez, R. A., Orgel, L. E. 1969. Phosphorylation of D-ribose in aqueous solution. *J. Org. Chem.* 34: 3702–3

Halvorson, H. O., Van Holde, K. E., eds. 1980. *The Origins of Life and Evolution*. New York: Alan R. Liss. 126 pp.

Harada, K. 1967. Formation of amino-acids by thermal decomposition of formamidine—oligomerization of hydrogen cyanide. *Nature* 214: 479–80

Hargreaves, W. R., Deamer, D. W. 1978a. Origin and early evolution of bilayer membranes. In *Light-Transducing Membranes: Structure, Function and Evolution*, ed. D. W. Deamer, pp. 23–59. New York: Academic. 358 pp.

Hargreaves, W. R., Deamer, D. W. 1978b. Liposomes from ionic, single-chain amphiphiles. *Biochemistry* 17: 3759–68

Hawker, J. R., Oró, J. 1981. Cyanamide mediated synthesis of Leu, Ala, Phe peptides under plausible primitive Earth conditions. In *Origin of Life*, ed. Y. Wolman, pp. 225–32. Dordrecht: Reidel. 613 pp.

Hayes, J. M. 1967. Organic constituents of meteorites—a review. *Geochim. Cosmochim. Acta* 31: 1395–1440

Herrera, A. L. 1942. A new theory of the origin and nature of life. *Science* 96: 14

Hinkle, K. W., Keady, J. J., Bernath, P. F. 1988. Detection of C_3 in the interstellar shell of IRC+10216. *Science* 241: 1319–22

Holden, J. T. 1968. Evolution of transport mechanisms. *J. Theor. Biol.* 21: 97–102

Holland, H. D. 1984. *The Chemical Evolution of the Atmosphere and Oceans*. Princeton: Princeton Univ. Press. 583 pp.

Holland, H. D., Schidlowski, M., eds. 1982. *Mineral Deposits and the Evolution of the Biosphere*. Berlin: Springer-Verlag. 333 pp.

Hollis, J. M., Snyder, L. E., Suenram, R. D., Lovas, F. J. 1980. A search for the lowest energy conformer of interstellar glycine. *Astrophys. J.* 241: 1001–6

Horowitz, N. H., Hobby, G. L., Hubbard, J. S. 1976. The Viking carbon assimilation experiments: interim report. *Science* 194: 1321–22

Hulshof, J., Ponnamperuma, C. 1976. Prebiotic condensation reactions in an aqueous medium: a review of condensing agents. *Origins Life* 7: 197–224

Inoue, T., Orgel, L. E. 1981. Substituent control of the poly(C)–directed oligomerization of guanosine 5′ phosphorimidazole. *J. Am. Chem. Soc.* 103: 7666–67

Inoue, T., Orgel, L. E. 1982. Oligomerization of (guanosine 5′-phosphor)-2-methylimidazolide on poly(C): an RNA polymerase model. *J. Mol. Biol.* 162: 201–17

Ip, W. H., Fernandez, J. A. 1988. Exchange of condensed matter among the outer terrestrial protoplanets and the effect on surface impact and atmospheric accretion. *Icarus* 74: 47–61

Irvine, W. 1989. Observational astrochemistry: recent results. *Adv. Space Res.* 9: 3–12

Jay, D. G., Gilbert, W. 1987. Basic protein enhances the incorporation of DNA into lipid vesicles: model for the formation of primordial cells. *Proc. Natl. Acad. Sci. USA* 84: 1978–80

Joyce, G. F. 1987. Nonenzymatic template-directed synthesis of informational macromolecules. *Cold Spring Harbor Symp. Quant. Biol.* 52: 41–51

Joyce, G. F. 1989. RNA evolution and the origins of life. *Nature* 338: 217–24

Joyce, G. F., Orgel, L. E. 1986. Non-enzymatic template directed synthesis on RNA random copolymers: poly(C,G) templates. *J. Mol. Biol.* 188: 433–41

Joyce, G. F., Inoue, T., Orgel, L. E. 1984. Non-enzymatic template-directed synthesis on RNA random copolymers: poly(C,U) templates. *J. Mol. Biol.* 176: 279–306

Joyce, G. F., Schwartz, A. W., Miller, S. L., Orgel, L. E. 1987. The case for an ancestral genetic system involving simple analogues of the nucleotides. *Proc. Natl. Acad. Sci. USA* 84: 4398–4402

Jukes, T. H. 1983. Evolution of the amino acid code. In *Evolution of Genes and Proteins*, ed. M. Nei, R. Kohen, pp. 191–207. Sunderland, Mass: Sinauer Assoc. 331 pp.

Jukes, T. H. 1984. Evolution of anticodons. *Life Sci. Space Res.* 21: 177–82

Kamminga, H. 1988. Historical perspective: the problem of the origin of life in the context of developments in biology. *Origins Life* 18: 1–11

Kanavarioti, A., White, D. H. 1987. Kinetic analysis of the template effect in ribooliogguanylate elongation. *Origins Life* 17: 333–49

Kanavarioti, A., Bernasconi, C. F., Doodokyan, D. L., Alberan, D. J. 1989. Magnesium ion catalysed P–N bond hydrolysis in imidazolide-activated nucleotides. Relevance to template-directed synthesis of polynucleotides. *J. Am. Chem. Soc.* 11: 7247–57

Kauffman, S. A. 1986. Autocatalytic sets of proteins. *J. Theor. Biol.* 119: 1–24

Kenyon, D. H., Steinman, G. 1969. *Biochemical Predestination.* New York: McGraw-Hill. 301 pp.

Kerridge, J. F., Anders, E. 1988. Boundary conditions for the origin of the Solar System. In *Meteorites and the Early Solar System*, ed. J. F. Kerridge, M. S. Matthews, pp. 1150–54. Tucson: Univ. Ariz. Press. 1269 pp.

Kissel, J., Krueger, F. R. 1987. The organic component in dust from Comet Halley as measured by the PUMA mass spectrometer on board *Vega 1*. *Nature* 326: 755–60

Knoll, A. H. 1985. The distribution and evolution of microbial life in the late Proterozoic era. *Annu. Rev. Microbiol.* 39: 391–417

Knoll, A. H. 1986. Geological evidence for early evolution. *Treballs Soc. Cat. Biol.* 39: 113–41

Knoll, A. H., Barghoorn, E. S. 1977. Archean microfossils showing cell division from the Swaziland system of South Africa. *Science* 198: 396–98

Kornberg, A. 1980. *DNA Replication.* San Francisco: Freeman. 724 pp.

Kuhn, H., Waser, J. 1981. Molecular self-organization and the origin of life. *Angew. Chem. Int. Ed. Engl.* 20: 500–20

Kvenvolden, K. A., ed. 1974. *Geochemistry and the Origin of Life.* Stroudsburg, Pa: Dowden, Hutchinson & Ross. 422 pp.

Kvenvolden, K. A., Lawless, J. G., Perking, K., Peterson, E., Flores, J., Ponnamperuma, C. 1970. Evidence for extraterrestrial amino acids and hydrocarbons in the Murchison meteorite. *Nature* 228: 923–26

Kvenvolden, K. A., Lawless, J. G., Ponnamperuma, C. 1971. Non-protein amino acids in carbonaceous meteorites. *Proc. Natl. Acad. Sci. USA* 68: 486–90

Lacey, J. C., Mullins, D. W. 1983. Experimental studies related to the origin of the genetic code and the process of protein synthesis—a review. *Origins Life* 13: 3–42

Lacey, J. C., Hall, L. M., Mullins, D. W. 1985. Rationalization of some genetic code assignments. *Origins Life* 16: 69–79

Lacey, J. C., Hawkins, A. F., Thomas, R. D., Watkins, C. L. 1988. Differential distribution of D and L amino acids between the 2′ and 3′ positions of the AMP residue at the 3′ terminus of transfer ribonucleic acid. *Proc. Natl. Acad. Sci. USA* 85: 4996–5000

Lammers, M., Follmann, H. 1983. The ribonucleotide reductases—a unique group of metalloenzymes essential for cell proliferation. *Struct. Bonding* 54: 27–91

Lazcano, A. 1986. Prebiotic evolution and the origin of cells. *Treballs Soc. Cat. Biol.* 39: 73–103

Lazcano-Araujo, A., Oró, J. 1981. Cometary material and the origins of life on Earth. In *Comets and the Origin of Life*, ed. C. Ponnamperuma, pp. 191–225. Dordrecht: Reidel. 282 pp.

Lazcano, A., Oró, J., Miller, S. L. 1983. Primitive Earth environments: organic syntheses and the origin and early evolution of life. *Precambrian Res.* 20: 259–82

Lazcano, A., Fastag, J., Gariglio, P., Ramirez, C., Oró, J. 1988a. On the early evolution of RNA polymerase. *J. Mol. Evol.* 27: 365–76

Lazcano, A., Guerrero, R., Margulis, L., Oró, J. 1988b. The evolutionary transition from RNA to DNA in early cells. *J. Mol. Evol.* 27: 283–90

Lazcano, A., Llaca, V., Fox, G. E., Oró, J. 1989. A classification of RNA polymerases based on their evolutionary relatedness. In *Abstr. 6th ISSOL Meet. and 9th Int. Conf. Origins of Life, Prague*, pp. 194–95 (Abstr.)

Levin, G. V., Straat, P. A. 1976. Viking labeled release biology experiment: interim report. *Science* 194: 1322–29

Llaca, V., Silva, E., Lazcano, A., Rangel, L. M., Gariglio, P., Oró, J. 1990. In search of the ancestral RNA polymerase: an experimental approach. In *Prebiological Self-Organization of Matter. College Park Colloq. Chem. Evol., 8th*, ed. C. Ponnamperuma, F. Eirich. Hampton, Va: Deepak-Sci. Technol. Corp. In press

Lohrmann, R., Orgel, L. E. 1977. Reactions of adenosine 5′-phosphorimidazole with adenosine analogs on a polyuridylic acid template. The uniqueness of the 2′-3′ unsubstituted beta-ribosyl system. *J. Mol. Biol.* 113: 193–98

Lohrmann, R., Bridson, P. K., Orgel, L. E. 1980. Efficient catalysis of polycytidylic acid-directed oligoguanylate formation by Pb^{++}. *J. Mol. Biol.* 142: 555–67

Lowe, C. U., Rees, M. W., Markham, R. 1963. Synthesis of complex organic compounds from simple precursors: formation of amino acids, amino-acid polymers, fatty acids and purines from ammonium cyanide. *Nature* 199: 219–22

Lunine, J. I., Stevenson, D. J. 1987. Clathrate and ammonia hydrates at high pressure: application to the origin of methane on Titan. *Icarus* 70: 61–77

Maizels, N., Weiner, A. M. 1987. Peptide-specific ribosomes, genomic tags, and the origin of the genetic code. *Cold Spring Harbor Symp. Quant. Biol.* 52: 743–49

Maloney, P. C. 1986. Evolution and ion pumps. In *Abstr., 5th ISSOL Meet. and 8th Int. Conf. Origins of Life, Berkeley, Calif.*, pp. 181–82 (Abstr.)

Maloney, P. C., Wilson, T. H. 1985. The evolution of ion pumps. *BioScience* 35: 43–48

Mar, A., Oró, J. 1990. Synthesis of the coenzymes ADPG, GDPG and CDP-ethanolamine under primitive Earth conditions. *J. Mol. Evol.* In press

Mar, A., Dworkin, J., Oró, J. 1987. Non-enzymatic synthesis of the coenzymes uridine diphosphate glucose and cytidine diphosphate choline, and other phosphorylated metabolic intermediates. *Origins Life* 17: 307–19

Margulis, L. 1970. *Origin of Eukaryotic Cells*. New Haven, Conn: Yale Univ. Press. 349 pp.

Margulis, L. 1982. *Symbiosis in Cell Evolution*. San Francisco: Freeman. 419 pp.

Margulis, L., Guerrero, R. 1986. Not "origins of life" but "evolution in microbes." *Treballs Soc. Cat. Biol.* 39: 105–12

Margulis, L., Barghoorn, E. S., Asherdorf, D., Banerjee, S., Chase, D., et al. 1980. The microbial community in the layered sediments at Laguna Figueroa, Baja California, Mexico: does it have Precambrian analogues? *Precambrian Res.* 11: 93–123

Matthews, C. N., Ludicky, R. 1987. The dark nucleus of Comet Halley: hydrogen cyanide polymer. *Proc. ESLAB Symp., 20th, Exploration of Halley's Comet. ESA Publ. SP-250*, 2: 273–77

Mazur, P., Barghoorn, E. S., Halvorson, H. O., Jukes, T. H., Kaplan, I. R., Margulis, L. 1978. Biological implications of the Viking mission to Mars. *Space Sci. Rev.* 22: 3–34

Mildvan, A., Loeb, L. 1979. The role of metal ions in the mechanisms of DNA and RNA polymerases. *CRC Crit. Rev. Biochem.* 6: 219–44

Miller, S. L. 1953. A production of amino acids under possible primitive Earth conditions. *Science* 117: 528–29

Miller, S. L. 1957. The mechanism of synthesis of amino acids by electric discharges. *Biochim. Biophys. Acta* 23: 480–87

Miller, S. L. 1982. Prebiotic synthesis of organic compounds. See Holland & Schidlowski 1982, pp. 155–76

Miller, S. L. 1987. Which organic compounds could have occurred on the prebiotic Earth? *Cold Spring Harbor Symp. Quant. Biol.* 52: 17–27

Miller, S. L., Bada, J. L. 1988. Submarine hot springs and the origin of life. *Nature* 334: 609–11

Miller, S. L., Orgel, L. E. 1974. *The Origins of Life on Earth*. Englewood Cliffs, NJ: Prentice-Hall. 229 pp.

Miller, S. L., Van Trump, J. E. 1981. The Strecker synthesis in the primitive ocean. In *Origin of Life*, ed. Y. Wolman, pp. 135–41. Dordrecht: Reidel. 613 pp.

Miller, S. L., Urey, H. C., Oró, J. 1976. Origin of organic compounds on the primitive Earth and in meteorites. *J. Mol. Evol.* 9: 59–72

Moorbath, S., O'Nions, R. K., Pankhurst, F. J. 1973. Early Archean age for the Isua Iron-Formation, West Greenland. *Nature* 245: 138–39

Morowitz, H. J., Heinz, B., Deamer, D. W. 1988. The chemical logic of a minimum protocell. *Origins Life* 18: 281–87

Mukhin, L. M., Gerasimov, M. V., Safonova, I. N. 1989. Origin of precursors of organic molecules during evaporation of meteorites and rocks. *Adv. Space Res.* 9: 95–97

Muller, H. J. 1955. Life. *Science* 121: 1–9

Nagy, B. 1975. *Carbonaceous Meteorites*. Amsterdam: Elsevier. 747 pp.

Nagy, B., Weber, R., Guerrero, J. C., Schidlowski, M., eds. 1983. *Developments and Interactions of the Precambrian Atmosphere, Lithosphere and Biosphere*. Amsterdam: Elsevier. 475 pp.

Nisbet, E. G. 1987. *The Young Earth: An Introduction to Archean Geology*. London: Allen & Unwin. 402 pp.

Nooner, D. W., Oró, J. 1979. Synthesis of fatty acids by a closed system Fischer-Tropsch process. In *Hydrocarbon Synthesis From Carbon Monoxide and Hydrogen*, ed. E. L. Kugler, F. W. Steffgen, pp. 159–71. Washington, DC: Am. Chem. Soc. 182 pp.

Oberbeck, V. R., Fogleman, G. 1989. On the possibility of life on early Mars. *Lunar Planet. Sci. XX*, pp. 800–1. Houston: Lunar Planet. Inst.

Odom, D. B., Yamrom, T., Oró, J. 1982. Oligonucleotide synthesis in a cycling system at low temperature by cyanamide or a water soluble carbodiimide. *Abstr., 24th COSPAR Plenary Meet., Ottawa, Can.*, p. 511. Paris: COSPAR

Oparin, A. I. 1924. *Proiskhodenie Zhizni*. Moscow: Moscoksky Rabotichii. 71 pp. Transl., 1967, as appendix in Bernal, J. D. *The Origin of Life*. Cleveland: World. 345 pp.

Oparin, A. I. 1936. *The Origin of Life*. New York: Dover. 270 pp.

Oparin, A. I. 1972. The appearance of life in the Universe. In *Exobiology*, ed. C. Ponnamperuma, pp. 2–14. Amsterdam: North-Holland. 485 pp.

Oparin, A. I. 1975. Sobre el origen de las primeras formas de vida. In *El Origen de la Vida*, ed. A. Lazcano-Araujo, A. Barrera, pp. 137–44. Mexico City: Univ. Nac. Auton. Mex. 146 pp.

Orgel, L. E. 1968. Evolution of the genetic apparatus. *J. Mol. Evol.* 38: 381–93

Orgel, L. E. 1987. Evolution of the genetic apparatus: a review. *Cold Spring Harbor Symp. Quant. Biol.* 52: 9–16

Orgel, L. E. 1989. The origin of polynucleotide-directed protein synthesis. *J. Mol. Evol.* 29: 465–74

Orgel, L. E., Lohrmann, R. 1974. Prebiotic chemistry and nucleic acid replication. *Acc. Chem. Res.* 7: 368–77

Oró, J. 1960. Synthesis of adenine from ammonium cyanide. *Biochem. Biophys. Res. Commun.* 2: 407–12

Oró, J. 1961. Comets and the formation of the biochemical compounds on the primitive Earth. *Nature* 190: 389–90

Oró, J. 1963a. Studies in experimental organic cosmochemistry. *Ann. NY Acad. Sci.* 108: 464–81

Oró, J. 1963b. Synthesis of organic compounds by electric discharges. *Nature* 197: 862–67

Oró, J. 1965. Stages and mechanisms of prebiological organic synthesis. In *The Origins of Prebiological Systems*, ed. S. W. Fox, pp. 137–62. New York: Academic. 482 pp.

Oró, J. 1976. In Levin & Straat 1976, p. 1328

Oró, J., ed. 1979. *The Viking Mission and the Question of Life on Mars. J. Mol. Evol. Spec. Iss.*, Vol. 14

Oró, J. 1980. Prebiological synthesis of organic molecules and the origin of life. See Halvorson & Van Holde 1980, pp. 47–63

Oró, J. 1988. Constraints imposed by cosmic evolution towards the origin of life. In *Bioastronomy*, ed. G. Marx, pp. 161–65. Dordrecht: Kluwer

Oró, J., Kamat, S. S. 1961. Amino acid synthesis from hydrogen cyanide under possible primitive Earth conditions. *Nature* 190: 442–43

Oró, J., Kimball, A. P. 1961. Synthesis of purines under possible primitive Earth conditions. *Arch. Biochem. Biophys.* 34: 217–27

Oró, J., Kimball, A. P. 1962. Synthesis of purines under possible primitive Earth conditions. II. Purine intermediates from hydrogen cyanide. *Arch. Biochem. Biophys.* 96: 293–313

Oró, J., Lazcano-Araujo, A. 1981. The role of HCN and its derivatives in prebiotic evolution. In *Cyanide in Biology*, ed. B. Vennesland, E. E. Conn, C. J. Knowles, J. Westley, F. Wissing, pp. 517–41. London: Academic. 548 pp.

Oró, J., Lazcano, A. 1984. A minimal living system and the origin of a protocell. *Adv. Space Res.* 4: 167–76

Oró, J., Mills, T. 1989. Chemical evolution of primitive solar system bodies. *Adv. Space Res.* 9: 105–20

Oró, J., Stephen-Sherwood, E. 1974. The prebiotic synthesis of oligonucleotides. In *Cosmochemical Evolution and the Origins of Life*, ed. J. Oró, S. L. Miller, C. Ponnamperuma, R. S. Young, 1: 159–72. Dordrecht: Reidel. 523 pp.

Oró, J., Gibert, J., Lichtenstein, H., Wikstrom, S., Flory, D. A. 1971a. Amino acids, aliphatic and aromatic hydrocarbons in the Murchison meteorite. *Nature* 230: 105–6

Oró, J., Nakaparsin, S., Lichtenstein, H., Gil-Av, E. 1971b. Configuration of amino acids in carbonaceous chondrites and a Precambrian chert. *Nature* 230: 107–8

Oró, J., Sherwood, E., Eichberg, J., Epps, D. 1978. Formation of phospholipids under primitive Earth conditions and the role of membranes in prebiological evolution. In *Light-Transducing Membranes: Structure, Function and Evolution*, ed. D. W. Deamer, pp. 1–21. New York: Academic. 358 pp.

Oró, J., Holzer, G., Lazcano, A. 1980. The contribution of cometary volatiles to the primitive Earth. *Life Sci. Space Res.* 18: 67–82

Oró, J., Rewers, K., Odom, D. 1982. Criteria for the emergence and evolution of life in the Solar System. *Origins Life* 12: 285–305

Oró, J., Basile, B., Cortes, S., Shen, C., Yamrom, T. 1984. The prebiotic synthesis and catalytic role of imidazoles and other condensing agents. *Origins Life* 14: 234–42

Oró, J., Squyres, S. W., Reynolds, R. T., Mills, T. M. 1990. Europa: prospects for an ocean and exobiological implications. *NASA Spec. Publ.* In press

Orpen, J. L., Wilson, J. F. 1981. Stromatolites at approximately 3500 Myr and a greenstone-granite unconformity in the Zimbabwean Archean. *Nature* 291: 218–20

Owen, T., Gautier, D. 1989. Titan: some new results. *Adv. Space Res.* 9: 73–78

Oyama, V. I., Berdahl, B. J. 1977. The Viking gas exchange experiment results from Chryse and Utopia surface samples. *J. Geophys. Res.* 82: 4669–76

Pace, N. R., Marsh, T. L. 1985. RNA catalysis and the origin of life. *Origins Life* 16: 97–116

Pace, N. R., Olsen, G. J., Woese, C. 1986. Ribosomal RNA phylogeny and the primary lines of evolutionary descent. *Cell* 45: 325–26

Peltzer, E. T., Bada, J. L., Schlesinger, G., Miller, S. L. 1984. The chemical conditions on the parent body of the Murchison meteorite: some conclusions based on amino-, hydroxy- and dicarboxylic acids. *Adv. Space Res.* 4: 69–74

Pflug, H. D. 1978. Yeast-like microfossils detected in the oldest sediments of the Earth. *Naturwissenschaften* 65: 611–15

Pflug, H. D., Jaeschke-Boyer, H. 1979. Combined structural and chemical analysis of 3800-Myr-old microfossils. *Nature* 280: 483–86

Pollack, J. P., Yung, Y. L. 1980. Origin and evolution of planetary atmospheres. *Annu. Rev. Earth Planet. Sci.* 8: 425–87

Pollack, J. P., Kasting, J. F. 1987. The case for a wet, warm climate on Mars. *Icarus* 71: 203–24

Ponnamperuma, C., ed. 1977. *Chemical Evolution of the Early Precambrian.* New York: Academic. 221 pp.

Ponnamperuma, C., Woeller, F. H. 1967. Alpha-amino nitriles formed by an electric discharge through a mixture of anhydrous methane and ammonia. *Curr. Mod. Biol.* 1: 156–58

Ponnamperuma, C., Lemmon, R. M., Marimer, R., Calvin, M. 1963. Formation of adenine by electron irradiation of methane, ammonia and water. *Proc. Natl. Acad. Sci. USA* 49: 737–40

Rao, M., Eichberg, J., Oró, J. 1982. Synthesis of phosphatidylcholine under possible primitive Earth conditions. *J. Mol. Evol.* 18: 196–202

Rao, M., Eichberg, J., Oró, J. 1987. Synthesis of phosphotidylethanolamine under possible primitive Earth conditions. *J. Mol. Evol.* 25: 1–6

Reanney, D. C. 1982. The evolution of RNA viruses. *Annu. Rev. Microbiol.* 36: 47–73

Reid, C., Orgel, L. E. 1967. Synthesis of sugars in potentially prebiotic conditions. *Nature* 216: 455

Ring, D., Wolman, Y., Friedmann, N., Miller, S. L. 1972. Prebiotic synthesis of hydrophobic and protein amino acids. *Proc. Natl. Acad. Sci. USA* 69: 765–68

Roy, P., Fukusho, A., Ritter, G. D., Lyon, D. 1988. Evidence for genetic relationship between RNA and DNA viruses from the sequence homology of a putative polymerase gene of Bluetongue virus with that of vaccinia virus: conservation of RNA polymerase genes from diverse species. *Nucleic Acids Res.* 16: 11,759–67

Rutten, M. G. 1971. *The Origin of Life by Natural Causes.* Amsterdam: Elsevier. 186 pp.

Sagan, C. 1989. What the Voyager spacecraft found at Neptune and what it means: at the frontier of the solar system. *Parade* Nov. 26: 3–7

Sagan (Margulis), L. 1989. On the origin of mitosing cells. *J. Theor. Biol.* 14: 225–75

Sanchez, R. A., Ferris, J. P., Orgel, L. E. 1966. Cyanoacetylene in prebiotic synthesis. *Science* 154: 784–86

Schidlowski, M. 1978. Evolution of the Earth's atmosphere: current state and exploratory concepts. In *Origin of Life*, ed. H. Noda, pp. 3–20. Tokyo: Cent. Acad. Publ. 637 pp.

Schidlowski, M., Appel, P. W. U., Eichmann, R., Junge, C. E. 1979. Carbon isotope geochemistry of the 3.7×10^9 yr old Isua sediments, West Greenland: implications for the Archean carbon and oxygen cycles. *Geochim. Cosmochim. Acta* 43: 189–99

Schidlowski, M., Hayes, J. M., Kaplan, I. R. 1983. Isotopic inferences of ancient biochemistries: carbon, sulfur, hydrogen and nitrogen. See Schopf 1983, pp. 149–86

Schlesinger, G., Miller, S. L. 1983a. Prebiotic synthesis in atmospheres containing CH_4, CO and CO_2. I. Amino acids. *J. Mol. Evol.* 19: 376–82

Schlesinger, G., Miller, S. L. 1983b. Prebiotic synthesis in atmospheres containing CH_4, CO and CO_2. II. Hydrogen cyanide, formaldehyde and ammonia. *J. Mol. Evol.* 19: 383–90

Schopf, J. W. 1975. Precambrian paleobiology: problems and perspectives. *Annu. Rev. Earth Planet. Sci.* 3: 213–49

Schopf, J. W. 1976. Are the oldest "fossils," fossils? *Origins Life* 7: 19–36

Schopf, J. W., ed. 1983. *The Earth's Earliest*

Biosphere: Its Origin and Evolution. Princeton: Princeton Univ. Press. 543 pp.

Schopf, J. W., Packer, B. M. 1987. Early Archean (3.3 billion to 3.5 billion years-old) microfossils from the Warrawoona Group, Australia. *Science* 237: 70–73

Schopf, J. W., Walter, M. R. 1983. Archean microfossils: new evidence of ancient microbes. See Schopf 1983, pp. 214–39

Schwartz, A. W., Orgel, L. E. 1985. Template-directed synthesis of novel, nucleic acid–like structures. *Science* 228: 585–87

Schwartz, A. W., Voet, A. B. 1984. Recent progress in the prebiotic chemistry of HCN. *Origins Life* 14: 91–98

Schwartz, A. W., Visscher, J., Van Der Woerd, R., Bakker, C. G. 1987a. In search of RNA ancestors. *Cold Spring Harbor Symp. Quant. Biol.* 52: 37–39

Schwartz, A. W., Visscher, J., Bakker, C. G., Niessen, J. 1987b. Nucleic acid–like structures. II. Polynucleotide analogues as possible primitive precursors of nucleic acids. *Origins Life* 17: 351–57

Shapiro, R. 1984. The improbability of prebiotic nucleic acid synthesis. *Origins Life* 14: 565–70

Shapiro, R. 1988. Prebiotic ribose synthesis: a critical analysis. *Origins Life* 18: 71–85

Shen, C., Yang, L., Miller, S. L., Oró, J. 1987. Prebiotic synthesis of imidazole-4-acetaldehyde and histidine. *Origins Life* 17: 295–305

Shen, C., Miller, S. L., Oró, J. 1990a. Prebiotic synthesis of histidine. *J. Mol. Evol.* In press

Shen, C., Mills, T., Oró, J. 1990b. Prebiotic synthesis of histidyl-histidine. *J. Mol. Evol.* In press

Shen, C., Lazcano, A., Oró, J. 1990c. Possible catalytic activities of histidyl-histidine in some prebiotic reactions. *J. Mol. Evol.* In press

Sill, G., Wilkening, L. 1978. Ice clathrate as a possible source of the atmospheres of the terrestrial planets. *Icarus* 33: 13–22

Simon, M. N., Simon, M. 1973. A search for interstellar acrylonitrile, pyrimidines and pyridines. *Astrophys. J.* 194: 757–61

Snyder, L. E., Buhl, D., Zuckerman, B., Palmer, P. 1969. Microwave detection of interstellar formaldehyde. *Phys. Rev. Lett.* 22: 679–81

Socha, R. F., Weiss, A. H., Sakharov, M. M. 1981. Homogeneously catalyzed condensation of formaldehyde to carbohydrates. VII. An overall formose reaction model. *J. Catal.* 67: 207–17

Spirin, A. S. 1976. Cell-free systems of polypeptide biosynthesis and approaches to the evolution of the translation apparatus. *Origins Life* 7: 109–18

Spirin, A. S. 1986. *Ribosome Structure and Protein Biosynthesis.* Menlo Park, Calif: Benjamin Cummings. 414 pp.

Sprengel, G., Follmann, H. 1981. Evidence for the reductive pathway of deoxyribonucleotide synthesis in an archaebacterium. *FEBS Lett.* 132: 207–9

Stephen-Sherwood, E., Oró, J., Kimball, A. P. 1971. Thymine: a possible prebiotic synthesis. *Science* 173: 446–47

Stillwell, W. 1980. Facilitated diffusion as a method for selective accumulation of materials from the primordial oceans by a lipid-vesicle protocell. *Origins Life* 10: 277–92

Stoks, P. G., Schwartz, A. W. 1981. Nitrogen heterocyclic compounds in meteorites: significance and mechanisms of formation. *Geochim. Cosmochim. Acta* 45: 563–69

Stribling, R., Miller, S. L. 1987. Energy yields for hydrogen cyanide and formaldehyde synthesis: the HCN and amino acid concentrations in the primitive oceans. *Origins Life* 17: 261–73

Suess, H., Urey, H. C. 1956. Abundances of the elements. *Rev. Mod. Phys.* 28: 53–62

Sulston, J. E., Orgel, L. E. 1971. Polynucleotide replication and the origin of life. In *Prebiotic and Biochemical Evolution*, ed. A. P. Kimball, J. Oró, pp. 89–94. Amsterdam: North-Holland. 296 pp.

Thomas, P. J., Chyba, C. F., Brookshaw, L., Sagan, C. 1989. Impact delivery of organic molecules to the early Earth and implications for the terrestrial origins of life. *Lunar Planet. Sci. XX*, pp. 1117–18. Houston: Lunar Planet. Inst. (Abstr.)

Troland, L. T. 1914. The chemical origin and regulation of life. *The Monist* 24: 92–133

Turner, B. E., Bally, J. 1987. Detection of interstellar PN: the first identified phosphorus compound in the interstellar medium. *Astrophys. J. Lett.* 321: L75–79

Tyler, S. A., Barghoorn, E. S. 1954. Occurrence of structurally preserved plants in Precambrian rocks of the Canadian Shield. *Science* 119: 606–8

Urey, H. C. 1952. On the early chemical history of the Earth and the origin of life. *Proc. Natl. Acad. Sci. USA* 38: 351–63

Usher, D. A. 1977. Early chemical evolution of nucleic acids: a theoretical model. *Science* 196: 311–13

Usher, D. A., McHale, A. H. 1976. Hydrolytic stability of helical RNA: a selective advantage for the natural 3′,5′-bond. *Proc. Natl. Acad. Sci. USA* 73: 1149–53

Usher, D. A., Needels, M. C. 1984. On the stereoselective aminoacylation of RNA. *Adv. Space Res.* 4: 163–66

von Kiedrowski, G. 1986. A self-replicating hexadeoxynucleotide. *Angew. Chem. Int. Ed. Engl.* 25: 932

Walker, J. G. C. 1977. *Evolution of the Atmosphere*. New York: Macmillan. 318 pp.

Walker, J. G. C. 1986. *Earth History: The Several Ages of the Earth*. Boston: Jones & Bartlett. 199 pp.

Walter, M. R. 1983. Archean stromatolites: evidence of the Earth's earliest benthos. See Schopf 1983, pp. 187–212

Weber, A. L., Lacey, J. C. 1978. Genetic code correlations: amino acids and their anticodon nucleotides. *J. Mol. Evol.* 11: 199–210

Welch, W. J., Vogel, S. N., Plambeck, R. L., Wright, M. H. C., Bieging, J. H. 1985. Gas jets associated with star formation. *Science* 228: 1389–95

Wetherill, G. W. 1980. Formation of the terrestrial planets. *Annu. Rev. Astron. Astrophys.* 18: 77–113

Whipple, F. L. 1976. A speculation about comets and the Earth. *Mem. Soc. R. Sci. Liège* 9: 101–11

White, D. H., Erickson, J. C. 1981. Enhancement of peptide bond formation of polyribonucleotides on clay surfaces in fluctuating environments. *J. Mol. Evol.* 17: 19–26

White, H. B. 1976. Coenzymes as fossils of an earlier metabolic state. *J. Mol. Evol.* 7: 101–4

White, H. B. 1982. Evolution of coenzymes and the origin of pyridine nucleotides. In *The Pyridine Nucleotide Coenzymes*, ed. J. Everse, B. Anderson, K. S. You, pp. 2–17. New York: Academic. 245 pp.

Wilson, T. H., Maloney, P. C. 1978. Speculations on the evolution of ion transport mechanisms. *Fed Proc.* 35: 2174–79

Woese, C. R. 1967. *The Genetic Code: The Molecular Basis for Gene Expression*. New York: Harper & Row. 200 pp.

Woese, C. R. 1980. Just so stories and Rube Goldberg machines: speculations on the origin of the protein synthetic machinery. In *Ribosomes: Structure, Function, and Evolution*, ed. G. Chambliss, G. R. Craven, J. Davies, K. Davis, L. Kahan, M. Nomura, pp. 357–73. Baltimore: Univ. Park Press. 984 pp.

Woese, C. R. 1987. Bacterial evolution. *Microbiol. Rev.* 51: 221–71

Woese, C. R., Fox, G. E. 1977. Phylogenetic structure of the prokaryotic domain: the primary kingdoms. *Proc. Natl. Acad. Sci. USA* 74: 5088–90

Wolman, Y., Haverland, W. H., Miller, S. L. 1972. Non-protein amino acids from spark discharges and their comparison with the Murchison meteorite amino acids. *Proc. Natl. Acad. Sci. USA* 69: 809–11

Wong, J. T. F. 1981. Coevolution of genetic code and amino acid biosynthesis. *Trends Biochem. Sci.* 6: 33–36

Zaug, A. J., Cech, T. R. 1986. The intervening sequence RNA of *Tetrahymena* is an enzyme. *Science* 231: 470–75

Zielinski, W. S., Orgel, L. E. 1985. Oligomerization of activated derivatives of 3′-amino-3′-deoxy-guanosine on poly(C) and poly(dC) templates. *Nucleic Acids Res.* 13: 2469–84

Zielinski, W. S., Orgel, L. E. 1987. Autocatalytic synthesis of a tetranucleotide analogue. *Nature* 327: 346–47

Zillig, W., Schnabel, R., Stetter, K. O. 1985. Archaebacteria and the origin of the eukaryotic cytoplasm. *Curr. Topics Microbiol. Immunol.* 114: 1–18

Ziurys, L. M. 1987. Detection of interstellar PN: the first phosphorus-bearing species observed in molecular clouds. *Astrophys. J. Lett.* 321: L81–85

Zuckerlandl, E., Pauling, L. 1965. Molecules as documents of evolutionary history. *J. Theor. Biol.* 8: 357–66

Annu. Rev. Earth Planet. Sci. 1990. 18:357–86
Copyright © 1990 by Annual Reviews Inc. All rights reserved

THE NATURE OF THE EARTH'S CORE

Raymond Jeanloz

Department of Geology and Geophysics, University of California, Berkeley, California 94720

INTRODUCTION

The Earth's interior consists of a rocky shell—the mantle and crust—surrounding a dense metallic core. Although most geological studies concern themselves only with the outermost segment of the planet, our understanding of the Earth's core is advancing rapidly. The most recent progress is especially interesting because it emphasizes the geological processes, and not just the structure, of the Earth's deepest interior. These advances come primarily from two developments: new geophysical observations, particularly in seismology and geomagnetism, and recent experimental achievements in studying materials at the ultrahigh pressures and temperatures existing in the core.

OVERVIEW OF CORE PROPERTIES

The outer core is by far the largest magma chamber inside the Earth, consisting of a liquid metallic alloy rather than the silicate compositions typical of magmas near the surface (Figure 1). It is nevertheless the main igneous body of our planet, forming a layer 2260 km thick between the crystalline lower mantle and the inner core. That the outer core is a liquid, with no elastic rigidity and a relatively low viscosity, is established from seismological and radio astronomical observations—specifically, the measured frequencies of the Earth's normal modes of free oscillation and of the forced nutations (see Figure 2 and Table 1). The finite elastic rigidity of the 1220-km radius inner core is also demonstrated by seismology (Dziewonski & Gilbert 1971). The observed seismic rigidity does not prove that the inner core is completely solid, in that it may be partially molten,

0084–6597/90/0515–0357$02.00

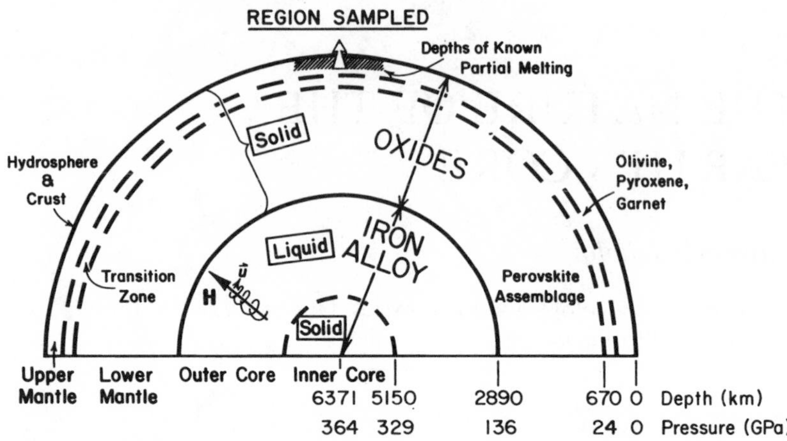

Figure 1 Schematic cross section of the Earth, with seismologically determined regions and pressures summarized as a function of depth at the bottom (100 GPa = 1 Mbar). The mantle and crust, consisting almost entirely of solid oxides, surround the outer (liquid) and inner (solid) core. Cyclonic motions (u) of the molten iron alloy making up the outer core generate the main geomagnetic field (**H**). Like the mantle, the inner core probably deforms by solid-state convection on geological time scales (from Jeanloz 1989a).

but the important conclusion is that the temperatures of the outer core and the inner core must be, respectively, above and below the effective melting temperature (solidus and liquidus) of the core-forming alloy.

The bulk composition of the core is considered to be nearly pure iron based on the following three lines of evidence (e.g. Stevenson 1981, Jacobs 1987a, Williams & Jeanloz 1990). First, the internal geomagnetic field must be produced by a dynamo mechanism, which is only possible in a liquid metallic region inside the Earth (e.g. Merrill & McElhinny 1983, Jacobs 1987b). As the fluidity and electrical conductivity of the mantle are clearly too low to sustain the geomagnetic field, it must be that the outer core is a liquid metal within which the geodynamo is created (Li & Jeanloz 1987, 1990, Courtillot & Le Mouël 1988). Second, the seismologically observed density and sound velocity of the core are close to those of iron measured at the appropriate pressures and temperatures (Birch 1964, McQueen & Marsh 1966, Jeanloz 1979, Brown & McQueen 1986). Third, iron is by far the most abundant element that has seismic properties resembling those of the core (e.g. Anders & Ebihara 1982, Clayton 1983). Thus, the conclusion that Fe is the main component of the Earth's core is based on a

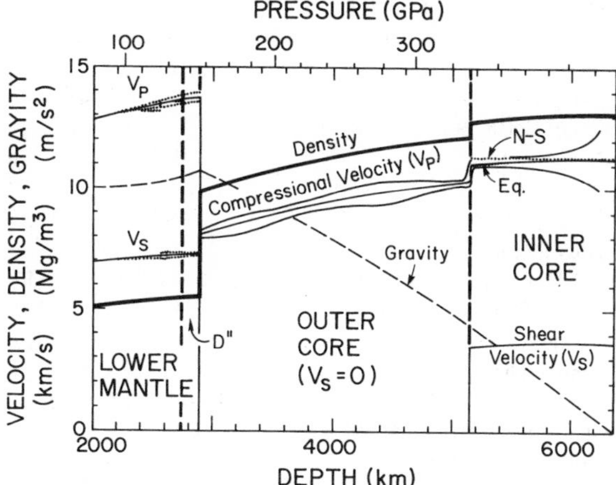

Figure 2 Seismologically measured density (in megagrams per cubic meter; *bold solid curve*), elastic wave velocities (compressional, V_P, and shear, V_S, in kilometers per second; *thin solid curves*), and gravitational acceleration (in meters per second squared; *thin dashed curve*) through the core and lowermost mantle are shown as functions of depth and corresponding pressure (*top scale*) (Dziewonski & Anderson 1981). Johnson & Lee's (1985) extremal bounds on the V_P profile through the core are included for comparison (*thin solid lines* about the average V_P curve). The difference between the polar (N–S) and equatorial (Eq.) compressional velocities through the inner core, as proposed by Morelli et al (1986) and Woodhouse et al (1986), are indicated by dotted lines. Note that the inner core equatorial velocity is barely distinguishable from the average V_P at the scale shown here, and that Morelli et al's proposed depth dependence of the anisotropy implies an essentially constant polar V_P throughout the inner core. Heterogeneity in the lowermost mantle (the D'' region) is illustrated by the variations in V_P and V_S profiles (*dotted curves*) observed at different locations above the core (Ruff & Helmberger 1982, Lay & Helmberger 1983, Garnero et al 1988, Baumgardt 1989, Young & Lay 1989).

combination of evidence from geomagnetism, seismology, high-pressure experimentation, and cosmochemistry.

Iron is an especially important constituent of dense planetary cores, in general, because of its large cosmic abundance among higher mass (or atomic number) elements. This natural abundance is well understood in terms of the stability of its nuclear structure, in that the binding energy per nucleon is at a maximum for ^{56}Fe (Clayton 1983). Therefore, considerable effort has gone into determining the phase equilibria of iron over a large range of pressures (P) and temperatures (T), mainly because of interest in geophysics and planetary science (Birch 1964, 1972, McQueen & Marsh 1966, Jeanloz 1979, Brown & McQueen 1982, 1986, Anderson 1986, Williams et al 1987, Williams & Jeanloz 1990). Not only is the high-pressure melting point required for estimating temperatures in the cores

Table 1 Summary of core properties[a]

Property	Outer core	Inner core
Size (fraction of Earth, %)		
Volume[b]	*15.7*	*0.7*
Mass[b]	*30.8*	*1.7*
Atomic[c]	*15.0*	*0.8*
Seismic quality factors[b,d]		
$Q(V_P)$ (at 1 Hz)	6×10^4	*250–600*
$Q(V_S)$ (at 10^{-3} Hz)	—	*100–4000*
Viscosity (Pa s)[e,f]	$<10^4$, $\sim 10^{-3}$	$10^{13\pm3}$
Electrical conductivity (S m^{-1})[g]	$6\pm3 \times 10^5$	$6\pm3 \times 10^5$
Specific heat (J kg^{-1} K^{-1})[h]	$0.5\pm0.3 \times 10^3$	$0.5\pm0.3 \times 10^3$
Thermal expansion coefficient (K^{-1})[h]	$8\pm6 \times 10^{-6}$	$7\pm4 \times 10^{-6}$
Thermal diffusivity (m^2 s^{-1})[i]	$1.5\pm1 \times 10^{-5}$	$1.5\pm1 \times 10^{-5}$

[a] Geophysically observed quantities are printed in italics, with the other values being derived either from laboratory measurements or from theoretical estimates.
[b] Dziewonski & Anderson (1981).
[c] Assumes a mean atomic weight of 56 g mol^{-1} and 54 g mol^{-1} for the inner and outer core, respectively (Jeanloz 1989a).
[d] Cormier (1981), Fukao & Suda (1989), T. G. Masters (personal communication, 1989).
[e] Gwinn et al (1986), Gans (1972), Stevenson (1981).
[f] Jeanloz & Wenk (1988).
[g] Keeler & Royce (1971), Knittle et al (1986).
[h] Jeanloz (1979), Brown & McQueen (1986), Williams et al (1987).
[i] Obtained from the Wiedemann-Franz relation (e.g. Shimoji 1977).

and deep interiors of terrestrial planets, but the solid-state transformations are equally important. For example, a current summary of the *P-T* phase diagram of iron shows that it is the hexagonal closest packed (hcp or ε) phase of iron that is stable at the conditions of the Earth's inner core (Figure 3).

INNER-CORE TECTONITE

The seismic rigidity of the inner core argues that this region is solid, in the conventional sense of the word. On geological time scales, however, the inner core may behave as a fluid, undergoing solid-state convection like the mantle. Indeed, estimates by Jeanloz & Wenk (1988) of the cooling rate and effective viscosity of the inner core (Table 1) yield a Rayleigh number approximately 6 ± 3 orders of magnitude larger than the critical Rayleigh number for an internally heated sphere (Figure 4; Chandrasekhar 1961). The implication is that the inner core is likely to be convectively unstable, or, in geological terms, tectonically active.

Support for this idea comes from recent seismological observations showing that the compressional-wave velocity through the inner core is

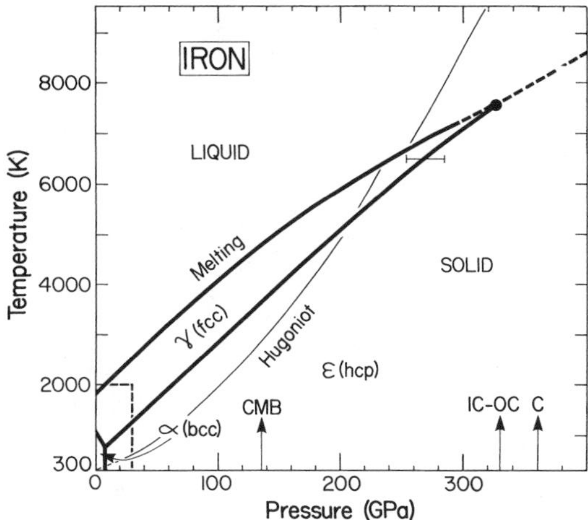

Figure 3 Pressure-temperature phase diagram of iron (*bold curves*, dashed where extrapolation is uncertain) as determined from the experiments of Brown & McQueen (1986), Akimoto et al (1987), Manghnani et al (1987), Mao et al (1987) and Williams at al (1987) (after Knittle & Jeanloz 1990). Boehler's (1986) preliminary results on the melting of iron are not included here because (*a*) they are not corrected for temperature gradients and therefore cannot be considered quantitatively reliable (Williams et al 1987, Bodea & Jeanloz (1989); (*b*) they are not supported by any published calibrations of melting temperatures (e.g. at zero pressure); and (*c*) the solid-state transition temperature measured in the same experiment disagrees with the results of Mao et al (1987), being systematically low and deviating increasingly as the transition temperature rises. The crystalline polymorphs of Fe include the α body-centered cubic (bcc) phase, the γ face-centered cubic (fcc) phase, and the ε hexagonal closest packed (hcp) phase. (The stability field for the δ bcc phase, near the zero-pressure melting temperature, is too small to show on the scale of this diagram.) The temperatures achieved along the Hugoniot shock-compression curve are indicated by the thin solid line. The pressures at the core-mantle boundary (CMB), at the inner core–outer core boundary (IC–OC), and at the center of the Earth (C) are shown for reference, as is the limited pressure-temperature range over which the phase diagram was known from experiments just 10 years ago (*dashed lines at lower left*; cf. Jamieson 1977).

systematically higher along the Earth's rotation axis (polar or north-south direction) than in the equatorial direction—that is, the inner core is acoustically birefringent (Morelli et al 1986, Woodhouse et al 1986, Shearer et al 1988). In particular, J. H. Woodhouse, A. M. Dziewonski, and coworkers were able to show by analyzing both body-wave travel times and frequencies of normal modes that the seismological data are best explained by anisotropy of the inner-core wave velocities (Figure 2), rather than by lateral heterogeneity.

The seismic anisotropy can, in turn, be explained by the presence of hcp crystals with a preferred orientation (Wenk et al 1988, Sayers 1989). The

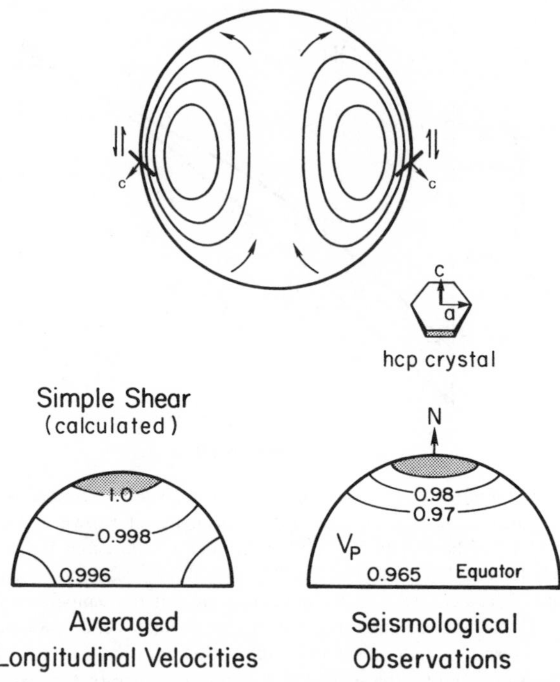

Figure 4 Solid-state convection of the inner core may induce a preferred orientation to the hcp crystallites comprising this region. (*Top*) The first unstable mode of convection for a self-gravitating sphere, the $l = 1$, $m = 0$ mode, is illustrated by streamlines and arrows depicting the flow. Double arrows indicate the resulting sense of simple shear, for this mode, that tends to orient the *c*-axes at angles of about 45° to the flow, as shown by schematic crystal orientations (arrow in the *c*-direction, perpendicular to the hexagonal unit-cell plane; the *a*- and *c*-axes of the hcp crystal structure are defined in the inset). (*Bottom left*) Equal-area stereographic projection of the calculated V_P anisotropy due to an hcp texture created by simple shearing during convection. The figure is oriented such that the plane of shear is contained in the equatorial plane of the projection. Contours give the ratio of the compressional velocity in a given direction to the maximum value (*shaded*), after averaging over the two opposite senses of simple shearing that are possible (cf. top figure). As discussed in the text, the values shown are derived from the elastic moduli of Ti measured at ambient conditions. The estimated moduli of hcp iron, extrapolated to the conditions of the inner core, are expected to yield an anisotropy about 5–10 times larger in magnitude. (*Bottom right*) A similar equal-area stereographic projection of the seismologically observed anisotropy of the inner core, with the polar direction oriented vertically (after Jeanloz & Wenk 1988).

predominant mineral phase of the inner core is expected to be hcp iron (Figure 3), a crystalline polymorph that is known to be anisotropic in its elastic properties (Jephcoat et al 1986). However, the complete set of elastic moduli are not known for hcp Fe, either at the pressures and temperatures of the inner core or at ambient conditions. Therefore, in order to quantify the results, calculations have been carried out using the measured elastic moduli of Ti because titanium is a close structural and elastic analog of hcp iron. That is, the ratio of unit-cell dimensions $c/a \leq 1.60$, and the c-axis is more compressible than the a-axis in both cases.

The results show that a V_P anisotropy of the order of 10^{-2} can be easily produced by way of a texture, or preferred orientation, with c-axes tending to point along the axis of highest velocity, on average (i.e. the polar direction in this instance). The texture can be thought of as a large-scale schistosity or foliation similar to that found in metamorphic rocks or glaciers. In the present case, the rock would be an iron tectonite.

If the inner core is undergoing solid-state convection, as inferred from the Rayleigh number estimate, both simple and pure-shear deformation are expected to produce a texture by way of slip, and perhaps by subsequent recrystallization. For illustrative purposes, the velocity anisotropy resulting from slip due to simple shear is shown in Figure 4 (*bottom left*). Similarly, pure shear would likely contribute to the anisotropy, but the overall pattern of shear and anisotropy cannot be determined because the inner-core flow field is unknown; it is probably more complex than the single $l = 1$, $m = 0$ mode depicted in the figure. Moreover, there is speculation, but as yet no satisfactory explanation, for why the convective pattern and hence the resulting anisotropy should be oriented relative to the Earth's rotation axis, as is observed (Jeanloz & Wenk 1988).

The seismological results do not conclusively prove that the inner core is convecting, in that other mechanisms could also produce anisotropy (e.g. crystal growth with a preferred orientation). Nevertheless, the occurrence and magnitude of seismic anisotropy are most simply explained by the effects of solid-state convection. The situation is analogous to the seismic anisotropy of the uppermost mantle, which is caused by olivine crystals being preferentially oriented in response to the large-scale shearing associated with plate tectonics (e.g. Francis 1969, Kirby & Kronenberg 1987, Nicolas & Christensen 1987). Thus, the observations suggest that tectonic activity extends all the way to the center of our planet, and that the inner-core anisotropy represents metamorphism at the most extreme conditions in the Earth.

IGNEOUS PETROLOGY AND GEOCHEMISTRY OF THE CORE

Composition of the Core Alloy

If the core consisted of pure iron, the temperature at the inner core–outer core boundary (solid-liquid interface) would be directly given by the high-pressure phase diagram of Fe (Figure 3). It has long been recognized, however, that although the sound velocity through the core is virtually identical to that of iron at equivalent *P-T* conditions, the density of the outer core is $10 \pm 2\%$ less than that of iron (Birch 1964, Jeanloz 1979, Brown & McQueen 1982). Therefore, the core cannot be composed of a single element; as far as can presently be ascertained, it must be an alloy. This means that there is not a single melting temperature at a given pressure, but that the core-forming substance melts over a finite temperature interval defined by the solidus and liquidus. [The possibility that the core composition coincides exactly with a eutectic point is not supported by existing data (Williams & Jeanloz 1990).]

A pessimistic view, but one that is most plausible in detail, is that iron in the core is alloyed with virtually all other elements (Birch 1952, Stevenson 1981). Yet, the important issue is to identify the most abundant alloying constituents, as their presence alters the melting-freezing equilibria of pure iron the most. This statement is made on the basis that the changes in liquidus and solidus temperatures are proportional to the concentrations of the alloying constituents for ideal thermodynamic solutions: ideal in the sense that an ideal liquid solution coexists either with an ideal solid solution or with solid phases that are completely immiscible (e.g. Moore 1972). That iron-alloy systems behave ideally on melting at high pressures is supported by the available data (Knittle & Jeanloz 1990, Williams & Jeanloz 1990; cf. Verhoogen 1980).

The main arguments relating to the composition of the core-forming alloy are intimately linked with specific scenarios that have been proposed for the origin and evolution of the core (Table 2). The two elements that are presently considered the likeliest major alloying components in the core are sulfur and oxygen, with nickel and hydrogen having also been proposed as possible or less significant constituents (Birch 1952, 1964, McQueen & Marsh 1966, Murthy & Hall 1970, Brett 1976, Ringwood 1977, 1979, Stevenson 1981, Fukai & Suzuki 1986, Knittle & Jeanloz 1986, 1989b, 1990, Ahrens & Jeanloz 1987, Suzuki et al 1989, Williams & Jeanloz 1990).

In brief, S and Ni have been proposed on the basis that iron-nickel and iron-sulfide phases are observed in meteorites, presumably the materials from which our planet accreted. In addition, if core-forming Fe equilibrates with other mineral phases at low pressures (e.g. early during accretion or core formation) it is likely to alloy with any sulfur and nickel that are available. Finally, the amounts of S and Ni present in the mantle appear to be less than expected from cosmic abundances, which suggests that these elements may be preferentially located in the core (e.g. Murthy & Hall 1970, Brett 1976). This last point is weakened, however, by the fact

Table 2 Candidate alloying constituents for Fe in the core

Element	Argument
H	*For—* 1. Abundant in the Universe and possibly at Earth's surface during accretion and core formation 2. Alloys with (or dissolves into) Fe at high pressures *Against—* 1. Cannot explain its depletion in the Earth—significant loss due to volatility is required in any case 2. Partitioning between silicates and iron and retention in liquid iron are unknown at high pressures and temperatures 3. Requires the presence of H deep inside the early (accreting) Earth, without subsequent loss
O	*For—* 1. Inevitably present in the core due to high-pressure reactions with the mantle 2. Abundant in the Earth (58 ± 2 atomic % of mantle) 3. Lowers the density of the alloy, satisfying geophysical constraints *Against—* 1. Limited to high pressures; assumes minor alloying prior to final core formation or Earth accretion if O is the main contaminant in the core 2. Raises the melting temperature of Fe, which leads to high core temperatures and possible complications with crystallization or formation of the inner core
Si, Mg	*For—* 1. Inevitably present in the core, at least in minor quantities, due to high-pressure reactions with the mantle 2. Relatively abundant in the Earth (18 ± 7 atomic % of mantle) *Against—* 1. Amounts present and apparent reactivity with liquid Fe (especially for Mg) are more limited than in the case of O
S	*For—* 1. Can help to explain its depletion in the mantle relative to cosmic abundances 2. Lowers the density of the alloy, satisfying geophysical constraints 3. Alloys readily with Fe, forming sulfides both at low and at high pressures 4. Forms a low-temperature alloy melt at low pressures, which (*a*) may help to precipitate core formation, and (*b*) makes alloying inevitable if S is present and equilibration (at least partial) occurs during accretion or core formation 5. Fe-sulfide phases are observed in meteorites *Against—* 1. Cannot explain total depletion; hence some loss due to volatility is required (*a*) if the observed iron meteoritic abundance is used (0.7 ± 0.5 wt%), or (*b*) if the full density deficit of the core is ascribed to S (10 ± 2 wt%) 2. Requires enrichment relative to less volatile elements (e.g. alkalis) if present in significant abundance in the core 3. Requires (at least) partial equilibration prior to or during Earth accretion and core formation
Ni	*For—* 1. Fe-Ni phases are observed in meteorites 2. Can explain its depletion in the mantle relative to cosmic abundances 3. Alloys readily with Fe, both at low and at high pressures 4. Forms a low-temperature alloy melt, which (*a*) may help to precipitate core formation, and (*b*) makes alloying inevitable if Ni is present and equilibration (at least partial) occurs during accretion or core formation *Against—* 1. Does not affect the high-pressure density of iron; therefore, none is needed on physical grounds

that the apparent depletion of sulfur in the mantle cannot be fully accounted for even if the maximum amount of S is put into the core, as constrained by the seismic density (Ahrens 1979, Ahrens & Jeanloz 1987). If one argues that sulfur is volatile enough to be lost during accretion, then there is no need to call upon S being present in the core on the basis of the cosmic abundances of the elements (Ringwood 1977, 1979, Brown et al 1984).

Unlike sulfur, the apparent depletion of Ni in the mantle can be accounted for by assuming that nickel is present in the core at a typical meteoritic abundance of 8 ± 7 wt% (Buchwald 1975). This difference with sulfur may simply be due to the fact that Ni is considerably less volatile than S and may therefore have been retained in its primordial abundance during accretion of the Earth (e.g. Ringwood 1979). It should be noted, however, that there is no geophysical reason for including any nickel in the core. Whereas alloying with S lowers the density of Fe, compatible with the observed density of the outer core being less than that of pure iron, the addition of Ni does not affect the seismologically observable density and wave velocity of iron at high pressures and temperatures (McQueen & Marsh 1966, Ahrens 1979). To summarize, the presence of significant amounts of S or Ni requires that the core metal has become alloyed at low pressures, prior to or early during core formation.

In contrast with sulfur and nickel, the possibility that O or H are dominant alloying constituents in the core is based entirely on high-pressure laboratory data; in particular, the relevant oxide and hydride alloy phases are not observed at low pressures, including in meteorite assemblages (Fukai & Akimoto 1983, Fukai & Suzuki 1986, Ohtani & Ringwood 1984, Ohtani et al 1984, Knittle & Jeanloz 1986, 1989b, 1990, Kato & Ringwood 1989, Suzuki et al 1989). The experiments demonstrate that both oxygen and hydrogen can alloy with (or be highly soluble in) liquid Fe at pressures above $\sim$5–30 GPa. Significant alloying is not observed at low pressures. Thus, alloying with oxygen and hydrogen is only expected to occur well after the core has begun to form, when the iron is already deep inside the planet, and perhaps well after accretion of the planet is complete.

Whereas hydrogen is the most abundant element in the Universe, it is greatly depleted in the Earth, and too little is currently known about its partitioning between metal and silicates to argue strongly for its abundance in the core. Qualitatively, one might expect that sufficient hydrogen could be retained in and around the Earth only if the planet remains cold (largely unmelted) during its accretion and earliest evolution, a scenario that is not favored at present (e.g. Hartmann et al 1986, Stevenson 1987a). Indeed, Ringwood (1977, 1979) has pointed out that elements much less volatile than H, or even S (e.g. alkalis), are clearly depleted in the Earth relative to cosmic abundances. The implication is that because less volatile elements are noticeably depleted, the more volatile species (H, S) may not be retained in sufficient abundance to be important alloying components for the core.

In contrast, oxygen is abundant inside the Earth, making up nearly 60% of the mantle on an atomic basis. Moreover, the addition of O to Fe lowers the density of the alloy, as required by the seismological data for the core. [Note, however, that the planet is depleted in oxygen relative to cosmic abundances (Jeanloz & Ahrens 1980).] Therefore, oxygen is a prime alloying candidate if the composition of the core has been evolving over geological time (Knittle & Jeanloz 1989b, 1990). In the same way, Si and Mg could also enter into the core due to high-pressure reactions with the mantle, though to a lesser extent than oxygen both because of their lower abundances [20 ± 1 and 15.8 ± 0.5 atomic % of the mantle, respectively (Jeanloz 1987a)] and because of their apparently lower solubility in liquid Fe (E. Knittle & R. Jeanloz, unpublished, 1990). Finally, the small concentrations of H, S, and Ni in the mantle imply that the abundances of these elements in the core could not increase noticeably throughout Earth history, after core formation was complete.

High-Pressure Melting Relations for Iron Alloys

The primary conclusion from ultrahigh *P-T* experiments is that alloying either oxygen or sulfur changes the melting temperature of Fe considerably, but that the effects of each are significantly different (Figure 5). Whereas the addition of S lowers the melting (liquidus) temperature of Fe, with a eutectic being present in the Fe-S system, the addition of O increases the solidus and liquidus temperatures, relative to pure iron, with no evidence of a eutectic (Knittle & Jeanloz 1990, Williams & Jeanloz 1990). The eutectic temperature for the Fe-S system calculated from a simple thermodynamic model for the liquid solution is in relatively good agreement with high-pressure liquidus measurements on intermediate compositions, such as $Fe_{0.84}S_{0.16}$, that correspond to the density of the outer core (Figure 5). If we extend the analysis further, the addition of oxygen is expected to increase the liquidus temperature of the Fe-S alloy (Williams & Jeanloz 1990).

As oxygen and sulfur, either alone or in combination, are likely to be the primary alloying constituents in the core, the difference between the melting relations in the Fe-O and Fe-S systems introduces a first-order uncertainty in modeling the temperature in the core. The smaller amounts of Ni and H that may be in the core are thought to pose less of a problem, with both being assumed to decrease the melting point of iron at high pressures; however, virtually no experimental data are available, especially for the Fe-H system.

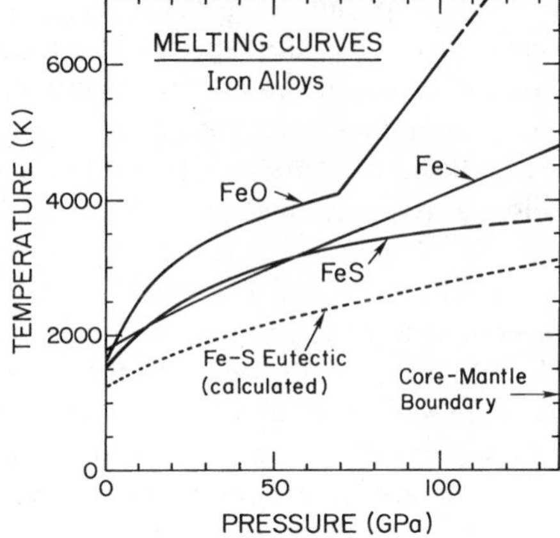

Figure 5 Pressure-temperature diagram summarizing the experimentally determined melting temperatures of Fe, FeO, and FeS (*solid lines*, dashed at pressures beyond the experimental range). For clarity, the raw experimental data are not shown. The dotted curve depicts the Fe-S eutectic temperatures calculated from a thermodynamic model that is derived from the high-pressure melting experiments. For alloy compositions that are compatible with the outer core density, experimentally determined melting temperatures in the Fe-O system (indistinguishable liquidus and solidus) lie between the melting curves of Fe and FeO, whereas experimental liquidus temperatures in the Fe-S system are observed to lie between the Fe melting curve and the calculated eutectic temperature (Knittle & Jeanloz 1986, 1990, Williams et al 1987, Williams & Jeanloz 1990).

The Core Geotherm

Limits on the temperature profile through the core are determined by considering the solid-liquid interface at the inner core–outer core boundary, but there are two difficulties that need to be addressed. First, there is no conclusive evidence that the inner core is in thermodynamic equilibrium with the outer core. The main problem is a lack of phase equilibrium data for plausible core compositions at the appropriate conditions, added to the fact that seismological observations do not yet offer a decisive constraint on the difference in composition between the inner and outer core [for example, the intrinsic density difference due to composition (Masters 1979, Jeanloz 1979, 1987b, Anderson 1986, Jephcoat & Olson 1987, Souriau & Souriau 1989)]. In particular, the inner core could be growing (crystallization from the outer-core liquid) or shrinking (dissolution into the outer core), regardless of whether or not it is at thermodynamic equilibrium with the liquid at the base of the outer core.

The second difficulty arises from the fact that the ε–γ–liquid triple point in the phase diagram of Fe occurs at the same pressure as the inner core–outer core boundary (Figure 3). Because of this coincidence, which may be fortuitous, the nature of the Fe alloy melting relations is expected to qualitatively change at about the pressure of the inner core–outer core boundary (Knittle & Jeanloz 1990). Thus, even if the inner core is at thermodynamic equilibrium with the outer-core liquid, solid-solution and possibly eutectic equilibria measured at lower pressures (e.g. Figure 5) are no longer reliable at inner-core pressures. In fact, it could be that just such changes in the iron alloy phase relations determine the pressure of the inner-core boundary, implying that the coincidence with the triple point is not fortuitous. That is, the size of the inner core may be set by the phase diagram of iron.

Keeping these uncertainties in mind, we proceed by assuming that the crystalline inner core is denser, and hence more iron rich, than the liquid outer core. The rationale is one of long-term stability, in that the configuration of the inner and outer core is not likely to be buoyantly unstable over geological history [see the discussions of Alder & Trigueros (1977) and Stevenson (1981, 1987b)]. In addition, the observed amplitudes of seismic waves reflected off of the inner core–outer core boundary indicate a density increase, in accord with normal mode data (Masters 1979, Souriau & Souriau 1989, Shearer & Masters 1990). Thus, the highest possible temperature for the inner core–outer core boundary is given by the melting point of pure iron (7600 K at 329 GPa). In contrast, the lowest possible temperature at the base of the outer core is set by the eutectic for the Fe-S system: Williams & Jeanloz (1990) estimate a value of 4700 K, which could be reduced to 4500 K by the addition of Ni. As the outer core is a convecting liquid, these values can be extrapolated along adiabatic gradients to obtain temperature limits of 3300–5800 K at the top of the core.

Though disparate, these absolute temperature limits do not depend on thermodynamic equilibrium at the bottom of the outer core but simply on the requirements that the outer core is liquid and that the inner core is crystalline and iron rich (compatible with the observed seismic density and anisotropy). As the upper temperature limit at the top of the core is well above the melting point of pure iron, this limit would allow considerable amounts of oxygen to be alloyed with the Fe. For comparison, although the lower temperature limit remains above the (calculated) eutectic temperature of the Fe-S system throughout the outer core, it requires that the liquid contains about 50–60% crystals by volume; that is, the outer core must be a liquid slurry if it is at the lowest temperatures considered here (Williams & Jeanloz 1990).

It is only by considering the possibility of a slurry that a temperature far below the liquidus can be invoked for a molten alloy system. As local equilibrium is likely to be achieved between coexisting crystals and liquid, the high-pressure melting data indicate that over 50% crystals by volume are present at temperatures of 3300–4500 K throughout the outer core, for a composition ($Fe_{0.84}S_{0.16}$) that is determined from the seismological density. It may seem implausible that such a slurry could be present for extended geological time periods, but there is no physical reason to reject this possibility as long as the crystallites remain small [radii of order 10^{-6} m (Williams & Jeanloz 1990)]. Alternatively, if one concludes that the outer core cannot be a slurry, then the geotherm must be close to (or above) the liquidus and well above the lower temperature limit quoted above.

A value of 5800 K for the temperature at the top of the core may be unrealistically high because it is based on considering only the melting relations for iron alloys. As the lowermost mantle is known from seismology to be solid, or at most partially molten, another estimate for the maximum temperature at the core-mantle boundary is about 4500 ± 500 K, as shown in Figure 6. This limit is derived from ultrahigh pressure measurements of the melting temperature of (Mg, Fe) silicate perovskite, the main constituent of the lower mantle (Knittle & Jeanloz 1989a). The resulting adiabatic temperature profile through the core yields temperatures of 5500–6600 K at the inner core–outer core boundary and 5700–6800 K at the Earth's center. It should be emphasized, however, that the thermal state and mineralogical constitution of the lowermost mantle are poorly known (Jeanloz & Richter 1979, Young & Lay 1987, Knittle & Jeanloz 1989b). As discussed below, the D'' layer at the base of the mantle is highly anomalous, so the melting temperature of silicate perovskite may yield an artificially low value for the maximum temperatures possible in the core.

Similar estimates of temperature (4900 ± 1000 K at the core-mantle boundary) are obtained if oxygen is considered to be an important alloying constituent. If we combine the effects of O and S, in an amount limited by the outer-core density (~ 10 wt% $\pm$ Ni in smaller proportions), the liquidus temperature at the inner core–outer core boundary is decreased by about 1000 ± 1000 K owing to the alloying of iron (Figure 6). This estimate is derived from melting point data at 15–80 GPa (Urakawa et al 1987, Knittle & Jeanloz 1990, Williams & Jeanloz 1990) and is compatible with a model of ideal melting point depression by the alloying constituent (Brown & McQueen 1982). The resulting adiabatic temperature profile through the core is shown in Figure 6: It is consistent with the requirements that the lowermost mantle and inner core are crystalline, and that the outer core is liquid and not far below the liquidus (i.e. a volume fraction <20–50% crystallites).

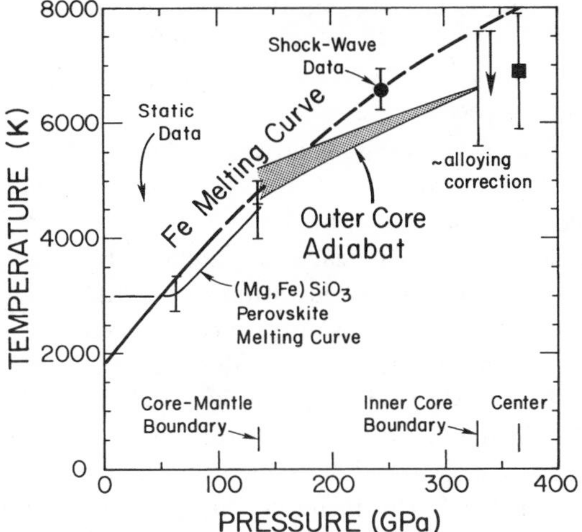

Figure 6 Pressure-temperature diagram summarizing current experimental constraints on the temperatures in the core. The melting curve of iron (*bold curve*, dashed beyond range of continuous measurements) is known from static and shock-wave measurements at high pressures [uncertainties shown at 136 and 243 GPa (Williams et al 1987)]. The expected correction for alloying of about 1000 ± 1000 K decrease in the liquidus (*arrow* near 340 GPa) is based on thermodynamic modeling of the limited data available on the melting of Fe alloys at high pressures (Brown & McQueen 1982, Williams et al 1987, Knittle & Jeanloz 1990, Williams & Jeanloz 1990). The resulting estimates of the temperature at the center of the Earth (*square*) and at the inner core–outer core boundary (*error bar without symbol*) are indicated, as is the calculated average temperaure profile through the outer core [*stippled band* (its width illustrates the uncertainty in the adiabatic gradient through the core)]. The melting curve of (Mg,Fe) silicate perovskite is known from the experiments of Heinz & Jeanloz (1987) and Knittle & Jeanloz (1989a), and provides an upper limit on the temperature in the lowermost mantle (*thin solid line*). Within the uncertainties, the melting point data are compatible with a temperature of about 4500–5000(±500–1000) K at the top of the core, and a central temperature of 6900 ± 1000 K.

D″ AND CONTACT METASOMATISM AT THE CORE-MANTLE BOUNDARY

High-Pressure Reactions Between Iron and Oxides

Having concluded that the core is at a high temperature (4500 ± 1200 K against the base of the mantle), one might wonder what happens when the hot liquid alloy of the outer core is placed in contact with the crystalline oxide and silicate phases of the lower mantle. The answer is revealed by direct experimental observation, in which the core-mantle boundary is simulated by placing liquid iron (or iron alloys such as FeS) against solid oxide minerals inside a laser-heated diamond cell. Invariably, the liquid metal is seen to react vigorously with the crystalline oxides and silicates at high pressures (Knittle & Jeanloz 1986, 1989b, 1990, Williams et al 1987, Williams & Jeanloz 1990).

In the case of the (Mg,Fe)SiO_3 perovskite phase, which dominates the lower mantle, liquid iron produces a reaction zone 20–30 μm wide into the perovskite when samples are heated for 2–3 s inside the diamond cell (Knittle & Jeanloz 1989b). The reaction is analogous to metasomatism, in that it appears to involve liquid infiltration along grain boundaries (but of iron rather than water), as well as diffusion and reaction across crystallite dimensions. The reaction of iron with oxides occurs only at high pressures [above 30–40 GPa for (Mg,Fe)SiO_3 perovskite], so it would not be expected to be significant in the Earth's upper mantle or in the mantles of small terrestrial planets (e.g. Mars). Consequently, this reaction is relevant to the deep mantle of the Earth, including the core-mantle boundary, only after accretion and core formation are largely complete and pressures at depth are sufficiently high.

The reason that metallic iron reacts with oxides at high pressures and temperatures can be understood by considering the properties of FeO. This compound is intermediate in composition between the iron of the core and the oxides of the mantle (Figure 1), and hence it may elucidate the nature of chemical processes at the core-mantle boundary. As summarized in Figure 7, experiments demonstrate that the Fe–O bond is dramatically altered by the combined effects of pressure and temperature. Specifically, FeO transforms from a nonmetallic ceramic at ambient conditions to a metal at combined pressures and temperatures above 70 GPa and 1000 K (Knittle & Jeanloz 1986, 1990, Sherman 1989, Jeanloz 1989c).

The qualitative result is that oxygen becomes a metallic alloying component at the conditions of the deep mantle, akin to sulfur (forming sulfides) but completely different from oxygen (forming ceramic oxides) at low pressures. One consequence of metallization is that the melting temperature of FeO increases rapidly with pressure above 70 GPa (Figures 5, 7). It is for this reason that the proposed alloying of oxygen with liquid iron (Table 2) leads to high values for the temperature in the outer core. Also, it is reasonable to conclude that elevated pressures are necessary for liquid iron to react significantly with (Mg,Fe)SiO_3 perovskite (and other oxides) because sufficient pressure is required to change the Fe–O bond in order that the chemical reaction be favored.

Analyses by Knittle & Jeanloz (1989b) confirm that the metallic alloys FeSi and (Fe,Mg)O wüstite are produced by the high-pressure reaction of liquid Fe with crystalline (Mg,Fe)SiO_3 perovskite. [Although the wüstite composition could not be uniquely distinguished in this study (between

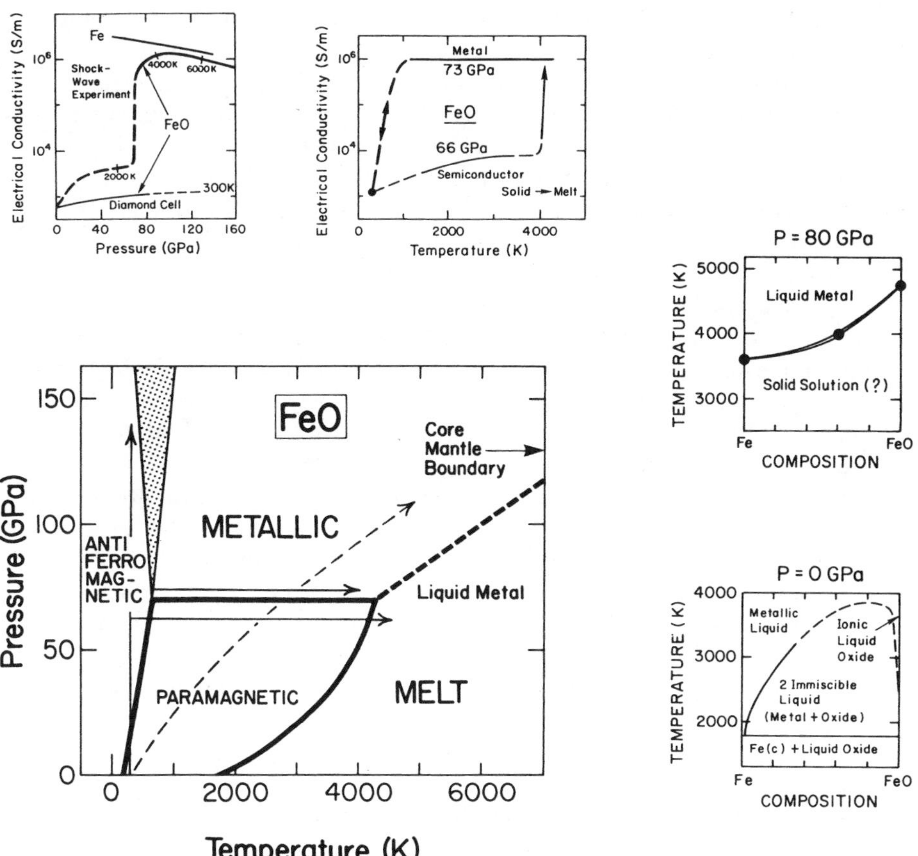

Figure 7 Summary of the phase diagram of FeO based on the experiments of Jeanloz & Ahrens (1980), Yagi et al (1985, 1988), Knittle et al (1986), and Knittle & Jeanloz (1990). The upper panels show the electrical conductivity measurements that help to locate the phase boundaries. Measurements were taken along paths indicated by arrows in the *P-T* phase diagram: Results under shock-wave loading (*dashed arrow*) and at 300 K in the diamond cell are shown in the left upper panel, and results as functions of temperature at 66 GPa and 73 GPa in the laser-heated diamond cell are displayed in the right upper panel. The isothermal boundary between the paramagnetic (nonmetallic) and metallic phases is determined by the data in the latter panel, whereas the boundary between the antiferromagnetic (nonmetallic) and metallic phases is not well constrained by measurements and is indicated by shading in the *P-T* diagram. The panels on the right illustrate the melting relations for the Fe-FeO system, as measured at zero pressure (*bottom right*) and as inferred from melting determinations for three compositions at 80 GPa (*closed symbols*: *top right*) (from Jeanloz 1989c). The melting behavior changes from that typical of a metal-salt system at zero pressure (e.g. with liquid immiscibility) to that of an alloy system at high pressures. Note that the electrical conductivity of metallic FeO is almost identical to that of Fe at elevated pressures and temperatures.

$Fe_{0.9}Mg_{0.1}O$ and nonstoichiometric $Fe_{0.9}O$), shock-wave experiments demonstrate that $Fe_{0.9}Mg_{0.1}O$ undergoes a high-pressure phase transformation similar to the shock-transformation of $Fe_{0.94}O$ to the metallic state (Marsh 1980, Jeanloz & Ahrens 1980, Yagi et al 1988).] In addition, the electrically insulating phases $MgSiO_3$ (iron-free) perovskite and SiO_2 stishovite are formed by the chemical reaction. Finally, some amount of Fe is observed among the reaction products after the ultrahigh *P-T* experiments are over, but it is unclear whether the iron is a reaction product or simply leftover (unconsumed) starting material.

If such experiments reliably simulate the core-mantle boundary, and not just the conditions of temperature and pressure at this depth, it seems inevitable that the mantle and core have reacted chemically over geological time. A cautionary note is required because the oxygen fugacities, defect states, and related thermodynamic parameters are uncertain both for the laboratory simulation and for the core-mantle boundary itself. Nevertheless, that crystalline silicates and liquid Fe react chemically at ultrahigh pressures and temperatures is a first-order observation from the experiments. In addition to the pressure effect on the bonding, the high temperatures enhance the solubility of mantle constituents in the liquid alloy of the outer core (Alder 1966). The implication is that oxygen and, to lesser degrees, Si and Mg are expected to enter into the core, and that the reaction-product mixture of metallic alloys and insulating oxides should be present at the base of the mantle (Figure 8).

D″ *and Lateral Heterogeneity at the Core-Mantle Boundary*

That such reactions actually occur inside our planet is vividly suggested by the heterogeneity of the *D″* layer at the base of the mantle. Seismologically, this $\sim$200–300($\pm$200–300)-km-thick layer is known to be characterized by large lateral variations in wave velocities, as well as by strong scattering (Young & Lay 1987, Lay 1989, Doornbos & Hilton 1989). The scattering of seismic waves can be attributed to heterogeneities that are $\sim$10–70 km in dimension, distributed over a 200 km thickness at the bottom of the mantle, and involve variations of less than 1% in V_P (Bataille & Flatté 1988). Large changes in elastic properties may be locally present, either on shorter length scales or in different parts of *D″* than those studied. Indeed, separate velocity profiles show differences of up to $\sim$2–5% in V_P and V_S at the bottom of the mantle (Figure 2).

The seismological heterogeneity of *D″* can be easily explained by local variations in the abundances of phases produced by the high-pressure chemical reaction of liquid iron with silicate perovskite. Table 3 summarizes the physical properties of the phases involved, and it is evident that local differences of 5–10% in the ratios of metallic to nonmetallic

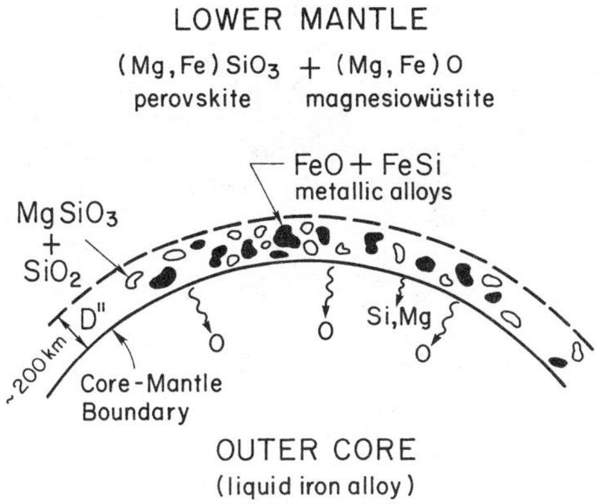

Figure 8 Schematic illustration of the core-mantle boundary, perhaps the most chemically active region inside the Earth. High-pressure experiments demonstrate that liquid iron reacts vigorously with crystalline (Mg,Fe)SiO_3 perovskite, and other silicates and oxides, at the conditions of the deep mantle; the reaction products include insulating phases [$MgSiO_3$ perovskite + SiO_2 stishovite or the closely related $CaCl_2$ phase (*clear blobs*)] and metallic alloy phases [(FeO + FeSi and unreacted Fe (*filled blobs*)], which coexist at the base of the mantle (Knittle & Jeanloz 1989b). Based on such experiments, the core is expected to react with the mantle over geological time scales, with iron permeating from the core into the lowermost mantle, and oxygen and to a lesser degree Si + Mg entering into the core from the mantle (*wavy arrows*). The resulting chemical reaction zone would be expected to exhibit heterogeneities similar to those observed seismologically in the D'' layer at the base of the mantle.

reaction products can readily yield several-percent variations in density and seismic wave velocities. Even though zero-pressure values are quoted for convenience, the differences between the values given should be roughly the same at the conditions of the core-mantle boundary.

Such differences in the relative abundances of the phases produced by the core-mantle reaction could simply reflect local variations in the degree of equilibration reached or in the amount of convective mixing suffered by the reaction products within D'' (see Sleep 1988). Alternatively, the heterogeneity could be due to variations in the amount of infiltration by liquid iron prior to reacting with the mantle rock (i.e. variations in the Fe : silicate ratio). Finally, it should be noted that although SiO_2 stishovite is not expected to be an important phase throughout most of the lower mantle, its occurrence in D'' due to mantle-core reactions could well produce lateral variations in shear-wave velocities. This is because stishovite becomes mechanically unstable with respect to the $CaCl_2$ structure at the conditions of the lowermost mantle, and the transition involves a soft shear mode (Tsuchida & Yagi 1989, Jeanloz 1989b).

Table 3 Physical properties of core-mantle boundary materials[a]

Composition phase	Density ρ (Mg m^{-1})	Seismic wave velocities: Compressional V_P (km s^{-1})	Shear V_S (km s^{-1})	Electrical conductivity[b] σ (S m^{-1})
Fe				
Liquid[c] (at 1810 K)	7.0	3.9	0	7×10^5
FeO				
Wüstite[d] (metallic)	6.1	7.6[e]	4.4[e]	$\sim 10^6$
FeSi				
ε (cubic)[f]	6.2	7.1[e]	4.0[e]	2×10^5
$MgSiO_3$				
Perovskite[g]	4.10	10.7	6.1	$< 10^{-4}$[e]
SiO_2				
Stishovite[h,i]	4.29	11.92	7.16	$< 10^{-6}$[e]

[a] At ambient conditions ($P = 0$, $T = 300$ K) unless otherwise indicated.
[b] At core conditions ($P \sim 100$–300 GPa, $T \sim 2000$–6000 K).
[c] Shimoji (1977), Steinemann & Keita (1988), Secco & Schloessin (1989).
[d] Jeanloz & Ahrens (1980), Knittle et al (1986), Knittle & Jeanloz (1986, 1990).
[e] Estimated value.
[f] Taylor & Kagle (1963), Keeler & Royce (1971), Marsh (1980).
[g] Yeganeh-Haeri et al (1989), Li & Jeanloz (1987, 1990).
[h] Weidner et al (1982).
[i] Ignores the probable distortion to the $CaCl_2$ structure at the conditions of the core-mantle boundary (see Tsuchida & Yagi 1989, Jeanloz 1989b).

The upshot is that the ultrahigh *P-T* experiments offer a plausible explanation for the seismological turbidity of D''. The heterogeneity in density and wave velocities at the bottom of the mantle would therefore be a direct result of chemical reactions having occurred between the mantle and core. If this explanation is valid, a necessary consequence is that the core-mantle system has been evolving over geological history.

EVOLUTION OF THE CORE-MANTLE SYSTEM

Geochemistry and Dynamics

Qualitatively, reaction between the core and mantle can be considered to be driven by the major difference in the bulk compositions of these two regions. Weathering at the Earth's surface offers an inverted analogy, with chemical reactions arising from the compositional contrast between the solid crust (corresponding to the mantle) and the fluid atmosphere and hydrosphere (corresponding to the outer core). In both instances, two very large reservoirs are locally equilibrating, and the reaction products are more or less dispersed in the fluid medium while remaining stagnant (on short geological time scales) within the solid region. One major difference, however, is that because of the very high temperatures, chemical reaction rates are fast and relatively unimpeded by kinetics at the core-mantle boundary in comparison with reactions at the Earth's surface. Thus, the high temperatures and the compositional contrast between the mantle and core suggest that the D'' layer is one of the most chemically active regions of our planet.

There are probably two time scales involved with the core-mantle reaction process—that of local equilibration (or approach thereof via chemical reaction), and that of large-scale dispersal of the resulting products and unreacted materials. Experimentally, the first time scale appears to be very rapid, in accord with previous suggestions that fluid infiltration and reaction may be an extremely effective process of melt-rock equilibration (e.g. Watson 1982, Stevenson 1986). As the liquid and crystalline surface energies are unknown for the relevant phases at deep-mantle conditions (Table 3), one can only speculate about the possible significance of capillary rise for bringing outer-core fluid up into the mantle rock. By extrapolation from the known behavior of capillary systems (e.g. Moore 1972, Shimoji 1977), it is possible that capillarity produces infiltration as far as $10^{\sim 1-3}$ m into the mantle. If so, the low viscosity of the outer-core fluid (Table 1) suggests that the infiltration could occur within about 10^{3-6} yr, a geologically short period (Stevenson 1986).

The second time scale, that of dispersal of the reaction products, is probably the longer of the two. At least this should be the case for the mantle because dispersal is controlled by the rate at which the overlying mantle convects (Davies & Gurnis 1986, Sleep 1988). The lower-mantle heterogeneities (or "dregs") are expected to be partly swept upward, away from the core-mantle boundary, thus leaving a D'' layer that varies laterally in thickness and degree of heterogeneity, as schematically indicated in Figure 8. This is entirely compatible with the seismological observations (e.g. Lay 1989, Doornbos & Hilton 1989). Given the density contrasts involved, of up to several tens of percent, it is unlikely that the heterogeneities are completely removed from the lowermost mantle: They may be recirculated within the D'' layer (see Table 3; Sleep 1988).

On the core side, widespread dispersal is likely to be much more rapid than in the mantle. The flow velocity in the outer core is usually thought to be of the order of 10^4 m yr^{-1}, the rate at which large-scale features of the geomagnetic field move, but the local fluid velocity may be greater yet, especially over distances of meters to kilometers (e.g. Merrill & McElhinny 1983; see also Braginsky 1978). Therefore, it is plausible that the oxygen, silicon, and magnesium are rapidly intermixed within the iron-alloy fluid as the core reacts with the mantle (Figure 8). This would imply that the composition of the core has been changing throughout geological history, becoming increasingly contaminated by the mantle, and that oxygen is probably the most important of the alloying constituents in the outer core (cf. Table 2). As oxygen appears to be fully miscible within liquid iron at high pressures (e.g. Figure 7), the seismologically measured density of the outer core implies that equilibration with the mantle is far from complete at this time. Otherwise, the density of the outer core would be much lower than is observed.

Speculations on the Geomagnetic Field

Because of the large variations in electrical conductivity among the phases involved, interpreting the seismological heterogeneity of the lowermost mantle in terms of the reaction products observed in ultrahigh *P-T* experiments leads to significant implications for the Earth's magnetic field. Indeed, the conductivity variations, of up to 11 orders of magnitude or more, far exceed the relative differences in elastic properties for these phases (Table 3). Therefore, the D'' region would be expected to exhibit substantial lateral heterogeneity in electrical conductivity.

Even if the relative abundances of the insulating and metallic phases vary by large amounts (e.g. tens of percent) from place to place, the rock conductivity undoubtedly does not change by so much as 10^{11}. Yet the plausible volume fraction of metal intermixed with silicate ranges from perhaps less than 10% for unreacted portions of mantle rock to over 50% for fully reacted material. (Variations in abundances among the metallic and insulating reaction products could also arise during the dispersal and remixing process, as noted above.) As this range spans the percolation threshold of 16% by volume, at which point metallic phases interconnect within a random three-dimensional medium (Scher & Zallen 1970, Zallen 1983), the lowermost mantle rock could easily change from relatively insulating ($\sigma < 10^0$ S m^{-1}) to metallic conductivity ($\sigma \sim 10^{5\text{–}6}$ S m^{-1}) as a function of location within the D'' region.

The reason that such variations in electrical conductivity are significant, if present, is because of their possible effect on the magnetic field lines emanating from the core (Figure 9). Due to induction, the field lines are expected to become pinned or blocked by the highly conductive regions at the bottom of the mantle. This is the exact analog of the "frozen-in-field" (or frozen-flux) effect that causes the magnetic field lines to shift around in response to the flow of the liquid metal in the outer core (e.g. Merrill & McElhinny 1983). It is because the metallic regions within D'' are dispersed over periods governed by mantle convection (the second, slower time scale of order $10^{6\text{–}7}$ yr or more) that these conductive zones are effectively stationary relative to the magnetic field lines. The latter move at the

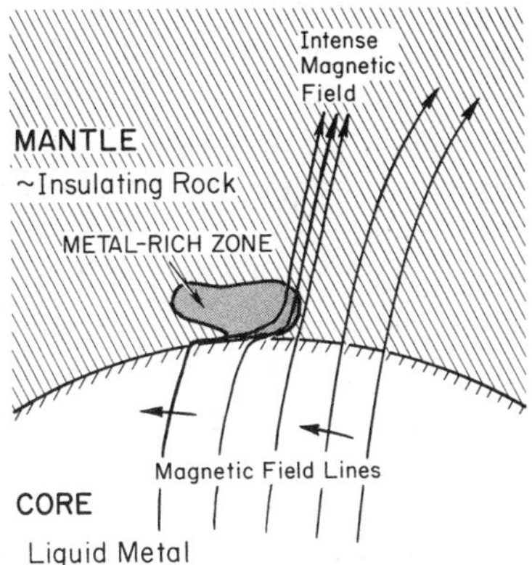

Figure 9 Qualitative illustration of the possible interaction between the Earth's magnetic field and the core-mantle boundary. Magnetic field lines dragged by the flow in the liquid outer core (*arrows*) intersect metal-rich regions created by reactions at the core-mantle boundary (*stippled*; see Figure 8). Pinning of the field lines in the metallic regions for periods of years to centuries can locally enhance the intensity of the magnetic field and result in an electromagnetic coupling between the mantle and the core. Depending on the configuration of the metallic zone in the third dimension, the field lines can either move around the anomaly (cross section shown through an equidimensional blob) or diffuse through it (cross section shown through an embayment in a larger volume of high conductivity). The piling up of the field lines leads to a flux-gradient force that could be correlated with the fluid flow and might induce an intrinsic time dependence to the observed field.

flow velocity of the outer-core liquid, which is estimated to be of order kilometers per years or more (cf. Braginsky 1978).

The pinning of the magnetic field lines by the lowermost mantle heterogeneities, schematically illustrated in Figure 9, can only occur for a modest period of time because of the effects of magnetic diffusion. Taking the heterogeneities to be ~ 100 km in dimension and assuming a conductivity of $\sigma \sim 10^6$ S m^{-1} for the metal-rich zones, as indicated by the seismological and experimental observations, respectively, lead to a magnetic diffusion time of order 10^{1-2} yr (see Merrill & McElhinny 1983).

There is now evidence for deviations from the frozen-flux approximation, and hence the occurrence of magnetic diffusion, on such time scales, a result that is based on recent analyses of the secular variation of the geomagnetic field over the past 200–300 yr (Bloxham & Gubbins 1985, Bloxham 1986). As the diffusion time varies with the square of the heterogeneity dimension (and linearly with conductivity), there is a wide range of possible time periods for pinning or deflecting of the magnetic field lines in the lowermost mantle. In particular, though the seismological data constrain the heterogeneities in D'' to vertical dimensions of tens to (at most) a few hundred kilometers (e.g. Figure 2), the lateral dimensions of heterogeneities at the base of the mantle could be up to thousands of kilometers. In addition, the effects of flux gradients and of field lines slipping around high-conductivity obstacles (in the third dimension) complicate the simple picture shown in Figure 9.

It is likely that the presence of strong lateral variations in the electrical conductivity of the lowermost mantle, as suggested here, leads to significant electromagnetic coupling between the mantle and core (cf. Merrill & McElhinny 1983). The coupling could be sufficient to locally affect the flow field near the top of the outer core. Indeed, the secular variation data have been interpreted as showing clear evidence of core-mantle coupling (Bloxham & Gubbins 1987). Although it was modeled in terms of a thermal rather than an electromagnetic mechanism, the basic conclusion from analyzing the secular variation of the geomagnetic field is that the mantle and core appear to be coupled. In fact, because thermal conduction through metals is dominated by electronic transport, as described by the Wiedemann-Franz relation, it is quite plausible that both thermal and electromagnetic coupling occur, and that they are at least partly correlated (see also Braginsky 1978, Braginsky & Roberts 1987, Braginsky & Meytlis 1987).

Thus, the geomagnetic observations appear to be compatible with, though they do not independently verify, the seismological and experimental evidence for heterogeneity in the D'' region. An obvious research direction that needs to be pursued at this time is to examine possible correlations, or lack thereof, between seismological heterogeneity at the base of the mantle and the recent secular variations of the geomagnetic field. In order to carry out such work, it will undoubtedly be necessary to characterize the seismological heterogeneity around the core-mantle boundary with a lateral resolution of hundreds of kilometers or less (cf. Bloxham 1986). For comparison, even the detailed velocity profiles of D'' shown in Figure 2 represent lateral averages over thousands of kilometers (Lay 1989).

An important reason for attempting to relate the structure of the geomagnetic field, as measured at the Earth's surface, to heterogeneities in D'' is that this might explain some of the peculiar features that have been observed in paleomagnetic records of field reversals. In particular, there is good evidence that successive geomagnetic reversals recorded over ~ 1 Myr in one location show strong similarities, and yet for any one reversal, records from different locations around the globe are generally not compatible with simple models of the reversal process inside the core (Bogue & Hoffman 1987).

This apparent contradiction between the temporally repeatable behavior and the spatially complex behavior of the transitional field recorded at the surface may simply reflect the presence of heterogeneities near the core-mantle boundary. For example, electromagnetic coupling with the heterogeneous lowermost mantle could significantly affect the fluid flow in the outer core during a reversal of the geomagnetic field.

Similarly, the evidence that the transitional field during a reversal (Coe & Prévot 1989), and perhaps even the recent geomagnetic field (Courtillot et al 1984, Courtillot & Le Mouël 1988), undergoes extremely rapid changes could be explained in terms of strong lateral variations in the electrical conductivity of the D'' layer. If so, the paleomagnetic record of transitional fields documented from several locations around the globe may offer especially important information: As a given field reversal represents an instant [i.e. is of short duration ($\sim 10^4$ yr)] relative to the time over which the D'' region changes ($>10^6$ yr), the paleomagnetic data for several reversals may allow us to deduce how the core-mantle boundary has evolved over geological history.

CONCLUSION

The Earth's core exhibits a wide range of geological processes, from tectonic deformation of the inner core, through igneous (magmatic) evolution of the outer core, to contact metamorphism and metasomatism at the core-mantle boundary. The processes and the detailed structure of the central region of our planet can now be studied because of modern capabilities in geophysical observation, primarily in seismology and geomagnetism, combined with recent developments in ultrahigh *P-T* experimentation and in modeling. Such studies are important in that the present thermal and geochemical state of the core reveals one of the main ways in which the Earth has evolved over geological history.

The high-pressure melting experiments clearly indicate that the temperature of the core is high, most probably ranging from 4500 ± 500 K at the top of the outer core to 6500 ± 1000 K at the center of the Earth (Figure 6). The absolute uncertainty for the core geotherm is as large as ± 1000–1500 K, however, mainly because of the uncertainties in the composition of the core alloy (Table 2). It is most likely that oxygen is a significant alloying component, based on the combined experimental and geophysical evidence that the crystalline mantle reacts with the liquid iron (or iron alloy) of the outer core. The alloying of oxygen with iron favors a high-temperature model for the core, but this could be offset by the presence of Ni, S, and H: Alloying by the latter elements allows for much of the liquid outer core to be at a temperature well below the melting point of pure iron (Figures 5, 6). Nevertheless, the high temperatures proposed here reduce the amount of alloying required to explain the density of the core, with a total of 8–10 wt% oxygen plus sulfur being adequate (Knittle & Jeanloz 1990, Williams & Jeanloz 1990).

A major implication of such high temperatures is that the tectonics and convection of the mantle, as well as the resulting geological processes observed at the surface, are to a large degree powered by heat from the core. Specifically, over 20% of the heat lost from the surface of the Earth is likely to originate from the core (Knittle & Jeanloz 1990). Both radioactive decay within the core and primordial heat, in particular from the hypothesized Moon-forming impact (Stevenson 1987a), could be the source of high temperatures in the core. Along with crystallization of the outer core, these heat sources are sufficient to power the dynamo action that creates the geomagnetic field (e.g. Gubbins et al 1979, Gubbins & Masters 1979).

The actual heat transfer from the core to the mantle is undoubtedly complicated by the chemical reactions that are thought to occur at the core-mantle boundary. First, the possible contributions of heats of reaction and partial melting in the lowermost mantle are highly uncertain and are therefore completely ignored so far. Second, the heat transfer across the D'' region is likely to be strongly modified by the presence of heterogeneities: The iron-rich reaction products (dregs) are expected to be sufficiently dense (Table 3) to stabilize the thermal and chemical boundary layer at the base of the mantle (Jeanloz & Richter 1979, Sleep 1988). It is therefore quite possible that the geotherm temperature increases by 1000–1500 K from the mantle to the core, across the depth of the D'' region.

As a result of the high temperatures, along with the compositional contrast between silicates and iron alloy, the core-mantle boundary is considered to be the most chemically active region of the Earth. The change in Fe–O bonding character induced by high pressures appears to facilitate the silicate-iron reactions observed in the laser-heated diamond cell (Figure 7). Both the high temperatures and this bonding change enhance the solubility of oxygen, and perhaps other mantle components, in the outer core alloy.

Thus, the high-pressure experiments provide a natural explanation for the lateral heterogeneity and strong scattering observed seismologically at the base of the mantle. Moreover, such core-mantle reactions appear to be inevitable between liquid iron alloy and deep-mantle silicates, with the implication being that the composition of the core has evolved over geological history. By identifying the specific phases making up the reaction products (Table 3), the experiments suggest that the geomagnetic field may reflect the geochemical evolution of the core-mantle boundary.

ACKNOWLEDGMENTS

Much of this article is based on research carried out in collaboration with E. Knittle and Q. Williams. I have benefitted from discussions with them, as well as with J. Bloxham, S. I. Braginsky, R. S. Coe, W. M. Kaula, M. H. Manghnani, T. G. Masters, P. H. Roberts, D. J. Stevenson, and H. R. Wenk. This work was supported by NASA.

Literature Cited

Ahrens, T. J. 1979. Equations of state of iron sulfide and constraints on the sulfur content of the Earth. *J. Geophys. Res.* 84: 985–98

Ahrens, T. J., Jeanloz, R. 1987. Pyrite: shock compression, isentropic release and composition of the Earth's core. *J. Geophys. Res.* 92: 10,363–75

Akimoto, S., Suzuki, T., Yagi, T., Shimomura, O. 1987. Phase diagram of iron determined by high-pressure/temperature X-ray diffraction using synchrotron radiation. In *High-Pressure Research in Mineral Physics*, ed. M. H. Manghnani, Y. Syono, pp. 149–54. Washington, DC: Am. Geophys. Union

Alder, B. J. 1966. Is the mantle soluble in the core? *J. Geophys. Res.* 71: 4973–79

Alder, B. J., Trigueros, M. 1977. Suggestion of a eutectic region between the liquid and solid core of the Earth. *J. Geophys. Res.* 82: 2535–39

Anders, E., Ebihara, M. 1982. Solar-system abundances of the elements. *Geochim. Cosmochim. Acta* 46: 2363–80

Anderson, O. L. 1986. Properties of iron at the Earth's core conditions. *Geophys. J. R. Astron. Soc.* 84: 561–79

Bataille, K., Flatté, S. M. 1988. Inhomogeneities near the core-mantle boundary inferred from short-period scattered *PKP* waves recorded at the Global Digital Seismogram Network. *J. Geophys. Res.* 93: 15,057–64

Baumgardt, D. R. 1989. Evidence for a *P*-wave anomaly in *D''*. *Geophys. Res. Lett.* 16: 657–60

Birch, F. 1952. Elasticity and constitution of the Earth's interior. *J. Geophys. Res.* 57: 227–86

Birch, F. 1964. Density and composition of the mantle and core. *J. Geophys. Res.* 69: 4377–88

Birch, F. 1972. The melting relations of iron and temperatures in the Earth's core. *Geophys. J. R. Astron. Soc.* 29: 373–87

Bloxham, J. 1986. Models of the magnetic field at the core-mantle boundary for 1715, 1777, and 1842. *J. Geophys. Res.* 91: 13,954–66

Bloxham, J., Gubbins, D. 1985. The secular variation of Earth's magnetic field. *Nature* 317: 777–81

Bloxham, J., Gubbins, D. 1987. Thermal core-mantle interactions. *Nature* 325: 511–13

Bodea, S., Jeanloz, R. 1989. Model calculations of the temperature distribution in the laser-heated diamond cell. *J. Appl. Phys.* 65: 4688–92

Boehler, R. 1986. The phase diagram of iron to 430 kbar. *Geophys. Res. Lett.* 13: 1153–56

Bogue, S. W., Hoffman, K. A. 1987. Morphology of geomagnetic reversals. *Rev. Geophys.* 25: 910–16

Braginsky, S. I. 1978. Nearly axially symmetric model of the hydromagnetic dynamo of the Earth. *Geomagn. Aeron.* 18: 225–31

Braginsky, S. I., Meytlis, V. P. 1987. Overheating instability in the lower mantle near the boundary with the core. *Izvest., Earth Physics* 23: 646–49

Braginsky, S. I., Roberts, P. H. 1987. A model-Z geodynamo. *Geophys. Astrophys. Fluid Dyn.* 38: 327–49

Brett, R. 1976. The current status of speculations on the composition of the core. *Rev. Geophys. Space Phys.* 14: 375–83

Brown, J. M., McQueen, R. G. 1982. The equation of state for iron and the Earth's core. In *High-Pressure Research in Geophysics*, ed. S. Akimoto, M. H. Manghnani, pp. 611–23. Tokyo: Cent. Acad. Publ. Jpn.

Brown, J. M., McQueen, R. G. 1986. Phase transition, Grüneisen parameter and elasticity for shocked iron between 77 GPa and 400 GPa. *J. Geophys. Res.* 91: 7485–94

Brown, J. M., Ahrens, T. J., Shampine, D. L. 1984. Hugoniot data for pyrrhotite and the Earth's core. *J. Geophys. Res.* 89: 6041–48

Buchwald, V. F. 1975. *Handbook of Iron Meteorites*, Vol. 1. Berkeley: Univ. Calif. Press. 243 pp.

Chandrasekhar, S. 1961. *Hydrodynamic and Hydromagnetic Stability*. New York: Dover. 654 pp.

Clayton, D. D. 1983. *Principles of Stellar Evolution and Nucleosynthesis*. Chicago: Univ. Chicago Press. 612 pp.

Coe, R. S., Prévot, M. 1989. Evidence suggesting extremely rapid field variation during a geomagnetic reversal. *Earth Planet. Sci. Lett.* 92: 292–98

Cormier, V. F. 1981. Short-period *PKP* phases and the anelastic mechanism of the inner core. *Phys. Earth Planet. Inter.* 24: 291–301

Courtillot, V., Le Mouël, J. L. 1988. Time variations of the Earth's magnetic field: from daily to secular. *Annu. Rev. Earth Planet. Sci.* 16: 389–476

Courtillot, V., Le Mouël, J. L., Ducruix, J. 1984. On Backus' mantle filter theory and the 1969 geomagnetic impulse. *Geophys. J. R. Astron. Soc.* 78: 619–25

Davies, G. F., Gurnis, M. 1986. Interaction of mantle dregs with convection: lateral heterogeneity at the core-mantle boundary. *Geophys. Res. Lett.* 13: 1517–20

Doornbos, D. J., Hilton, T. 1989. Models of the core-mantle boundary and the travel times of internally reflected core phases. *J. Geophys. Res.* 94: 15,741–51

Dziewonski, A. M., Anderson, D. L. 1981. Preliminary reference Earth model. *Phys. Earth Planet. Inter.* 25: 297–356

Dziewonski, A. M., Gilbert, F. 1971. Solidity of the inner core of the Earth inferred from normal mode observations. *Nature* 234: 465–66

Francis, T. J. G. 1969. Generation of seismic anisotropy in the upper mantle under mid-oceanic ridges. *Nature* 221: 162–65

Fukai, Y., Akimoto, S. 1983. Hydrogen in the Earth's core: experimental approach. *Proc. Jpn. Acad.* 59B: 158–62

Fukai, Y., Suzuki, T. 1986. Iron-water reaction under high pressure and its implication in the evolution of the Earth. *J. Geophys. Res.* 91: 9222–30

Fukao, Y., Suda, N. 1989. Core modes of the Earth's free oscillations and structure of the inner core. *Geophys. Res. Lett.* 16: 401–4

Gans, R. F. 1972. Viscosity of the Earth's core. *J. Geophys. Res.* 77: 360–66

Garnero, E., Helmberger, D., Engen, G. 1988. Lateral variations near the core-mantle boundary. *Geophys. Res. Lett.* 15: 609–12

Gubbins, D., Masters, T. G. 1979. Driving mechanisms for the Earth's dynamo. *Adv. Geophys.* 21: 1–50

Gubbins, D., Masters, T. G., Jacobs, J. A. 1979. Thermal evolution of the Earth's core. *Geophys. J. R. Astron. Soc.* 59: 57–99

Gwinn, C. R., Herring, T. A., Shapiro, I. I. 1986. Geodesy by radio interferometry: studies of the forced nutations of the Earth 2. Interpretation. *J. Geophys. Res.* 91: 4755–65

Hartmann, W. K., Phillips, R. J., Taylor, G. J., eds. 1986. *Origin of the Moon.* Houston: Lunar Planet. Inst. 781 pp.

Heinz, D. L., Jeanloz, R. 1987. Measurement of the melting curve of $Mg_{0.9}Fe_{0.1}SiO_3$ perovskite at lower mantle conditions and its geophysical implications. *J. Geophys. Res.* 92: 11,437–44

Jacobs, J. A. 1987a. *The Earth's Core.* New York: Academic. 304 pp. 2nd ed.

Jacobs, J. A., ed. 1987b. *Geomagnetism*, Vols. 1, 2. New York: Academic. 450 pp., 586 pp.

Jamieson, J. C. 1977. Phase transitions in rutile-type structures. In *High-Pressure Research*, ed. M. H. Manghnani, S. Akimoto, pp. 209–17. New York: Academic

Jeanloz, R. 1979. Properties of iron at high pressure and the state of the core. *J. Geophys. Res.* 84: 6059–69

Jeanloz, R. 1987a. Earth's mantle. In *Encyclopedia of Physical Science and Technology*, ed. R. A. Meyers, 4: 486–502. New York: Academic

Jeanloz, R. 1987b. Composition of the Earth's inner core. *Nature* 325: 303

Jeanloz, R. 1989a. High pressure chemistry of the Earth's mantle and core. In *Mantle Convection*, ed. W. R. Peltier, pp. 203–59. New York: Gordon & Breach

Jeanloz, R. 1989b. Phase transitions in the mantle. *Nature* 340: 184

Jeanloz, R. 1989c. Physical chemistry at ultrahigh pressures and temperatures. *Annu. Rev. Phys. Chem.* 40: 237–59

Jeanloz, R., Ahrens, T. J. 1980. Equations of state of FeO and CaO. *Geophys. J. R. Astron. Soc.* 62: 505–28

Jeanloz, R., Richter, F. M. 1979. Convection, composition and the thermal state of the lower mantle. *J. Geophys. Res.* 84: 5497–5504

Jeanloz, R., Wenk, H. R. 1988. Convection and anisotropy of the inner core. *Geophys. Res. Lett.* 15: 72–75

Jephcoat, A. P., Mao, H. K., Bell, P. M. 1986. Static compression of iron to 78 GPa with rare gas solids as pressure-transmitting media. *J. Geophys. Res.* 91: 4677–84

Jephcoat, A., Olson, P. 1987. Is the inner core of the Earth pure iron? *Nature* 325: 332–35

Johnson, L. R., Lee, R. C. 1985. Extremal bounds on the *P* velocity in the Earth's core. *Bull. Seismol. Soc. Am.* 75: 115–30

Kato, T., Ringwood, A. E. 1989. Melting relationships in the system Fe–FeO at high pressures: implications for the composition and formation of the Earth's core. *Phys. Chem. Miner.* 16: 524–38

Keeler, R. N., Royce, E. B. 1971. Shock waves in condensed media. In *Physics of High Energy Density*, ed. P. Caldirola, pp. 51–150. New York: Academic

Kirby, S. H., Kronenberg, A. K. 1987. Rheology of the lithosphere: selected topics. *Rev. Geophys.* 25: 1219–44

Knittle, E., Jeanloz, R. 1986. High-pressure metallization of FeO and implications for the Earth's core. *Geophys. Res. Lett.* 13: 1541–44

Knittle, E., Jeanloz, R. 1989a. Melting curve of (Mg,Fe)SiO_3 perovskite to 96 GPa: evidence for a structural transition in lower mantle melts. *Geophys. Res. Lett.* 16: 421–24

Knittle, E., Jeanloz, R. 1989b. Simulating the core-mantle boundary: an experimental study of high-pressure reactions between silicates and liquid iron. *Geophys. Res. Lett.* 16: 609–12

Knittle, E., Jeanloz, R. 1990. The high pressure phase diagram of $Fe_{0.94}O$, a possible constituent of the Earth's outer core. *J. Geophys. Res.* In press

Knittle, E., Jeanloz, R., Mitchell, A. C., Nellis, W. 1986. Metallization of $Fe_{0.94}O$ at elevated pressures and temperatures observed by shock-wave electrical resistivity measurements. *Solid State Commun.* 59: 513–15

Lay, T. 1989. Structure of the core-mantle transition zone: a chemical and thermal boundary layer. *Eos, Trans. Am. Geophys. Union* 70: 54–55, 58–59

Lay, T., Helmberger, D. V. 1983. A lower mantle *S*-wave triplication and the shear velocity structure of *D''*. *Geophys. J. R. Astron. Soc.* 75: 799–838

Li, X., Jeanloz, R. 1987. Electrical conductivity of (Mg,Fe)SiO_3 perovskite and a perovskite-dominated assemblage at lower mantle conditions. *Geophys. Res. Lett.* 14: 1075–78

Li, X., Jeanloz, R. 1990. Laboratory studies of the electrical conductivity of silicate perovskites at high pressures and temperatures. *J. Geophys. Res.* In press

Manghnani, M. H., Ming, L. C., Nakagiri, N. 1987. Investigation of the α-Fe $\leftrightarrow$ ε-Fe phase transition by synchrotron radiation. In *High-Pressure Research in Mineral Physics*, ed. M. H. Manghnani, Y. Syono, pp. 155–63. Washington, DC: Am. Geophys. Union

Mao, H. K., Bell, P. M., Hadidiacos, C. 1987. Experimental phase relations of iron to 360 kbar, 1400°C, determined in an internally heated diamond-anvil apparatus. In *High-Pressure Research in Mineral Physics*, ed. M. H. Manghnani, Y. Syono, pp. 135–38. Washington, DC: Am. Geophys. Union

Marsh, S. P., ed. 1980. *LASL Shock Hugoniot Data*. Berkeley: Univ. Calif. Press. 658 pp.

Masters, G. 1979. Observational constraints on the chemical and thermal structure of the Earth's deep interior. *Geophys. J. R. Astron. Soc.* 57: 507–34

McQueen, R. G., Marsh, S. P. 1966. Shock wave compression of iron-nickel alloys and the Earth's core. *J. Geophys. Res.* 71: 1751–56

Merrill, R. T., McElhinny, M. W. 1983. *The Earth's Magnetic Field*. New York: Academic. 401 pp.

Moore, W. J. 1972. *Physical Chemistry*. Englewood Cliffs, NJ: Prentice-Hall. 977 pp. 4th ed.

Morelli, A., Dziewonski, A. M., Woodhouse, J. H. 1986. Anisotropy of the inner core inferred from *PKIKP* travel times. *Geophys. Res. Lett.* 13: 1545–48

Murthy, V. R., Hall, H. T. 1970. The chemical composition of the core: possibility of sulfur in the core. *Phys. Earth Planet. Inter.* 2: 276–82

Nicolas, A., Christensen, N. I. 1987. Formation of anisotropy in upper mantle peridotites—a review. In *The Composition, Structure and Dynamics of the Lithosphere-Asthenosphere System. Geodyn. Ser.*, ed. C. Froidevaux, K. Fuchs, 16: 111–23. Washington, DC: Am. Geophys. Union

Ohtani, E., Ringwood, A. E. 1984. Composition of the core, I. Solubility of oxygen in molten iron at high temperatures. *Earth Planet. Sci. Lett.* 71: 85–93

Ohtani, E., Ringwood, A. E., Hibberson, W. 1984. Composition of the core, II. Effect of high pressure on solubility of FeO in molten iron. *Earth Planet. Sci. Lett.* 71: 94–103

Ringwood, A. E. 1977. Composition of the core and implications for the origin of the Earth. *Geochem. J.* 11: 111–35

Ringwood, A. E. 1979. *Origin of the Earth and Moon*. New York: Springer-Verlag. 295 pp.

Ruff, L. J., Helmberger, D. V. 1982. The structure of the lowermost mantle determined by short-period *P*-wave amplitudes. *Geophys. J. R. Astron. Soc.* 68: 95–119

Sayers, C. M. 1989. Seismic anisotropy of the inner core. *Geophys. Res. Lett.* 16: 267–70

Scher, H., Zallen, R. 1970. Critical density in percolation processes. *J. Chem. Phys.* 53: 3759–61

Secco, R. A., Schloessin, H. H. 1989. The electrical resistivity of solid and liquid Fe at pressures up to 7 GPa. *J. Geophys. Res.* 94: 5887–94

Shearer, P. M., Masters, G. 1990. The density and shear velocity contrast at the inner core boundary. *Geophys. J. Int.* In press

Shearer, P. M., Toy, K. M., Orcutt, J. A. 1988. Axi-symmetric Earth models and inner-core anisotropy. *Nature* 333: 228–32

Sherman, D. M. 1989. The nature of the pressure-induced metallization of FeO and its implication to the core-mantle boundary. *Geophys. Res. Lett.* 16: 515–18

Shimoji, M. 1977. *Liquid Metals*. New York: Academic. 391 pp.

Sleep, N. H. 1988. Gradual entrainment of a chemical layer at the base of the mantle by overlying convection. *Geophys. J.* 95: 437–47

Souriau, A., Souriau, M. 1989. Ellipticity and density at the inner core boundary from sub-critical *PKiKP* and *PcP* data. *Geophys. J. Int.* 98: 39–54

Steinemann, S. G., Keita, N. M. 1988. Compressibility and internal pressure anomalies of liquid *3d* transition metals. *Helv. Phys. Acta* 61: 557–65

Stevenson, D. J. 1981. Models of the Earth's core. *Science* 241: 611–19

Stevenson, D. J. 1986. On the role of surface tension in the migration of melts and fluids. *Geophys. Res. Lett.* 13: 1149–52

Stevenson, D. J. 1987a. Origin of the Moon—the collision hypothesis. *Annu. Rev. Earth Planet. Sci.* 15: 271–315

Stevenson, D. J. 1987b. Limits on the lateral density and velocity variations in the Earth's outer core. *Geophys. J. R. Astron. Soc.* 88: 311–19

Suzuki, T., Akimoto, S., Yagi, T. 1989. Metal-silicate-water reaction under high pressure. I. Formation of metal hydride and implications for composition of the core and mantle. *Phys. Earth Planet. Inter.* 56: 377–88

Taylor, A., Kagle, B. J., eds. 1963. *Crystallographic Data on Metal and Alloy Structures*. New York: Dover. 263 pp.

Tsuchida, Y., Yagi, T. 1989. A new post-stishovite high-pressure polymorph of silica. *Nature* 340: 217–20

Urakawa, S., Kato, M., Kumazawa, M. 1987. Experimental study on the phase relations in the system Fe-Ni-O-S to 15 GPa. In *High-Pressure Research in Mineral Physics*, ed. M. H. Manghnani, Y. Syono, pp. 95–111. Washington, DC: Am. Geophys. Union

Verhoogen, J. 1980. *Energetics of the Earth*. Washington, DC: Natl. Acad. Sci. 135 pp.

Watson, E. B. 1982. Melt infiltration and magma evolution. *Geology* 10: 236–40

Weidner, D. J., Bass, J. D., Ringwood, A. E., Sinclair, W. 1982. The single-crystal elastic moduli of stishovite. *J. Geophys. Res.* 87: 4740–46

Wenk, H. R., Takeshita, T., Jeanloz, R., Johnson, G. C. 1988. Texture development and elastic anisotropy during deformation of ε iron. *Geophys. Res. Lett.* 15: 76–79

Williams, Q., Jeanloz, R. 1990. Melting relations in the iron-sulfur system at ultrahigh pressures: implications for the thermal state of the Earth. *J. Geophys. Res.* In press

Williams, Q., Jeanloz, R., Bass, J., Svendsen, B., Ahrens, T. J. 1987. The melting curve of iron to 250 GPa: a constraint on the temperature at Earth's center. *Science* 236: 181–82

Woodhouse, J. H., Giardini, D., Li, X. D. 1986. Evidence for inner core anisotropy from free oscillations. *Geophys. Res. Lett.* 13: 1549–52

Yagi, T., Fukuoka, K., Takei, H., Syono, Y. 1988. Shock compression of wüstite. *Geophys. Res. Lett.* 15: 816–19

Yagi, T., Suzuki, T., Akimoto, S. 1985. Static compression of wüstite ($Fe_{0.98}O$) to 120 GPa. *J. Geophys. Res.* 10: 8784–88

Yeganeh-Haeri, A., Weidner, D. J., Ito, E. 1989. Elasticity of $MgSiO_3$ in the perovskite structure. *Science* 243: 787

Young, C. J., Lay, T. 1987. The core-mantle boundary. *Annu. Rev. Earth Planet. Sci.* 15: 25–46

Young, C. J., Lay, T. 1989. The core shadow zone boundary and lateral variations of the *P* velocity structure of the lowermost mantle. *Phys. Earth Planet. Inter.* 54: 64–81

Zallen, R. 1983. *The Physics of Amorphous Solids*. New York: Wiley. 304 pp.

Annu. Rev. Earth Planet. Sci. 1990. 18:387–447
Copyright © 1990 by Annual Reviews Inc. All rights reserved

SYNCHROTRON RADIATION: APPLICATIONS IN THE EARTH SCIENCES

W. A. Bassett

Department of Geological Sciences, Snee Hall, Cornell University, Ithaca, New York 14853

G. E. Brown, Jr.

Department of Geology and Center for Materials Research, Stanford University, Stanford, California 94305-2115

1. INTRODUCTION

A revolutionary advance in X-ray studies of materials occurred in the mid-1970s when the first synchrotron X-ray sources became available for general scientific use (for background information, see Winick & Doniach 1980, Koch 1983, Winick 1987a). X rays have played an important role in the Earth sciences since 1915, when W. H. and W. L. Bragg first used this energetic form of electromagnetic radiation to determine the atomic-level structures of minerals. This basic structural information is now an essential part of our knowledge about the Earth and its dynamic processes. During the past 20 years, major advances in experimental and theoretical methods, in X-ray instrumentation, in X-ray sources, and in computer automation of experiments have resulted in significant gains in our ability to study the structure, bonding, and composition of Earth materials using X rays. Earth scientists now use X-ray scattering methods (both diffraction and non-Bragg scattering) in a variety of applications, including (*a*) rapid identification and unit cell characterization of minerals; (*b*) in situ study of mineral structures at high temperatures and pressures; (*c*) derivation of structural information about amorphous silicates, including melts; (*d*) determination and refinement of the atomic coordinates and site occu-

0084–6597/90/0515–0387$02.00

pancies of relatively complex minerals by analysis of diffraction data from single crystals as well as from fine-grained mineral powders; and (*e*) study of bonding in minerals through accurate determinations of electron density distributions. X-ray textural analysis of rocks has also become routine, and the recent application of three-dimensional X-ray tomography methods to porous or fractured rocks offers the promise of rapid, quantitative imaging of drill-core samples, including nondestructive measurements of porosity, permeability, and phase distributions.

X-ray spectroscopy methods have also had a major impact on the study of Earth materials. Electron- and X-ray-induced characteristic X-ray emission (electron microprobe analysis and X-ray fluorescence analysis) are now the primary means of rapid, quantitative compositional analysis of minerals, rocks, natural glasses, and synthetic experimental run products. High-energy-resolution X-ray emission and photoelectron spectroscopies have been used to obtain information on chemical bonding in minerals, including oxidation states. Most recently, X-ray absorption spectroscopy has made it possible to obtain short- and intermediate-range structural information (bond distances and coordination numbers) on cation and anion environments in Earth materials of all types (minerals, glasses, melts, gels, coal, aqueous solutions, and mineral surfaces and interfaces).

All of these applications benefit significantly from the 10^6–10^8 increase in brilliance of synchrotron X-ray sources relative to stationary and rotating anode X-ray tubes (see Figure 1). Here, we define brilliance as the number

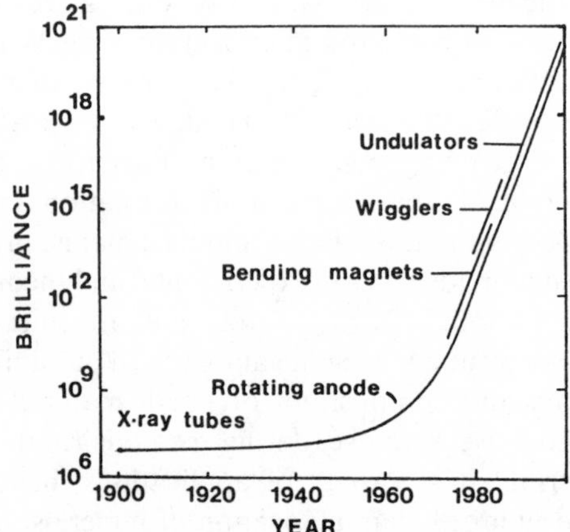

Figure 1 Available brilliance (number of photons per second per square millimeter per square milliradian per 0.1% bandwidth) of X-ray sources as a function of time. Reproduced with permission of Theophrastus Publications (Augustithis 1988).

of photons per second per unit solid angle per unit source area per unit of spectral bandwidth. These sources produce highly collimated beams of X rays over a broad spectral range (from a few electron volts to >50 keV; see Figure 2) that can be wavelength tuned using movable crystal monochromators. This remarkable improvement in X-ray sources has extended greatly the range of structural and chemical investigations possible with X rays, permitting studies of very dilute samples, samples at extreme conditions of pressure and temperature, samples with extremely small volumes, and phenomena of very short duration. Synchrotron radiation sources also satisfy the need for data of high-energy resolution and high signal-to-noise ratio, such as those required for profile fitting analysis of powder diffraction data from low-symmetry crystal structures. In addition, the spectral characteristics of synchrotron radiation have made possible new research areas, such as structural studies of reaction sites at mineral/water interfaces and of trace-element environments in silicate minerals and glasses, as well as kinetic studies of phase transitions. The high intensity of synchrotron X rays has permitted viscosity measurements of silicate melts at high temperatures and pressures using radiography methods. When synchrotron radiation sources are operated in a timing mode, extremely intense X rays are produced with a pulsed time structure (on the order of a few tenths of a nanosecond), making possible time-resolved X-ray diffraction and spectroscopy studies of transient phenomena, such as solid-solid phase transitions. Research in these areas was not feasible before the availability of high-intensity synchrotron X rays, and each of these areas offers the potential for new discoveries about Earth materials and Earth processes.

In the 15 years that synchrotron radiation has been utilized for research in the physical and biological sciences, it has become a remarkably versatile and widely used analytical tool for probing the structure, composition, and bonding of materials, including those of interest in the Earth sciences. Further development of this field will closely parallel the development of new synchrotron sources, particularly insertion devices (multipole wiggler and undulator magnets), which greatly enhance X-ray flux and brilliance. New research opportunities will undoubtedly be provided by these next-generation sources, which will produce X rays 1000 to 10,000 times more brilliant than existing synchrotron X-ray sources.

An organization of Earth scientists who use synchrotron radiation (Geo-Sync) has been formed to plan the development of synchrotron radiation beamlines and to facilitate their use by Earth scientists.

This paper presents an overview of synchrotron radiation sources and their characteristics. It also reviews recent synchrotron-based research on Earth materials and discusses future Earth sciences applications using the

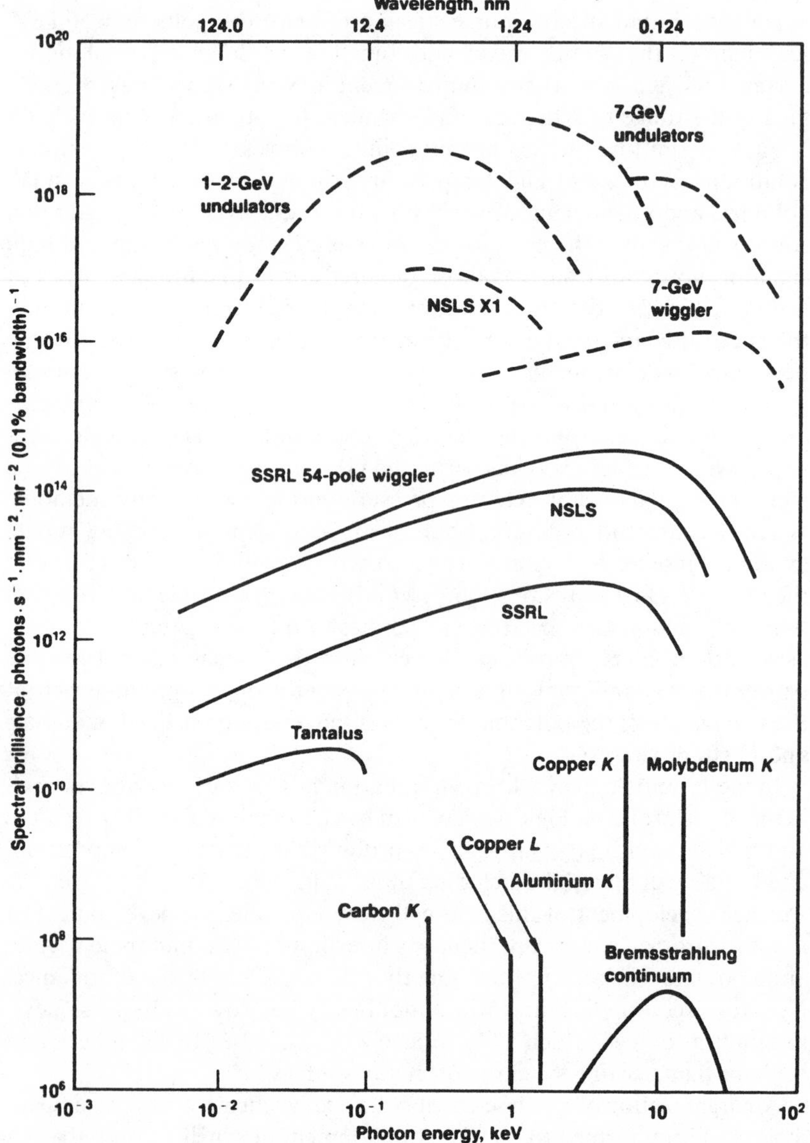
Wavelength, nm
124.0
12.4
1.24
0.124
7-GeV undulators
1–2-GeV undulators
NSLS X1
7-GeV wiggler
SSRL 54-pole wiggler
NSLS
SSRL
Tantalus
Copper K
Molybdenum K
Copper L
Aluminum K
Carbon K
Bremsstrahlung continuum
Spectral brilliance, photons · s⁻¹ · mm⁻² · mr⁻² (0.1% bandwidth)⁻¹
Photon energy, keV

next generation of synchrotron radiation sources, which are presently under construction. Because of length limitations, not all recent work could be covered. Therefore, we have provided references to the literature where other applications can be found.

2. PRODUCTION OF SYNCHROTRON RADIATION

2.1 *Principles*

Electromagnetic radiation is produced whenever a charged particle is accelerated. Deflection of a beam of charged particles is a form of acceleration; thus, a beam of charged particles that is made to circulate in a ring through use of bending magnets is accelerated radially and constantly loses energy in the form of electromagnetic radiation. Radiation from such a source is called synchrotron radiation, and it is necessary to continually provide energy to the beam of charged particles to make up for the loss. Radio-frequency cavities are placed in the ring to accelerate the particles as they move around the ring. This makes it possible to maintain a beam of charged particles in a storage ring for hours.

At low velocity, a beam of electrons or positrons produces dipole radiation in the shape of a toroid around the acceleration vector. As the velocity of the particles approaches that of light, the increasing effect of relativity causes the electromagnetic radiation to be emitted in the forward direction like a headlight. The divergence of the beam decreases as the energy of the particle beam is increased. In the mid-1970s, the potential of such intense, well-directed radiation was recognized and tangential pipes or beamlines were added to storage rings to provide a means for directing the radiation into experimental areas where it could be used.

Although synchrotron radiation can be produced wherever there is a beam of charged particles, the source most widely used for the production of synchrotron X rays is the electron or positron storage ring. The electrons

---

Figure 2 Comparison of spectral brilliances for a variety of synchrotron radiation sources and laboratory X-ray sources. The range of brilliance for the laboratory sources reflects the values possible for stationary anode tubes and for rotating anode tubes, with and without microfocussing. Tantalus is the synchrotron storage ring at the University of Wisconsin, SSRL is the Stanford Synchrotron Radiation Laboratory, and NSLS is the National Synchrotron Light Source at Brookhaven National Laboratory. The dashed lines represent calculated spectral brilliances for the planned 7-GeV Advanced Photon Source at Argonne National Laboratory (7-GeV wiggler and 7-GeV undulators—these lines show the envelope of the possible output spectrum); the 1–2 GeV soft X-ray vacuum ultraviolet Advanced Light Source, which is now under construction at Lawrence Berkeley Laboratory (1–2 GeV undulators—this line shows the envelope of the output spectrum); and the proposed beamline X1 at Brookhaven National Laboratory (NSLS X1). From Lear & Weber (1987).

or positrons are typically accelerated in a linear accelerator, transferred to a booster synchrotron where they are further accelerated, and then transferred to the storage ring. In the storage ring the electrons travel in bunches. Each bunch produces a burst or pulse of radiation as it passes through a bending magnet or an insertion device consisting of a series of magnets. The bunches may be varied in number and spacing to produce sequential bursts of radiation through a beamline. The pulse length is a few tenths of a nanosecond, and the time interval between bunches is tens of nanoseconds to microseconds.

2.2 *Characteristics*

X rays produced by a synchrotron source differ from those produced by a conventional X-ray tube in several significant ways. The properties that make synchrotron radiation unique are described below. A more detailed discussion of these properties can be found in Winick (1980).

2.2.1 HIGH INTENSITY Synchrotron radiation produced by a bending magnet has a brilliance that is about three orders of magnitude greater than the characteristic radiation and four to six orders of magnitude greater than the bremsstrahlung produced by a rotating anode source (Figure 2). Here, the units for brilliance are photons per second per 0.1% spectral bandwidth per square milliradian per square millimeter. Synchrotron radiation produced by a wiggler insertion device has a brilliance that is six or seven orders of magnitude greater than the characteristic radiation from a rotating anode source (Figure 2). Synchrotron radiation from an undulator source can be as much as eight orders of magnitude more brilliant. However, the brilliance of radiation from an undulator is much greater for some energies than for others, as defined in Section 2.3.2. Another unit of measurement that is commonly used in discussing synchrotron radiation from storage rings is emittance, which is defined as the product of the size and angular spread of the electron or positron beam and has units of nanometer-radians. Low-emittance rings produce narrow beams of X rays and thus focus a small beam size on a sample. The "third generation" storage rings currently under construction or in the design stage will have emittances of 50 to 100 nm-rad.

2.2.2 BROAD ENERGY RANGE The high brilliance of radiation from bending magnets and wigglers typically extends over a range of energies (or wavelengths) from less than 1 keV (~1.4 nm) to more than 70 keV (~0.02 nm) (Figure 2). The brilliance of the X rays produced by bending magnets or wigglers is a smooth function of energy, which is not the case for the brilliance of radiation produced by undulators. However, undulators are being designed that can be scanned through a range of energies without large fluctuations in brilliance.

2.2.3 EXCELLENT COLLIMATION Synchrotron radiation from a bending magnet is highly collimated in the vertical direction, having a divergence of only a fraction of a milliradian. This property makes it possible to produce a very intense beam of X rays by placing a pinhole some distance upstream from a sample. It also allows very high resolution in diffraction studies. Even though the divergence of the X-ray beam is small, the cross-sectional area at the location of the experiment is in many cases several millimeters high by several centimeters wide.

2.2.4 PULSED TIME STRUCTURE The radiation arrives in pulses, each pulse being produced as a bunch of electrons or positrons passes through a bending magnet, a wiggler, or an undulator. The frequency of these pulses is determined by the spacing and number of bunches in the storage ring.

2.2.5 POLARIZATION The radiation is highly polarized. The electric vector of the radiation lies in the plane defined by the direction of deflection of the particle beam. For bending magnets, this is the horizontal plane of the storage ring. However, there are insertion devices that deflect the particle beam vertically and therefore produce polarized radiation with the electric vector vertical.

2.3 *Devices for Generating Synchrotron Radiation*

2.3.1 BENDING MAGNET The most commonly used device for generating synchrotron radiation in a storage ring containing electrons or positrons is the bending magnet. Such magnets are typically electromagnets rather than superconducting and are generally constructed of steel. As stated above, they are used to deflect the beam of positrons or electrons in a circular arc, thus producing synchrotron radiation.

2.3.2 INSERTION DEVICES Over the last decade, two types of devices have been developed as alternatives to bending magnets for producing synchrotron radiation. These are the wiggler and the undulator, and they are often referred to collectively as insertion devices because they are inserted in straight sections of a storage ring. Each of these devices is a string of permanent (e.g. $SmCo_5$ or Nd-Fe) or superconducting magnets designed to deflect the beam first in one direction and then in the other. Because the net deflection is zero, these devices are inserted in straight stretches of the ring. The principle is the same as that of the bending magnet, but the radiation from each magnet in the string is added to the radiation from the others, thus producing an even greater intensity. Although the wiggler and the undulator both operate on the same principle, the undulator is different in that there is interaction between the radiation produced by one magnet and the radiation produced by the other magnets along the string. There is constructive and destructive interference, which leads to enhanced intensity of some wavelengths and diminished intensity of others. A more detailed description of these insertion devices can be found in a recent Advanced Photon Source workshop report (Rivers 1988).

3. FACILITIES

At present, 12 electron or positron storage rings in seven countries produce synchrotron radiation in the hard X-ray range (~ 3 keV to > 50 keV), and 13 storage rings in the same countries (plus Sweden) produce synchrotron radiation in the soft X-ray/vacuum ultraviolet (VUV) range (1 eV to ~ 3 keV) (Table 1). The most important hard X-ray sources in the US, Europe, and Asia are discussed in more detail below. Very little work on Earth materials has been carried out using available soft X-ray/VUV sources. Therefore, they are not discussed in much detail herein. In addition, brief overviews of future US and European synchrotron sources are given below, with emphasis on those features that will permit new classes of experiments on Earth materials.

3.1 *United States*

In the United States there are four operating high-energy synchrotron sources that produce hard X rays: (*a*) the Stanford Positron Electron Accumulation Ring (SPEAR) at the Stanford Synchroron Radiation Laboratory (SSRL); (*b*) the Positron-Electron Project (PEP) storage ring at SSRL; (*c*) the Cornell High Energy Synchrotron Source (CHESS); and (*d*) the National Synchrotron Light Source (NSLS) at Brookhaven National Laboratory. The total number of experimental stations available on these sources is about 50. Several soft X-ray/VUV sources are also available, including nine beamlines at SPEAR, a separate soft X-ray/VUV ring at NSLS, and smaller facilities in Wisconsin (TANTALUS) and at the National Institute for Standards and Technology (formerly National Bureau of Standards) in Gaithersburg, Maryland (SURF II). The facilities available and experimental work possible at each of the four major US synchrotron radiation sources are discussed below.

3.1.1 SSRL (STANFORD SYNCHROTRON RADIATION LAB) The first general user synchrotron source in the US is located at Stanford University. It was developed in 1974 as a parasitic facility on the SPEAR storage ring, which was used primarily for high-energy physics research at the Stanford Linear

Accelerator Center (SLAC). This facility became a Department of Energy-(DOE)-supported national-level laboratory in 1977 and has since been widely used by scientists from all disciplines as a major synchrotron radiation source. Currently this storage ring is injected from the Stanford Linear Accelerator, but it will have an independent injector in the early 1990s when it becomes a fully dedicated synchrotron radiation source. The PEP storage ring at SLAC is also used in parasitic fashion for synchrotron radiation research but presently in a much more limited way than SPEAR.

3.1.1.1 *SPEAR (Stanford Positron Electron Accumulation Ring)* On SPEAR at SSRL there are nine beamlines, four with bending magnets (critical energy of approximately 4.7 keV), four with wiggler insertion devices, and one with an undulator insertion device. A total of 22 experimental stations are available on these beamlines; 13 provide hard X rays (3–50 keV), and 9 provide photons in the soft X-ray/vacuum ultraviolet range (1–3000 eV). The soft X-ray/VUV stations provide experimental capabilities for ultra-high-vacuum compatible studies [X-ray photoelectron and ultraviolet photoelectron spectroscopy, low-energy electron diffraction, surface extended X-ray absorption fine structure (SEXAFS), X-ray lithography/microscopy]. The hard X-ray stations provide a variety of experimental capabilities, including small- and large-angle X-ray scattering, X-ray anomalous scattering, surface scattering, single-crystal goniometers and rotation cameras for single-crystal diffraction studies, X-ray absorption spectroscopy (EXAFS and XANES), and X-ray topography. All 22 stations except one have double-crystal monochromators or ruled gratings that enable continuous tuning of photon energies over a wide range of values. The one white-radiation station on a bending magnet beamline can be used for energy-dispersive X-ray scattering, including high-pressure/high-temperature diamond anvil cell studies, dispersive EXAFS/XANES spectroscopy, and X-ray topography. The 54-pole wiggler beamline provides an exceptionally bright source of hard X rays that have been used for a variety of purposes, including X-ray fluorescence microprobe studies, EXAFS/XANES studies of dilute species and low-atomic-number elements, and glancing-angle X-ray scattering studies from surfaces. Approximately half of the experimental stations on SPEAR have been used by Earth scientists, although none of them has been available for Earth sciences applications for more than a few months each year because SPEAR is oversubscribed by users in all disciplines. The types of Earth sciences applications that have been carried out on SPEAR beamlines include X-ray absorption spectroscopy, X-ray scattering (including anomalous scattering), structure refinement by single-crystal and powder diffraction, X-ray fluorescence microprobe studies, soft X-ray total elec-

Table 1 Synchrotron radiation sources (after Wong 1986)

Machine	Location	Energy (GeV)	Current (mA)	Bending radius (m)	Critical energy (keV)	Remarks
PETRA	Hamburg, FRG	15	50	192	39.0	Possible future use for synchrotron radiation research
PEP (SSRL)	Stanford, CA, USA	15	50	165.5	45.2	Synchrotron radiation facility planned
		12	45	(23.6)	(163)	(From 17-kG wiggler)
CHESS (Cornell)	Ithaca, NY, USA	8	50	32.5	35.0	Used parasitically
VEPP-4	Novosibirsk, USSR	7	10	16.5	46.1	Initial operation at 4.5 GeV
		4.5		(18.6)	(10.9)	(From 8-kG wiggler)
DESSY	Hamburg, FRG	5	50	12.1	22.9	Partly dedicated
		2.5	300		2.9	
SPEAR (SSRL)	Stanford, CA, USA	4.0	50	12.7	11.1	50% dedicated
		3.0	100		4.7	
		3.0		(5.5)	(10.8)	(From 18-kG wiggler)
SRS	Daresbury, UK	2.0	500	5.55	3.2	Dedicated
				(1.33)	(13.3)	(For 50-kG wiggler)
VEPP-3	Novosibirsk, USSR	2.25	100	6.15	4.2	Partly dedicated
				(2.14)	(11.8)	(From 35-kG wiggler)
DCI	Orsay, France	1.8	500	4.0	3.63	Partly dedicated

ADONE	Frascati, Italy	1.5	60	5.0	1.5	Partly dedicated
				(2.8)	(2.7)	(From 18-kG wiggler)
VEPP-2M	Novosibirsk, USSR	0.67	100	1.22	0.54	Partly dedicated
ACO	Orsay, France	0.54	100	1.1	0.32	Dedicated
SOR Ring	Tokyo, Japan	0.40	250	1.1	0.13	Dedicated
SURF II	Gaithersburg, MD, USA	0.25	25	0.84	0.041	Dedicated
TANTALUS I	Stoughton, WI, USA	0.24	200	0.64	0.048	Dedicated
PTB	Braunschweig, FRG	0.14	150	0.46	0.013	Dedicated
N-100	Kharkov, USSR	0.10	25	0.50	0.004	
Photon Factory	Tsukuba, Japan	2.5	500	8.33	4.16	Dedicated
				(1.67)	(20.5)	(For 50-kG wiggler)
NSLS	Brookhaven National Laboratory, Upton, NY, USA	2.5	200	6.88	5.01	Dedicated
				(1.67)	(20.5)	(For 50-kG wiggler)
BESSY	West Berlin, FRG	0.80	500	1.83	0.62	Dedicated; industrial use planned
NSLS	Brookhaven National Laboratory, Upton, NY, USA	0.70	750	1.90	0.40	Dedicated
ETL	Electrotechical Laboratory, Tsukuba, Japan	0.66	100	2	0.32	Dedicated
UVSOR	Institute of Molecular Science, Okatabi, Japan	0.60	500	2.2	0.22	Dedicated
MAX	Lund, Sweden	0.50	100	1.2	0.23	Dedicated
KURCHATOV	Moscow, USSR	0.45		1.0	0.21	Dedicated

tron yield spectroscopy, and high-pressure/high-temperature diffraction studies. Many of these studies were funded by the National Science Foundation (NSF). SPEAR normally operates in a fully dedicated synchrotron mode (3 GeV and 60–100 mA beam current) about 300 eight-hour shifts each year and in a parasitic mode [during high-energy physics experiments (1.5–2.0 GeV and 10–30 mA beam current)] for a smaller number of shifts each year. When operating in the dedicated mode, typical fill lifetimes of 8 to 12 hr are achieved. Several new beamlines and side stations on SPEAR are presently under construction.

3.1.1.2 *PEP (Positron Electron Project)* The PEP storage ring is currently equipped with two synchrotron beamlines, each utilizing a 52-pole undulator insertion device. This ring operates at 8 GeV in parasitic mode and at up to 16 GeV in dedicated mode. Extremely low emittances and exceptionally high brilliances have been achieved in limited parasitic runs during the past year. These runs have been devoted to surface scattering studies that were not possible before the PEP facility was developed for synchrotron radiation. Because the emittances and brilliances achieved are comparable to those hoped for at the Advanced Photon Source (APS), PEP will serve as the most advanced synchrotron source during the next 7 to 9 years while the APS is under construction and will also function as a test bed for developing new types of insertion devices to be used at APS. To date, no Earth sciences projects have been carried out on the PEP storage ring; however, its characteristics would make it an excellent choice for X-ray fluorescence microprobe studies of Earth materials. Fill lifetimes of up to 20 hr have been achieved during the limited operation of PEP, which is currently funded by the DOE.

3.1.2 CHESS (CORNELL HIGH ENERGY SYNCHROTRON SOURCE) At CHESS, there are three bending magnet beamlines (A, B, C) with critical energies of approximately 8.7 keV, resulting in higher X-ray intensities at higher X-ray energies than at the SSRL or at the NSLS (see Section 3.1.3). In addition, beamline A is equipped with a wiggler so that either bending magnet or wiggler radiation can be used. Beamline A has three stations that offer a variety of types of radiation, from fixed wavelength to variable wavelength to white radiation. Beamline B has one station and offers white radiation from a bending magnet. Beamline C has two stations that are each able to provide white radiation or variable monochromatic radiation. The radiation at CHESS is produced by the storage ring electron beam. Each fill lasts for 2 to 4 hr.

The operation of CHESS, which started up in 1980, is funded by the NSF. Most of the X-ray diffraction studies at high pressures and temperatures using diamond anvil cells have been conducted with white radiation available on beamline B using energy-dispersive techniques. Some runs have been made that utilized the intense white radiation as well as monochromatic radiation from the wiggler on line A. Some preliminary studies of element-specific X-ray radiography have been conducted on beamline C using monochromatic radiation selected to show the distribution of iron in fossils. The NSF has been the major source of funding for these studies.

Two experimental stations are being constructed at CHESS for conducting high-pressure/high-temperature X-ray diffraction experiments using diamond anvil cells. One of these stations will be on a bending magnet beamline, and the other will be on a wiggler beamline. Both will utilize white radiation and the energy-dispersion method. Each station will be equipped with a solid-state detector, multichannel analyzer, Microvax III computer, an apparatus for remote alignment, and provisions for remote application of pressure and temperature. Experienced staff members familiar with both synchrotron radiation and high-pressure/high-temperature experiments will be available to help users set up and run their experiments. Each of these experimental facilities will be on side stations, necessitating in some cases a special experimental design that will permit working within a few inches of a throughgoing beam pipe.

The new experimental stations will be equipped with an optical system that will permit direct observation of the sample during the experiment, laser heating of the sample, temperature measurement by blackbody curve fitting, and pressure measurement by the ruby fluorescence method.

3.1.3 NSLS (NATIONAL SYNCHROTRON LIGHT SOURCE) The NSLS consists of two storage rings: a VUV ring that operates at 750 mA and 750 MeV, and an X-ray ring that operates at 200 mA and 2.5 GeV. The VUV ring began operations in 1983, and the X-ray ring in 1985. On the X-ray ring there are 28 ports that correspond to the beamlines described above for SSRL and CHESS. Each port is split into beamlines that correspond to experimental stations as described above for SSRL and CHESS. There is the capacity for 56 such stations at NSLS, of which approximately 37 are currently operational. The critical energy of the radiation produced by bending magnets on the X-ray ring is 5 keV. Beam lifetimes of 10 hr are common, and the facility operates 24 hr per day for 45 weeks each year. Currently, the NSLS provides more photons on target than all other synchrotron light sources in the world combined. A number of insertion devices (wigglers and undulators) are in use or planned. The operation of NSLS is funded by the DOE.

Nine beamlines at the NSLS play important roles in Earth sciences research. They are X2, X7A, X7B, X17B, X17C, X19A, X26A, X26C, and

U3. Beamline X2 is used for X-ray tomography. X7A is used for powder diffraction and diamond anvil cell research. X7B is used for single-crystal X-ray diffraction. The X17 beamline uses a 5-T superconducting wiggler magnet with a critical energy of 22 keV. X17B is used for studies with a cubic-anvil large-volume press and for high-energy X-ray fluorescence and tomography studies. X17C is dedicated to diamond anvil cell high-pressure research. X19 is used for X-ray absorption spectroscopy. X26A and X26C are used for the X-ray microprobe facility and for microtomography. U3 is available for photoelectron spectroscopy with ultraviolet light. In addition, many other NSLS X-ray beamlines are available to the Earth sciences community for diffraction and spectroscopy experiments.

3.1.4 FUTURE SYNCHROTRON RADIATION SOURCES Upgrading of existing synchrotron radiation sources will continue over the next few years prior to completion of new national facilities. These changes will enhance the general capabilities of each source; some of these changes are being implemented by the Earth sciences community.

At SSRL, several new hard and soft X-ray beamlines are currently being planned or are under construction on SPEAR. A new multipole undulator beamline is also planned for PEP. These projects should be completed by early 1990.

At Cornell, construction of CHESS II began in March 1988 and is expected to be completed in July 1989. It will be the mirror image of CHESS I and will be located on the east side of the interaction area. CHESS II will utilize X rays produced by the positron beam. There will be two X-ray beamlines. A straight section will support a multipole permanent magnet wiggler capable of providing a flux that is five times greater than that produced by the CHESS I wiggler. The CHESS II experimental area will be approximately equal in size to that of CHESS I.

At the NSLS, significant upgrades are underway or planned on several beamlines. X25 will be upgraded with a 30-pole hybrid wiggler with a 5-keV critical energy. It will be optimized for high-angular-resolution scattering but will also be available for other experiments that require the same spectrum as the bending magnets but with more flux. X21 will also be upgraded with a 30-pole wiggler and will be available for high-energy-resolution scattering as well as other experiments. Beamline X13 is being fitted with a soft X-ray undulator for electron spectroscopy.

Over the next decade, several third-generation synchrotron sources will be constructed in the US and Europe that will have significant impact on the study of Earth materials. These machines will utilize wiggler and undulator insertion devices to produce photons ranging from 50 eV to 100 keV in energy, with brilliances several orders of magnitude greater than existing sources (Figure 1). Currently, the PEP ring at Stanford University (Table 1) serves as a test bed for these new high-brilliance/low-emittance sources, and extremely low emittances (high brightnesses) have already been achieved using an undulator insertion device on this synchrotron radiation source (Coisson & Winick 1987). The US synchrotron radiation community is planning two new sources: One will operate in the 7–8 GeV range, producing photons in the X-ray to hard X-ray energy range (1–100 keV), at Argonne National Laboratory (the Advanced Photon Source, or APS); and the other will operate in the 1–2 GeV range, producing photons in the soft X-ray to vacuum ultraviolet energy range (50–3000 eV), at Lawrence Berkeley Laboratory (the Advanced Light Source, or ALS). The APS will accommodate 34 insertion devices and 35 bending magnets along its 1060-m circumference, providing up to 100 experimental stations (Shenoy et al 1988). In this ring, positrons will circulate in 1 to 60 bunches at a total current of about 100 mA, with a revolution time of about 3.5 μs. The European synchrotron radiation community is also planning a major facility (European Synchrotron Radiation Facility, or ESRF) (Buras & Tazzari 1984). These new sources of X rays of enormous brilliance will undoubtedly create new opportunities for research in the Earth sciences.

In summary, 30 to 40 new experimental stations are planned for the next 5 to 7 years in the US. That number could easily rise to 50 or 60 as more stations are added to existing beamlines and more beamlines are added at PEP. When APS starts up in the mid-1990s, it will have 69 beamlines with multiple stations. GeoSync has recommended that four insertion device beamlines and four bending magnet beamlines be constructed at the APS. These beamlines would be funded and supervised by the Earth sciences community and dedicated to the study of Earth materials. Some of the potential applications of these next-generation synchrotron beamlines are discussed at the end of this review.

3.2. *Synchrotron Radiation Facilities in Other Countries*

Reviews of synchrotron radiation facilities in countries other than the US are provided by Winick (1987b) and Kulipanov (1987). As of March 1989, there are 33 storage ring synchrotron radiation sources either in operation or under construction/design outside the US. These include seven rings in Japan; six rings in the USSR; five rings in Germany; three rings in France; two rings each in Brazil, China (PRC), India, and Italy; and one ring each in Taiwan (ROC), England, Korea, and Sweden. Table 1 provides a listing of some of these facilities.

4. APPLICATIONS

The great variety of research problems in the Earth sciences that require structural and/or compositional information for their solutions provides a rich context for synchrotron radiation. To date, Earth sciences applications of synchrotron radiation have been made in the following areas: (*a*) X-ray scattering studies of Earth materials (crystalline and noncrystalline) under ambient conditions, (*b*) diffraction studies of Earth materials at high pressures and/or temperatures, (*c*) spectroscopic studies of Earth materials (primarily X-ray absorption spectroscopy), and (*d*) spatially resolved X-ray fluorescence (microprobe) studies of compositional variations in Earth materials. X-ray tomography and topography are other synchrotron-based methods that may become important in characterizing Earth materials. Soft X-ray/vacuum ultraviolet radiation (1–∼3000 eV) from synchrotron sources has not been widely used in Earth sciences research, but there is great potential, however, for applying the intense radiation in this energy range to problems involving the structural environments of low-atomic-number elements (e.g. Na, F, Al, Si, S, Cl) and the characterization of surface reactions of minerals with liquids and gases. These and other applications are discussed below.

5. X-RAY SCATTERING STUDIES

Powder diffraction techniques are no longer restricted to phase identification and unit cell measurement. They now can be used to refine the crystal structure of a powdered phase using the full diffraction pattern fitting method devised by Rietveld (1969). These tasks are made much easier using synchrotron radiation because of the dramatic improvements in resolution, peak shapes, and time required for data collection relative to conventional laboratory X-ray sources. In addition, using intensity information obtained from synchrotron-based powder diffraction, one can now solve crystal structures by Patterson or direct methods (Lehmann et al 1987, Prewitt et al 1987, Thompson et al 1987). Figure 3 shows an example of the improved resolution produced with synchrotron-based diffraction on a powdered mixture of the minerals quartz (SiO_2), orthoclase ($KAlSi_3O_8$), and albite ($NaAlSi_3O_8$) (Parrish et al 1986). Although it is possible to identify phases in the pattern produced by conventional powder diffraction, it is not possible to extract separate peak intensities for Rietveld analysis. The pattern produced by synchrotron radiation overcomes this difficulty.

For single-crystal diffraction, synchrotron X rays have been used successfully to collect diffraction intensities from extremely small crystals. For

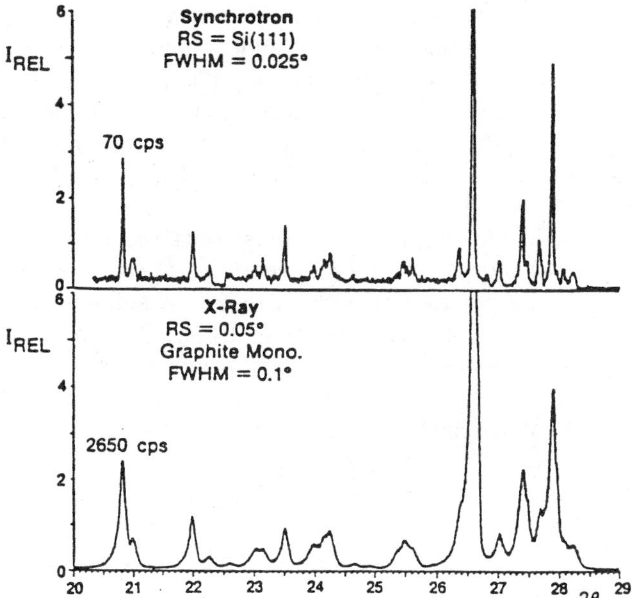

Figure 3 Comparison of a powder diffraction pattern of a mixture of quartz, orthoclase, and albite produced by monochromatized synchrotron radiation (*top*) with a powder diffraction pattern of the same mixture produced by a monochromatized conventional source (*bottom*). The synchrotron-produced pattern required one hour. From Parrish et al (1986).

example, Kvick & Rohrbaugh (1987) were able to refine the structure of a borosodalite zeolite from data collected with a 50-μm-diameter crystal, and they could have used an even smaller sample. Bachman et al (1983) collected a full-intensity data set from a 6-μm-diameter fluorite crystal, and Sueno et al (1987) collected intensity data from very small single crystals of MnO (2.5 μm $\times$ 2.5 μm $\times$ 2.5 μm) and tremolite (2 μm $\times$ 2 μm $\times$ 30 μm), with a factor of 19 gain in diffracted intensities relative to an 8-kW rotating anode generator. Such work demonstrates the feasibility of structural analysis of small particle materials, a class of solids very important in Earth and material sciences. The particle sizes of such materials are often limited by thermodynamic or kinetic factors, such as difficulty of nucleation or unfavorable growth rates, and previously they could not be well characterized using either single-crystal or powder diffraction methods.

Another important type of diffraction experiment in the Earth sciences is Bragg scattering from crystalline materials under high-pressure and/or high-temperature conditions. A number of techniques have been developed by Earth scientists for making in situ diffraction studies of crystalline samples while they are under pressure and at high temperatures. These techniques are used to make measurements that are of fundamental importance to investigations of the Earth's interior, including the effects of pressure and temperature on density, phase relationships, rheologic properties, the nature of phase transitions, etc. This research area is discussed below in Section 6.

The structure of noncrystalline Earth materials, such as silicate liquids and their quenched glasses, ore-forming solutions, hydrous gels, and glassy extrusive rocks, can be structurally characterized by non–Bragg scattering methods and by X-ray spectroscopic methods (see Section 7). Scattering from such materials is about two orders of magnitude weaker than regular Bragg scattering (diffraction) and hence is very time intensive or of limited usefulness without extremely intense X-ray sources. Because of the lack of long-range atomic-scale ordering in these materials, it is possible to obtain only radial distribution information over a limited radial distance ($<$6–8 Å) with most scattering experiments. Although synchrotron X-ray scattering has not yet been used to obtain structural information on noncrystalline Earth materials, it should result in scattering data of significantly higher signal-to-noise ratio and higher resolution due to the higher momentum-space range sampled with high-energy synchrotron radiation. In addition, X-ray scattering studies of liquids, such as silicate melts, under high-temperature/high-pressure conditions will benefit greatly from synchrotron radiation.

Surface diffraction studies of crystal and mineral grain surfaces, solid-liquid interfaces, and melt-solid interfaces are now feasible with high-intensity, highly collimated synchrotron radiation. The types of studies that have been undertaken to date can be divided into two categories: glancing angle diffraction from surfaces, surface adsorbates, and thin films; and X-ray standing wave field-excited X-ray fluorescence. Examples of these studies include the reconstructed surface structure of single-crystal Au (110) (Robinson 1983) and Ge (001) (Eisenberger & Marra 1981), the structure of graphite-adsorbed two-dimensional liquid crystal layers (Moncton & Brown 1983), and study of the GaAs-Al interface by glancing angle surface diffraction (Marra et al 1979). In addition, the non–Bragg glancing angle scattering study of Fischer-Colbrie (1986) revealed the atomic-level structure of amorphous thin films (250 Å thick) of $GeSe_2$. The thin-film atomic structure is almost identical to that in the bulk. Standing wave field studies have included characterization of Ge adsorbed on Si (111) (Dev et al 1986) and the $NiSi_2$-Si (111) interface (Vlieg et al 1986). Other studies in progress are designed to study single-crystal surfaces in contact with electrolyte solutions.

Although surface diffraction and standing wave field studies have not been attempted on geological materials, they will be of great importance in characterizing mineral surfaces and the mechanisms of surface reactions. Such information is of fundamental significance to many Earth processes, such as weathering and diagenesis, crystal growth, ore genesis, and nuclear waste and pollutant transport in groundwaters. The power of such techniques is opening up the field of study of two-dimensional systems in many areas of physics, chemistry, and engineering. The use of synchrotron X-ray sources can lead to similar advances in the study of Earth materials (see also Section 7.5.1).

Time-dependent diffraction studies have been made possible by the very high intensity of synchrotron radiation. Data can be collected fast enough to measure kinetics of reactions, mechanisms of phase transitions, and other transient phenomena. To date, this approach has been used to study solid-solid phase transitions at high pressures and temperatures. This application of synchrotron radiation is discussed below in Section 6.

Synchrotron X-ray scattering could also be useful in studies of other geochemical and mineralogical processes, including unmixing and precipitation reactions in depressurized and devolatilized mineral solid solutions, weathering and absorption/adsorption processes on surfaces of mineral grains, recrystallization reactions in metamorphic rocks and minerals, and melting processes in minerals and rocks under a wide range of geological conditions.

Most of the developments described above would have been impossible without synchrotron radiation. The availability of even more intense and more brilliant X-ray sources will lower limits on sample size and increase energy resolution beyond the limit imposed by current synchrotron sources. Such improvements will make possible new classes of scattering experiments, particularly those involving time-resolved measurements of transient phenomena.

6. IN-SITU HIGH-PRESSURE/ HIGH-TEMPERATURE STUDIES

6.1 *Introduction*

High-pressure/high-temperature research is vitally important for studying the composition, thermal state, and processes in the Earth and other planets. Measurements of physical and chemical properties of rocks and plausible mantle mineral phases under appropriate conditions of pressure and temperature, together with interrelationships of such data with geophysical and geochemical observations, are indispensible for a better understanding of and a basis for the modeling of planetary interiors.

Among the most exciting geophysical problems to be resolved are the dynamics, rheology, compositional stratification, and heterogeneity of the Earth's mantle and core and the nature of lithosphere-asthenosphere coupling. Recent analysis of available physical property data on mantle mineral phases have led to a number of useful geophysical models (Jeanloz & Thompson 1983, Anderson & Bass 1986, Weidner & Ito 1987). However, comparisons of the seismic properties of hypothetical mineral assemblages with corresponding properties inferred from seismic observations are unable to discriminate, for example, among various petrological models for pyrolite and piclogite. This inability to discriminate is primarily due to the lack of experimental data on the physical properties of relevant mineral phases under appropriate pressure-temperature conditions. It is also due to the fact that variations in composition or temperature often cause similar effects on physical properties of minerals. Because of the lack of physical property data, nagging questions remain about the nature of the seismic discontinuities at ~400 km and ~670 km. For example, it is not clear whether these discontinuities are isochemical phase boundaries or chemical boundaries or both (Anderson 1984, Anderson & Bass 1986). Furthermore, the scale of mantle convection is an unsolved problem.

To resolve such problems, we need to know the elastic, thermodynamic, and rheological properties as well as the crystal structures and phase equilibria of candidate assemblages in the system Mg-Fe-Al-Ca-Na-Si-O-H at high pressures and temperatures. Such knowledge depends on precise measurements of properties of candidate minerals and on extension of pressure-temperature ranges of experimental facilities to those found in the Earth's interior. It also requires that the measurements be made under in situ pressure-temperature conditions. Figure 4 shows the ranges of pressure and temperature achieved by different methods.

6.2 *Diamond Anvil Cell*

Earth scientists have been particularly active in developing high-pressure methods, such as the diamond anvil cell (DAC), which can be used to generate pressures in excess of those at the Earth's center (>360 GPa, or 3.6 Mbar). When coupled with laser heating, the DAC makes it feasible to study materials under the extreme conditions of the Earth's core ($T > 6500$ K) and of the interiors of the giant planets. During the past decade, considerable progress has been made in high-pressure/high-temperature diffraction studies of crystalline materials (Furnish & Bassett 1983, Skelton et al 1985, Mao et al 1988a). The extreme *P-T* conditions in a laser-heated DAC exist in a very small volume (picoliters), which places severe constraints on scattering or spectroscopic experiments on such samples. Synchrotron radiation plays a valuable role in making studies at

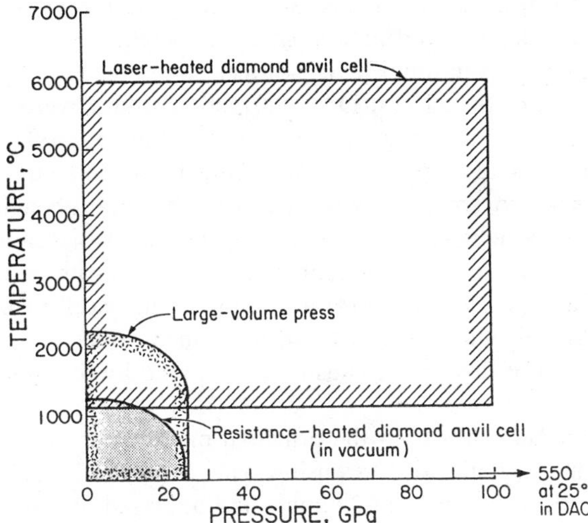

Figure 4 The ranges of pressure and temperature that can be achieved with different pressure-temperature techniques.

these extreme conditions possible. Further progress in this area is important for an understanding of the dynamic processes within the Earth.

6.2.1 AMBIENT TEMPERATURE DIFFRACTION AT PRESSURE The DAC consists of two gem-quality, faceted diamonds as anvils. Anvil faces are ground and polished on the points of the diamonds. These are typically 0.3–0.6 mm in diameter and are carefully aligned so that the opposing faces are parallel to each other. A sample placed in a gasket is squeezed between the two anvils. A well-collimated X-ray beam (10–100 μm in diameter) is directed through one of the diamond anvils and through the sample at the center of the anvil face. Diffracted X rays scatter out through the other diamond anvil to a detector. DACs are now capable of generating pressures of up to several megabars (Xu et al 1986).

The first DACs were developed and used at the National Bureau of Standards in the early 1960s (Piermarini & Weir 1962, Van Valkenburg 1963). A few years later, Bassett & Takahashi (1965) added the capability of heating the sample while at high pressure by placing a resistance heater around the sample and anvils. Since then, DACs have been used with conventional X-ray sources to collect a wealth of data on phase relationships, crystal structures, and equations of state of high-pressure phases.

Samples in DACs are typically in the nanogram mass range. Because of this and the strong attenuation of X rays as they pass through the diamond anvils, collecting a single diffraction pattern may require tens to hundreds of hours using a conventional X-ray source.

In 1980, some of the earliest DAC diffraction studies were made using synchrotron radiation (Manghnani et al 1980, Will et al 1980), resulting in reductions in data acquisition times from hours to minutes or even seconds. This improvement provided two very important benefits: Higher pressures could be achieved because of the small sample size, and real-time studies could be made of transient phenomena, such as phase transitions. Another advantage of synchrotron radiation is the excellent collimation of the X rays. This allows the beam to be defined rather far upstream of the sample without divergence causing it to spread by the time it reaches the sample.

Energy dispersion has been the principal method for obtaining X-ray diffraction data at high pressures using synchrotron radiation. Scattered X rays are collected at a fixed 2θ angle and analyzed according to their energy (Figure 5). The method takes advantage of the high intensity of

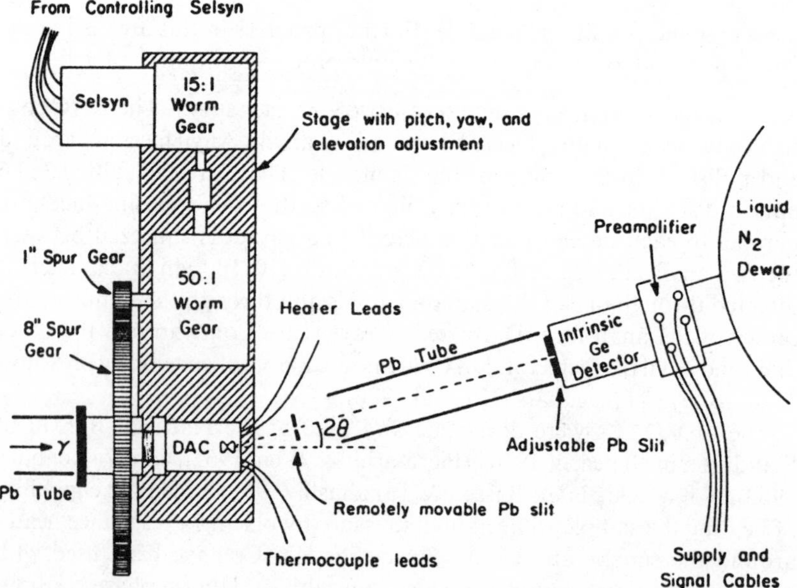

Figure 5 Diagram of the apparatus used by Furnish & Bassett (1983) at CHESS for energy-dispersive X-ray diffraction at high pressures and temperatures in the diamond anvil cell.

radiation over a broad range of energies to yield diffraction data very rapidly. It is well suited for observing changes in diffraction patterns due to phase transitions or for measuring lattice parameter changes due to pressure and/or temperature changes. In contrast, angle dispersion, because of its higher resolution, is a better method for solving or observing changes in complex structures, which produce closely spaced diffraction peaks.

6.2.2 HIGH-TEMPERATURE DIFFRACTION AT PRESSURE Since the first use of heated DACs (Bassett & Takahashi 1965), the design of internal heaters has been greatly improved. It is now possible to produce static temperatures as high as 1500 K in DACs (Schiferl et al 1987). In most high-temperature DACs, the heaters consist of resistance wire wound around the diamond anvils or the diamond supports. Vacuum or inert atmosphere is required to prevent oxidation of the diamonds as well as the metal parts when temperatures over 1200 K are used. DACs equipped with resistance heaters have been used for a number of synchrotron radiation studies, including the mapping of phase relationships, studies of the kinetics and mechanisms of phase transformations, and full equation-of-state measurements.

Much higher temperatures (of order 6000–7000 K) have been achieved in DACs using a focused coherent infrared beam from a continuous yttrium aluminum garnet (YAG) laser. The first high-pressure/high-temperature diffraction study using synchrotron radiation on a laser-heated sample in a DAC was carried out by Boehler et al (1988). This method holds considerable promise for future diffraction studies at very high pressures and temperatures.

6.2.3 SPECTROSCOPY AT PRESSURE In addition to diffraction, the DAC has also been used for spectroscopic analyses of samples under pressure by the EXAFS method (e.g. Ingalls et al 1978). Although these types of studies are few, they have significant potential for providing detailed structural information on Earth materials at very high pressures. This approach is discussed in more detail in Section 7.4.4.

6.3 *Large-Volume Presses*

Two types of large-volume presses have been used in high-pressure/high-temperature diffraction studies with synchrotron radiation: the MAX-80 type press, which was developed in Japan; and the belt design, which has been used in Germany.

6.3.1 MAX-80 The Max-80 cubic anvil apparatus employs an assembly in which the ram force (500 t) applied along the vertical axis is converted to six equivalent components of force that act on six faces of the cubic sample assembly along three orthogonal axes. It consists of upper and lower pyramidal guide blocks (bolsters) installed on the heads of the hydraulic rams, four trapezoid end blocks (thrust blocks), and six anvil holders. The inner surfaces of the guide blocks form a tetragonal pyramid with faces at 45° with the load axis. Two of the six anvils (made of tungsten carbide) are along the centerline of the pyramid and are fixed opposite to each other on each guide block. The other four anvils are positioned on the midpoints of the square edges of a bipyramid. For high-temperature studies, the sample is surrounded by a graphite heater. The X-ray beam is introduced into the MAX-80 through one of the gaskets, which are made of low-atomic-number materials so as to be transparent to X rays. The scattered X rays then emerge through a gasket on the other side of the sample and are collected by a solid-state detector.

The MAX-80 system has the following attributes: (*a*) Pressure is controlled by oil pressure in the rams, which can be accurately monitored using a digital readout; (*b*) state-of-the-art systems are used for controlling the pressure and temperature; (*c*) the hydraulic pumps, movement mechanism, and control system are easy to maintain; (*d*) the pressure distribution is very nearly hydrostatic up to pressures of ~25 GPa; and (*e*) the large sample allows efficient, uniform internal heating to temperatures of ~2200°C [see Smith & Manghnani (1988) for more detail].

To date, the MAX-80 apparatus has been used with synchrotron radiation primarily for in situ diffraction measurements at high pressures and temperatures. Another application is observation of a sample while at high pressure and temperature by X-ray radiography or the shadow method. X rays passing through the sample cast a shadow of the sample on a vidicon tube. Changes taking place in the position or shape of the sample can then be observed on a TV monitor (Kanzaki et al 1987).

6.3.2 BELT APPARATUS The other type of large-volume, high-pressure apparatus for synchrotron X-ray diffraction studies is the belt apparatus (Hinze et al 1983). It consists of a belt that is a steel ring supported by a tungsten carbide die with an inner diameter of 6 mm. The sample, which is 1 mm high, is placed in the cylindrical part of the belt and is squeezed by two pistons driven by a hydraulic press. Once the desired pressure is achieved, the load can be maintained by two screw caps. The device is then removed from the hydraulic press and placed on a synchrotron X-ray beamline. The X-ray beam is introduced through an axial hole in one of the anvils. A corresponding hole in the opposing anvil permits the direct beam to pass out of the sample chamber. Another hole in the downstream anvil is slanted at an angle that permits the diffracted X rays to pass out of the sample chamber to a solid-state detector. Because the energy-dispersive method is used, a complete X-ray diffraction pattern can be obtained at a fixed 2θ angle. All three holes are closed with sapphire or diamond disks. High temperatures are produced by an internal graphite or coaxial resistance wire heater. This apparatus has achieved pressures up to 8.0 GPa and temperatures up to 1000°C.

6.4 *Results*

6.4.1 PHASE RELATIONSHIPS Synchrotron radiation has proved to be a valuable tool for rapidly exploring the phase relationships in pressure-temperature diagrams for various materials. This information is vitally important for determining the properties of these materials within the interiors of Earth and other planets.

6.4.1.1 *Iron* Several synchrotron radiation studies have been made on iron to determine phase relationships (Akimoto et al 1987, Boehler et al 1987, Huang et al 1987, Manghnani et al 1987), equation-of-state information (Boehler et al 1987, 1988, 1989, Manghnani et al 1987, Huang et al 1988a), and phase transition mechanisms (Bassett & Huang 1987). Diffraction data were collected on samples at high pressures and temperatures by in situ energy-dispersive diffraction techniques.

Four major phases of iron exist at different pressures and temperatures. They are body-centered cubic (bcc), face-centered cubic (fcc), hexagonal close packed (hcp), and melt. In situ X-ray diffraction has been used to define the solid-solid phase boundaries in the iron phase diagram. Boehler et al (1987) determined the bcc-fcc and the fcc-hcp boundaries. Their results are in good agreement with earlier measurements based on electrical resistance measurements. For the bcc-hcp phase boundary, however, a great deal of disagreement exists. Three of the investigations using synchrotron radiation (Akimoto et al 1987, Huang et al 1987, Manghnani et al 1987) found that the transformation from bcc to hcp occurs over a wide range of pressures at room temperature. This range can be attributed to the mechanism of the phase transition (Bassett & Huang 1987), the rapidity with which data are collected with synchrotron radiation, and (especially) the contrast between the properties of the iron and the medium surrounding it (N. von Bargen, personal communication). When these effects are taken into consideration, the results are found to be fairly consistent with an equilibrium phase transition at about 13.0 GPa. Observations on the bcc-hcp transformation mechanism are described in Section 6.4.3.

6.4.1.2 *Iron-nickel* Huang et al (1988b) used synchrotron radiation with a resistance-heated DAC to investigate the diffusionless phase diagram of Fe-Ni alloys as a function of pressure (up to 25 GPa), temperature (up to

600°C), and composition (up to 35 wt% Ni). The Fe-Ni phase diagrams are topologically similar to that of iron. However, the triple point (bcc-fcc-hcp) decreases in temperature with increasing Ni content, and the slope (dT/dP) of the fcc-hcp boundary decreases with increasing Ni content. They conclude that an inner core of Fe-Ni may consist of two phases (fcc and hcp) of different Fe : Ni ratios.

6.4.1.3 *Magnetite* Huang & Bassett (1986) determined the high-pressure/high-temperature phase relationships in Fe_3O_4 up to 34 GPa and 600°C in a resistance-heated DAC using energy-dispersive X-ray diffraction at CHESS. Gold was used as the pressure calibrant, and temperature was measured with a thermocouple. The most probable value of the equilibrium slope (dT/dP) of the transformation from magnetite to its high-pressure phase was found to be $-68°C\ GPa^{-1}$.

6.4.1.4 *Hydrogen and helium* The ultrahigh-pressure behavior of hydrogen, including the insulator-metal transition, is one of the major areas of investigation in planetary and condensed matter physics. In addition, metallic hydrogen may have novel properties such as very high temperature superconductivity, which makes the material potentially of great interest in applied physics. Experiments on hydrogen represent a great experimental challenge owing to the high compressibility of the material and the limited number of techniques that can be used to study the small volumes of the sample at ultrahigh pressures. Despite these difficulties, hydrogen has been compressed to pressures near 250 GPa, at which point solid hydrogen becomes opaque while remaining molecular (Mao & Hemley 1989). Mao & Hemley attribute this behavior to band overlap and conclude that it is the initial stage of metallization.

Synchrotron X-ray diffraction has played a major role in the study of the structure and equation of state of hydrogen (Mao et al 1988a). Currently, X-ray diffraction measurements of hydrogen and deuterium under pressure in the DAC are performed using a synchrotron X-ray beam 30–50 μm in diameter. From such measurements, the structure and pressure-volume relationships of hydrogen have been obtained to 26.5 GPa. Hydrogen remains in the hexagonal close packed structure up to 26.5 GPa and exhibits increasing structural anisotropy with pressure. The pressure-volume curve determined by Mao et al (1988a) is the most accurate to date and removes the discrepancy between earlier indirect determinations. It also provides new experimental constraints on the molecular-to-atomic transition predicted to take place at high pressure.

Single-crystal X-ray diffraction measurements have been performed on solid He from 15.6 to 23.3 GPa at 300 K with synchrotron radiation (Mao et al 1988b). The diffraction patterns demonstrate that the structure of the solid is hexagonal close packed over this pressure-temperature range, contrary both to the interpretation of high-pressure optical studies and to theoretical predictions. The solid is more compressible than indicated by equations of state calculated with recently determined helium pair potentials. These results suggest that a significant revision of current views of the phase diagram and of the energetics of dense solid helium is in order.

6.4.1.5 *Other low-atomic-number planetary materials* Although pressures at the center of the giant planets [e.g. Jupiter: ~10 TPa (100 Mbar); Saturn: ~7.5 TPa (75 Mbar)] are well in excess of the maximum pressure that can be achieved with present static methods, much of the volume of these planets lies within the experimentally accessible pressure regime. It is within this pressure regime that such important transitions as the molecular to monatomic transition in solid hydrogen (described above) are predicted to occur. The mantles and outer envelopes of these planets are characterized by abundant light elements (H, He, C, N, O). Current advances in X-ray diffraction of low-atomic-number elements with narrow (~10-μm) X-ray beams at high pressure indicate that we will be able to tackle major problems concerning the phase equilibrium, equation of state, and structural properties of the materials that comprise the large gaseous planets. Recent synchrotron X-ray studies on materials such as solid N_2, Ne, and H_2O-ice indicate that crystal structures and equations of state can be constrained in the 100-GPa range at room temperature (Hemley et al 1987, 1989, Jephcoat et al 1989).

6.4.2 EQUATIONS OF STATE Synchrotron radiation is a valuable tool for measuring the effects of pressure and temperature on molar volume. When both high pressures and high temperatures are required, the lifetime of the parts of the experimental apparatus may be limited; therefore, rapid acquisition of data is essential.

6.4.2.1 *Iron* There have been several investigations of the equations of state of the phases of iron using synchrotron radiation, including measurements of pressure-volume and pressure-temperature-volume relationships in the bcc, fcc, and hcp phases (Boehler et al 1987, 1988, 1989, Manghnani et al 1987, Huang et al 1988b; A. P. Jephcoat et al, personal communication). Manghnani et al (1987) report a thermal expansion α of

$$\alpha_V(\text{deg}^{-1} \times 10^5) = 5.53 \pm 0.04 - (0.070 \pm 0.006)P\,(\text{GPa})$$

for the bcc phase, and that

$$\alpha_V(\text{deg}^{-1} \times 10^5) = 5.29 \pm 0.13 - (0.090 \pm 0.009)P\,(\text{GPa})$$

for the hcp phase. Boehler et al (1989) give the following equation for the bulk modulus K of the fcc phase:

$$K = 127 \pm 8\,\text{GPa} + 2.2(P/\text{GPa}).$$

A. P. Jephcoat et al (personal communication) give the following values for zero-pressure volume V_0, zero-pressure bulk modulus K_0, and pressure derivative of bulk modulus K_0' for the hcp phase at ambient temperature, based on their diffraction measurements made to 270 GPa:

$$V_0 = 22.34 \pm 0.03\ \text{Å}^3, \quad K_0 = 174 \pm 4\,\text{GPa}, \quad \text{and} \quad K_0' = 5.0 \pm 0.1.$$

A. P. Jephcoat et al (personal communication) conclude, as a result of their measurement of its equation of state, that iron has a density that is approximately 10% greater than that of the Earth's core.

They also conclude that the equation of state of Fe(80%)-Ni(20%) is so similar to that of pure Fe that the density difference between the Fe-Ni alloy and Fe is within 1% over the entire pressure range up to 270 GPa, so that Ni is essentially an invisible component in the core.

6.4.2.2 *Majorite* Equation-of-state measurements on majorite were made in a large-volume press by Yagi et al (1987) using synchrotron radiation. The compositions studied were 58% enstatite–42% pyrope and 18% ferrosilite–82% almandine. They found that dissolution of the pyroxene component in garnet decreases the bulk modulus but has little effect on thermal expansion. The garnet-structured $MgSiO_3$ appears to be much more compressible than previously thought. The volume dependence of thermal expansion can be expressed by $\alpha/\alpha_0 = (V/V_0)^5$ for pyrope and $\alpha/\alpha_0 = (V/V_0)^7$ for the almandine-rich composition.

6.4.2.3 *Silicate perovskite* The stability range and pressure-volume equation of state of the cubic perovskite phase of $CaSiO_3$ have been investigated by in situ DAC X-ray diffraction techniques (Mao et al 1989a). The study was carried out using both energy-dispersive synchrotron radiation techniques and conventional angle-dispersive film techniques and a sealed-tube X-ray source. Measurements were made up to 134 GPa, a pressure equivalent to that at the Earth's core-mantle boundary. Pressures were determined by the use of a platinum internal X-ray standard and by the ruby fluorescence method. The data were fit to a third-order Birch-Murnaghan equation with zero-pressure parameters: $V_0 = 45.31 \pm 0.08$ Å^3, $K_0 = 273 \pm 6$ GPa, $K_0' = 4.3 \pm 0.2$, and $\rho_0 = 4.258 \pm 0.008$ Mg m^{-3}. Because these values are close to those of $(Mg_{0.88}Fe_{0.12})SiO_3$-perovskite and to those inferred from the Preliminary Reference Earth Model (PREM) for the lower mantle, the relative amounts of Ca-silicate and Mg/Fe-silicate perovskites in the lower mantle cannot easily be distinguished.

The effect of pressure, temperature, and composition on the lattice parameters and density of $(Fe,Mg)SiO_3$-perovskite to 30 GPa have been measured using monochromatic synchrotron radiation with the photographic film technique (Mao et al 1989b). This study provides the first direct measurements of unit cell distortions and equation-of-state parameters of orthorhombic perovskite as functions of composition and simultaneous high pressure and temperature. The results demonstrate that with increasing pressure, the orthorhombic perovskite undergoes further distortion away from the ideal cubic structure. The zero-pressure bulk modulus ($K_0 = 272.5 \pm 2.4$ GPa) is very close to the value mentioned above for $CaSiO_3$-perovskite, but it is nearly double the bulk modulus of magnesiowüstite.

6.4.3 PHASE TRANSITION KINETICS AND MECHANISMS Time-dependent diffraction studies enable the kinetics of crystal growth and phase changes to be studied directly. An example of such work is the NaCl (B1) to CsCl (B2) phase transition in BaS, studied in the MAX-80 high-pressure apparatus at the Photon Factory over the pressure range 6.3–6.77 GPa (Hamaya et al 1985). Energy-dispersive powder diffraction patterns were obtained in 100-s intervals for this study. Work by Skelton et al (1985) indicates that energy-dispersive diffraction patterns can be collected from a DAC apparatus in periods of less than a second, depending on the data quality desired, at present synchrotron radiation sources. If we assume that the data collection speed of detection systems improves in the future (see Clarke et al 1988), the rates of chemical and physical changes that can be measured should vary approximately linearly with X-ray intensity. In the Earth sciences, kinetic studies are important in a variety of contexts, including phase transitions, crystal growth, mineral dissolution, and surface chemical reactions.

6.4.3.1 *Olivine-spinel* High-intensity synchrotron radiation has been used in time-resolved measurements at high pressures and temperatures for analyzing the nature of the olivine-spinel phase transition. Furnish & Bassett (1983) used white radiation and energy-dispersive DAC techniques to examine the mechanism of the olivine-spinel phase transition in Fe_2SiO_4. The transition proceeds in two stages: first, a restacking of the anions from the distorted hexagonal close packing of the olivine structure to the cubic close packing of the spinel structure, followed by diffusion of the cations to their proper locations for the spinel structure. These results indicate the existence of an intermediate, metastable phase in which the anions are in the positions of the spinel structure and the cations are in motion from the olivine positions to the spinel positions. Will & Lauterjung (1987) made similar observations for the olivine-spinel transition in Mg_2GeO_4 using synchrotron radiation with a belt-type high-pressure apparatus. They also measured the kinetics of this transition. These results are important in understanding both the behavior of olivine in subducting lithosphere and the role that phase transitions may play in the origin of deep-focus earthquakes.

6.4.3.2 *Iron (bcc-hcp)* In another synchrotron X-ray diffraction DAC experiment, Bassett & Huang (1987) observed a lattice distortion in the hcp phase of iron that occurs during the transformation from bcc to hcp. While both phases coexist, the hcp crystal lattice is stretched along the *c*-axis and compressed along the *a*-axis. Hence, they conclude that the bcc-to-hcp transition takes place by distortion of the structure without the breaking of bonds. When the transition is partially completed, the atoms in the bcc arrangement remain attached to the atoms in the hcp arrangement, thus producing the distortion.

To date, time-resolved investigations of Earth materials have been made in minutes with bending magnet radiation. Wigglers and undulators will reduce the time required for the acquisiton of diffraction patterns by orders of magnitude, thus making it possible to obtain significant data in seconds or less.

6.4.4 VISCOSITY OF SILICATE MELTS Direct recording of images by shadow radiography, which has played such a major role in the medical profession, also has important applications in high-pressure/high-temperature research. Kanzaki et al (1987) made radiographs of a platinum sphere dropping through a melt of albite composition at 3 GPa and 1100°C. They used synchrotron radiation from a bending magnet and recorded the shadowgraphs with a high-resolution TV camera. From the rate of falling, they calculated the viscosity. The more intense radiation from a wiggler or an undulator source might make it possible to reach still higher pressures and temperatures, where thicker and/or denser container walls would be required.

7. SPECTROSCOPY

7.1 *Introduction*

A key element in understanding geological processes on all size scales is the dependence of the physical and chemical properties of Earth materials on molecular-level structure, chemical bonding, and composition. Spectroscopic and scattering studies of natural materials and their synthetic analogues provide this information. However, because of a number of limiting factors (e.g. low-concentration levels of certain important elements in Earth materials, small sample volumes imposed by experimental restrictions, the need for high signal-to-noise ratio data, low dimensionality of reaction sites at interfaces and grain boundaries), some of which act in concert, extremely high photon fluxes are needed for many of these studies. Spectroscopic methods that make use of conventional laboratory X-ray sources can provide some of the information needed but only for relatively high concentration samples, which may not duplicate natural conditions. Synchrotron-based methods have significantly extended the range of studies possible to include Earth materials with low elemental concentrations and/or small volumes. The most important of these are X-ray absorption spectroscopy (XAS) methods (EXAFS and XANES), which provide information on the local structure (≤ 6 Å radius around a selected atom) and bonding for atoms in all types of materials.

The availability of synchrotron radiation has resulted in other new structure-sensitive methods, including differential anomalous scattering (DAS) (Fuoss et al 1981). This method involves the collection of X-ray scattering data above and just below the absorption edge of an element and thus takes advantage of the differences in the real and imaginary contributions to anomalous scattering near the absorption edge of the scattering element. The two data sets are normalized and subtracted to yield scattering data with greatly enhanced contributions from the chosen element. Because DAS data are not limited in the low-momentum-space range (<3 Å^{-1}) as are EXAFS data, distance information from DAS pair correlations extend to greater distances than those from EXAFS, and they are not as strongly affected by large degrees of static disorder (Kortright et al 1983). The major limitation of DAS is its inapplicability to elements with atomic number less than about 25 because of the small momentum-space region accessible for these elements. In contrast, EXAFS spectroscopy can be applied to almost all elements, but the distance information is typically limited to less than 5 Å around the absorber and may suffer if static or thermal disorder is severe. Therefore, DAS provides complementary information to EXAFS spectroscopy, but its use is more limited. To the best of our knowledge, no DAS studies of Earth materials have been published; however, the method should prove valuable in studying polymeric metal complexes in aqueous solutions and metal sites in noncrystalline silicates.

The characteristics of synchrotron radiation have resulted in significantly enhanced resolution and sensitivity for several well-established X-ray spectroscopy methods, including X-ray photoelectron spectroscopy (XPS) and X-ray emission spectroscopy (XES). Synchrotron-based XPS has been utilized extensively for studies of many types of materials (Smith & Himpsel 1983), particularly chemisorption and physisorption on metal surfaces, but it has not been used in major ways on Earth materials. Similarly, synchrotron-based XES has not been used for studies of Earth materials. These methods provide direct information on bonding (e.g. electron binding energies from XPS) but only indirect structural information.

7.2 *EXAFS (Extended X-Ray Absorption Fine Structure)*

The theory and practice of EXAFS spectroscopy are discussed extensively in a number of review papers and books, including Teo & Joy (1980), Brown & Doniach (1980), Lee et al (1981), Hayes & Boyce (1982), Stern & Heald (1983), Wong (1986), Teo (1986), Koningsberger & Prins (1988), and Brown et al (1988). Here, a brief overview is presented.

A typical X-ray absorption experiment consists of passing a beam of X rays through a sample of thickness x cm and of linear absorption coefficient μ. The transmitted (I) and incident (I_0) intensities are measured using gas-filled ionization chambers and are related by $\ln(I_0/I) = \mu x$. Figure 6 shows a typical X-ray absorption spectrum, plotted as $\ln(I_0/I)$ versus energy (in elecron volts). When the concentration of an element is low or a sample is very small, the X-ray fluorescence yield method is preferred to the transmission method. Fluorescent X rays from the sample are detected with a large-area gas-filled ion chamber or an array of scintillation detec-

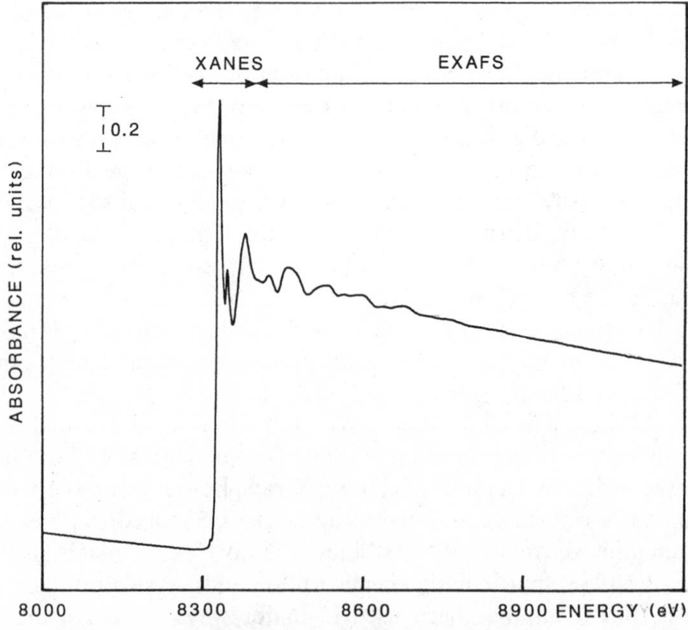

Figure 6 K-edge X-ray absorption spectrum of Ni in Ni-kerolite (= nickel-antigorite) [$Ni_6Si_4O_{10}(OH)_8$], showing the XANES and EXAFS regions. From Manceau & Calas (1986).

tors. The fluorescence EXAFS method is based on the principle that each absorption event involving the excitation of a core-level electron produces X-ray fluorescence.

When the energy $h\nu$ of the incident X-ray beam is less than the binding energy E_b of a core-level electron of an element in the sample, little absorption occurs except for the type that increases smoothly with decreasing X-ray energy. When $h\nu \approx E_b$, there is a high probability that a core-level electron will be excited, and a strong, abrupt increase in μx occurs with increasing incident X-ray energy, producing an absorption edge. If the excited electron (photoelectron) is from the K-shell, the resulting absorption edge is referred to as the K-edge. The energy region extending from about 10 eV below the absorption edge to about 50 eV above the edge is referred to as the XANES region and is discussed in the next section. When $h\nu$ is 50 to 100 eV greater than E_b, the photoelectron travels from the absorber element to the nearest-neighbor atoms surrounding the absorber and is singly backscattered to the vicinity of the absorber. The constructive and destructive interferences between the outgoing and backscattered photoelectron waves produce oscillations in μx with increasing incident X-ray energy, which are referred to as extended X-ray absorption fine structure (EXAFS).

Analysis of the EXAFS spectrum of a selected element in a sample can yield the distance, number, and identity of first- and second-nearest neighbors surrounding an absorber. Distance is inversely related to the frequency of the EXAFS oscillations produced by an absorber-backscatterer pair, and the amplitude of EXAFS oscillations is directly related to the number and type of backscatterers. Key points about EXAFS spectroscopy can be summarized as follows: (*a*) It is an element-specific structural method that can be used to study most elements in different states of matter (solid, liquid, gas) at concentration levels of parts per million to high weight percentages; (*b*) it is a local structural probe that "sees" only the two or three closest shells of neighbors around an absorbing atom ($\leq$5–6 Å); and (*c*) it is capable of yielding average interatomic distances accurate to $\pm$0.02 Å and average coordination numbers to $\pm$20%, assuming that systematic errors have been minimized in the experiment and that static and thermal disorder are small. The details of EXAFS data analysis are presented in Brown et al (1988) and Teo (1986).

7.3 *XANES (X-Ray Absorption Near-Edge Structure)*

XANES spectra are produced by a combination of electronic transitions from deep core levels to bound energy levels and by transitions to continuum (unbound) levels. The former type of transition can produce spectral features below the absorption edge in a region commonly referred to as the preedge. Their intensity and energy are related to the symmetry of an absorber's local environment and the absorber's oxidation state (if variable), respectively. The latter type of transition results in multiple scattering of the photoelectron among atoms surrounding the absorber. The interference of these scattered waves produces the fine structure from a few electron volts to as much as 50 eV above an absorption edge. The details of this fine structure are very sensitive to the geometry of the absorber's local environment and bonding. Although XANES spectroscopy provides qualitative structure and bonding information, it is complementary to the more quantitative EXAFS spectroscopy (see Petiau et al 1987, Lytle et al 1988, Bianconi 1988).

7.4 *Mineralogical Applications of XAS*

Applicatons of EXAFS and XANES spectroscopies to mineralogical and geochemical problems have been reviewed by Waychunas & Brown (1984), Calas et al (1984a, 1986, 1987), Brown et al (1986, 1988), Petiau (1986), Brown & Waychunas (1988), and Brown & Parks (1989). The discussion below presents representative examples of these applications.

7.4.1 STRUCTURE AND BONDING OF CATIONS IN MINERALS X-ray absorption spectroscopy (XAS) analysis of local structural details in minerals complements X-ray diffraction analysis, particularly in compositionally complex minerals where more than one element can occupy one type of crystallographic site. For example, an XAS study of Co and Ni in supergene manganese oxides (asbolane and lithiophorite) (Manceau et al 1987) revealed different oxidation states and local environments for these elements. Cobalt K-edge XANES spectra indicate trivalent, 6-coordinated cobalt in the phases studied. EXAFS study has shown these ions to have the same local environment as manganese, which suggests that Co is randomly distributed in layers occupied by Mn. The short Co-O bond length (1.92 Å) and the unique XANES spectrum suggest that Co^{3+} occurs as a low-spin octahedral cation, with high crystal-field stabilization energy. This suggestion could explain the well-known selective uptake of Co by Mn oxides. Unlike Co, Ni exhibits different local surroundings in both lithiophorite and asbolane. In the former, Ni atoms are located in hydrargillite $Al(OH)_3$ layers, and in the latter they occupy sites in separate $Ni(OH)_2$ layers of unknown extent. The local environments of Mn and Fe in oxides and oxyhydroxides have also been studied by EXAFS spectroscopy (Manceau & Combes 1988).

A survey of Fe-bearing minerals and oxides using XANES spectroscopy (Waychunas et al 1983) shows the richness of structure and bonding information on cation local environments that can be obtained in this way (see also Calas & Petiau 1984). Information from XANES spectra is complementary to that from EXAFS spectroscopy (see Waychunas et al 1986a). The intensities of preedge features in a XANES spectrum are particularly sensitive to site distortion, and the energies of these features are sensitive to oxidation state, if variable (see Wong et al 1984). The information content of XANES spectra can be increased by collecting polarized XANES spectra, as has been done for selected Fe- and Ti-bearing minerals (Waychunas & Brown 1988). This latter study used oriented single crystals of minerals like gillespite, rutile, and epidote and took advantage of the highly polarized nature of the synchrotron X-ray beam in the plane of the storage ring. The study of polarization dependence of XANES spectra can enable one to sort out shape resonances from bound-state transitions. Polarized EXAFS spectra can yield bond distances between absorber and ligand atoms as a function of crystallographic orientation (Waychunas & Brown 1990). Manceau et al (1988) have carried out polarized EXAFS work on Fe in biotite and chlorite with similar success. Such studies offer the potential for obtaining information about the local environment(s) of cations at low concentrations at grain boundaries and defects.

The study of Ti in silicates and oxides has been hindered by the lack of definitive information on its formal oxidation states and site geometry. The site geometry of Ti^{4+} cannot be studied directly by optical absorption spectroscopy or most other spectroscopic methods, but it can be characterized by XAS, as demonstrated by Waychunas (1987). The main conclusions from this work are (*a*) that Ti^{3+} is relatively rare, even for samples predicted to have significant Ti^{3+} from previous indirect spectroscopic studies; and (*b*) that tetrahedral Ti^{4+} occurs only at very small concentration levels in silicates, with the majority of Ti^{4+} ions occurring in 6-coordinate sites with varying degrees of distortion.

Other examples of the use of XAS to determine site geometry and/or oxidation states of elements in minerals include the location of Fe atoms in natural kaolinites (Bonnin et al 1982); Fe-site occupancy in Garfield nontronite (Bonnin et al 1985); Fe-Ni segregation in smectites (Decarreau et al 1987); crystal chemistry of chromium in phyllosilicates (Calas et al 1984b); site occupancies in chromites and Cu oxidation states in sulfides (Calas et al 1986); location of calcium in phosphates (Harries et al 1987); modification of the Ca-site geometry in diopside-jadeite solid solution (Davoli et al 1987); oxidation state and location of copper in vermiculites (Ildefonse et al 1986); location of Tc, Er, and Lu in epidote (Cressey & Steel 1988); characterization of the Cu and Ag sites in argentian tetrahedrites (Charnock et al 1988); and a study of the bonding in chalcopyrite (Petiau et al 1988).

7.4.2 CATION DISTRIBUTION IN A SILICATE PEROVSKITE Because of the high intensity of synchrotron X-ray sources, it is possible to carry out XAS studies of extremely small samples, such as those produced in ultrahigh-pressure DAC experiments. One geophysically relevant example of this type of study is the investigation of the partitioning of Fe within a high-pressure silicate perovskite ($Mg_{0.88}Fe_{0.12}SiO_3$) synthesized at 50 GPa and ~2000 K (Jackson et al 1987a). This phase is thought to dominate in the Earth's lower mantle, yet we have only limited knowledge about its structure and properties (e.g. Knittle et al 1986). Only 30 μg of sample were available from a total of 30 separate synthesis runs under these conditions. This amount precludes many types of spectroscopic characterization; however, XAS study is possible using fluorescence detection techniques. The Fe K-XANES spectrum of this sample shows a very weak preedge feature caused by 1*s* to 3*d* transitions. This observation is consistent with a relatively regular site with a center of symmetry, such as an octahedron. Fitting of the EXAFS spectrum resulted in an average Fe-O distance of 2.14 ± 0.04 Å and a coordination number near six. Based on the stoichiometry of this silicate perovskite, there is sufficient Si to fill the 6-coordinated *B* site, and the $Fe_{0.12}$ (and $Mg_{0.88}$) would be expected to occupy the large, distorted, 8- to 12-coordinated A site of the perovskite structure. However, if Fe^{2+} is present in the *B* site, as indicated by XANES and EXAFS spectral analysis, a small amount of Si (~0.1 Si atoms per formula unit) should occupy the *A* site, where it might preferentially bond to four nearest-neighbor oxygens in the irregular site. An in situ high-pressure EXAFS study of Mg-Fe-silicate perovskite (D. Andrault, personal communication, 1989) and a single-crystal diffraction study using synchrotron radiation (Kudoh et al 1989) have found evidence for Fe in the *A* site rather than the *B* site. However, their samples were prepared at lower temperatures than that used by Jackson et al (1987). Additional structural work on single crystals of high-pressure silicate perovskites of adequate size is required to determine Fe and Si site occupancies in silicate perovskites as a function of temperature.

7.4.3 SHORT-RANGE ORDER IN MINERALS One of the best uses of EXAFS spectroscopy takes advantage of its ability to distinguish among types of nearest neighbors and to provide quantitative estimates of the extent of short-range ordering (SRO) around a selected element in solid solutions. In this application, the inability of the EXAFS probe to detect long-range ordering is a distinct advantage. The basic assumption made when studying SRO in minerals is that the local structure around the absorbing atom is known. If this structure is not known, the model cannot separate structural from compositional effects resulting in phase cancellation. In addition, this assumption may not be valid when the absorber is a dilute species because local structural relaxation often accompanies element substitutions, and this relaxation is not easy to characterize by spectroscopic or scattering methods other than EXAFS.

An EXAFS study of SRO of Fe^{2+} in a series of magnesiowüstites that were synthesized at and rapidly quenched from a temperature of 1140°C indicates that iron has a random rather than ordered arrangement of second-neighbor Fe and Mg atoms (Waychunas et al 1986c). A similar study of Fe^{3+}-doped MgO samples, which were annealed at low temperature, shows evidence for the clustering of Fe^{3+} (Waychunas 1983). Evidence for partial ordering of Fe^{2+} in Ca-Fe pyroxenes was also found through analysis of the Fourier transform magnitude of the second atomic shell around iron (Waychunas et al 1986c).

7.4.4 HIGH-PRESSURE STUDIES One of the few mineralogical examples in this area is the high-pressure EXAFS study of FeS_2 (pyrite) by Ingalls et al (1978), in which Fe-Fe and Fe-S distances were determined as a function of pressure to ~5 GPa using a DAC. Another is the DAC EXAFS study of the local environment of iron in cubanite ($CuFe_2S_3$) up to a pressure of 9.4 GPa (Sueno et al 1986a, 1989); these authors found an average Fe-S bond length of 2.18 Å, which is 0.1 Å shorter than in cubanite at low pressure and considerably shorter than ordinary 6-coordinated Fe-S bond lengths in sulfides (2.4–2.5 Å). Other examples on nongeological materials include dispersive XAS studies of solid bromine to pressures of 57.5 GPa (Itie et al 1986) and crystalline and vitreous GeO_2 to pressures up to 29.1 GPa (Itie et al 1989), and transmission EXAFS studies of NaBr and Ge to several gigapascals (Ingalls et al 1980), of ZnSe and CuBr to pressures of 8.7 GPa (Tranquada & Ingalls 1984), and of Ni-dimethylglyoxime to 5.6 GPa (Sueno et al 1986b), all using the DAC. The EXAFS study of Ge and GaAs to 13 GPa and 22 GPa, respectively, using the MAX-80 device (Shimomura & Kawamura 1987) examined the structural changes accompanying the semiconductor-to-metal transitions in these materials. The study by Itie et al (1989) observed a change in Ge coordination number in GeO_2 from 4 to 6 between 7 and 9 GPa. In addition to studies of the pressure dependence of metal-anion bond lengths, high-pressure XAS offers exciting possibilities for studying bonding changes with increasing pressure (e.g. Rohler et al 1984, Rohler 1984) as well as phase transitions (Tranquada et al 1984). Even higher pressure XANES/EXAFS spectroscopy studies using the DAC should be possible with next-generation synchrotron X-ray sources. In addition, high-pressure EXAFS studies of light elements in minerals may be possible using X-ray Raman scattering, which produces EXAFS-like spectra (see Tohji

& Udagawa 1987). X rays from light elements like Si and Al are strongly absorbed by the diamonds of a DAC; thus, conventional EXAFS methods will not work for these elements in a DAC.

7.5 *Geochemical Applications of XAS*

7.5.1 SORPTION REACTIONS AT MINERAL/WATER INTERFACES Many geochemical processes occur at solid/liquid interfaces, but there are few ways to characterize these environments in situ. One of the more novel applications of XAS is the in situ study of cation sorption complexes at water/mineral interfaces (Roe et al 1987, Hayes et al 1987, Brown et al 1988, 1989). Direct structural characterization of such complexes is not generally possible using most spectroscopic or scattering methods because of (*a*) the low concentration levels of the complexes (coverages are typically less than one monolayer), (*b*) the interference caused by water, and (*c*) the need to dry samples for surface-sensitive, high-vaccuum spectroscopy or diffraction methods. However, these difficulties can be overcome in large part with XAS.

To characterize the average local environment and composition of sorption complexes at mineral/water interfaces, fluorescence-yield X-ray absorption measurements have been made on high-surface-area, wet oxide samples with less than monolayer sorbate coverages. Samples that have been studied include Co^{2+}(aq) on γ-Al_2O_3 and TiO_2 (rutile) (Chisholm-Brause et al 1989a), Pb^{2+}(aq) on γ-Al_2O_3 (Chisholm-Brause et al 1989b), and aqueous selenate and selenite ions on α-FeOOH (goethite) (Hayes et al 1987). Direct, in situ measurements of the average distances, numbers, and identities of first- and second-neighbors surrounding the sorbed atoms in these samples were derived from the experimental EXAFS data. Results for Co^{2+} on γ-Al_2O_3 and TiO_2, at pH 6.8, indicate that Co^{2+} forms inner-sphere complexes, that it does not sorb as a three-dimensional precipitate or diffuse into the oxide, and that it forms multinuclear complexes on γ-Al_2O_3 but smaller polymers or monomeric complexes on TiO_2. Results for Co^{2+} on γ-Al_2O_3 provide the first, direct structural evidence for multinuclear metal sorption complexes on an oxide surface. These results also provide the first direct structural data to suggest that the nature of the oxide surface influences the type of adsorption complex formed. Sorption complexes formed by Pb^{2+} on γ-Al_2O_3 at pH 6.0 are similar; Pb^{2+} probably forms dimeric or a combination of monomeric and small multinuclear complexes. Selenate is sorbed as an outer-sphere complex (pH 3.5) and selenite is sorbed as an inner-sphere, bidentate complex (pH 5.6) on α-FeOOH. These structural results provide the first molecular-level explanation for the weak binding of selenate and the strong binding of selenite at the goethite/water interface.

This application of XAS has opened up a new research area involving in situ structural studies of reactions at mineral/water interfaces. Further work in this area should provide unique information about the geochemistry of mineral surfaces, particularly when the next-generation synchrotron sources make real-time, structural studies of chemical reactions on mineral surfaces feasible.

Another class of surface-sensitive XAS measurements, referred to as surface EXAFS (SEXAFS), provides quantitative information on local structure and bonding of atoms chemisorbed or physisorbed on clean single-crystal surfaces [see Citrin (1986) for a review]. Measurements of this type are carried out under very high vacuum conditions and use electron yield methods. Thus, they differ from the surface-sensitive fluorescence EXAFS work described above, which is done in air and is used to characterize wet surfaces. Yet another type of surface-sensitive EXAFS measurement makes use of an incident X-ray beam that intercepts a flat sample at a very small glancing angle. This technique is referred to as ReflEXAFS and can be used with samples in air or vacuum (Bosio et al 1984). To date, no SEXAFS or ReflEXAFS studies have been carried out on Earth materials; however, such measurements are potentially very important for studies of gas and liquid reactions with mineral surfaces.

7.5.2 NONCRYSTALLINE MATERIALS Disordered natural phases constitute an important class of Earth materials and include natural glasses, melts, aqueous solutions, gels and the poorly crystallized phases that form from them, and metamict minerals that are disordered because of radiation damage. The geochemical behavior and phase relations of most elements are controlled in part by the types of sites each element occupies in these disordered phases. Until recently, little has been known about the structural variations present in these materials because few methods can yield structural data on these structurally and chemically complex materials. XAS provides one of the few quantitative probes of local structure that is also element specific. Thus, it is particularly well suited for characterization studies of such materials, and significant results have been obtained using it. Some of these applications are reviewed below.

7.5.2.1 *Cation environments in silicate glasses* One of the major applications of XAS spectroscopy has been the study of local structural environments of cations in glasses. Glasses possess little structural order beyond a few angstroms around a cation or anion, i.e. beyond the first coordination shell. Thus the local nature of the structural information provided by XAS, the fact that this information is essentially one dimensional, and the method's element selectivity make XAS well suited for the study of multicomponent glasses. Several recent reviews of XAS studies of amorphous silicates present reasonably up-to-date treatments of these studies (Calas & Petiau 1983a, Greaves et al 1984, Brown et al 1986, Calas et al 1987). Therefore, the discussion below is limited to a few representative applications of XAS to glasses.

Aluminum in natural silicate melts and in industrial glasses normally acts as a network-forming element. However, for sodium aluminosilicate melts or glasses with peraluminous compositions (where the Al/Na atom ratio exceeds 1.0), some have suggested that a fraction of the Al occurs in octahedral coordination and acts as a network modifier as an explanation for anomalous changes in physical properties, such as viscosity and electrical conductivity, as the Al/Na ratio is varied (e.g. Hunold & Bruckner 1980). This possibility has been assessed by XAS study of Al in a series of sodium aluminosilicate glasses using high-vacuum, total electron yield detection methods. Such methods are necessary because of the very strong air and matrix absorption problems caused by the low energy of the Al K-edge (1560 eV). Analysis of Al K-XANES (Brown et al 1983) and Al K-EXAFS (McKeown et al 1985a) for a series of glasses in the Na_2O-Al_2O_3-SiO_2 system (with variable Al/Na and constant Si) clearly shows that Al remains in tetrahedral coordination, even in peraluminous glasses with Al/Na ≤ 1.6. Thus, another cause besides octahedral aluminum must be sought to explain the effect of Al/Na ratio on the transport properties of these glasses and melts.

The most important alkalis and alkaline earth elements in natural melts and in industrial glasses are Na, K, Ca, and Mg. However, the structural roles of Na, K, and Ca are difficult to assess in these materials by X-ray or neutron scattering studies because of the range of alkali- or alkaline Earth distances and the overlapping of these distances with O-O distances in radial distribution functions. XAS measurements have provided some constraints on the local site geometries of these elements in glasses, including those in the system Na_2O-K_2O-Al_2O_3-SiO_2 (McKeown et al 1985b, Jackson et al 1987b) and glasses of composition $CaAl_2Si_2O_8$ and $CaMgSi_2O_6$ (Binsted et al 1985). In the latter study, Ca was found to occupy 7-coordinate sites in both glasses, which are similar to the Ca site in crystalline anorthite and unlike the 8-coordinate Ca site in crystalline diopside. This result is consistent with the similarity in densities of anorthite, anorthite glass, and diopside glass, and the higher density of diopside. A similar study has also been made on Ca in a model basaltic glass by Hardwick et al (1985).

The structural environments and oxidation states of 3*d*-transition metal ions in silicate glasses have been studied intensively by many spectroscopic methods, including XAS. One of the first of these studies (Brown et al 1978) examined the structural role of Fe^{3+} in glasses of composition

$KFeSi_3O_8$, $NaFeSi_2O_6$, and $NaAl_{0.85}Fe_{0.15}Si_2O_6$ quenched from melts at 1450°C and 1 atm pressure. Analysis of Fe K-XANES and EXAFS spectra for these glasses indicates that Fe^{3+} is teterahedrally coordinated by oxygens. This conclusion was confirmed by ^{57}Fe Mössbauer spectroscopy, optical absorption spectroscopy, and X-ray radial distribution analysis on the same glass samples. The study of Fe^{3+} in sodium disilicate glasses (Calas & Petiau 1983b) also found iron to be tetrahedrally coordinated.

The Ti^{4+} ion is known to occur in 4-, 5-, and 6-coordinated sites in crystalline oxide structures; however, its coordination environment in most glasses is not known. Ti K-XAS studies of TiO_2-SiO_2 glasses (Sandstrom et al 1980, Greegor et al 1983) concluded that Ti^{4+} is primarily 4-coordinated, with a small amount being 6-coordinated on the basis of multiple-shell fits to the EXAFS and interpretation of the intensities of the preedge feature. These studies also found that the proportion of 6- to 4-coordinated Ti^{4+} increases with increasing TiO_2 content in these glasses.

Synchrotron-based EXAFS methods have also been used to study the local environments of selected rare earth elements (REE) in silicate glasses at trace-level concentrations (<2000 ppm). L_{III} EXAFS spectra were collected for La, Gd, and Yb in silicate glasses of the following compositions: albite ($NaAlSi_3O_8$), sodium trisilicate ($Na_2Si_3O_7$), the peralkaline composition ($Na_{3.3}AlSi_7O_{17}$) approximately halfway between the two, and each of these compositions with 1–2 wt% fluorine or chlorine added (Ponader & Brown 1989a,b). EXAFS data on La, Gd, and Yb in the halogen-free glasses provide the first quantitative measures of REE-oxygen distances for trace levels of these elements in quenched melts of geochemical relevance. Significant systematic changes were found in each REE's coordination environment with variations in melt polymerization. In general, as polymerization decreases, the REE's environment is more strongly influenced by its bonding requirements and less influenced by the melt's network topology. For the halogen-bearing glasses, Yb and Gd form complexes with fluorine in F-bearing melts, and La forms mixed fluorine-oxygen complexes. In the chlorine-bearing glasses, none of these rare earth elements was found to form complexes with Cl. These results have a bearing on the transport of rare earth elements in silicic magma chambers.

7.5.2.2 *Iron in silicate melts* As with other disordered materials, structural study of multicomponent silicate melts is limited to only a few direct methods. The added complications of high-temperature sample containment and control of atmosphere present significant experimental difficulties. Synchrotron radiation-based XAS spectroscopy overcomes some of these difficulties. Waychunas et al (1986b, 1988) obtained the first structural data on the coordination geometry of Fe^{2+} in silicate melts of compositions $M_2FeSi_3O_8$(M = Na and K) up to 1200 K and under a vacuum of about 10^{-4} torr. EXAFS-derived Fe-O distances are ~1.94 Å, and coordination numbers average ~4. Perhaps the most surprising result was the finding that Debye-Waller factors for Fe in the melts are no larger than for crystalline model compounds, such as $FeAl_2O_4$ (hercynite), at similar high temperatures. This result suggests that the static disorder around the average Fe site in these melts is no greater than that in the well-crystallized oxides studied. The site geometry of Fe in the glasses quenched from these melts was also studied by EXAFS, XANES, and Mössbauer spectroscopies. The resulting Fe-O distances are ~2.00 Å, and average coordination numbers are similar to those for the melts. The XANES spectra of the melts and glasses are characterized by strong preedge features, consistent with tetrahedral Fe^{2+}. Finally, ^{57}Fe Mössbauer spectra gave isomer shift values near 0.98 mm s^{-1} relative to metallic Fe and showed no significant levels of Fe^{3+} or metallic Fe. Hence, little local structural relaxation occurred around the Fe site during quenching.

These and other results (Waseda & Toguri 1978, Matsui & Kawamura 1980, Jackson et al 1987c; C. M. B. Henderson, personal communication, 1988), which have found evidence that Fe^{2+} and Mg occur in 4-coordinated sites in basic silicate melts, led Waychunas et al (1988) to suggest that both Fe^{2+} and Mg may undergo pressure-induced coordination changes from 4- to 6-coordinated sites in silicate melts in the Earth's upper mantle. This type of pressure-induced coordination change was suggested for Fe^{3+} in silicate melts (Waff 1975). Such coordination changes should have significant effects on transport properties of silicate melts under upper-mantle conditions.

7.5.2.3 *Metal complexes in aqueous solutions* Direct structural studies of dilute solutions (<1 M) containing transition element complexes are difficult for non-element-specific methods because of the weak signal from the complexes and the large background contribution from the solvent. That these difficulties are largely overcome by XAS spectroscopy has led to numerous XAS studies of the local structure of transition element complexes in dilute and concentrated aqueous electrolyte solutions (reviewed by Brown et al 1988).

An example is the EXAFS study of Zn chloride solutions (Parkhurst et al 1984) that examined the types of Zn complexes as a function of concentration (5.6 M to 0.001 M $ZnCl_2$) and Cl : Zn ratio (2 : 1, 4 : 1, 10 : 1, 100 : 1). The lowest concentrations approach the levels of Zn and Cl in natural hydrothermal solutions. A transition occurs from an inner-sphere tetra-chloro to an outer-sphere aquo complex (with nearest-neighbor water molecules) between 5.6 M and 1.0 M. As the concentration is lowered from 1.0 M to 0.001 M, the average number of water molecules in the solvation sphere around Zn increases from 5 to 8. As the Cl : Zn ratio increases, the number of water molecules in the hydration sphere is reduced. A clear extension of this work would involve direct study of metal complexes in hydrothermal solutions at temperature and pressure, where current aqueous thermodynamic data bases rely either on indirect data concerning complex type or on assumption.

7.5.2.4 *Iron in hydroxide gels* Gels are common in low-temperature geological environments and represent an intermediate stage in the formation of many sedimentary and supergene minerals, including silica, aluminum, ferric-iron oxides, and oxy-hydroxides. The nature of the resulting crystalline phases strongly depends on the physico-chemical conditions (pH, temperature, activity of the component, etc) of formation and evolution ("aging") of these gels. Unfortunately, differences among gels at various stages of evolution may not be distinguishable with conventional spectroscopic and diffraction methods because they are X ray amorphous and the radial distribution function would consist of all possible pair correlations. However, XAS methods are well suited for the study of such gels because of their element selectivity, which permits derivation of pair correlations involving the element of interest. An Fe K-EXAFS study by Combes et al (1989a,b) on ferric-iron hydroxide gels has shown them to have short-range order, with shorter Fe-O radial distances (six oxygens at 1.96 Å) than in the derivative crystalline phases (three oxygens at 1.95 Å and three at 2.10 Å). Iron-oxygen distances in the gels are similar to those encountered in the aqueous solutions. The most interesting result is that the local structure around iron is dependent on formation conditions. In gels precipitated from Fe^{2+}-containing solutions, FeO_6 octahedra were found to be edge shared as in γ-FeOOH; however, in gels precipitated from Fe^{3+}-containing solutions, the local structure around iron is similar to that in goethite (chains of edge-sharing FeO_6 octahedra linked by corners). EXAFS study of the evolution of these gels to hematite at 92°C revealed an intermediate phase that had not been detected by other methods. This intermediate amorphous phase is characterized by the appearance of face-sharing octahedra, as in the structure of hematite; its appearance precedes the formation of hematite nuclei, which can be detected by X-ray diffraction. The formation of ferric-silicate precipitates from basic solutions at room temperature and their evolution into the clay mineral nontronite have also been studied through XAS (Decarreau et al 1987). A strong local disorder around iron was observed, although a local smectitelike structure appears in "amorphous" precipitates. Although no changes in local order were observed at 75 and 100°C, a significant evolution to a more ordered iron environment was observed at 150°C.

7.5.2.5 *Metamict minerals* Natural metamict minerals have received renewed interest recently because of the information they can provide about stable containment, over geologic time periods, of radioactive waste elements such as ^{137}Cs in refractory phases. At radiation doses near the saturation level, these minerals are amorphous and XAS analysis is very useful in the study of local structural environments of certain elements. Several studies on complex Ca-Ti-Nb-Ta oxides of the pyrochlore group by Greegor et al (1984, 1986a,b) and Ewing et al (1987) generally show that local order beyond the first coordination shell of Ta and Ti is lost with increasing radiation damage. This local ordering reappears on annealing the samples. XAS results on metamict materials of several initial structure types show the same type of nearest-neighbor structure around the major network-forming cations, leading to the suggestion that the final local environments of these cations in the metamict state is similar for a wide range of materials (Ewing et al 1987).

8. X-RAY FLUORESCENCE MICROCHEMICAL ANALYSIS

8.1 *Introduction*

There are a number of microanalytical techniques, many of which have been developed in the past couple of decades. Electron microprobe analysis (EMPA), which is one of the earliest and most widely used, is now supplemented by such techniques as secondary ion mass spectrometry (SIMS), particle-induced X-ray emission (PIXE), particle-induced gamma-ray emission (PIGE), nuclear reaction analysis (NRA), Rutherford backscattering (RBS), and synchrotron radiation-induced X-ray emission (SRIXE). These techniques have extended sensitivities to less than 1 part per million (ppm) and in some cases have preserved the high spatial resolution possible with the electron microprobe. The SRIXE method is attractive for several reasons: (*a*) It is nondestructive (low-energy deposition for a given sensitivity compared with electron or proton microprobes); (*b*) it can analyze for a wide range of elements, from potassium to the heaviest elements; (*c*) it has high sensitivity (<0.1–5 ppm); (*d*) it has good spatial resolution (<10 μm); and (*e*) the physics of photon interactions with matter are well understood, so that quantification is relatively straightforward. Additional advantages are simple sample preparation (a polished surface or grain-mount on a low-z surface is all that is required) and rapidity of data collection (~ 5 min per sample).

The depth of penetration, which ranges from ~ 0.01 mm to 1 mm depending on the energy of the X rays, can be an advantage or a disadvantage. The penetration can make it possible to analyze the contents of inclusions at depth, and it can give a better indication of the bulk composition of a grain if the surface of the grain is contaminated. However, the penetration may lead to "seeing through" to material lying under the grain being analyzed. SRIXE analyses to date have been made in air; therefore, analyses of elements lighter than potassium cannot be made because of the absorption by air of the lower energy X rays characteristic of the lighter elements. This problem can be avoided by placing the X-ray path in helium or vacuum.

8.2 *Methods*

At the National Synchrotron Light Source a collimated X-ray microprobe (CXRM) has been set up on beamline X26A for making SRIXE measurements. This facility became operational in 1986 and uses white radiation from a bending magnet source with spot sizes as small as 8 μm $\times$ 8 μm at 9 m from the 2.5-GeV electron beam. Fluorescence from major elements is minimized by filtering out the low-energy radiation with the aluminum or Kapton absorbers. The photo flux at the sample is approximately 1.5×10^8 photons s^{-1} μm^{-2}. Samples are mounted on a stepper-motor-driven x-y-z-θ stage with 1 μm resolution. Single-spot analyses and one-dimensional and two-dimensional scanning are available. Technical upgrades are in progress to improve spatial resolution, energy resolution, and elemental sensitivity. An 8 : 1 critical reflectance, ellipsoidal, focusing mirror has been installed, and initial tests show an increase in photon flux of about a factor of 30 over the direct beam. This enhancement will allow wavelength-dispersive detectors to be used on the microprobe. Installation of a monochromator is being planned. Fixed tantalum collimators have been constructed that have produced beams down to 3-μm diameters. A frame grabber and freeze-frame peripheral are used to display, manipulate, print, and store elemental maps and optical images of the sample as seen through a camera-equipped, petrographic microscope. Fluorescence emissions are observed with a Si(Li) energy-dispersive detector system at 90° to the incident beam and within the electron orbital plane to minimize Compton scattering (Figure 7). A representative X-ray fluorescence spectrum is shown in Figure 8.

A focused X-ray microprobe (FXRM) has also been developed using a Kirkpatrick-Baez focusing system (Kirkpatrick & Baez 1948). This device can produce a photon spot 2–5 μm across with a flux that is very similar to that of the CXRM at an energy of 10 keV and a bandwidth of 10%. The lower backgrounds of the FXRM produce lower detection liimits than the CXRM, but the CXRM can be used for detection of elements with

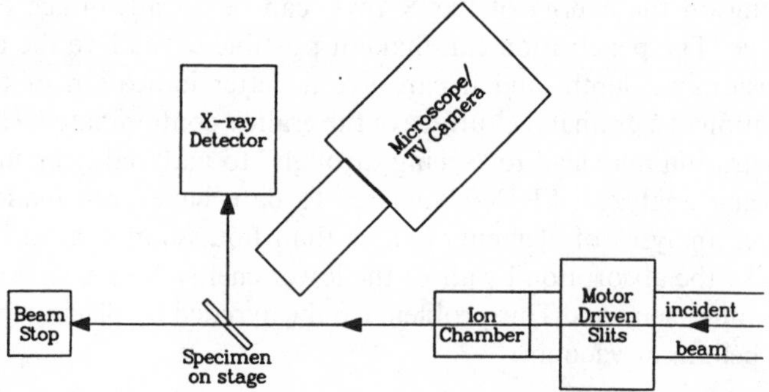

Figure 7 Schematic of X-ray microprobe on beamline X26C at NSLS using collimated synchrotron radiation as the excitation source. From Sutton et al (1988).

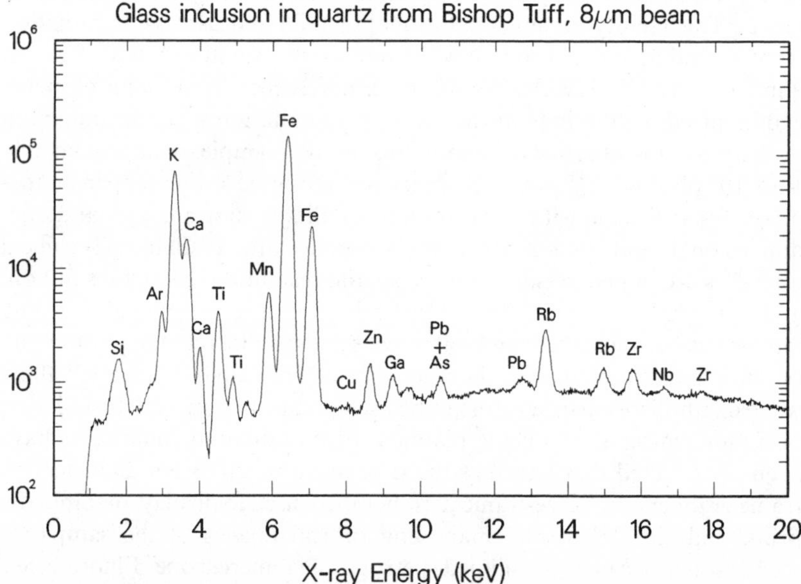

Figure 8 Synchrotron radiation-induced X-ray fluorescence spectrum for a glass inclusion in quartz from the Bishop Tuff using an apparatus similar to that shown in Figure 7 (S. R. Sutton, personal communication, 1989).

higher absorption edge energies ($\leq$30 keV). The usefulness of both approaches has been demonstrated.

8.3 *Results*

8.3.1 PLATINUM-GROUP ELEMENT PARTITIONING IN SULFIDES Pyrrhotite, pentlandite, and chalcopyrite samples from Sudbury (Canada), Merensky Reef (South Africa), and Stillwater (Montana) have been analyzed for Pd and Ru distributions (Cabri et al 1985). The dominant carrier of Pd in all three ores was found to be pentlandite, with levels in the range 100–300 ppm for Merensky and Stillwater and 1–5 ppm for Sudbury. Ruthenium and palladium were detected only in the pentlandite and pyrrhotite from Merensky (10 ppm at a counting rate of 3 s $pixel^{-1}$. In the other samples it was below the detection limit. Imaging of these samples has been achieved by collecting data point by point across the samples. Images consisting of 40 pixels by 40 pixels have been formed, with each pixel being 10 μm across. In these images, it is possible to directly compare the intensities of the Pd emission from the different mineral phases and to detect any variations within the phases. Chen et al (1987) have used SRIXE to detect gold in pyrite grains in unoxidized Carlin-type ore deposits.

8.3.2 FELDSPARS Lu et al (1989) made an extensive comparison of the synchrotron X-ray fluorescence analysis (SXRFA) method with other microanalysis methods, including EMPA, PIXE, PIGE, NRA, SIMS, and RBS, as well as atomic absorption spectrophotometry (AAS). Twenty-one feldspar and synthetic glass samples were analyzed with white X radiation from a bending magnet at NSLS. They were analyzed for Ba, Ti, Mn, Fe, Cu, Zn, Ga, Ge, Rb, and Sr. The resulting standardless analyses agreed with analyses made by EMPA and AAS within $\pm 20\%$ with two exceptions, Rb and Ba. Lu et al attribute the discrepancies for the latter to a systematic error in the fundamental parameters used to obtain theoretical yields for the fluorescence emission and conclude that an external standard or empirical correction is desirable.

8.3.3 ELEMENT DISTRIBUTIONS IN IRON METEORITES Different phases of iron meteorites have been examined by the SRIXE method for various elements. One of the most significant observations is that the Cu distribution coefficients are less than 1 for Ni-rich meteorites. Such Cu enrichments in metal are unlikely to have resulted from eutectic crystallization alone, since laboratory experiments demonstrate that Cu exhibits a chalcophilic character under such conditions. These results provide evidence that subsolidus exsolution can "fractionate" Cu and Ni tenfold. Such subsolidus reactions may have played important roles in establishing the present distributions of elements in the Earth.

Scans across phase interfaces have revealed some unusually high elemental concentrations. For instance, at grain boundaries between troilite (FeS) and kamacite (Fe-Ni) a remarkable accumulation of several different elements has been observed. Among the elements occurring at this grain boundary are Ge, Ga, and Cu. These observations were made using radiation from a prototype undulator being tested at CHESS for use at APT (Rivers et al 1989).

8.3.4 FLUID INCLUSION STUDIES The focused X-ray microprobe (FXMP) has been used to analyze synthetic fluid inclusions in quartz (Frantz et al 1988). Inclusions containing known amounts of calcium chloride, manganese chloride, and zinc chloride were introduced into quartz by equilibrating fractured quartz prisms with aqueous fluids using standard hydrothermal techniques. Elements contained within fluid inclusions with dimensions of only 17 μm $\times$ 7 μm $\times$ 3 μm were identified, and semiquantitative analyses of their concentrations were obtained. The focused X-ray spot (5–10 μm in diameter) was scanned across individual inclusions in 1-μm-size steps. Contour maps based on these measurements show exceptional spatial resolution in that the shapes of the inclusions are delineated. In one case, even the position of the vapor bubble can be seen. The sensitivity and accuracy of the measurements depend on the depth of the inclusion beneath the quartz surface and on X-ray energy. For instance, half of the intensity of the K_α radiation for Zn is absorbed in 40 μm of quartz, while half of the intensity of the K_α radiation for Ca is absorbed in 5 μm of quartz.

Results from these experiments show that FXMP is potentially a powerful nondestructive and noninvasive method for studying high-Z elements in fluid inclusions. This technique, when applied to natural specimens, can provide valuable information on the conditions of formation of rock systems and can be particularly valuable for studying the formation of ore minerals.

8.3.5 MICROMETEORITES Micrometeorites collected in the stratosphere are of great importance in studying the nature of the interplanetary dust cloud and primitive bodies, such as comets and some asteroids. Studies on these particles have shown the presence of solar flare tracks (Bradley et al 1984), solar wind–implanted noble gases (Rajan et al 1977, Hudson et al 1981), and nonterrestrial isotopic ratios (Esat et al 1979, Zinner et al 1983), all indicative of an extraterrestrial origin. The determination of chemical compositions is of fundamental interest in understanding the histories of the particles and their relationships with conventional meteorites. The nondestructive nature of the synchrotron radiation–induced X-ray emission technique allows subsequent investigation by TEM, ion microprobe, infrared transmission spectroscopy, and Raman spectroscopy. Element sensitivities better than 10 ppm have been achieved for 20-μm particles. The first-order conclusion from these analyses is that particles that have "chondritic" major-element abundances also have "chondritic" trace-element abundances (Sutton & Flynn 1988). However, in detail, each micrometeorite has a unique composition. The most striking chemical differences from conventional meteorites are common enrichments of volatile elements (e.g. Zn, Se, and Br) in the particles. Bromine enrichments as large as 40 times chondrite abundances have been observed, suggesting that atmospheric injection of halogens by extraterrestrial material may affect atmospheric chemistry and enhance ozone depletion (Sutton & Flynn 1989). Two particles exhibited extremely low Ni concentrations and abundance patterns remarkably similar to that of mafic igneous rocks. SRIXE has also been used to estimate cosmic dust particle densities, an important parameter in calculating orbital evolution time scales and atmospheric entry heating (Sutton & Flynn 1989).

8.3.6 SRIXE ANALYSES USING INSERTION DEVICES Insertion devices have the potential to improve the SRIXE technique in several ways. Undulators on high-energy storage rings provide fluxes 1000 times greater than those available from bending magnet sources. This permits the use of much smaller beam diameters, the use of monochromatic X rays for better sensitivity and/or micro-EXAFS capability, and the use of detection systems with better resolution and peak/background ratio but poorer efficiency. Rivers et al (1989) made measurements with a prototype APS undulator installed at CHESS. With beam currents of only 2 mA, they were able to demonstrate parts-per-million sensitivity in a beam less than 10 μm in size.

Another type of insertion device that will extend the SRIXE technique is high-field wigglers, such as the superconducting wiggler X17 at the NSLS. This device extends the useful energy range of the synchrotron out beyond 100 keV, permitting excitation of the K lines of all of the elements. This is particularly important for analyses of REE, whose L lines are complex and have interferences from the K lines of the much more abundant transition elements. Preliminary experiments at X17, with only 1% of the normal NSLS beam current, have yielded detection limits of 4 ppm for the REE in beam diameters of 30 μm.

9. X-RAY TOMOGRAPHY

X-ray tomography provides a means of making nondestructive, noninvasive studies of the internal structures of specimens.

First-generation tomography experiments use a pencil beam of X rays, collimated to sizes of 2–3 μm. The sample is scanned through the beam pixel by pixel, and the absorption is measured at each point. The sample is then rotated by a small angle, and the sample is scanned again. The resulting data set can then be computer processed to reconstruct a cross-section image of the X-ray attenuation coefficient. This type of tomography has been done on beamline X26 at the NSLS with spatial resolutions as small as 2 μm (Spanne & Rivers 1987). It is also possible to collect a fluorescence signal at each point and, thus, to reconstruct a cross section of the concentration of some element (Spanne & Rivers 1988, Jones et al 1989).

Third-generation tomography experiments use a fan beam and a one-dimensional or two-dimensional positron-sensitive detector to measure X-ray absorption. It is then not necessary to translate the sample during the measurement; only rotation is required. This can lead to a large reduction in data collection time relative to older methods. With a two-dimensional detector, one can collect the data for hundreds of slices simultaneously. Flannery et al (1987) have developed a two-dimensional phosphor plate that is coupled optically to a CCD detector. With this equipment they have produced tomographic reconstructions with spatial resolutions better than 10 μm at the EXXON beamline at the NSLS. They determined the attenuation coefficient to better than 1%. Flannery et al (1987) have applied the technique to specimens of coal and predict that it will have a wide range of applications in materials science, biology, and medicine, as well as geology. When absorption data are collected just below and just above the absorption edge for a particular element in a specimen, the distribution of that element in the specimen can be imaged, thus providing even more detailed information.

Although some very interesting results have been obtained on the structures of fossils using conventional X-ray sources (Cisne 1981, Conroy & Vannier 1984, Fraser & Shelton 1988, Haubitz et al 1988), synchrotron radiation clearly offers the advantages of excellent collimation, energy selection, and intensity for making this kind of study.

10. X-RAY TOPOGRAPHY

X-ray topography is a technique that produces an image using diffracted X rays. The intensity at any point in the image is extremely sensitive to small differences in lattice spacings and orientations. Therefore, it is very useful for studying dislocations, planar defects, stacking faults, domain walls, and growth defects. The method is explained in detail by Sauvage & Petroff (1980).

X-ray topography has been successfully applied to the study of growth defects in synthetic quartz (Zarka & Lin 1983) and to the observation of the incommensurate phase in berlinite (Zarka et al 1986). Because it is able to show minor variations in atomic spacings and orientations, X-ray topography is a valuable way to observe the effects of variations in parameters such as temperature and composition during growth of a crystal. The potential for studying growth in natural crystals makes this an important technique for mineralogists and petrologists.

11. FUTURE APPLICATIONS

Over the next five to seven years, two new high-brilliance synchrotron radiation sources will be constructed in the US, one devoted to hard X-ray production (the Advanced Photon Source) and one to soft X-ray and vacuum ultraviolet light production (the Advanced Light Source) (see Section 3.1.4). Their brilliances will be 1000 to 10,000 times greater than existing synchrotron sources and will greatly extend the range of structural and chemical problems that can be addressed.

The most obvious research areas that will benefit from these next-generation X-ray sources are those involving low elemental concentrations, very small sample volumes, extremely high pressures and temperatures, and transient phenomena during which very rapid data acquisition is required. Many elements of interest in Earth materials (e.g. Ar, Nd, Sm) occur at trace concentration levels (<1000 ppm), which makes conventional scattering or spectroscopic studies of their structural environment and bonding virtually impossible. Consequently, little is known about the distributions of these elements in mineral structures (crystallographic sites versus defect sites) or about their crystal chemical behavior (bonding, site partitioning, short-range ordering, complex formation, diffusion mechanisms, etc). Such information is useful in a variety of geological contexts; for example, it is needed to provide an understanding of the "blocking temperatures" of radiogenic elements and their daughter products in minerals that have undergone metamorphism. Moreoever, with the gains being made in synchrotron X-ray microprobe analyses of trace elements in minerals at the NSLS, there is a growing need for parallel information on the crystal chemistry of the minerals. X-ray absorption spectroscopy studies of such elements in bulk geologic samples are currently difficult or impossible because of insufficient flux from existing synchrotron X-ray beamlines. The new high-brilliance X-ray sources that will become available in the mid-1990s should permit XAS studies of trace elements in minerals, gels, glasses, melts, and electrolyte solutions at concentration levels approaching those in natural systems and at spatial resolution levels of 1 μm^2 or less.

The study of compositional variations of minor and trace elements in Earth materials, including fluid inclusions, using the X-ray fluorescence microprobe will be greatly enhanced by next-generation synchrotron sources. The high brilliance of these sources will permit higher spatial resolution than now possible and will lower detection limits for most elements. This method will complement the next-generation ion microprobe.

Another research area that will benefit from these new high-brilliance X-ray sources involves reactions, element partitioning, structure, and bonding at mineral surfaces and interfaces. Most chemical reactions involving minerals and fluids occur at these sites of low dimensionality. Furthermore, the strength of minerals and rocks and the transport of elements in Earth materials are ultimately related to sites at grain boundaries, surfaces, dislocations, stacking faults, twin boundaries, and domain boundaries. For example, hydrolytic weakening of minerals at crack tips is of great significance in the fracturing and faulting of crustal rocks. Very little is known about these types of sites in minerals because few methods have the ability to spatially resolve and isolate such sites or to provide element-specific structure and bonding information for elements in such sites at low concentrations. Polarized X-ray absorption spectroscopy studies of cations at these types of sites using high-brilliance sources will provide the first opportunity to tackle some of these problems in surface science related to Earth materials. Moreover, these photon sources will permit studies of chemisorption and physisorption on geologically relevant single crystals for the first time. Such studies are virtually impossible using conventional spectroscopies or scattering methods because of the typically very low concentration levels of elements at submonolayer coverages and the need for extremely sensitive methods that require high-vacuum environments.

A fourth research area that is well-suited to utilize high-brilliance X rays is the study of Earth materials as functions of temperature and pressure. Knowledge of the structure-property relationships of minerals and fluids under extreme conditions is of critical importance in developing models of many geological processes, particularly dynamical processes occurring in the Earth's mantle and in the core-mantle boundary region. As experimental temperatures and pressures have been pushed to the extreme values typical of those in the Earth's lower mantle and core ($T > 3000$ K, $P > 250$ GPa) using laser-heated DACs, new information on phase stability, phase transitions, and equations of state has been gained. At these extreme *P-T* conditions, the need for high-brilliance photon sources has increased because of the large pressure and temperature gradients in the DAC (Jeanloz & Heinz 1984) and the very small sample volume (on the order of picoliters) at maximum *P* and *T*. Both X-ray diffraction and X-ray absorption experiments carried out under such *P-T* conditions could benefit greatly from the next-generation, high-brilliance X-ray sources, which will produce very high X-ray fluxes in a very small spot size.

Information on bonding can be derived from synchrotron-based spectroscopies (X-ray absorption and emission, ultraviolet and X-ray photoelectron emission), providing inputs and checks for the band structure calculations needed in ab initio equation-of-state calculations. Further, they provide energy level information needed to calculate point defect energies for diffusion, deformation, and electrical conductivity of Earth materials.

These new sources will permit new classes of time-resolved X-ray diffraction and X-ray spectroscopy experiments (e.g. structural studies of transient phenomena such as first-order solid-solid phase transitions, melting, and higher order transitions such as changes in the spin state of iron). Such studies will require rapid detection methods, such as position-sensitive detectors employed in dispersive EXAFS studies (e.g. Itie et al 1986). Development of rapid detection devices for time-resolved synchrotron studies is underway (Clarke et al 1988). Time-dependent X-ray diffraction and X-ray absorption spectroscopy studies have been carried out on a few materials, including EXAFS study of flash-melted Al films using a single nanosecond pulse of X rays emitted from a laser-produced plasma (Epstein et al 1983). Time-resolved EXAFS methods are also used to study structural changes during electrochemical reactions, such as the changes in structure of $Ni(OH)_2$ as it is being electrochemically oxidized in concentrated alkali electrolytes (McBreen et al 1987). Clearly, much groundwork is needed to fully utilize time-resolved X-ray diffraction and X-ray spectroscopy in the study of transient phenomena; however, the potential is great for major breakthroughs in understanding the kinetics and mechanisms of reactions and phase changes of Earth materials using existing and next-generation synchrotron X-ray sources.

As new methods are developed in response to increases in X-ray brilliance of new synchrotron radiation sources, new applications will undoubtedly result. Some of these methods will prove useful in Earth sciences research and could well be developed by Earth scientists. For example, synchrotron-produced far-infrared spectroscopy of minerals will yield valuable information on lattice vibrational modes, and X-ray Raman scattering (Tohji & Udagawa 1987) may make possible in situ high-pressure spectroscopic studies of low-atomic-number elements (e.g. Si) in important mantle phases. One can only guess what new knowledge and methods may come from synchrotron radiation research in the years ahead.

ACKNOWLEDGMENTS

Material for this review came from a number of sources. We drew liberally from chapters we wrote or cowrote in *Synchrotron X-ray Sources and*

New Opportunities in the Earth Sciences: Workshop Report (Smith & Manghnani 1988) and wish to acknowledge the participants of that workshop for their contributions to this review. We wish to thank the National Science Foundation (NSF) for generous support of our synchrotron radiation studies of Earth materials through grants EAR-8706997 (Bassett), EAR-8513488 (Brown), and EAR-8805440 (Brown). Preparation of this manuscript was supported by NSF grants EAR-8706997 (Bassett) and EAR-8805440 (Brown). We also wish to thank the National Aeronautics and Space Administration (NASA) for their support of the manuscript preparation through grant NAGW-963 (Bassett). Mark L. Rivers, Joseph V. Smith, and Steven R. Sutton provided very helpful reviews of the manuscript. Mark and Steve are thanked in particular for providing significant input into Sections 3.1.3, 8, and 9 and for providing Figure 8.

Literature Cited

Akimoto, S., Suzuki, T., Yagi, T., Shimomura, O. 1987. Phase diagram of iron determined by high-pressure/temperature X-ray diffraction using synchrotron radiation. In *High-Pressure Research in Mineral Physics*, ed. M. H. Manghnani, Y. Syono, pp. 149–54. Tokyo/Washington, DC: TERRAPUB/Am. Geophys. Union. 486 pp.

Anderson, D. L. 1984. The Earth as a planet: paradigms and paradoxes. *Science* 223: 347–55

Anderson, D. L., Bass, J. D. 1986. Transition region of the Earth's upper mantle. *Nature* 320: 321–28

Augustithis, S. S. 1988. *Synchrotron Radiation Applications in Mineralogy and Petrology*. Athens: Theophrastus Publ. 216 pp.

Bachman, R., Kohler, H., Schulz, H., Weber, H.-P., Kupcik, V., et al. 1983. Structure analysis of a CaF_2 single crystal with an edge length of only 6 μm: an experiment using synchrotron radiation. *Angew. Chem. Int. Ed. Engl.* 22: 1011–12

Bassett, W. A., Huang, E. 1987. Mechanism of the bcc-hcp phase transition in iron. *Science* 238: 780–83

Bassett, W. A., Takahashi, T. 1965. Silver iodide polymorphs. *Am. Mineral.* 50: 1576–94

Bianconi, A. 1988. XANES spectroscopy. See Koningsberger & Prins 1988, pp. 573–662

Binsted, N., Greaves, G. N., Henderson, C. M. B. 1985. An EXAFS study of glassy and crystalline phases of composition $CaAl_2Si_2O_8$ (anorthite) and $CaMgSi_2O_6$ (diopside). *Contrib. Mineral. Petrol.* 89: 103–9

Boehler, R., Nicol, M., Johnson, M. L. 1987. Internally-heated diamond-anvil cell: phase diagram and *P-V-T* of iron. In *High-Pressure Research in Mineral Physics*, ed. M. H. Manghnani, Y. Syono, pp. 173–76. Tokyo/Washington, DC: TERRAPUB/Am. Geophys. Union. 486 pp.

Boehler, R., von Bargen, N., Hoffbauer, W., Huang, E. 1988. $d\alpha/dP$ at high P and T from synchrotron measurements and systematics in αK. *Eos, Trans. Am. Geophys. Union* 69: 1461 (Abstr.)

Boehler, R., Besson, J. M., Nichol, M., Nielsen, M., Itie, J. P., et al. 1989. X-ray diffraction of γ-Fe at high temperatures and pressures. *J. Appl. Phys.* 65: 1795–97

Bonnin, D., Muller, S., Calas, G. 1982. Le fer dans les kaolins. Etude par spectrométries RPE, Mossbauer, EXAFS. *Bull. Minéral.* 105: 467–75

Bonnin, D., Calas, G., Suquet, H., Pezerat, H. 1985. Site occupancy of Fe^{3+} in Garfield nontronite: a spectroscopic study. *Phys. Chem. Miner.* 12: 55–64

Bosio, L., Cortes, R., Froment, M. 1984. ReflEXAFS studies of protective oxide formation on metal surfaces. In *EXAFS and Near-Edge Structure III. Springer Proc. Phys. 2*, ed. K. O. Hodgson, B. Hedman, J. E. Penner-Hahn, pp. 484–86. New York/Berlin: Springer-Verlag. 533 pp.

Bradley, J. P., Brownlee, D. E., Fraundorf, P. 1984. Discovery of nuclear tracks in interplanetary dust. *Science* 226: 1432–34

Brown, G. E. Jr., Parks, G. A. 1989. Synchrotron-based X-ray absorption studies of cation environments in Earth materials. *Rev. Geophys.* 27(4): 519–33

Brown, G. E. Jr., Waychunas, G. A. 1988. Synchrotron-based spectroscopic and anomalous scattering studies of Earth materials. See Smith & Manghnani 1988, pp. 69–82

Brown, G. E. Jr., Dikmen, F. D., Waychunas, G. A. 1983. Total electron yield K-XANES and EXAFS investigation of aluminum in amorphous aluminosilicates. *Rep. 83/01*, pp. 146–47. Stanford Synchrotron Radiat. Lab., Stanford, Calif.

Brown, G. E. Jr., Keefer, K. D., Fenn, P. M. 1978. Extended X-ray absorption fine structure (EXAFS) study of iron-bearing silicate glasses: iron coordination environment and oxidation state. *Geol. Soc. Am. Abstr. With Programs* 10: 373 (Abstr.)

Brown, G. E. Jr., Waychunas, G. A., Ponader, C. W., Jackson, W. E., McKeown, D. A. 1986. EXAFS and NEXAFS studies of cation environments in oxide glass. *J. Phys.* 47(C8): 661–68

Brown, G. E. Jr., Calas, G., Waychunas, G. A., Petiau, J. 1988. X-ray absorption spectroscopy and its applications in mineralogy and geochemistry. In *Spectroscopic Methods in Mineralogy and Geology. Rev. Mineral.*, ed. F. C. Hawthorne, 18: 431–512. Washington, DC: Mineral. Soc. Am. 698 pp.

Brown, G. E. Jr., Parks, G. A., Chisholm-Brause, C. J. 1989. In-situ X-ray absorption spectroscopic studies of ions at oxide-water interfaces. *Chimia* 43: 248–56

Brown, G. S., Doniach, S. 1980. The principles of X-ray absorption spectroscopy. See Winick & Doniach 1980, pp. 353–85

Buras, B., Tazzari, S., eds. 1984. *Report of the European Synchrotron Radiation Facility*. Geneva: CERN, LEP Div.

Cabri, L., Rivers, M. L., Smith, J. V., Jones, K. W. 1985. Trace elements in sulfide minerals by milliprobe X-ray fluorescence using white synchrotron radiation. *Eos, Trans. Am. Geophys. Union* 66: 1150 (Abstr.)

Calas, G., Petiau, J. 1983a. Structure of oxide glasses: spectroscopic studies of local order and crystallochemistry: geochemical implications. *Bull. Minéral.* 106: 33–55

Calas, G., Petiau, J. 1983b. Coordination of iron in oxide glasses through high-resolution K-edge spectra: informations from the pre-edge. *Solid State Commun.* 48: 625–29

Calas, G., Petiau, J. 1984. X-ray absorption spectra at the K-edges of 3d transition elements in minerals and reference compounds. *Bull Minéral* 107: 85–91

Calas, G., Bassett, W. A., Petiau, J., Steinberg, D., Tchoubar, D., Zarka, A. 1984a. Mineralogical applications of synchrotron radiation. *Phys. Chem. Miner.* 121: 17–36

Calas, G., Manceau, A., Novikoff, A., Boukili, H. 1984b. Comportement du chrome dans les mineaux d'alteration du gisement de Campo Formoso (Bahia, Bresil). *Bull Minéral.* 107: 755–66

Calas, G., Petiau, J., Manceau, A. 1986. X-ray absorption spectroscopy of geological materials. *J. Phys.* 47(C8): 813–18

Calas, G., Brown, G. E. Jr., Waychunas, G. A., Petiau, J. 1987. X-ray absorption spectroscopic studies of silicate glasses and minerals. *Phys. Chem. Miner.* 15: 19–29

Charnock, G. N., Garner, C. G., Patrick, P. A. D., Vaughan, D. G. 1988. Investigation into the nature of copper and silver sites in argentian tetrahedrites using EXAFS spectroscopy. *Phys. Chem. Miner.* 15: 296–99

Chen, J. R., Chao, E. C. T., Minkin, J. A., Back, J. M., Bagby, W. C., et al. 1987. Determination of the occurrence of gold in an unoxidized Carlin-type ore sample using synchrotron radiation. *Nucl. Instrum. Methods Phys. Res. Sect. B* 22: 394–400

Chisholm-Brause, C. J., Brown, G. E. Jr., Parks, G. A. 1989a. EXAFS investigation of aqueous Co(II) adsorbed on oxide surfaces in-situ. *Physica B* 158: 674–76

Chisholm-Brause, C. J., Roe, A. L., Hayes, K. F., Brown, G. E. Jr., Parks, G. A., Leckie, J. O. 1989b. XANES and EXAFS study of aqueous Pb(II) adsorbed on oxide surfaces. *Geochim. Cosmochim. Acta.* In press

Cisne, J. L. 1981. *Triarthrus eatoni* (Trilobita): anatomy of its exoskeletal, skeletomuscular, and digestive system. *Paleontographica Americana*, Vol. 9. Ithaca, NY: Paleontol. Res. Inst. 140 pp.

Citrin, P. 1986. An overview of SEXAFS during the past decade. *J. Phys.* 47(C8): 437–72

Clarke, R., Sigler, P., Mills, D. 1988. Time-resolved studies and ultrafast detectors. *Rep. ANL/APS-TM-2*, Argonne Natl. Lab., Argonne, Ill. 51 pp.

Coisson, R., Winick, H., eds. 1987. *Workshop on PEP as a Synchrotron Radiation Source*. Stanford, Calif: Stanford Synchrotron Radiat. Lab. 558 pp.

Combes, J. M., Manceau, A., Calas, G., Bottero, J. Y. 1989a. Formation of ferric oxides from aqueous solutions: a polyhedral approach by X-ray absorption spectroscopy: I. Hydrolysis and formation of ferric gels. *Geochim. Cosmochim. Acta* 53: 583–94

Combes, J. M., Manceau, A., Calas, G. 1989b. Formation of ferric oxides from aqueous solutions: a polyhderal approach by X-ray absorption spectroscopy: II.

Hematite formation from ferric gels. *Geochim. Cosmochim. Acta.* In press

Conroy, G. C., Vannier, M. W. 1984. Noninvasive three-dimensional computer imaging of matrix-filled fossil skulls by high-resolution computed tomography. *Science* 226: 456–58

Cressey, G., Steel, A. T. 1988. On EXAFS studies of Tc, Er, and Lu site location in the epidote structure. *Phys. Chem. Miner.* 15: 304–12

Davoli, I., Paris, E., Mottana, A., Marcelli, A. 1987. Xanes analysis on pyroxenes with different Ca concentration in M2 site. *Phys. Chem. Miner.* 14: 21–25

Decarreau, A., Colin, F., Herbillon, A., Manceau, A., Nahon, D., et al. 1987. Domain segregation in Ni-Mg-Fe smectites. *Clays Clay Miner.* 35: 1–10

Dev, B. N., Materlik, G., Johnson, R. L., Kranz, W., Funke, P. 1986. X-ray standing wave studies of germanium adsorbed on Si(111) surfaces. *Surf. Sci.* 178: 1–9

Eisenberger, P., Marra, W. C. 1981. X-ray diffraction study of Ge(001) reconstructed surface. *Phys. Rev. Lett.* 46: 1081–84

Epstein, H. M., Schwerzel, R. E., Mallozzi, P. J., Campbell, B. E. 1983. Flash-EXAFS for structural analysis of transient species: rapidly melting aluminum. *J. Am. Chem. Soc.* 105: 1466–68

Esat, T. M., Brownlee, D. E., Papanastassiou, D. A., Wasserburg, G. J. 1979. Magnesium isotopic composition of interplanetary dust particles. *Science* 206: 190–97

Ewing, R. C., Chakoumakos, B. C., Lumpkin, G. R., Murakami, T. 1987. The metamict state. *Mater. Res. Soc. Bull.* 12: 58–66

Fischer-Colbrie, A. 1986. Grazing incidence X-ray studies of thin amorphous layers. *Rep. 86/05*, Stanford Synchrotron Radiat. Lab., Stanford, Calif. 235 pp.

Flannery, B. P., Deckman, H. W., Roberge, W. G., D'Amico, K. L. 1987. Three-dimensional X-ray microtomography. *Science* 237: 1439–44

Frantz, J., Mao, H., Zhang, Y., Wu, Y., Thompson, A., et al. 1988. Analysis of fluid inclusions by X-ray fluorescence using synchrotron radiation. *Chem. Geol.* 69: 235–44

Fraser, N. C., Shelton, C. G. 1988. Studies of tooth implantation in fossil tetrapods using high-resolution X-radiography. *Geol. Mag.* 125: 117–22

Fuoss, P. H., Eisenberger, P., Warburton, W. K., Bienenstock, A. 1981. Application of differential anomalous X-ray scattering to structural studies of amorphous materials. *Phys. Rev. Lett.* 46: 1537–40

Furnish, M. D., Bassett, W. A. 1983. Investigation of the mechanism of the olivine-spinel transition in fayalite by synchrotron radiation. *J. Geophys. Res.* 88: 10,333–41

Greaves, G. N., Binstead, N., Henderson, C. M. B. 1984. The environment of modifiers in glasses. In *EXAFS and Near-Edge Structure III. Springer Proc. Phys. 2*, ed. K. O. Hodgson, B. Hedman, J. E. Penner-Hahn, pp. 297–301. New York/Berlin: Springer-Verlag. 533 pp.

Greegor, R. B., Lytle, F. W., Sandstrom, D. R., Wong, J., Schultz, P. 1983. Investigation of TiO_2-SiO_2 glasses by X-ray absorption spectroscopy. *J. Non-Cryst. Solids* 55: 27–43

Greegor, R. B., Lytle, F. W., Chakoumakos, B. C., Lumpkin, G. R., Ewing, R. C. 1986a. Structural investigation of metamict minerals using X-ray absorption spectroscopy. *Abstr. Program. Int. Mineral. Assoc. Gen. Meet., 14th, Stanford, Calif.* p. 114 (Abstr.)

Greegor, R. B., Lytle, F. W., Ewing, R. C., Haaker, R. F. 1984. EXAFS/XANES studies of metamict materials. In *EXAFS and Near-Edge Structure III. Springer Proc. Phys. 2*, ed. K. O. Hodgson, B. Hedman, J. E. Penner-Hahn, pp. 343–48. New York/Berlin: Springer-Verlag. 533 pp.

Greegor, R. B., Lytle, F. W., Chakoumakos, B. C., Lumpkin, G. R., Ewing, R. C., et al. 1986b. Investigation of the Ta site in alpha-recoil damaged natural pyrochlores by XAS. *Rep. 86/01*, p. 58–59. Stanford Synchrotron Radiat. Lab., Stanford, Calif.

Hamaya, N., Yagi, T., Yamaoka, S., Shimomura, O., Akimoto, S. 1985. Time-resolved X-ray diffraction analysis of the phase transition in BaS at high pressures and temperatures. In *Solid State Physics Under Pressure*, ed. S. Minomura, pp. 357–62. Tokyo/Dordrecht: KTK Sci. Publ./D. Reidel. 382 pp.

Hardwick, A., Whittaker, E. J. W., Diakun, G. P. 1985. An extended X-ray absorption fine structure (EXAFS) study of the calcium site in a model basaltic glass, $Ca_3Mg_4Al_2Si_7O_{24}$. *Mineral. Mag.* 49: 25–29

Harries, J. E., Irlam, J. C., Holt, C., Hasnain, S. S., Hukins, D. W. L. 1987. Analysis of EXAFS spectra from the brushite and monetite forms of calcium phosphate. *Mater. Res. Bull.* 22: 1151–57

Haubitz, B., Prokop, M., Dohring, W., Ostrom, J. H., Wellnhofer, P. 1988. Computed tomography of *Archaeopteryx*. *Paleobiology* 14: 206–13

Hayes, K. F., Roe, A. L., Brown, G. E. Jr., Hodgson, K. O., Leckie, J. O., Parks, G. A. 1987. In-situ X-ray absorption study of surface complexes at oxide/water interfaces: selenium oxyanions on α-FeOOH. *Science* 238: 783–86

Hayes, T. M., Boyce, J. B. 1982. Extended X-ray absorption fine structure spectroscopy. *Solid State Phys.* 37: 173–365

Hemley, R. J., Jephcoat, A. P., Mao, H. K., Finger, L. W., Zha, C. S., Cox, D. E. 1987. Static compression of H_2O-ice to 128 GPa (1.28 Mbar). *Nature* 330: 737–40

Hemley, R. J., Jephcoat, A. P., Zha, C. S., Mao, H. K., Finger, L. W., Cox, D. E. 1989. Equation of state of solid neon from X-ray diffraction measurements to 110 GPa. *Proc. AIRAPT Int. Conf., 11th, Kiev.* In press

Hinze, E., Lauterjung, J., Will, G. 1983. A new belt apparatus for energy-dispersive X-ray diffraction under high pressure and temperature. *Nucl. Instrum. Methods* 208: 569–72

Huang, E., Bassett, W. A. 1986. Rapid determination of Fe_3O_4 phase diagram by synchrotron radiation. *J. Geophys. Res.* 91: 4697–4703

Huang, E., Bassett, W. A., Tao, P. 1987. Study of bcc-hcp iron phase transition by synchrotron radiation. In *High-Pressure Research in Mineral Physics*, ed. M. H. Manghnani, Y. Syono, pp. 165–72. Tokyo/Washington, DC: TERRAPUB/Am. Geophys. Union. 486 pp.

Huang, E., Bassett, W. A., Tao, P. 1988a. Pressure-temperature-volume relationship for hexagonal close packed iron determined by synchrotron radiation. *J. Geophys. Res.* 92: 8129–35

Huang, E., Bassett, W. A., Weathers, M. S. 1988b. Phase relationships in Fe-Ni alloys at high pressures and temperatures. *J. Geophys. Res.* 93: 7741–46

Hudson, B., Flynn, G. J., Fraundorf, P., Hohenberg, C. M., Shirck, J. 1981. Noble gases in stratospheric dust particles: confirmation of extraterrestrial origin. *Science* 211: 383–86

Hunold, K., Bruckner, R. 1980. Physikalische Eigenschaften und struktureller Feinbau von natrium-aluminosilikat Glasern und Schmelzen. *Glastech. Ber.* 53: 149–61

Ildefonse, P., Manceau, A., Prost, D., Toledo-Groke, M. C. 1986. Hydroxy Cu-vermiculite formed by the weathering of Fe-biotites at Salobo, Carajas, Brazil. *Clays Clay Miner.* 34: 338–45

Ingalls, R., Garcia, G. A., Stern, E. A. 1978. X-ray absorption at high pressure. *Phys. Rev. Lett.* 40: 334–36

Ingalls, R., Crozier, E. D., Whitmore, J. E., Seary, A. J., Tranquada, J. M. 1980. Extended X-ray absorption fine structure of NaBr and Ge at high pressure. *J. Appl. Phys.* 51: 3158–63

Itie, J. P., Jean-Louis, M., Dartyge, E., Fontaine, A., Jucha, A. 1986. High pressure XAS on bromine in the dispersive mode. *J. Phys.* 47(C8): 897–900

Itie, J. P., Polian, A., Calas, G., Fontaine, A., Tolentino, H. 1989. Pressure-induced coordination changes in crystalline and vitreous GeO_2. *Phys. Rev. Lett.* 63: 398–401

Jackson, W. E., Knittle, E., Brown, G. E. Jr., Jeanloz, R. 1987a. Partitioning of Fe within high-pressure silicate perovskite: evidence for unusual geochemistry in the lower mantle. *Geophys. Res. Lett.* 14: 224–26

Jackson, W. E., Brown, G. E. Jr., Ponader, C. W. 1987b. EXAFS and XANES study of the potassium coordination environment in glasses from the $NaAlSi_3O_8$-$KAlSi_3O_8$ binary: structural implications for the mixed alkali-effect. *J. Non-Cryst. Solids* 93: 311–22

Jackson, W. E., Cooney, T., Ponader, C. W., Brown, G. E. Jr., Waychunas, G. A. 1987c. Coordination environment of Fe^{2+} in fayalite glasses and in glasses from the fosterite-fayalite binary by EXAFS spectroscopy. *Abstr. Vol., NATO Adv. Study Inst. Phys. Prop. and Thermodyn. Behav. of Miner., Cambridge, Engl.* (Abstr.)

Jeanloz, R., Heinz, D. L. 1984. Experiments at high temperature and high pressure: laser heating through the diamond cell. *J. Phys.* 45(C8): 83–92

Jeanloz, R., Thompson, A. B. 1983. Phase transitions and mantle discontinuities. *Rev. Geophys. Space Phys.* 21: 51–74

Jephcoat, A. P., Hemley, R. J., Zha, C. S., Mao, H. K. 1989. Phase transitions and equation of state of solid nitrogen from X-ray diffraction measurements to 80 GPa. *Bull. Am. Phys. Soc.* In press

Jones, K. W., Gordon, B. M., Hanson, A. L., Pounds, J. G., Rivers, M. L., et al. 1989. X-ray microscopy using synchrotron radiation. In *Microbeam Analysis—1989*, ed. P. E. Russell, pp. 191–95. San Francisco: San Francisco Press

Kanzaki, M., Kurita, K., Fujii, T., Kato, T., Shimomura, O., Akimoto, S. 1987. A new technique to measure the viscosity and density of silicate melts at high pressure. In *High Pressure Research in Mineral Physics*, ed. M. H. Manghnani, Y. Syono, pp. 195–200. Tokyo/Washington, DC: TERRAPUB/Am. Geophys. Union. 486 pp.

Kirkpatrick, P., Baez, A. V. 1948. Formation of optical images by X-rays. *J. Opt. Soc. Am.* 39: 766

Knittle, E., Jeanloz, R., Smith, G. L. 1986. The thermal expansion of silicate perovskite and stratification of the Earth's mantle. *Nature* 319: 214–16

Koch, E. E., ed. 1983. *Handbook on Synchrotron Radiation*, Vols. 1a,b. Amsterdam/New York/Oxford: North-Holland. 605 pp., 560 pp.

Koningsberger, D. C., Prins, R., eds. 1988. *X-ray Absorption: Principles, Applications, Techniques of EXAFS, SEXAFS and XANES*. New York: Wiley. 673 pp.

Kortright, J., Warburton, W., Bienenstock, A. 1983. Anomalous X-ray scattering and its relationship to EXAFS. In *EXAFS and Near Edge Structure. Springer Ser. Chem. Phys. 27*, ed. A. Bianconi, L. Incoccia, S. Stipcich, pp. 362–72. Berlin/New York: Springer-Verlag. 485 pp.

Kudoh, Y., Prewitt, C. T., Finger, L. W., Darovskikh, A., Ito, E. 1989. Effect of iron on the crystal structure of (Mg,Fe,Cr)SiO_3 perovskite. *Eos, Trans. Am. Geophys. Union* 70: 1386 (Abstr.)

Kulipanov, G. N. 1987. The status of synchrotron research in the Soviet Union. *Nucl. Instrum. Methods Phys. Res. Sect. A* 261: 1–7

Kvick, A., Rohrbaugh, S. J. 1987. In *National Synchrotron Light Source Annu. Rep. 52045*, ed. S. M. White-Pace, N. F. Gmur, p. 340. Brookhaven Natl. Lab., Upton, N.Y.

Lear, R. D., Weber, M. J., eds. 1987. *Synchrotron Radiation. Rep. UCRL-52000-87-11.12*, Lawrence Livermore Natl. Lab., Livermore, Calif. 54 pp.

Lee, P. A., Citrin, P. H., Eisenberger, P. M. 1981. Extended X-ray absorption fine structure—its strengths and limitations as a structural tool. *Rev. Mod. Phys.* 53: 769–806

Lehmann, M. S., Christensen, A. N., Fjellvag, H., Feidenhans'l, R., Nielsen, M. 1987. Structure determination by use of pattern decomposition and the Rietveld method on synchrotron X-ray and neutron powder data: the structures of $Al_2Y_4O_9$ and I_2O_4. *J. Appl. Crystallogr.* 20: 123–29

Lu, F.-Q., Smith, J. V., Sutton, S. R., Rivers, M. L., Davis, A. M. 1989. Synchrotron X-ray fluorescence analysis of rock-forming minerals: Part I. Comparison with other techniques; Part II. White beam energy dispersive procedures for feldspars. *Chem. Geol.* 75: 123–43

Lytle, F. W., Greegor, R. B., Panson, A. J. 1988. Discussion of X-ray-absorption near-edge structure: application to Cu in the high-T_c superconductors $La_{1.8}Sr_{0.2}CuO_4$ and $YBa_2Cu_3O_7$. *Phys. Rev. B* 37: 1550–62

Manceau, A., Calas, G. 1986. Nickel-bearing clay minerals. 2. Intracrystalline distribution of nickel: an X-ray absorption study. *Clay Miner.* 21: 341–60

Manceau, A., Combes, J. M. 1988. Structure of Mn and Fe oxides and oxyhydroxides: a topological approach by EXAFS. *Phys. Chem. Miner.* 15: 283–95

Manceau, A., Llorca, S., Calas, G. 1987. Crystal chemistry of cobalt and nickel in lithiophorite and asbolane from New Caledonia. *Geochim. Cosmochim. Acta* 51: 105–13

Manceau, A., Bonnin, D., Kaiser, P., Fretigny, C. 1988. Polarized EXAFS of biotite and chlorite. *Phys. Chem. Miner.* 16: 180–85

Manghnani, M. H., Ming, L. C., Jamieson, J. C. 1980. Prospects of using synchrotron radiation facilities with diamond-anvil cells: high pressure research applications in geophysics. *Nucl. Instrum. Methods* 177: 219–26

Manghnani, M. H., Ming, L. C., Nakagiri, N. 1987. Investigation of the α-Fe to ε-Fe phase transition by synchrotron radiation. In *High Pressure Research in Mineral Physics*, ed. M. H. Manghnani, Y. Syono, pp. 155–63. Tokyo/Washington, DC: TERRAPUB/Am. Geophys. Union. 486 pp.

Mao, H. K., Hemley, R. J. 1989. Optical studies of hydrogen above 200 gigapascals: evidence for metallization by band overlap. *Science* 244: 1462–65

Mao, H. K., Jephcoat, A. P., Hemley, R. J., Finger, L. W., Zha, C. S., et al. 1988a. Synchrotron X-ray diffraction measurements of single-crystal hydrogen to 26.5 GPa. *Science* 239: 1131–34

Mao, H. K., Hemley, R. J., Wu, Y., Jephcoat, A. P., Finger, L. W., et al. 1988b. High-pressure phase diagram and equation of state of solid helium from a single-crystal X-ray diffraction to 23.3 GPa. *Phys. Rev. Lett.* 60: 2649–52

Mao, H. K., Chen, L. C., Hemley, R. J., Jephcoat, A. P., Wu, Y., Bassett, W. A. 1989a. Stability and equation of state of $CaSiO_3$-perovskite to 134 GPa. *J. Geophys. Res.* In press

Mao, H. K., Hemley, R. J., Shu, J. F., Chen, L. C., Jephcoat, A. P., et al. 1989b. The effect of pressure, temperature, and composition on lattice parameters and density of (Fe,Mg)SiO_3-perovskite to 30 GPa with implications to the chemical composition of the lower mantle. In preparation

Marra, W. C., Eisenberger, P., Cho, A. Y. 1979. X-ray total-external-reflection-Bragg diffraction: a structural study of the GaAs-Al interface. *J. Appl. Phys.* 50: 6927–33

Matsui, Y., Kawamura, K. 1980. Instantaneous structure of a $MgSiO_3$ melt simulated by molecular dynamics. *Nature* 285: 648–49

McBreen, J., O'Grady, W. E., Pandya, K. I., Hoffman, R. W., Sayers, D. E. 1987. EXAFS study of the nickel oxide electrode. *Langmuir* 1987 (3): 428–33

McKeown, D. A., Waychunas, G. A., Brown, G. E. Jr. 1985a. EXAFS study of the coordination environment of aluminum in a series of silica-rich glasses and selected minerals within the Na_2O-Al_2O_3-SiO_2 system. *J. Non-Cryst. Solids* 74: 349–71

McKeown, D. A., Waychunas, G. A., Brown, G. E. Jr. 1985b. EXAFS and XANES study of the local coordination environment of sodium in a series of silica-rich glasses and selected minerals within the Na_2O-Al_2O_3-SiO_2 system. *J. Non-Cryst. Solids* 74: 325–48

Moncton, D. E., Brown, G. S. 1983. High-resolution X-ray scattering. *Nucl. Instrum. Methods* 208: 579–86

Parkhurst, D. A., Brown, G. E. Jr., Parks, G. A., Waychunas, G. A. 1984. Structural study of zinc complexes in aqueous chloride solutions by fluorescence EXAFS spectroscopy. *Geol. Soc. Am. Abstr. With Programs* 16: 618 (Abstr.)

Parrish, W., Hart, M., Huang, T. C. 1986. Synchrotron X-ray polycrystalline diffractometry. *J. Appl. Crystallogr.* 19: 92–100

Petiau, J. 1986. La spectrometrie des rayons X. In *Methodes Spectroscopiques Appliquées aux Minéraux*, ed. G. Calas, pp. 200–30. Paris: Soc. Fr. Minéral. Cristalogr. 380 pp.

Petiau, J., Calas, G., Sainctavit, Ph. 1987. Recent developments in the experimental studies of XANES. *J. Phys., Colloq. C9* 48: 1085–96

Petiau, J., Sainctavit, Ph., Calas, G. 1988. K X-ray absorption spectra and electronic structure of chalcopyrite $CuFeS_2$. *Mater. Sci. Eng.* 81: 237–49

Piermarini, G. J., Weir, C. E. 1962. A diamond cell for X-ray diffraction studies at high pressures. *J. Res. Natl. Bur. Stand. Sect. A* 66: 325–31

Ponader, C. W., Brown, G. E. Jr. 1989a. Rare earth elements in silicate glass/melt systems: I. Effects of composition on the coordination environments of La, Gd, and Yb. *Geochim. Cosmochim. Acta.* In press

Ponader, C. W., Brown, G. E. Jr. 1989b. Rare earth elements in silicate glass/melt systems: II. Interactions of La, Gd, and Yb with halogens. *Geochim. Cosmochim. Acta.* In press

Prewitt, C. T., Coppens, P., Phillips, J. C., Finger, L. W. 1987. New opportunities in synchrotron X-ray crystallography. *Science* 238: 312–19

Rajan, R. S., Brownlee, D. E., Tomandl, D., Hodge, P. W., Farrar, H., Britten, R. A. 1977. Detection of ^{4}He in stratospheric particles gives evidence of extraterrestrial origin. *Nature* 267: 133–34

Rietveld, H. M. 1969. A profile refinement method for nuclear and magnetic structures. *J. Appl. Crystallogr.* 2: 65–71

Rivers, M. L. 1988. Characteristics of the Advanced Photon Source and comparison with existing synchrotron facilities. See Smith & Manghnani 1988, pp. 5–22

Rivers, M. L., Sutton, S. R., Gordon, B. M. 1989. X-ray fluorescence microprobe imaging with undulator radiation. In *Synchrotron Radiation in Materials Research. Proc. Mater. Res. Soc.*, ed. R. Clark, J. Gland, J. Weaver. 143: 285–90

Robinson, I. K. 1983. Direct determination of the Au(110) reconstructed surface by X-ray diffraction. *Phys. Rev. Lett.* 50: 1145–48

Roe, A. L., Hayes, K. F., Chisholm, C. J., Brown, G. E. Jr., Hodgson, K. O., et al. 1987. XAS study of ion adsorption at aqueous/oxide interfaces. *Rep. 87/01*, pp. 142–44. Stanford Synchrotron Radiat. Lab., Stanford, Calif.

Rohler, J. 1984. High pressure L_{III} absorption in mixed valent cerium. In *EXAFS and Near-Edge Structure III. Springer Proc. Phys. 2*, ed. K. O. Hodgson, B. Hedman, J. E. Penner-Hahn, pp. 379–84. New York/Berlin: Springer-Verlag. 533 pp.

Rohler, J., Keulerz, K., Dartyge, E., Fontaine, A., Jucha, A., Sayers, D. 1984. High-pressure energy dispersive X-ray absorption of EuO up to 300 kbar. In *EXAFS and Near-Edge Structure III. Springer Proc. Phys. 2*, ed. K. O. Hodgson, B. Hedman, J. E. Penner-Hahn, pp. 385–87. New York/Berlin: Springer-Verlag. 533 pp.

Sandstrom, D. R., Lytle, F. W., Wei, P. S. P., Greegor, R. B., Wong, J., Shultz, P. 1980. Coordination of Ti in TiO_2-SiO_2 glass by X-ray absorption spectroscopy. *J. Non-Cryst. Solids* 41: 201–7

Sauvage, M., Petroff, J. F. 1980. Application of synchrotron radiation to X-ray topography. See Winick & Doniach 1980, pp. 607–38

Schiferl, D., Fritz, J. N., Katz, A. I., Schaeffer, M., Skelton, E. F., et al. 1987. Very high temperature diamond anvil cell for X-ray diffraction: application to the comparison of the gold and tungsten high-temperature–high pressure internal standards. In *High Pressure Research in Mineral Physics*, ed. M. H. Manghnani, Y. Syono, pp. 75–83. Tokyo/Washington, DC: TERRAPUB/Am. Geophys. Union. 486 pp.

Shenoy, G. K., Viccaro, P. J., Mills, D. M.

1988. Characteristics of the 7-GeV advanced photon source: a user's guide. *Rep. ANL-88-9*, Argonne Natl. Lab., Argonne, Ill. 52 pp.

Shimomura, O., Kawamura, T. 1987. EXAFS and XANES under pressure. In *High Pressure Research in Mineral Physics*, ed. M. H. Manghnani, Y. Syono, pp. 187–93. Tokyo/Washington, DC: TERRA PUB/Am. Geophys. Union. 486 pp.

Skelton, E. F., Qadri, S. B., Elam, W. T., Webb, A. W. 1985. A review of high pressure research with synchrotron radiation and recent kinetic studies. In *Solid State Physics Under Pressure*, ed. S. Minomura, pp. 329–34. Tokyo/Dordrecht: KTK Sci. Publ./D. Reidel. 382 pp.

Smith, J. V., Manghnani, M., eds. 1988. *Synchrotron X-ray Sources and New Opportunities in the Earth Sciences: Workshop Report. Rep. ANL/APS-TM-3*. Argonne Natl. Lab., Argonne, Ill. 127 pp.

Smith, N. V., Himpsel, F. J. 1983. Photoelectron spectroscopy. In *Handbook on Synchrotron Radiation*, ed. E.-E. Koch, 2: 905–54. Amsterdam/New York/Oxford: North Holland. 1165 pp.

Spanne, P., Rivers, M. L. 1987. Computerized microtomography using synchrotron radiation from the NSLS. *Nucl. Instrum. Methods Phys. Res. B.* 24/25: 1063–67

Spanne, P., Rivers, M. L. 1988. Microscopy and elemental analysis in tissue samples using computed microtomography with synchrotron X rays. *BioSci. Abstr.* 1: 101–3 (Abstr.)

Stern, E. A., Heald, S. M. 1983. Basic principles and applications of EXAFS. In *Handbook on Synchrotron Radiation*, ed. E.-E. Koch, 2: 955–1014. Amsterdam/New York/Oxford: North-Holland. 1165 pp.

Sueno, S., Nakai, I., Imafuku, M., Ohsumi, K., Morikawa, H., et al. 1986a. EXAFS measurements under high pressure conditions using a diamond anvil cell (II). *Photon Fact. Act. Rep.* 1986(4): 181 (Abstr.)

Sueno, S., Nakai, I., Ohsumi, K., Morikawa, H. 1989. EXAFS measurement under high pressure using diamond anvil cell. *Int. Geol. Congr., 28th Abstr. Vol.*, 3: 193 (Abstr.)

Sueno, S., Nakai, I., Imafuku, M., Morikawa, H., Kimata, M., et al. 1986b. EXAFS measurements under high pressure conditions using a combination of a diamond anvil cell and synchrotron radiation. *Chem. Lett.* 1986 (10): 1663–66

Sueno, S., Kimata, M., Nakai, I., Ohmasa, M., Ohsumi, K., Sasaki, S. 1987. Recent activities of the mineralogy group in Photon Factory, KEK, Japan. *Geol. Soc. Am. Abstr. With Programs* 19: 859 (Abstr.)

Sutton, S. R., Flynn, G. J. 1988. Stratospheric particles: synchrotron X-ray fluorescence determination of trace element contents. *Proc. Lunar Planet. Sci. Conf., 18th*, pp. 607–14

Sutton, S. R., Flynn, G. J. 1989. Density estimates for eleven cosmic dust particles based on synchrotron X-ray fluorescence analyses. *Abstr. Annu. Meet. Meteorit. Soc., 52nd*, p. 235 (Abstr.)

Sutton, S. R., Rivers, M. L., Jones, K. W., Smith, J. V. 1988. X-ray fluorescence microprobe analysis. See Smith & Manghnani 1988, pp. 93–112

Teo, B. K. 1986. *EXAFS: Basic Principles and Data Analysis*. New York/Berlin: Springer-Verlag. 349 pp.

Teo, B. K., Joy, D. C., eds. 1980. *EXAFS Spectroscopy*. New York: Plenum. 275 pp.

Thompson, P., Cox, D. E., Hastings, J. B. 1987. Rietveld refinement of Debye-Scherrer synchrotron X-ray data from Al_2O_3. *J. Appl. Crystallogr.* 20: 79–83

Tohji, K., Udagawa, Y. 1987. X-ray Raman scattering as a tool for structure determination. *Photon Fact. Act. Rep.* 1987 (5): 266 (Abstr.)

Tranquada, J. M., Ingalls, R. 1984. An EXAFS study of the instability of the zincblende structure under pressure. In *EXAFS and Near-Edge Structure III. Springer Proc. Phys. 2*, ed. K. O. Hodgson, B. Hedman, J. E. Penner-Hahn, pp. 388–90. New York/Berlin: Springer-Verlag. 533 pp.

Tranquada, J. M., Ingalls, R., Crozier, E. D. 1984. High pressure X-ray absorption studies of phase transitions. In *EXAFS and Near-Edge Structure III. Springer Proc. Phys. 2*, ed. K. O. Hodgson, B. Hedman, J. E. Penner-Hahn, pp. 374–78. New York/Berlin: Springer-Verlag. 533 pp.

Van Valkenburg, A. 1963. High pressure microscopy. In *High Pressure Measurement*, ed. A. Giardini, E. C. Lloyd, pp. 87–94. Washington, DC: Butterworth

Vlieg, E., Fischer, A. E. M. J., van der Veen, J. F., Dev, B. N., Materlik, G. 1986. Geometric structure of the $NiSi_2$-Si(111) interface: an X-ray standing-wave analysis. *Surf. Sci.* 178: 36–46

Waff, H. S. 1975. Pressure-induced coordination changes in magmatic liquids. *Geophys. Res. Lett.* 2: 193–96

Waseda, Y., Toguri, J. M. 1978. The structure of the molten FeO-SiO_2 system. *Metall. Trans. B* 9: 595–601

Waychunas, G. A. 1983. Mössbauer, EXAFS and X-ray diffraction study of Fe^{3+} clusters in MgO:Fe and magnesiowüstite $(Mg,Fe)_{1-x}O$—evidence for specific cluster geometries. *J. Mater. Sci.* 18: 195–207

Waychunas, G. A. 1987. Synchrotron radiation XANES spectroscopy of Ti in minerals: effects of Ti bonding distances, Ti valence and site geometry on absorption edge structure. *Am. Mineral.* 72: 89–101

Waychunas, G. A., Brown, G. E. Jr. 1984. Applications of EXAFS and XANES spectroscopy to problems in mineralogy and geochemistry. In *EXAFS and Near-Edge Structure III. Springer Proc. Phys. 2*, ed. K. O. Hodgson, B. Hedman, J. E. Penner-Hahn, pp. 336–42. New York/Berlin: Springer-Verlag. 533 pp.

Waychunas, G. A., Brown, G. E., Jr. 1990. Polarized X-ray absorption spectroscopy of metal ions in minerals: applications to site geometry and electronic structure determination. *Phys. Chem. Miner.* Submitted for publication

Waychunas, G. A., Apted, M. J., Brown, G. E. Jr. 1983. X-ray K-edge absorption spectra of Fe minerals and model compounds: near-edge structure. *Phys. Chem. Miner.* 10: 1–9

Waychunas, G. A., Brown, G. E. Jr., Apted, M. 1986a. X-ray K-edge absorption spectra of Fe minerals and model compounds: II. EXAFS. *Phys. Chem. Miner.* 13: 31–47

Waychunas, G. A., Brown, G. E. Jr., Ponader, C. W., Jackson, W. E. 1986b. High temperature X-ray absorption study of iron sites in crystalline, glassy and molten silicates and oxides. *Rep. 87/01*, pp. 139–41. Stanford Synchrotron Radiat. Lab., Stanford, Calif.

Waychunas, G. A., Dollase, W. A., Ross, C. R. 1986c. Determination of short range order (SRO) parameters from EXAFS pair distribution functions of oxide and silicate solid solutions. *J. Phys.* 47(C8): 845–48

Waychunas, G. A., Brown, G. E. Jr., Ponader, C. W., Jackson, W. E. 1988. Evidence from X-ray absorption for network-forming Fe^{2+} in molten alkali silicates. *Nature* 332: 251–53

Weidner, D. J., Ito, E. 1987. Mineral physics constraints on a uniform mantle composition. In *High Pressure Research in Mineral Physics*, ed. M. H. Manghnani, Y. Syono, pp. 439–46. Tokyo/Washington, DC: TERRAPUB/Am. Geophys. Union. 486 pp.

Will, G., Lauterjung, J. 1987. The kinetics of the pressure induced olivine-spinel phase transition in Mg_2GeO_4. In *High Pressure Research in Mineral Physics*, ed. M. H. Manghnani, Y. Syono, pp. 177–86. Tokyo/Washington, DC: TERRAPUB/Am. Geophys. Union. 486 pp.

Will, G., Hinze, E., Nuding, W. 1980. The compressibility of FeO measured by energy dispersive X-ray diffraction in a diamond anvil squeezer up to 200 kbar. *Phys. Chem. Miner.* 6: 157–67

Winick, H. 1980. Properties of synchrotron radiation. See Winick & Doniach 1980, pp. 11–25

Winick, H. 1987a. Synchrotron radiation. *Sci. Am.* 255: 88–99

Winick, H. 1987b. Status of synchrotron radiation facilities outside the USSR. *Nucl. Instrum. Methods Phys. Res. A* 261: 9–17

Winick, H., Doniach, S., eds. 1980. *Synchrotron Radiation Research.* New York/London: Plenum. 754 pp.

Wong, J. 1986. Extended X-ray absorption fine structure: a modern structural tool in materials science. *Mater. Sci. Eng.* 80: 107–28

Wong, J., Lytle, F. W., Messmer, R. P., Maylotte, D. H. 1984. K-edge absorption spectra of selected vanadium compounds. *Phys. Rev. B* 30: 5596–5610

Xu, J. A., Mao, H. K., Bell, P. M. 1986. High pressure ruby and diamond fluorescence observations at 0.21 to 0.55 TPa. *Science* 232: 1404–6

Yagi, T., Akaogi, M., Shimomura, O., Tamai, H., Akimoto, S. 1987. High pressure and high temperature equation of state of majorite. In *High-Pressure Research in Mineral Physics*, ed. M. H. Manghnani, Y. Syono, pp. 141–47. Tokyo/Washington, DC: TERRAPUB/Am. Geophys. Union. 486 pp.

Zarka, A., Lin, L. 1983. Study of local variations in spacing and orientation in a Z-cut plate of a synthetic quartz crystal by X-ray topography. *J. Cryst. Growth* 61: 397–405

Zarka, A., Capelle, B., Philippot, E., Jumas, J. C. 1986. Observation of defects and incommensurate phases in berlinite crystals by high temperature X-ray topography. *J. Appl. Crystallogr.* 19: 477–81

Zinner, E. K., McKeegan, K. D., Walker, R. M. 1983. Laboratory measurements of D/H ratios in interplanetary dust. *Nature* 305: 119–21

SUBJECT INDEX

A

B

T

CUMULATIVE INDEXES

CONTRIBUTING AUTHORS VOLUMES 1–18

CHAPTER TITLES VOLUMES 1–18

GEOPHYSICS AND PLANETARY SCIENCE

OCEANOGRAPHY, METEOROLOGY, AND PALEOCLIMATOLOGY

PALEONTOLOGY, STRATIGRAPHY, AND SEDIMENTOLOGY

ANNUAL REVIEWS INC.

A NONPROFIT SCIENTIFIC PUBLISHER

4139 El Camino Way
P.O. Box 10139
Palo Alto, CA 94303-0897 • USA

ORDER FORM

ORDER TOLL FREE
1-800-523-8635
(except California)

Annual Reviews Inc. publications may be ordered directly from our office; through booksellers and subscription agents, worldwide; and through participating professional societies. Prices subject to change without notice. ARI Federal I.D. #94-1156476

- **Individuals:** Prepayment required on new accounts by check or money order (in U.S. dollars, check drawn on U.S. bank) or charge to credit card—American Express, VISA, MasterCard.
- **Institutional buyers:** Please include purchase order.
- **Students:** $10.00 discount from retail price, per volume. Prepayment required. Proof of student status must be provided (photocopy of student I.D. or signature of department secretary is acceptable). Students must send orders direct to Annual Reviews. Orders received through bookstores and institutions requesting student rates will be returned. You may order at the Student Rate for a maximum of 3 years.
- **Professional Society Members:** Members of professional societies that have a contractual arrangement with Annual Reviews may order books through their society at a reduced rate. Check with your society for information.
- **Toll Free Telephone orders:** Call 1-800-523-8635 (except from California) for orders paid by credit card or purchase order and customer service calls only. California customers and all other business calls use 415-493-4400 (not toll free). Hours: 8:00 AM to 4:00 PM, Monday-Friday, Pacific Time. **Written confirmation** is required on purchase orders from universities before shipment.
- **FAX: 415-855-9815 Telex: 910-290-0275**

Regular orders: Please list below the volumes you wish to order by volume number.
Standing orders: New volume in the series will be sent to you automatically each year upon publication. Cancellation may be made at any time. Please indicate volume number to begin standing order.
Prepublication orders: Volumes not yet published will be shipped in month and year indicated.
California orders: Add applicable sales tax.
Postage paid (4th class bookrate/surface mail) **by Annual Reviews Inc.** Airmail postage or UPS, extra.

ANNUAL REVIEWS SERIES		Prices Postpaid per volume USA & Canada/elsewhere	Regular Order Please send: Vol. number	Standing Order Begin with: Vol. number
Annual Review of **ANTHROPOLOGY**				
Vols. 1-16	(1972-1987)	**$31.00/$35.00**		
Vols. 17-18	(1988-1989)	**$35.00/$39.00**		
Vol. 19	(avail. Oct. 1990)	**$39.00/$43.00**	Vol(s). ______	Vol. ______
Annual Review of **ASTRONOMY AND ASTROPHYSICS**				
Vols. 1, 4-14, 16-20	(1963, 1966-1976, 1978-1982)	**$31.00/$35.00**		
Vols. 21-27	(1983-1989)	**$47.00/$51.00**		
Vol. 28	(avail. Sept. 1990)	**$51.00/$55.00**	Vol(s). ______	Vol. ______
Annual Review of **BIOCHEMISTRY**				
Vols. 30-34, 36-56	(1961-1965, 1967-1987)	**$33.00/$37.00**		
Vols. 57-58	(1988-1989)	**$35.00/$39.00**		
Vol. 59	(avail. July 1990)	**$39.00/$44.00**	Vol(s). ______	Vol. ______
Annual Review of **BIOPHYSICS AND BIOPHYSICAL CHEMISTRY**				
Vols. 1-11	(1972-1982)	**$31.00/$35.00**		
Vols. 12-18	(1983-1989)	**$49.00/$53.00**		
Vol. 19	(avail. June 1990)	**$53.00/$57.00**	Vol(s). ______	Vol. ______
Annual Review of **CELL BIOLOGY**				
Vols. 1-3	(1985-1987)	**$31.00/$35.00**		
Vols. 4-5	(1988-1989)	**$35.00/$39.00**		
Vol. 6	(avail. Nov. 1990)	**$39.00/$43.00**	Vol(s). ______	Vol. ______

ANNUAL REVIEWS SERIES		**Prices Postpaid per volume USA & Canada/elsewhere**	**Regular Order Please send: Vol. number**	**Standing Order Begin with: Vol. number**
Annual Review of **COMPUTER SCIENCE**				
Vols. 1-2	(1986-1987)	**$39.00/$43.00**		
Vols. 3-4	(1988, 1989-1990)	**$45.00/$49.00**	Vol(s). ______	Vol. ______
Annual Review of **EARTH AND PLANETARY SCIENCES**				
Vols. 1-10	(1973-1982)	**$31.00/$35.00**		
Vols. 11-17	(1983-1989)	**$49.00/$53.00**		
Vol. 18	(avail. May 1990)	**$53.00/$57.00**	Vol(s). ______	Vol. ______
Annual Review of **ECOLOGY AND SYSTEMATICS**				
Vols. 2-18	(1971-1987)	**$31.00/$35.00**		
Vols. 19-20	(1988-1989)	**$34.00/$38.00**		
Vol. 21	(avail. Nov. 1990)	**$38.00/$42.00**	Vol(s). ______	Vol. ______
Annual Review of **ENERGY**				
Vols. 1-7	(1976-1982)	**$31.00/$35.00**		
Vols. 8-14	(1983-1989)	**$58.00/$62.00**		
Vol. 15	(avail. Oct. 1990)	**$62.00/$66.00**	Vol(s). ______	Vol. ______
Annual Review of **ENTOMOLOGY**				
Vols. 10-16, 18	(1965-1971, 1973)			
20-32	(1975-1987)	**$31.00/$35.00**		
Vols. 33-34	(1988-1989)	**$34.00/$38.00**		
Vol. 35	(avail. Jan. 1990)	**$38.00/$42.00**	Vol(s). ______	Vol. ______
Annual Review of **FLUID MECHANICS**				
Vols. 2-4, 7-19	(1970-1972, 1975-1987)	**$32.00/$36.00**		
Vols. 20-21	(1988-1989)	**$34.00/$38.00**		
Vol. 22	(avail. Jan. 1990)	**$38.00/$42.00**	Vol(s). ______	Vol. ______
Annual Review of **GENETICS**				
Vols. 1-21	(1967-1987)	**$31.00/$35.00**		
Vols. 22-23	(1988-1989)	**$34.00/$38.00**		
Vol. 24	(avail. Dec. 1990)	**$38.00/$42.00**	Vol(s). ______	Vol. ______
Annual Review of **IMMUNOLOGY**				
Vols. 1-5	(1983-1987)	**$31.00/$35.00**		
Vols. 6-7	(1988-1989)	**$34.00/$38.00**		
Vol. 8	(avail. April 1990)	**$38.00/$42.00**	Vol(s). ______	Vol. ______
Annual Review of **MATERIALS SCIENCE**				
Vols. 1, 3-12	(1971, 1973-1982)	**$31.00/$35.00**		
Vols. 13-19	(1983-1989)	**$66.00/$70.00**		
Vol. 20	(avail. Aug. 1990)	**$70.00/$74.00**	Vol(s). ______	Vol. ______
Annual Review of **MEDICINE**				
Vols. 9, 11-15	(1958, 1960-1964)			
17-38	(1966-1987)	**$31.00/$35.00**		
Vols. 39-40	(1988-1989)	**$34.00/$38.00**		
Vol. 41	(avail. April 1990)	**$38.00/$42.00**	Vol(s). ______	Vol. ______

FROM

NAME ______________________

ADDRESS ______________________

______________ ZIP CODE ____________

PLACE
STAMP
HERE

Annual Reviews Inc. A NONPROFIT SCIENTIFIC PUBLISHER

4139 El Camino Way
P.O. Box 10139
Palo Alto, CA 94303-0897

ANNUAL REVIEWS SERIES		Prices Postpaid per volume USA & Canada/elsewhere	Regular Order Please send:	Standing Order Begin with:
			Vol. number	Vol. number
Annual Review of **PSYCHOLOGY**				
Vols. 4, 5, 8, 10, 13-24, 26-30, 33-38	(1953, 1954, 1957, 1959, 1962-1973) (1975-1979, 1982-1987)	**$31.00/$35.00**		
Vols. 39-40	(1988-1989)	**$34.00/$38.00**		
Vol. 41	(avail. Feb. 1990)	**$38.00/$42.00**	Vol(s). ______	Vol. ______
Annual Review of **PUBLIC HEALTH**				
Vols. 1-8	(1980-1987)	**$31.00/$35.00**		
Vols. 9-10	(1988-1989)	**$39.00/$43.00**		
Vol. 11	(avail. May 1990)	**$43.00/$47.00**	Vol(s). ______	Vol. ______
Annual Review of **SOCIOLOGY**				
Vols. 1-13	(1975-1987)	**$31.00/$35.00**		
Vols. 14-15	(1988-1989)	**$39.00/$43.00**		
Vol. 16	(avail. Aug. 1990)	**$43.00/$47.00**	Vol(s). ______	Vol. ______

Note: Volumes not listed are out of print.

SPECIAL PUBLICATIONS			Prices Postpaid per volume USA & Canada/elsewhere	Regular Order Please Send:
The Excitement and Fascination of Science				
Volume 1	(published 1965)	Clothbound	**$25.00/$29.00**	______ Copy(ies).
Volume 2	(published 1978)	Hardcover	**$35.00/$39.00**	______ Copy(ies).
		Softcover	**$15.00/$19.00**	______ Copy(ies).
Volume 3	(published 1990)	Clothbound	**$90.00/$95.00**	______ Copy(ies).

(Volume 3 is published in two parts with complete indexes for Volumes 1, 2, and both parts of Volume 3. Sold in two-part set only.)

Intelligence and Affectivity:
The Relationship During Child Development, by Jean Piaget
(published 1981) Hardcover .... **$8.00/$9.00** ______ Copy(ies).

TO: **ANNUAL REVIEWS INC.,** a nonprofit scientific publisher
4139 El Camino Way • P.O. Box 10139
Palo Alto, CA 94303-0897 USA

Please enter my order for the publications checked above. **California orders, add sales tax.**
Prices subject to change without notice.

Institutional purchase order No. ______

Amount of remittance enclosed $ ______

Charge my account ☐ VISA

☐ MasterCard ☐ American Express

INDIVIDUALS: Prepayment required in U.S. funds or charge to bank card below. Include card number, expiration date, and signature.

Acct. No. ______

Exp. Date ______ ______
Signature

Name ______
Please print

Address ______
Please print

______ Zip Code ______ Date ______

______ Send free copy of current **Prospectus** ☐
Area(s) of Interest

ARI Federal I.D. #94-1156476